BOTANY: AN INTRODUCTION TO PLANT BIOLOGY

Botany: An Introduction to Plant Biology

Editor:

Sarika Garg

Delve Publishing

www.delvepublishing.com

Botany: An Introduction to Plant Biology

Editor: **Sarika Garg**

708 3rd Avenue, 6th Floor
New York, NY 10017
United States of America
www.delvepublishing.com

ISBN: 978-1-68095-753-2

Library of Congress Control Number: 2016951509

About the Editor

Sarika Garg

Ph.D. Montreal, Canada

Sarika Garg procured her doctoral degree at the Max Planck Unit For Structural Molecular Biology, University of Hamburg, Germany in 2011. This was followed by Postdoctoral Scientist position at University of Saskatchewan, Canada. Later, she moved to CRCHUM, University of Montreal for her second Postdoc. She has diverse research interests and has contributed immensely in many research fields viz. Enzyme Technology, Environmental Science, and Neuroscience. She has numerous research papers and review articles to her credit. She has coauthored two books 'Enzymes & Enzyme Technology'; and 'Biochemical Tests: Principles and Protocols'. She has also contributed several chapters in various books including a chapter on Regulation of Gene Expression in Encyclopedia of Molecular Cell Biology and Molecular Medicine. She is a recipient of 2012 SHRF Postdoctoral Fellowship Award and 2012 Laura E. Chapman Research Award. She is currently working as a Career Counselor in her very own venture 'higherstudiescounseling.com', Montreal, Canada and as a freelancer science writer.

Contents

List of Contributors

Kumar A.
School of Biotechnology, Devi Ahilya University, Indore, India.

Zhi-Yao Ma
Key Laboratory of Horticultural Plant Biology, College of Horticulture and Forestry Science, Huazhong Agricultural University, Ministry of Education, Wuhan, China

Jun Wen
Department of Botany, National Museum of Natural History, MRC166, Smithsonian Institution, Washington, DC, USA

Stefanie M. Ickert-Bond
UA Museum of the North Herbarium and Department of Biology and Wildlife, University of Alaska Fairbanks, Fairbanks, AK, USA

Long-Qing Chen
Key Laboratory of Horticultural Plant Biology, College of Horticulture and Forestry Science, Huazhong Agricultural University, Ministry of Education, Wuhan, China

Xiu-Qun Liu
Key Laboratory of Horticultural Plant Biology, College of Horticulture and Forestry Science, Huazhong Agricultural University, Ministry of Education, Wuhan, China

Linnea Hesse
Institut für Botanik, Technische Universität Dresden, D-01062 Dresden, Germany
Institut für Spezielle Botanik, Johannes Gutenberg—Universität Mainz, D-55112 Mainz, Germany
Present address: Plant Biomechanics Group Freiburg, University of Freiburg, Schänzlestrasse 1, D-79104 Freiburg im Breisgau, Germany

Sarah T. Wagner
Institut für Botanik, Technische Universität Dresden, D-01062 Dresden, Germany

Christoph Neinhuis
Institut für Botanik, Technische Universität Dresden, D-01062 Dresden, Germany

Lisa S. Spirko
Department of Biology, University of North Florida, 1 UNF Drive, Jacksonville, FL 32224, USA

Anthony M. Rossi
Department of Biology, University of North Florida, 1 UNF Drive, Jacksonville, FL 32224, USA

Yan Cheng
Institute of Vegetable Research, Shanxi Academy of Agricultural Sciences, China

Rong-li Ma
Institute of Vegetable Research, Shanxi Academy of Agricultural Sciences, China

Yan-sheng Jiao
Institute of Vegetable Research, Shanxi Academy of Agricultural Sciences, China

Ning Qiao
Institute of Vegetable Research, Shanxi Academy of Agricultural Sciences, China

Ting-ting Li
Horticulture College, Shanxi Agricultural University, China

V. Douhovnikoff
Bowdoin College, 6500 College Station, Brunswick, ME 04011, USA

S. H. Taylor
Bowdoin College, 6500 College Station, Brunswick, ME 04011, USA

E. L. G. Hazelton
Department of Watershed Sciences, Ecology Center, Utah State University, Logan, UT 84322, USA

C. M. Smith
Bowdoin College, 6500 College Station, Brunswick, ME 04011, USA

J. O'Brien
Bowdoin College, 6500 College Station, Brunswick, ME 04011, USA

S.T. Maseko
Department of Crop Sciences, Tshwane University of Technology, Private Bag X680, Pretoria 0001, South Africa

F.D. Dakora
Chemistry Department, Tshwane University of Technology, Private Bag X680, Pretoria 0001, South Africa

Paraskevi Tavladoraki
Laboratory of Biochemistry, Physiology and Biotechnology of Plants, Department of Science, University "Roma Tre", Rome, Italy

Alessandra Cona
Laboratory of Biochemistry, Physiology and Biotechnology of Plants, Department of Science, University "Roma Tre", Rome, Italy

Riccardo Angelini
Laboratory of Biochemistry, Physiology and Biotechnology of Plants, Department of Science, University "Roma Tre", Rome, Italy

Julien Le Roy
Centre National de la Recherche Scientifique, UMR 8576-Unité de Glycobiologie Structurale et Fonctionnelle, Université de Lille, Lille, France

Brigitte Huss
Centre National de la Recherche Scientifique, UMR 8576-Unité de Glycobiologie Structurale et Fonctionnelle, Université de Lille, Lille, France

Anne Creach
Centre National de la Recherche Scientifique, UMR 8576-Unité de Glycobiologie Structurale et Fonctionnelle, Université de Lille, Lille, France

Simon Hawkins
Centre National de la Recherche Scientifique, UMR 8576-Unité de Glycobiologie Structurale et Fonctionnelle, Université de Lille, Lille, France

Godfrey Neutelings
Centre National de la Recherche Scientifique, UMR 8576-Unité de Glycobiologie Structurale et Fonctionnelle, Université de Lille, Lille, France

Marta Francisco
Department of Plant Sciences, University of California, Davis, CA, USA
Group of Genetics, Breeding and Biochemistry of Brassicas, Department of Plant Genetics, Misión Biológica de Galicia, Spanish Council for Scientific Research, Pontevedra, Spain

Bindu Joseph
Department of Plant Sciences, University of California, Davis, CA, USA

Hart Caligagan
Department of Plant Sciences, University of California, Davis, CA, USA

Baohua Li
Department of Plant Sciences, University of California, Davis, CA, USA

Jason A. Corwin
Department of Plant Sciences, University of California, Davis, CA, USA

Catherine Lin
Department of Plant Sciences, University of California, Davis, CA, USA

Rachel Kerwin
Department of Plant Sciences, University of California, Davis, CA, USA

Meike Burow
DynaMo Center of Excellence, Copenhagen Plant Science Centre, University of Copenhagen, Frederiksberg, Denmark

Daniel J. Kliebenstein
Department of Plant Sciences, University of California, Davis, CA, USA
DynaMo Center of Excellence, Copenhagen Plant Science Centre, University of Copenhagen, Frederiksberg, Denmark

Rongjun Fang
State Key Laboratory of Pharmaceutical Biotechnology, NJU-NJFU Joint Institute of Plant Molecular Biology, School of Life Sciences, Nanjing University, Nanjing 210046, People's Republic of China.
Jiangsu University of Science and Technology, Zhenjiang 212003, People's Republic of China.
Co-Innovation Center for Sustainable Forestry in Southern China, Nanjing Forestry University, Nanjing 210037, People's Republic of China.

Ailan Zou
State Key Laboratory of Pharmaceutical Biotechnology, NJU-NJFU Joint Institute of Plant Molecular Biology, School of Life Sciences, Nanjing University, Nanjing 210046, People's Republic of China.
Co-Innovation Center for Sustainable Forestry in Southern China, Nanjing Forestry University, Nanjing 210037, People's Republic of China.

Hua Zhao
State Key Laboratory of Pharmaceutical Biotechnology, NJU-NJFU Joint Institute of Plant Molecular Biology, School of Life Sciences, Nanjing University, Nanjing 210046, People's Republic of China.
Co-Innovation Center for Sustainable Forestry in Southern China, Nanjing Forestry University, Nanjing 210037, People's Republic of China.

Fengyao Wu
State Key Laboratory of Pharmaceutical Biotechnology, NJU-NJFU Joint Institute of Plant Molecular Biology, School of Life Sciences, Nanjing University, Nanjing 210046, People's Republic of China.
Co-Innovation Center for Sustainable Forestry in Southern China, Nanjing Forestry University, Nanjing 210037, People's Republic of China.

Yu Zhu
State Key Laboratory of Pharmaceutical Biotechnology, NJU-NJFU Joint Institute of Plant Molecular Biology, School of Life Sciences, Nanjing University, Nanjing 210046, People's Republic of China.
Co-Innovation Center for Sustainable Forestry in Southern China, Nanjing Forestry University, Nanjing 210037, People's Republic of China.

Hu Zhao
State Key Laboratory of Pharmaceutical Biotechnology, NJU-NJFU Joint Institute of Plant Molecular Biology, School of Life Sciences, Nanjing University, Nanjing 210046, People's Republic of China.
Co-Innovation Center for Sustainable Forestry in Southern China, Nanjing Forestry University, Nanjing 210037, People's Republic of China.

Yonghui Liao
State Key Laboratory of Pharmaceutical Biotechnology, NJU-NJFU Joint Institute of Plant Molecular Biology, School of Life Sciences, Nanjing University, Nanjing 210046, People's Republic of China.
Co-Innovation Center for Sustainable Forestry in Southern China, Nanjing Forestry University, Nanjing 210037, People's Republic of China.

Ren-Jie Tang
Department of Plant and Microbial Biology, University of California, Berkeley, CA 94720, USA.

Yanjun Pang
State Key Laboratory of Pharmaceutical Biotechnology, NJU-NJFU Joint Institute of Plant Molecular Biology, School of Life Sciences, Nanjing University, Nanjing 210046, People's Republic of China.
Co-Innovation Center for Sustainable Forestry in Southern China, Nanjing Forestry University, Nanjing 210037, People's Republic of China.

Rongwu Yang
State Key Laboratory of Pharmaceutical Biotechnology, NJU-NJFU Joint Institute of Plant Molecular Biology, School of Life Sciences, Nanjing University, Nanjing 210046, People's Republic of China.
Co-Innovation Center for Sustainable Forestry in Southern China, Nanjing Forestry University, Nanjing 210037, People's Republic of China.

Xiaoming Wang
State Key Laboratory of Pharmaceutical Biotechnology, NJU-NJFU Joint Institute of Plant Molecular Biology, School of Life Sciences, Nanjing University, Nanjing 210046, People's Republic of China.
Co-Innovation Center for Sustainable Forestry in Southern China, Nanjing Forestry University, Nanjing 210037, People's Republic of China.

Jinliang Qi
State Key Laboratory of Pharmaceutical Biotechnology, NJU-NJFU Joint Institute of Plant Molecular Biology, School of Life Sciences, Nanjing University, Nanjing 210046, People's Republic of China.
Co-Innovation Center for Sustainable Forestry in Southern China, Nanjing Forestry University, Nanjing 210037, People's Republic of China.
Department of Plant and Microbial Biology, University of California, Berkeley, CA 94720, USA.

Guihua Lu
State Key Laboratory of Pharmaceutical Biotechnology, NJU-NJFU Joint Institute of Plant Molecular Biology, School of Life Sciences, Nanjing University, Nanjing 210046, People's Republic of China.
Co-Innovation Center for Sustainable Forestry in Southern China, Nanjing Forestry University, Nanjing 210037, People's Republic of China.

Yonghua Yang
State Key Laboratory of Pharmaceutical Biotechnology, NJU-NJFU Joint Institute of Plant Molecular Biology, School of Life Sciences, Nanjing University, Nanjing 210046, People's Republic of China.
Co-Innovation Center for Sustainable Forestry in Southern China, Nanjing Forestry University, Nanjing 210037, People's Republic of China.

Thangasamy Saminathan
Department of Biology, Gus R. Douglass Institute, West Virginia State University, Institute, WV 25112-1000, USA.

Abiodun Bodunrin
Department of Biology, Gus R. Douglass Institute, West Virginia State University, Institute, WV 25112-1000, USA.

Nripendra V. Singh
ICAR-National Research Center on Pomegranate, Kegaon, Solapur, Maharashtra 413255, India.

Ramajayam Devarajan
ICAR-Indian Institute of Oil Palm Research, Pedavegi, West Godavari, Andhra Pradesh 534450, India.

Padma Nimmakayala
Department of Biology, Gus R. Douglass Institute, West Virginia State University, Institute, WV 25112-1000, USA.

Moersfelder Jeff
National Clonal Germplasm Repository, USDA-ARS, University of California, Davis, CA 95616, USA.

Mallikarjuna Aradhya
National Clonal Germplasm Repository, USDA-ARS, University of California, Davis, CA 95616, USA.

Umesh K. Reddy
Department of Biology, Gus R. Douglass Institute, West Virginia State University, Institute, WV 25112-1000, USA.

Bo Zhou
College of Life Science, Northeast Forestry University, 26 Hexing Road, Harbin 150040, China.
State Key Laboratory of Tree Genetics and Breeding, Northeast Forestry University, 26 Hexing Road, Harbin 150040, China.

Pengzhen Fan
College of Life Science, Northeast Forestry University, 26 Hexing Road, Harbin 150040, China.

Yuhua Li
College of Life Science, Northeast Forestry University, 26 Hexing Road, Harbin 150040, China.
State Key Laboratory of Tree Genetics and Breeding, Northeast Forestry University, 26 Hexing Road, Harbin 150040, China.

Haifang Yan
College of Life Science, Northeast Forestry University, 26 Hexing Road, Harbin 150040, China.

Qijiang Xu
College of Life Science, Northeast Forestry University, 26 Hexing Road, Harbin 150040, China.
State Key Laboratory of Tree Genetics and Breeding, Northeast Forestry University, 26 Hexing Road, Harbin 150040, China.

X.J. Zhang
College of Horticulture, Northwest A&F University, Yangling, Shaanxi 712100, PR China

L.X. Wang
College of Horticulture, Northwest A&F University, Yangling, Shaanxi 712100, PR China

Y.L. Liu
College of Horticulture, Northwest A&F University, Yangling, Shaanxi 712100, PR China

X.X. Chen
College of Natural Resources and Environment, Northwest A&F University, Yangling, Shaanxi 712100, PR China

Y.Z. Yang
College of Horticulture, Northwest A&F University, Yangling, Shaanxi 712100, PR China

Z.Y. Zhao
College of Horticulture, Northwest A&F University, Yangling, Shaanxi 712100, PR China
Apple E&T Research Centre of Shaanxi Province, Yangling, Shaanxi 712100, PR China

Christoph Jedmowski
Department of Ecology, Evolution and Diversity, University of Frankfurt, Max von Laue Strasse 13, 60438 Frankfurt, Germany

Ahmed Ashoub
Department of Ecology, Evolution and Diversity, University of Frankfurt, Max von Laue Strasse 13, 60438 Frankfurt, Germany
Agricultural Genetic Engineering Research Institute (AGERI), ARC, Giza 12619, Egypt

Osama Momtaz
Agricultural Genetic Engineering Research Institute (AGERI), ARC, Giza 12619, Egypt

Wolfgang Brüggemann
Department of Ecology, Evolution and Diversity, University of Frankfurt, Max von Laue Strasse 13, 60438 Frankfurt, Germany
[3]Biodiversity and Climate Research Center, Senckenberganlage 25, 60325 Frankfurt, Germany

Ganesh K. Jaganathan
Institute of Biothermal Technology, University of Shanghai for Science and Technology, Shanghai 20009, China

Sarah E. Dalrymple
School of Natural Sciences and Psychology, Liverpool John Moores University, James Parsons Building, Byrom Street, Liverpool L3 3AF, UK

Issouf Zerbo
Laboratory of Plant biology and Ecology, University of Ouagadougou, 03 BP 7021, Ouagadougou 03, Burkina Faso

Markus Bernhardt-Römermann
Institute of Ecology, Friedrich Schiller University Jena, Dornburger Strasse 159, 07743 Jena, Germany

Oumarou Ouédraogo
Laboratory of Plant biology and Ecology, University of Ouagadougou, 03 BP 7021, Ouagadougou 03, Burkina Faso

Karen Hahn
Department of Ecology and Geobotany, Institute for Ecology, Evolution and Diversity, J. W. Goethe University, Max-von-Laue-Strasse 13, 60438 Frankfurt am Main, Germany

Adjima Thiombiano
Laboratory of Plant biology and Ecology, University of Ouagadougou, 03 BP 7021, Ouagadougou 03, Burkina Faso

Poornima D. Vijendra
Department of Biochemistry, Kuvempu University, P.G. Centre, Shivagangothri, Davangere, Karnataka 577002, India
Department of Biochemistry, Davangere University, Shivagangothri, Davangere, Karnataka 577002, India

Kavitha M. Huchappa
Department of Biochemistry, Kuvempu University, P.G. Centre, Shivagangothri, Davangere, Karnataka 577002, India

Roopa Lingappa
Department of Biochemistry, Kuvempu University, P.G. Centre, Shivagangothri, Davangere, Karnataka 577002, India

Giridhara Basappa
Department of Biochemistry, Davangere University, Shivagangothri, Davangere, Karnataka 577002, India

Sathisha G. Jayanna
Department of Biochemistry, Kuvempu University, Shankaraghatta, Shivamogga District, Karnataka 577451, India

Vadlapudi Kumar
Department of Biochemistry, Kuvempu University, P.G. Centre, Shivagangothri, Davangere, Karnataka 577002, India
Department of Biochemistry, Davangere University, Shivagangothri, Davangere, Karnataka 577002, India

Jingbin Chen
Institute of Vegetable Crops, Jiangsu Academy of Agricultural Sciences, Nanjing, China
Department of Agronomy, Faculty of Agriculture at Kamphaeng Saen, Kasetsart University, Nakhon Pathom, Thailand

Prakit Somta
Department of Agronomy, Faculty of Agriculture at Kamphaeng Saen, Kasetsart University, Nakhon Pathom, Thailand

Xin Chen
Institute of Vegetable Crops, Jiangsu Academy of Agricultural Sciences, Nanjing, China

Xiaoyan Cui
Institute of Vegetable Crops, Jiangsu Academy of Agricultural Sciences, Nanjing, China

Xingxing Yuan
Institute of Vegetable Crops, Jiangsu Academy of Agricultural Sciences, Nanjing, China

Peerasak Srinives
Department of Agronomy, Faculty of Agriculture at Kamphaeng Saen, Kasetsart University, Nakhon Pathom, Thailand

Line Jensen
Department of Plant and Environmental Sciences, Faculty of Science, University of Copenhagen, Højbakkegård Allé 9-13, 2630 Taastrup, Denmark

Josefine Nymark Hegelund
Department of Plant and Environmental Sciences, Faculty of Science, University of Copenhagen, Højbakkegård Allé 9-13, 2630 Taastrup, Denmark

Andreas Olsen
Department of Plant and Environmental Sciences, Faculty of Science, University of Copenhagen, Højbakkegård Allé 9-13, 2630 Taastrup, Denmark

Henrik Lütken
Department of Plant and Environmental Sciences, Faculty of Science, University of Copenhagen, Højbakkegård Allé 9-13, 2630 Taastrup, Denmark

Renate Müller
Department of Plant and Environmental Sciences, Faculty of Science, University of Copenhagen, Højbakkegård Allé 9-13, 2630 Taastrup, Denmark

Erna Laere
Department of Applied Biomedical Sciences and Biotechnology, School of Health Sciences, International Medical University, 57000 Bukit Jalil, Kuala Lumpur, Malaysia

Anna Pick Kiong Ling
Department of Applied Biomedical Sciences and Biotechnology, School of Health Sciences, International Medical University, 57000 Bukit Jalil, Kuala Lumpur, Malaysia

Ying Pei Wong
Department of Applied Biomedical Sciences and Biotechnology, School of Health Sciences, International Medical University, 57000 Bukit Jalil, Kuala Lumpur, Malaysia

Rhun Yian Koh
Department of Applied Biomedical Sciences and Biotechnology, School of Health Sciences, International Medical University, 57000 Bukit Jalil, Kuala Lumpur, Malaysia

Mohd Azmi Mohd Lila
Department of Veterinary Pathology and Microbiology, Faculty of Veterinary Medicine, Universiti Putra Malaysia (UPM), 43400 Serdang, Selangor, Malaysia

Sobri Hussein
Agrotechnology and Bioscience Division, Malaysian Nuclear Agency, Bangi, 43000 Kajang, Selangor, Malaysia

Preface

Botany is a biological science involving the study of plants. Plants are distinctive in manipulating energy from the sunlight alongside other chemical elements for growth. This process of photosynthesis supplies food and energy necessary for life on earth.

Because of the vast botanical research interests, several plant biologists are working in different domains to exploit botany for the benefits of living being. Some biologists centralize on the whole plant pattern. Some study the plant structure and organization. Many biologists are curious in unveiling the mechanism behind conversion of simple chemical compounds to more complex chemicals by plants. Some botanists combine plant biology with ecology by studying the interconnection of plants with other life forms and the habitat. Some even deal with the plants hereditary transmission and variation.

The botanical research outcomes ameliorate living being and environment in several ways. They expand our supply of foods, medicines, fibres, building materials and other plant products. Botanical knowledge helps people to maintain trees, flowering plants, forests, parks, lands etc thus contributing towards a healthy environment.

In the preparation of this book, several research studies are collated with the hope that they will aid the student in understanding the basics as well as the applications of botany. The first chapter accounts the Introduction and basics of botany. In Chapters 2-4, the structure and organization of plants is discussed. Plant physiology and development is explored in the following four chapters. Chapters 9-11 describe the plant metabolism. Plant regulation is presented in chapters 12-14. Study of the interaction of plants with other organisms and the environment is reported in chapters 15-18. Genetics and genetic engineering of the plants is talked about in the following two chapters. In the last chapter, light is thrown on the application part of botany.

I anticipate that this book will prove to be convenient for teaching, learning and research.

Editor
Sarika Garg, Ph.D.
Montreal, Canada

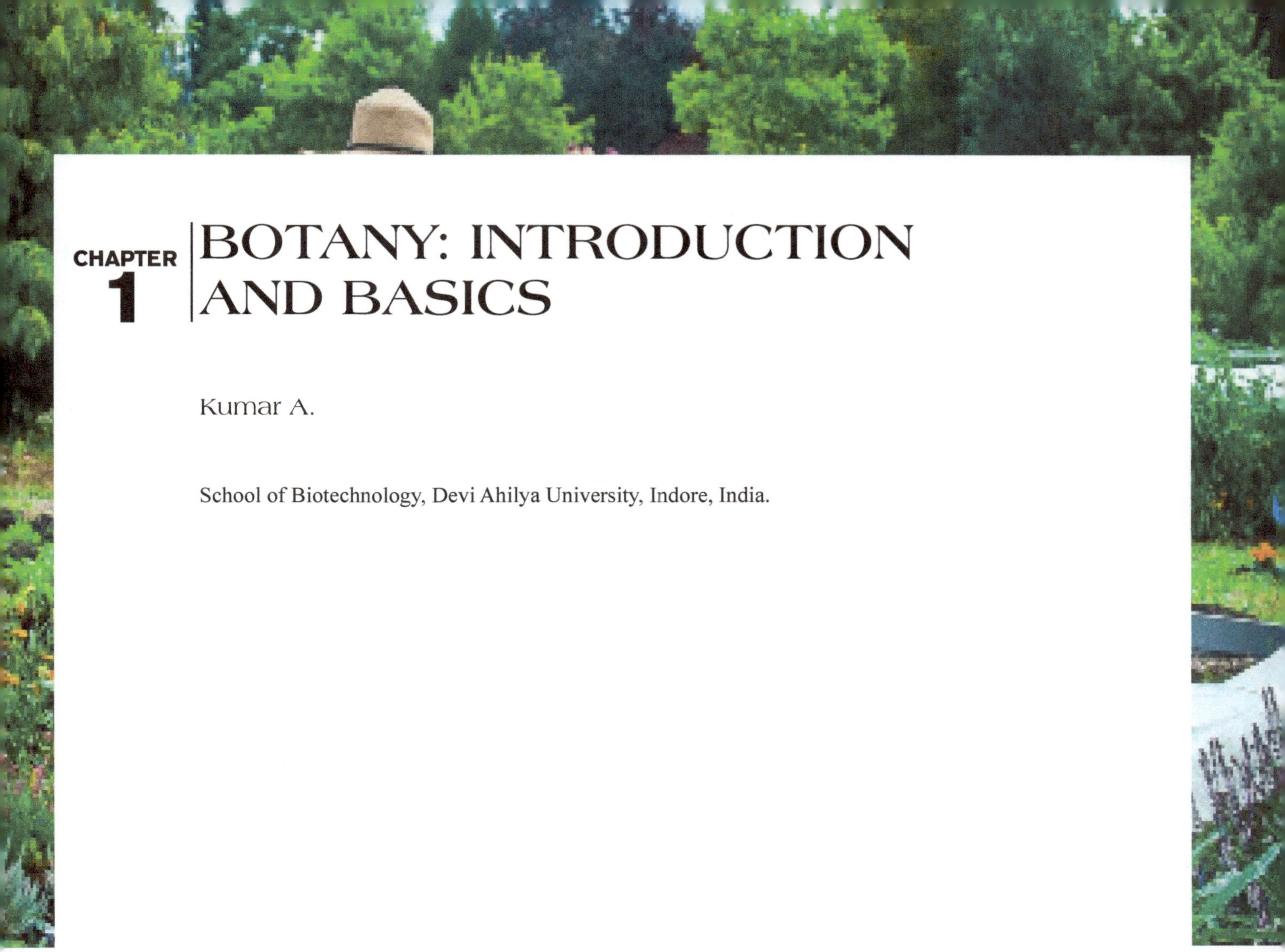

CHAPTER 1 BOTANY: INTRODUCTION AND BASICS

Kumar A.

School of Biotechnology, Devi Ahilya University, Indore, India.

INTRODUCTION

The science of plant is called as Botany. It includes study of plants of all kinds and applies their knowledge to support basic daily functions of human life by providing food and nutrition, and to supplement medicines, replenish earth's oxygen supply, prevent soil erosion, slow down wind movement, cool the atmosphere and provide wildlife habitat. Plants affect most aspects of life in many different ways by providing products such as food, medicine, rubber, wood, fabrics, building material, alcohol and many other plant products. Botanical research is also a key to the development of biofuels such as bio-ethanol, bio-diesel and methane gas which are used as alternative to fossil fuels. This science has increased the economic productivity as it helps farmers to increase the production of their crops. It is also important in the area of environmental protection as plants can solve the pollution problem. Botany influences the studies of other academic disciplines such as life science, science communication, ecology, and evolutionary biology.

As described above, plants are essential for survival on earth. These are unique as they make their own food unlike any other living organism by the process called photosynthesis. Photosynthesis is a process to convert light energy into chemical energy. This chemical energy is stored in carbohydrate molecules such as sugars which are synthesized from carbon dioxide and water. Photosynthesis maintains atmospheric oxygen level and supplies the organic compounds energy necessary for life on Earth.

As we know, there are wide varieties of plants. Plants are found on land, in oceans, and in fresh water. Plants have been on Earth for millions of years. Plants were on Earth before animals and currently number about 260,000 species. The morphology or external structure of plants are characterized by the presence of roots, stems, leaves, flowers and fruits. To study the different plants, botanists have sorted and classified them based on their similarities and differences. This system of classification is also called as taxonomy.

MORPHOLOGY OF PLANTS

Plant morphology includes both the vegetative structures and reproductive structures of plant. Vegetative structure includes the shoot system and root system. A shoot system is composed of stems and leaves and a root system includes roots. Reproductive structures are involved in the production of seeds. They include flowers, fruits and seeds. Each structure has certain characteristics and function(s):

1. ***Roots***: The root or radicle is the first plant part to appear from a seed and roots are usually underground. Their principal functions are to absorb nutrients and moisture, anchor the plant in the soil, support the stem and store food.

2. ***Stems***: Stems are the most basic organs on plants. Stems can range from hard and woody to soft and green but

Citation: Kumar A (2016). Botany: Introduction and Basics.

all stems serve a supportive function on a plant. Stems place the leaves in favorable positions for exposure to light.

3. ***Leaves***- The principal function of leaves is photosynthesis in which chlorophylls absorb sunlight to convert it into chemical energy and carbon dioxide is converted into sugars using enzymes by a cyclic pathway called as Calvin Cycle. Leaf surfaces are flattened to present a large area for efficient light absorption. In many plants, leaves have been modified to a sharp point to serve as protection spines.

4. ***Flowers***- Flowers are beautiful and showiest part of a plant. The main function of flower is the sexual reproduction. Fragrance and colors attract pollinators (insects or birds) that play an important role in the reproductive process.

5. ***Fruits***- Fruit consists of fertilized, mature ovules (seeds) plus the ovary wall. Fruit may be fleshy or dry and hard. In some fruits, seeds are enclosed within the ovary as in apple or may be situated outside of a fruit tissue as in corn.

6. ***Seeds***- A seed is formed by the process of reproduction and is enclosed in a protective outer covering. A seed contains all of the genetic information required to develop into an entire plant.

ANATOMY OF PLANT

Plant anatomy is the term used for the study of the internal structure of plants. Plants have cells as the basic unit, and cells are organized into tissues and tissues are arranged to form the organs. Different organs show different internal structures.

Plant Tissues

Plant tissues are classified into two main groups- meristematic and permanent tissues. A meristem tissue containing undifferentiated cells (meristematic cells) is found in zones of the plant where growth can take place. Meristematic cells give rise to various organs of the plant and keep the plant growing. The shoot apical meristem (SAM) gives rise to organs like the leaves and flowers while the root apical meristem (RAM) provides the meristematic cells for the future root growth. Based on location in the plant, meristematic tissues fall into two categories: primary and secondary. Primary meristematic tissue helps the plant to grow vertically. It means that it helps the plant grow up towards the sun and down into the soil. Secondary meristematic tissue helps the plant to increase the girth or lateral growth of its stems, branches, and roots. Secondary meristematic growth is also responsible for producing the bark on woody trees. Cells of meristematic tissue gradually divide and get differentiated to form permanent tissues.

Permanent tissue is composed of cells that have lost the power of dividing and has attained a definite form and shape. Permanent tissue can be divided into Simple and Complex tissues. Simple tissues are homogenous and composed of structurally and functionally similar cells. These are of three types: Parenchyma, Collenchyma and Sclerenchyma. Parenchyma forms the major component within organs. Parenchyma is the most common tissue which is morphologically and physiologically simple and unspecialized. These cells are found in epidermis, cortex, pericycle, pith etc. These are responsible for photosynthesis, storage of food, secretion etc. Collenchyma cells are elongated cells with irregularly thick cell walls that provide support and structure. Their thick cell walls are composed of the compounds, cellulose and pectin. These cells are often found under the epidermis, or in the outer layer of cells in young stems and in leaf veins. Sclerenchyma serves as the support in plants and contains various kinds of hard woody cells. Mature sclerenchyma cells are dead cells that have heavily thickened walls containing lignin. Such cells occur in many different shapes and sizes. Fibres and sclereids are the two main types of mature sclerenchyma cells to occur. Fibres are greatly elongated cells whose long, tapering ends interlock providing maximum support to a plant. These are found almost everywhere in the plant body including the stem, root and the vascular bundles in leaves. Sclereids are extremely variable in shape and are present in various tissues of the plant such as the periderm, cortex, pith, xylem, and phloem. These also occur in leaves and fruits and constitute the hard shell of nuts and the outer hard coat of many seeds.

Complex tissues are made up of more than one type of cells and these work together as a unit. These tissues transport water, salt and prepared food material to various parts of the plant body. Complex tissues are of two types: xylem and phloem. Xylem or wood is a conducting tissue and is composed of different kinds of elements namely Tracheids, Vessels or tracheae, Xylem parenchyma and Xylem sclerenchyma. Phloem is also a conducting tissue and is composed of elements of different kinds: sieve tubes, companion cells, phloem parenchyma, phloem fibres or bast fibres.

Root Anatomy

Cells in a root are arranged as root hair, epidermis, epiblema, cortex, endodermis, pericycle and vascular tissue which is in the centre of a root to transport the water absorbed by the root to other places of the plant.

Stem Anatomy

Stem usually consists of three tissues: dermal tissue, ground tissue and vascular tissue. The dermal tissue covers the outer surface of the stem and its functions are waterproofing, protection and control of gas exchange. The ground tissue usually consists of parenchyma cells and fills in around the vascular tissue. Vascular tissue provides long distance transport and structural support.

Leaf Anatomy

A leaf is made of many layers that are sandwiched between two layers of tough skin cells called epidermis. The epidermis also secretes a waxy substance called as cuticle. These layers protect the leaf from insects, bacteria, and other pests. There are also pairs of sausage-shaped guard cells among the epidermal cells. Each pair of guard cells forms a pore called stomata. Gases enter and exit the leaf through the stomata. Most food production takes place in elongated cells called palisade mesophyll. Gas exchange occurs in the air spaces between the oddly-shaped cells of the spongy mesophyll. Veins support the leaf and are filled with vessels that transport food, water, and minerals to the plant.

Chloroplast

Chloroplast is an organelle found in the cells of the mesophyll in plant leaves and is one of the three types of plastids. Its main role is to conduct photosynthesis, where the photosynthetic pigment chlorophyll captures the energy from sunlight and converts it and stores it in the energy-storage molecules ATP and NADPH while freeing oxygen from water. There are usually 30-40 chlorophyll molecules per mesophyll cell. The chloroplast has an inner and outer membrane with an empty intermediate space in between. Inside the chloroplast are stacks of thylakoids, called as grana and stroma. These thylakoids contain the chlorophyll that is necessary for the plant to go through photosynthesis. The space filled by the chlorophyll is called the thylakoid space.

Flower Anatomy

The flower consists of many different parts. Some of the most important parts being separated into both male and female parts are shown in Figure 1. The salient features of flower anatomy are:

- Petals give a flower its unique shape and are often bright colored to attract insects.
- Sepals are the small leaf-like parts growing at the base of the petals. They serve to protect the flower before it blossoms.
- Peduncle refers to the stem or stalk of a flower.
- Receptacle is the thickened part at the bottom of the flower which holds its major organs.
  - Pistil is the female organ of the flower. It consists of four major parts:
  - Stigma is the head of the pistil. The stigma receives pollen which will begin the process of fertilization.
  - Style is the stalk of the pistil. When pollen reaches the stigma, it begins to grow as a tube through the style called as pollen tube which will reach up to the ovary. Therefore, style acts as a buffer against pollen contamination since only compatible pollen is able to grow as a pollen tube.
  - Ovary is the base of the pistil. It holds the ovules awaiting fertilization.
  - Ovules are the flower's eggs located inside the ovary. Upon fertilization by pollen, they will eventually grow into a seed.
- Stamen is the male organ of the flower, consisting of two major parts:
  - Anther is the head of the stamen. The anther is responsible for the production of pollen, which will hopefully be transported to the pistil by animals or insects, such as bees.
  - Filament is the stalk that holds the anther and attaches it to the flower.

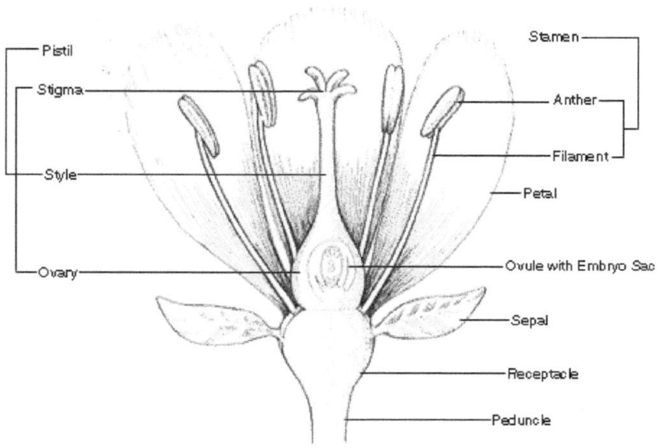

Figure 1: Anatomy of a flower (www.amnh.org).

TAXONOMY OF PLANT

The most basic divisions are vascular and non-vascular. Plants are non-vascular if they cannot circulate rainwater through their stems and leaves but must absorb it from the environment that surrounds them. The remaining plant species are all vascular that is they have a system for circulating fluids. The vascular system consists of xylem, phloem and vascular cambium. The vascular system can be thought as a plant's plumbing system. Xylem tubes conduct water and dissolved minerals and phloem tubes carry food such as sugars. The cambium is a layer of meristematic tissue that separates the xylem and phloem and continuously produces new xylem and phloem cells. This new tissue is responsible for increase in stem's girth.

Non-vascular plants are very small because they lack a vascular system. Another characteristic of non-vascular plants is that they don't have roots. In place of roots, non- vascular plants have rhizoids which are small hairs that are inserted into the substrate to keep the plant in place. Non-vascular plants are commonly found in moist environments so that they can absorb water right into the main part of the plant without relying on roots. Non-vascular plants are also called as bryophytes and are divided into three different types-mosses, liverworts, and hornworts. Mosses are small flowerless plants composed of short central stem with wiry branches and very small leaf-like structures. There are approximately 12,000 species of mosses classified in the Bryophyta. Hornwort is flowerless horn-like or needle-like capsule which is developed from a flattish, green sheet. They prefer very moist environments and some are even found submerged in ponds and lakes. The hornworts constitute a small group comprised of probably only 100-150 species. Liverworts are the simplest plants commonly identified by flattened leaves that often grow in two distinctive rows. These are found in terrestrial and semi-aquatic environments. These are considered as weeds as they can easily take over farm if it is moist.

Vascular plants such as trees are also called as Tracheophyta and flowering plants have vascular vessels to transport water and food throughout the plant. The vascular vessels are divided into two types based on the material they transport. Phloems are vessels on the outer layer of the stem that transport food materials such as sugars from the leaves where they are produced or from storage tissues to the rest of the plant. The second type of vascular vessel is the xylem that transports water throughout the plant. The xylem vessels transport water from the roots up to the leaves. Vascular vessels not only transport the food and water, but also help the plant to grow larger. Common examples of vascular plants include trees, shrubs, grasses, flowering plants and ferns.

Vascular plants include a broad range of plants and these are divided further into more specific categories. Vascular plants are divided by their method of reproduction. Vascular plants that reproduce by the use of spores (not by seeds) are characterized as ferns. This type of vascular plant is often referred as the seedless vascular plant. In contrast, there are vascular plants that reproduce by creating seeds rather than spores and are classified as gymnosperms and angiosperms. Gymnosperms meaning "naked seeds" are the plants in which the ovules or seeds are not enclosed in an ovary. There are around 1000 species of gymnosperms and these include conifers, fir trees, pine trees. Angiosperms differ from gymnosperms in few characteristics including flowers, endosperm within the seeds, and the production of fruits that contain the seeds. Angiosperms include the most diverse group of land plants which are found in most habitats *viz.* forests, grasslands, sea margins and even deserts. Examples of Angiosperms are grasses, apples and bananas.

Angiosperms or flowering plants are further divided into Monocotyledons and Dicotyledons. Monocots consist of plants having seeds with one cotyledon. Examples of monocots are grains (wheat, corn, rice, millet), lilies, daffodils, sugarcane, banana, palm, ginger, onions, bamboo, sugar, cone, palm tree, and grass. Dicots consist of plants having seeds with two cotyledons. Examples of dicots are Legumes (pea, beans, lentils, peanuts), daisies, mint, lettuce, tomato and oak. The whole classification has been enumerated in Figure 2.

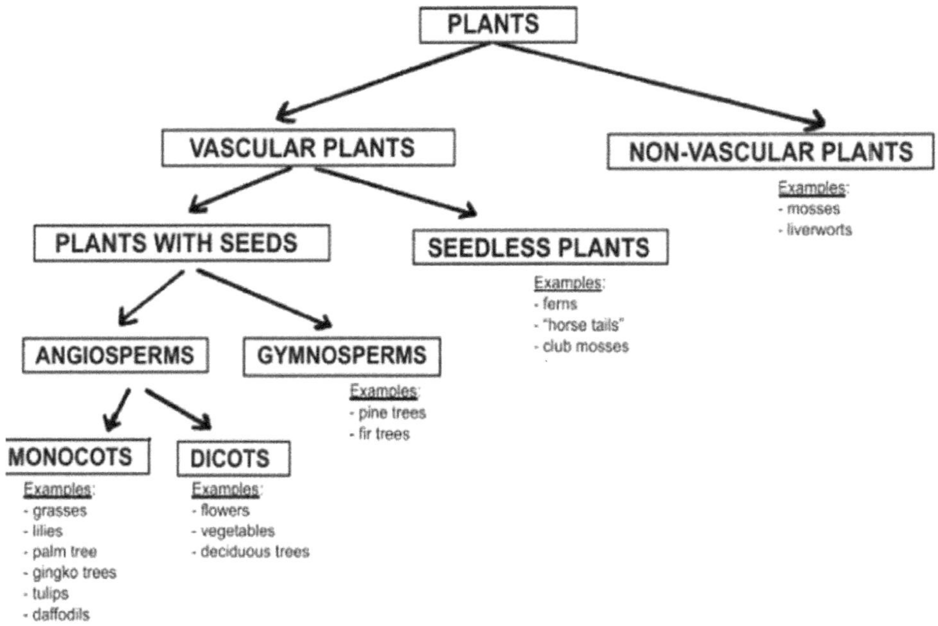

Figure 2: Taxonomy of plants (www.easypacelearning.com).

PHYSIOLOGY OF PLANT

Plant physiology is the study of the functions and processes occurring in plants. It includes the vital processes *viz.* photosynthesis, photorespiration, transpiration, plant nutrition, photoperiodism, photomorphogenesis, circadian rhythms, environmental stress physiology, seed germination, dormancy and floral induction. Some of the processes are described below:

Photosynthesis

Photosynthesis is a process in which plants use energy from the sun to transform water and carbon dioxide into oxygen and organic compounds. Photosynthesis is a two stage process. The first process is the Light Reaction that requires the direct energy of light to make energy carrier molecules that are used in the next process. The second stage is Dark Reactions that occur when the products of the Light Reaction are used to form C-C covalent bonds of carbohydrates.

Light Reaction

Light reactions occur in the thylakoid membranes of the chloroplasts and are of two types: cyclic and non-cyclic. In the non-cyclic light reaction, one molecule of the pigment chlorophyll absorbs one photon and loses one electron. This electron is passed to a modified form of chlorophyll called pheophytin which passes the electron to a quinone molecule starting the flow of electrons down an electron transport chain that leads to the ultimate reduction of $NADP^+$ to NADPH. In addition, this creates a proton gradient (energy gradient) across the chloroplast membrane, which is used for the synthesis of ATP by ATP synthase. The chlorophyll molecule ultimately regains the lost electron when a water molecule is split in a process called photolysis, which releases a dioxygen (O_2) molecule as a waste product (Figure 3). The overall equation for the light-dependent reactions under the conditions of non-cyclic electron flow in green plants is as follows:

$2\ H_2O + 2\ NADP^+ + 3\ ADP + 3\ Pi + light \rightarrow 2\ NADPH + 2\ H^+ + 3\ ATP + O_2$

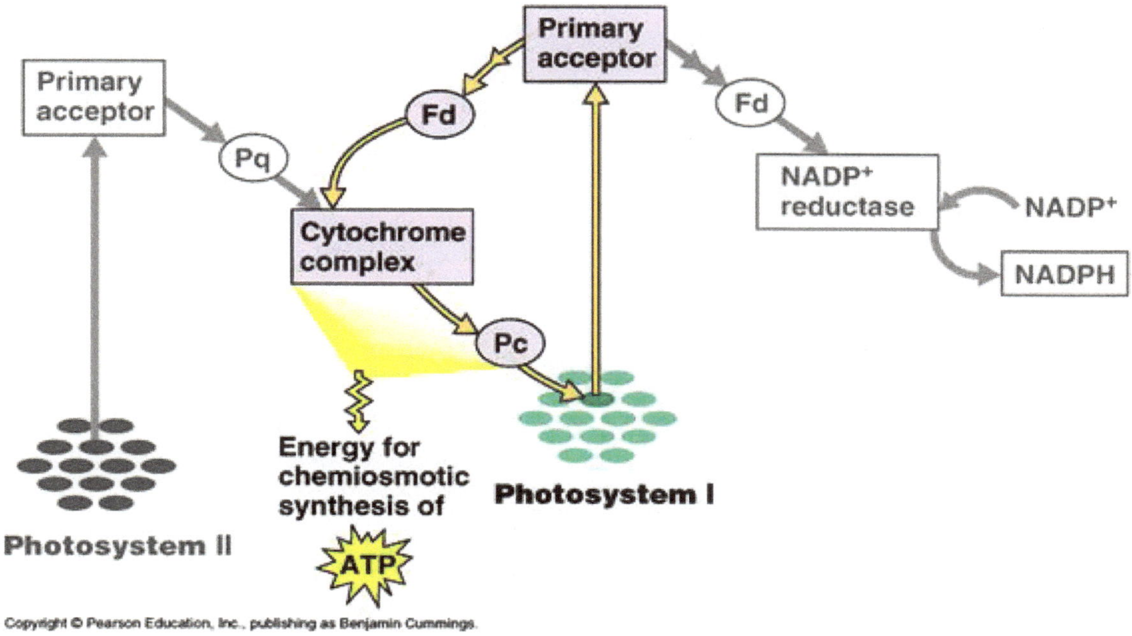

Figure 3: Non Cyclic Light reaction (ww2.methuen.k12.ma.us).

In cyclic light reaction, light energy is captured by the antenna complex and transferred to the Photosystem I reaction centre, which contributes two high energy electrons to the primary electron receptor. They are passed to ferrodoxin (Fd), an iron containing protein which acts as an electron carrier. A second electron carrier plastoquinone (Pq) carries the electrons to a complex of two cytochromes. In the process, energy is provided to produce a proton gradient across the membrane which can be used for ADP to ATP conversion. The electrons are returned by plastocyanin (Pc) to the P700 pigment in the reaction centre to complete the cycle (Figure 4).

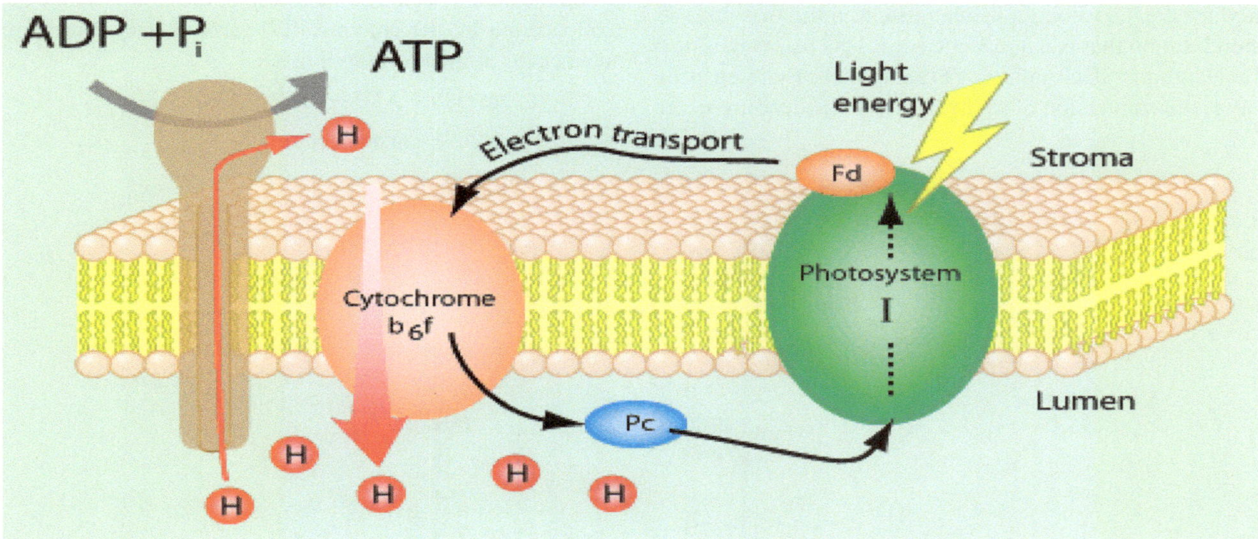

Figure 4: Cyclic Light reaction (hyperphysics.phy-astr.gsu.edu).

Dark Reaction

Dark reaction makes use of ATP and NADPH. This reaction cycle is also called Calvin Benison Cycle, and it occurs in the stroma. ATP provides the energy while NADPH provides the electrons required to fix CO_2 into carbohydrates. The carbon fixation reactions are actually a cycle which can be broken down into three stages. In the first step CO_2 is fixed by Rubisco (ribulose 1,5-bisphosphate carboxylase/oxygenase) to convert one molecule of ribulose 1,5-bisphosphate (R1,5BP) into two molecules of 3-phosphoglycerate (3PG). In the next stage, 3PG is converted to hexose sugars (sugars that contain 6 carbons, like glucose, fructose) via the molecule glyceraldehyde-3-phosphate (G3P). In the last stage, R1,5BP is regenerated so that cycle can be started again (Figure 5).

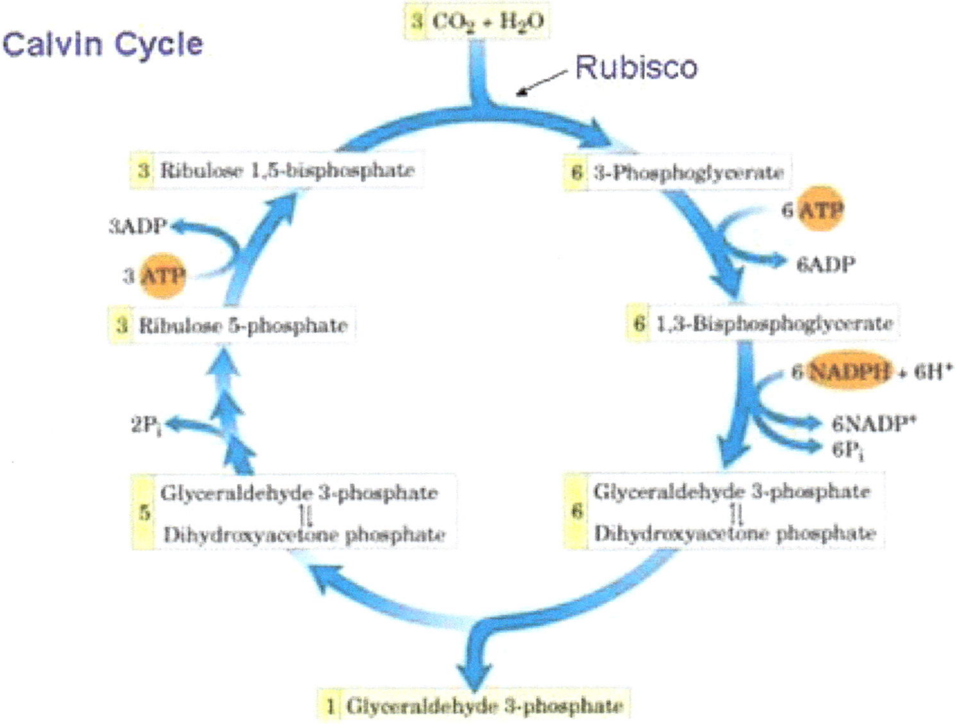

Figure 5: Dark Reaction (preib-biology.blogspot.com).

C4 Pathway

C4 pathway is the alternative pathway of Calvin cycle (C3 cycle) taking place during the dark phase of photosynthesis. It is named for the 4-carbon molecule present in the first product of carbon fixation that is oxaloacetic acid. This pathway is also called as Hatch and Slack pathway (Figure 6). The first step in the pathway is the conversion of pyruvate to phosphoenolpyruvate (PEP), by the enzyme pyruvate orthophosphate dikinase. This reaction requires inorganic phosphate and ATP plus pyruvate, producing phosphoenolpyruvate, AMP, and inorganic pyrophosphate (PPi). The next step is the fixation of CO_2 into oxaloacetate by the enzyme PEP carboxylase. Both of these steps occur in the mesophyll cells:

Pyruvate + Pi + ATP → PEP + AMP + PPi

PEP + CO2 → oxaloacetate

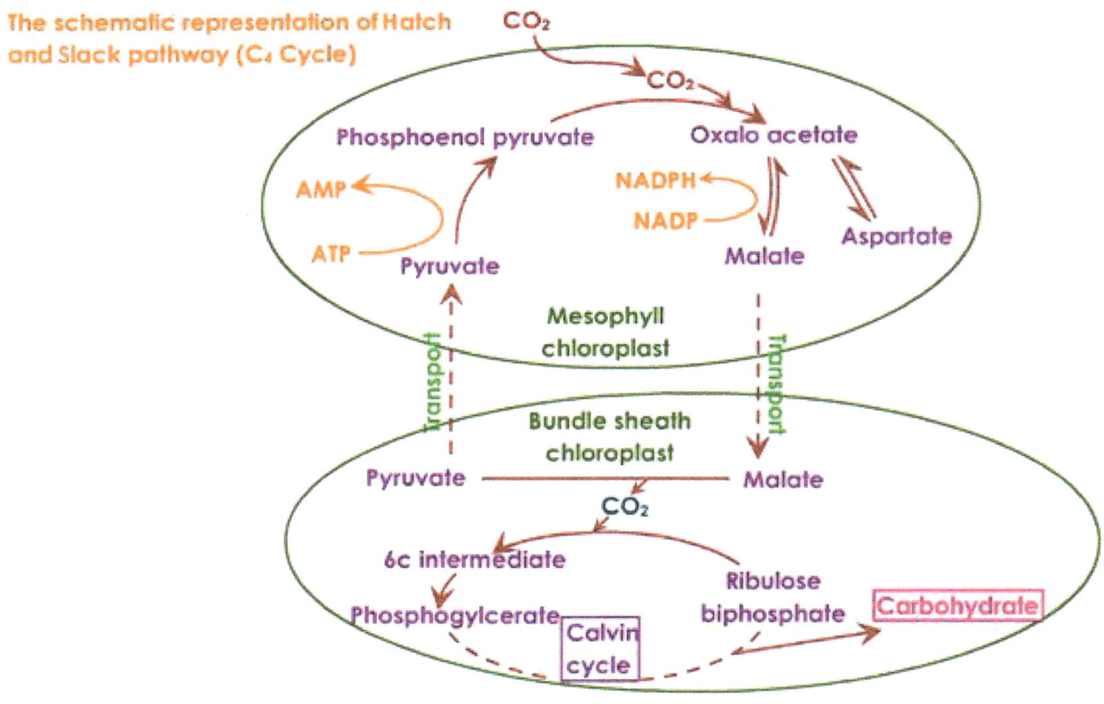

Figure 6 – C4 Pathway in Plants (www.tutorvista.com).

CAM (Crassulacean Acid Metabolism) Pathway

Crassulacean acid metabolism also known as CAM photosynthesis is a carbon fixation pathway that is evolved as an adaptation to arid conditions (Figure 7). This metabolism was first studied in plants of the Crassulaceae family. In a plant using full CAM, the stomata in the leaves remain shut during the day time to reduce transpiration but open at night to collect carbon dioxide (CO_2). The CO_2 is stored as the four carbon acid malate in vacuoles at night, and then in the daytime, the malate is transported to chloroplasts where it is converted back to CO_2, which is then used during photosynthesis. The pre-collected CO_2 is concentrated around the enzyme RuBisCO increasing photosynthetic efficiency. It is consisted of two parts:

Night

During the night time, plant opens stomata allowing CO_2 to enter and fixed it as organic acids that are stored in vacuoles. The carbon dioxide is fixed in the cytoplasm of mesophyll cells by a PEP reaction similar to that of C4 pathway.

Day

During the day, the CO_2 storing organic acids are released from the vacuoles of the mesophyll cells and enter the stroma of the chloroplasts where an enzyme releases the CO_2, which then enters into the Calvin cycle.

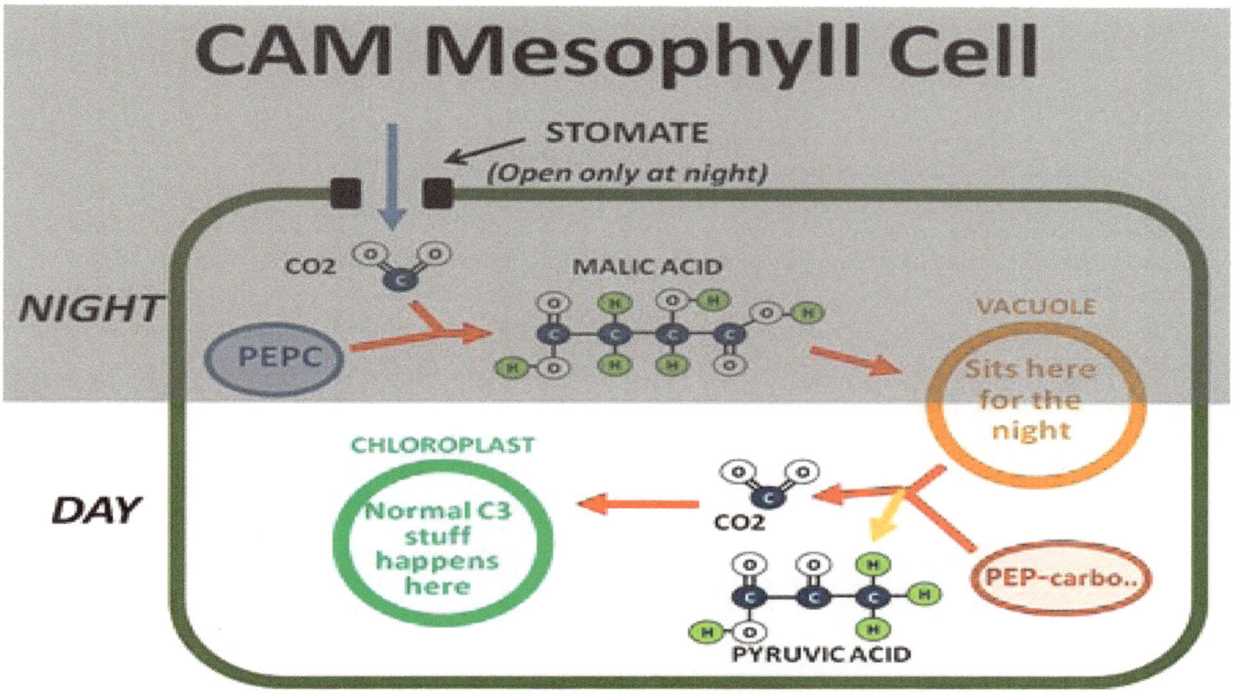

Figure 7: CAM Pathway in Plants (www.cais.uga.edu).

Photorespiration

Photorespiration is a biochemical process where oxygen inhibits the Calvin cycle especially under conditions of water stress. Photorespiration results in the light-dependent uptake of oxygen and release of carbon dioxide and is associated with the synthesis and metabolism of a small molecule called glycolate. In photosynthesis, carbon dioxide is taken in and in photorespiration carbon dioxide is given off. These two processes work against each other. Photorespiration is defined as a light dependent uptake of O_2 and output of CO_2. Photorespiration occurs in three orgenelles viz, chloroplast, peroxisomes and mitochondria. In the chloroplast, phosphoglycolate is unstable and is converted to glycolate. PGA is used up in the calvin cycle. The peroxisomes present in the cell convert the glycolate into glyoxylate and then into glycine. Glycine is converted to serine and CO_2 in the mitochondria. The serine thus formed is converted to glycerate through a series of reactions which occur in peroxisome (Figure 8).

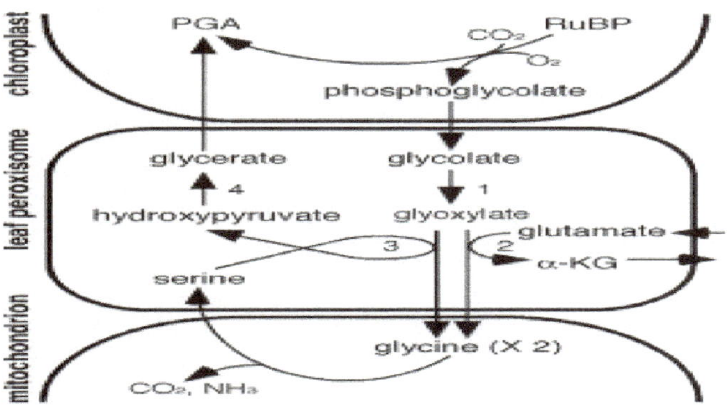

Figure 8: Photorespiration in Plants (www.tutorhelpdesk.com).

REGULATION OF PLANT GROWTH

Plant Growth Factors

Phytohormones or plant growth regulators are signal molecules produced within the plant and occur in low concentrations.

There are five classes of phytohormones: auxins, cytokinins, gibberellins, abscisic acid and ethylene.

Auxins

In nature, auxins bring about elongation of stem and internodes, apical dominance, abscission etc. Auxins were the first class of growth regulators discovered. 3-Indole butyric acid (IBA) and indole acetic acid (IAA) are commonly used auxins.

Cytokinins

These hormones are reported to cause cell division, modification of apical dominance, shoot differentiation. Cytokinins have a highly synergistic effect in concert with auxins, and the ratios of these two groups of plant hormones affect most major growth periods during a plant's lifetime.

Gibberellins

These hormones are reported to cause stem elongation, germination, dormancy, flowering, sex expression, enzyme induction, and leaf and fruit senescence. Gibberellins comprise more than twenty compounds, of which GA3 is the most frequently used gibberellin. Abscisic acid is a strong antagonist of Gibberellins action.

Abscisic acid

Abscisic acid (ABA) is often referred to as inhibitory rather than stimulatory hormone. Abscisic acid is involved in several plant functions. It is involved in the closure of stomata, bud and seed dormancy and is known to inhibit other hormonal actions. The plant hormone abscisic acid (ABA) is the major role in mediating the adaptation of the plant to stress.

Ethylene

Ethylene is a gaseous plant hormone. It is also called as fruit ripening gas. The effects of ethylene include fruit ripening, loss of chlorophyll, abortion of plant parts, stem shortening, abscission of plant parts, and epinasty (bending of stems). Ethylene also plays a role in response to water stress, chilling and mechanical injury.

REFERENCES

1. Bold, H. C., Alexopoulos, C. J., & Delevoryas, T. (1987). Morphology of plants and fungi. Harper and Row.
2. Farquhar, G. D., & Sharkey, T. D. (1982). Stomatal conductance and photosynthesis. Annual review of plant physiology, 33(1), 317-345.
3. Sage, R. F., Sage, T. L., & Kocacinar, F. (2012). Photorespiration and the evolution of C4 photosynthesis. Annual review of plant biology, 63, 19-47.
4. Miransari, M., & Smith, D. L. (2014). Plant hormones and seed germination. Environmental and Experimental Botany, 99, 110-121.

CHAPTER 2

MORPHOLOGY, STRUCTURE, AND ONTOGENY OF TRICHOMES OF THE GRAPE GENUS (VITIS, VITACEAE)

Zhi-Yao Ma[1], Jun Wen[2], Stefanie M. Ickert-Bond[3], Long-Qing Chen[1], and Xiu-Qun Liu[1]

[1]Key Laboratory of Horticultural Plant Biology, College of Horticulture and Forestry Science, Huazhong Agricultural University, Ministry of Education, Wuhan, China

[2]Department of Botany, National Museum of Natural History, MRC166, Smithsonian Institution, Washington, DC, USA

[3]UA Museum of the North Herbarium and Department of Biology and Wildlife, University of Alaska Fairbanks, Fairbanks, AK, USA

ABSTRACT

Trichomes are widely distributed on surfaces of different organs in the grape genus Vitis and are of taxonomic utility. To explore the morphology, structure and ontogeny of Vitis trichomes, we investigated the diversity and distribution of trichomes in 34 species of Vitis. Two main types of trichomes in Vitis are documented: non-glandular and glandular. Within non-glandular trichomes, ribbon and simple trichomes are found on different vegetative plant organs. The morphology and ontogeny of these types of trichomes are further examined with light microscopy and scanning electron microscopy. The ultrastructure of the glandular trichomes is explored with transmission electron microscopy. The ribbon trichomes are twisted, greatly elongated and unicellular, and this trichome type may be a morphological synapomorphy of Vitis and its closest tropical relative Ampelocissus and Pterisanthes in Vitaceae. The simple trichomes are documented in most species sampled in the genus. The glandular trichomes are multicellular, non-vascularized and composed of both epidermis and subjacent layers. We show that prickles occurring along the stems and petioles of Vitis davidii are modified glandular trichomes. We observed that glandular trichomes of V. romanetii secrete mucilage and volatile substances which trap insectes on the glands. Transmission electron microscopy indicates that metabolic products accumulate in vacuoles, the cytoplasm and intercellular spaces. We infer that glandular trichomes and young prickles are involved in the secretion of these metabolic products and the intercellular spaces may be the places of temporary storage of these secretions.

INTRODUCTION

Trichomes usually originate from epidermal cells and are ubiquitous in many plant families, showing great diversity in morphology, cellular structure and function (Uphof, 1962). These epidermal appendages are unicellular or multicellular, branched or unbranched, glandular or non-glandular (Levin, 1973; Werker, 2000; Yang and Ye, 2013). They have been observed on all vegetative and reproductive organs in angiosperms (Werker, 2000; Wagner et al., 2004). These special structures can protect from herbivory by large mammals, insect and pathogens attacks, increase light reflectance, regulate temperature and decrease water loss (Levin, 1973; Wagner et al., 2004).

Grapes have been widely recognized for their agronomic and economic importance as fresh fruits, and sources of wine and raisins (Wen, 2007). Other species of the grape genus (*Vitis* L.) are important germplasm resources for the wine grape (*V. vinifera* L.), and have been bred to improve resistance

Citation: Ma Z-Y, Wen J, Ickert-Bond SM, Chen L-Q, Liu X-Q. Morphology, Structure, and Ontogeny of Trichomes of the Grape Genus (Vitis, Vitaceae). Frontiers in Plant Science. 2016; 7:704. doi:10.3389/fpls.2016.00704.

of cultivated varieties against many fungal diseases and to enhance cold tolerance (Staudt and Kassemeyer, 1995; Wang et al., 1995; Staudt, 1997; Brown et al., 1999; Wan et al., 2007; Zhang et al.,2012; Gerrath et al., 2015).

Vitis includes ca. 70 species mostly in the temperate regions of the Northern Hemisphere (Chen et al.,2007; Wen, 2007; Zecca et al., 2012; Moore and Wen, 2016). *Vitis* has been shown to be part of the*Ampelocissus–Vitis* clade, one of the five major clades in Vitaceae (Soejima and Wen, 2006; Liu et al.,2013, 2016; Wen et al., 2013). The genus *Vitis* has been widely studied over the past 10 years and confirmed to include two subgenera (subgenus *Muscadinia* and subgenus *Vitis*) (Mullins et al., 1992; Soejima and Wen, 2006; Aradhya et al., 2008; Tröndle et al., 2010; Péros et al., 2011; Zecca et al.,2012; Miller et al., 2013; Wan et al., 2013; Liu et al., 2016). Subgenus *Muscadinia* includes only two species: *V. popenoei* J. H. Fennel from Mexico and *V. rotundifolia* Michx. from the south-eastern U.S.A. (Brizicky, 1965; Moore, 1991; Wen, 2007), while, subgenus *Vitis* includes the Eurasian and the New World clades (Péros et al., 2011; Zecca et al., 2012; Wan et al., 2013; Liu et al., 2016). The phylogeny and taxonomy within the subgenus *Vitis* have not been well resolved due to the highly similar morphology, overlapping distribution, introgression and interspecific hybridization among species within the clade (Aradhya et al., 2008; Moore and Wen, 2016). Many *Vitis* species are differentiated and delimited mainly by the types, distribution and density of trichomes (Moore, 1991; Chen et al., 2007).

However, trichomes of *Vitis* have rarely been studied systematically. Detailed documents of trichome types and distribution in vegetative parts of *Vitis* have not been shown in previous studies (Galet, 1988; Moore, 1991; Chen et al., 2007). He et al. (1994) and Cheng et al. (2013) have explored trichome diversity and some anatomy in *Vitis* trichomes. Nonetheless, they did not show clear and detailed anatomical structures and distribution of trichomes and did not perform any developmental studies of trichomes. The lack of structural and ontogenetic studies on *Vitis* trichomes has limited the exploration of their function and the significance as the basis for inferring the evolutionary relationships among the species in *Vitis*.

Vitis davidii is the only species with prickles on stems and petioles in *Vitis*. The unique character was considered to distinguish *V. davidii* from other species in *Vitis* (Galet, 1988; Chen et al., 2007). However, very little is known about its structure, ontogeny, function and the relationship with other trichome types.

The primary purpose of this study is to: (1) document types and distribution of *Vitis* trichomes to provide available and reliable information for taxonomy, and (2) to explore the anatomical structure, ultrastructure and ontogeny of *Vitis* trichomes, especially prickles of *V. davidii*, to infer the relationship among trichomes and prickles.

MATERIALS AND METHODS

Plant Materials

This study sampled 34 *Vitis* species covering three different distribution areas: 27 from Asia, six from North America and one from Europe (Table 1).

Table 1: The distribution of trichomes on different vegetative organs in *Vitis*.

Species		Plant parts									Number of plants
		Current year's branchlets	Young leaves	Mature leaves							
				Petioles	Leaf surfaces					Leaf margins	
					Adaxial		Abaxial				
					Leaf epidermis	Leaf veins	Leaf epidermis	Leaf veins	Leaf vein axils		
ASIAN SPECIES											
1	*V. adenoclada*[123]	*G+ R*	*G R+ S*	*G R*		*R S*	*R+*	*R S*			3
2	*V. amurensis*	*R*	*R+ S*	*S**				*R* S**	S		4
3	*V. balansana*	*R*	*R+*				R*				3
4	*V. bellula* var. *pubigera*	*R*	*R+ S*	*S*		*S*	*R+*	*R S*		*S*	3
5	*V. betulifolia*	*R*	*R+ S*	*R*		*R S*	*R*	*R S*			1
6	*V. bryoniifolia*	*R*	*R+ S*	*R S*		*R S*	*R*	*R S*		*S*	4
7	*V. chunganensis*[123]	*G* R**	*G* R**	*G**				*G**			3
8	*V. chungii*		*R* S**			*S**		*S**		*S**	5

9	*V. davidii*[123]	*P+*	*P* *R* *S*	*P*		S		*R* *S*	*S*	*S*	3
10	*V. erythrophylla*		*R* *S*	*S*		*R* *S*		*R* *S*		*S*	3
11	*V. flexuosa*[1]	*S*	*R* *S*	*S*		S		S	S		5
12	*V. hancockii*[1]	*S+*	*R* *S+*	*R* *S+*		S		*R* *S+*	*S*	*S*	4
13	*V. heyneana*	*R*	*R+* *S*	*R*		*R* *S*	*R+*	*R* *S*			3
14	*V. heyneana* subsp. *ficifolia*	*R*	*R+* *S*	*R*		*R* *S*	*R+*	*R* *S*			3
15	*V. hui*	*R* *S*	*R+* *S*	*R* *S*	*S+*	*R* *S*	*R+*	*R* *S*		*S*	3
16	*V. lanceolatifoliosa*	*R*	*R+* *S*	*R*		*R* *S*	*R+*	*R* *S*			3
17	*V. piasezkii*	*R** *S**	*R* *S*	*R** *S**	*S+*	*S*	*R**	*S+*		*S*	3
18	*V. piasezkii* var. *pagnucii*		*R* *S*	*S*		S		*R* *S*	S		3
19	*V. pseudoreticulata*	*R*	*R+* *S*	*R* *S*		*R* *S*		*R* *S*	*S*	*S**	5
20	*V. retordii*[12]	*R*	*R+* *S+*	*R* *S**	*S+*	*S*	*R+*	*R* *S*		*S*	3
21	*V. romanetii*[123]	*G+* *R* *S+*	*G+* *R+* *S+*	*G* *R* *S+*		*S*	*R*	*G* *R* *S+*		*S*	3
22	V. ruyuanensis		S	S		S		S		*S*	4
23	*V. shenxiensis*[12]	*G* *R*	*G* *R* *S*	*G* *R* *S*		S		S	*S*	*S*	1
24	*V. sinocinerea*	*R* *S*	*R+* *S*	*R* *S*		*S*	*R+*	*R* *S*		*S*	3
25	*V. tsoii*	*S*	*R* *S*	*R* *S*		*R* *S*	*R**	*R* *S*		*S*	3
26	*V. wilsoniae*[12]	*R*	*R+* *S*	*R* *S*		S		*R* *S*	*S*	*S*	3
27	*V. wuhanensis*	*R*	*R+* *S*	*R* *S*		*R* *S*		*R* *S*		*S*	3
EUROPEAN SPECIE											
28	*V. vinifera*	*R*	*R* *S**	*R* *S**			*R**	*R* *S**			4
NORTH AMERICAN SPECIES											
29	*V. labrusca*	*G** *R*	*R+* *S*	*G** *R*		*S*	*R+*	*R* *S*		*S*	1
30	*V. mustangensis*	*R*	R+	R			R+	R			1
31	*V. riparia*		S	S		S		*S*	*S*	*S*	1
32	V. rupestris		*R* *S*	*S*		S		*S*	*S*	*S*	1
33	*V. shuttleworthii*	*R*	*R+*	*R*			R+	R			1
34	*V. rotundifolia*	*G** *S*	*R* *S*	*S*		S		S	S		1

Abbreviations in table: R, ribbon trichomes; G, glandular trichomes; P, prickles; S, simple trichomes. The blank spot indicate that the organ is glabrous. The numbers "1," "2," "3" indicate the species were examined with light, scanning electron and transmission electron microscopy respectively. "+" indicate that the trichome type is stable present and forms remarkable dense indumentum on the vegetative organ.

""indicate that the trichome type is variable and not stable present on the vegetative organ. The unmarked indicate that the trichome type is stable present, but not forms remarkable dense indumentum on the vegetative organ.*

Stereomicroscope

Herbarium specimens and fresh vegetative parts (current year's branchlets, young leaves, petioles, adaxial and abaxial leaf surfaces of mature leaves, and leaf margins) of *Vitis* were examined to characterize the types and distribution of trichomes in 34 *Vitis* species using a Nikon SMZ800 stereomicroscope. At least three plants were sampled for most species, except *V. betulifolia*, *V. shenxiensis* and all North American species for which there was only one individual sampled. Young leaves were taken from shoot tips, while mature ones were taken from nodes after the 9th node (counting from the top) to assure their mature condition. Five young and five mature leaves from each plant were sampled respectively. The presence/absence and types of trichomes on different vegetative organs were documented.

detailed anatomy, ontogeny and ultrastructure studies of *Vitis* trichomes were conducted using light microscopy (LM), scanning electron microscopy (SEM) and transmission electron microscopy (TEM).

Light Microscopy

To examine the anatomical structure of trichomes, 33 samples of stems, petioles and leaf mid-veins, from nine *Vitis* species (*V. adenoclada*, *V. chunganensis*, *V. davidii*, *V. flexuosa*, *V. hancockii*, *V. retordii*,*V. romanetii*, *V. shenxiensis*, and *V. wilsoniae*), were fixed in 70% FAA for 24 h. Samples were dehydrated in an ethanol series (Johansen, 1940) and embedded in paraffin wax. Transversal and longitudinal sections of 6–8 μm were cut with a Leica RM2235 rotary microtome, mounted on slides and stained with 1% safranin O and 0.5% fast green.

Scanning Electron Microscopy

Thirty-eight samples including stems, petioles and leaves from representatives of eight species (*V. adenoclada*, *V. chunganensis*, *V. davidii*, *V. heyneana*, *V. retordii*, *V. romanetii*, *V. shenxiensis*, and *V. wilsoniae*) were fixed in 2.5% glutaraldehyde in 0.1M phosphate buffer, pH7.2. The fixed samples were dehydrated in an ethanol series and critical point dried. Samples were placed with double-sided tape on stubs and sputter-coated with a thin gold layer (Robards, 1978). The types and morphology of trichomes were examined with a JEOL JSM-6390LV Scanning Electron Microscope. Based on the SEM images, we measured the heights/lengths of different trichome types and the sizes of parts of glandular trichomes or prickles.

Transmission Electron Microscopy

Four samples of glandular trichome or young prickle from *V. adenoclada*, *V. chunganensis*, *V. davidii*, and *V. romanetii* were fixed in 1% osmium tetroxide, in 0.1M phosphate buffer, pH7.2, and processed following standard techniques (Roland and Vian, 1991). Ultra-thin sections of both glandular trichomes and the prickles were made and viewed under a Hitachi H-7650 Transmission Electron Microscope.

Histochemical Lignin Estimation

The histochemical lignin tests for prickles of different growth stages of *V. davidii* were performed in fresh hand-cut sections. The sections were soaked in a fresh phloroglucinol solution, and then a few drops of 35% HCl were added (Pillonel et al., 1991). The positive result will shows an intense pink color.

RESULTS

Trichome Types

We used the terminology of Payne (1978) to describe the trichomes of *Vitis* in this study. Two general trichome types in *Vitis* are documented: non-glandular and glandular. Non-glandular trichomes include ribbon and simple trichomes.

Non-Glandular Trichomes

In *Vitis* the distribution and densities of non-glandular trichomes vary among different vegetative organs including current year's branchlets, young leaves and mature leaves, as well as on petioles, adaxial and abaxial leaf surfaces and leaf margins (Figures 1A–I). Almost all species sampled bear the ribbon and simple trichomes on different plant organs (Table 1).

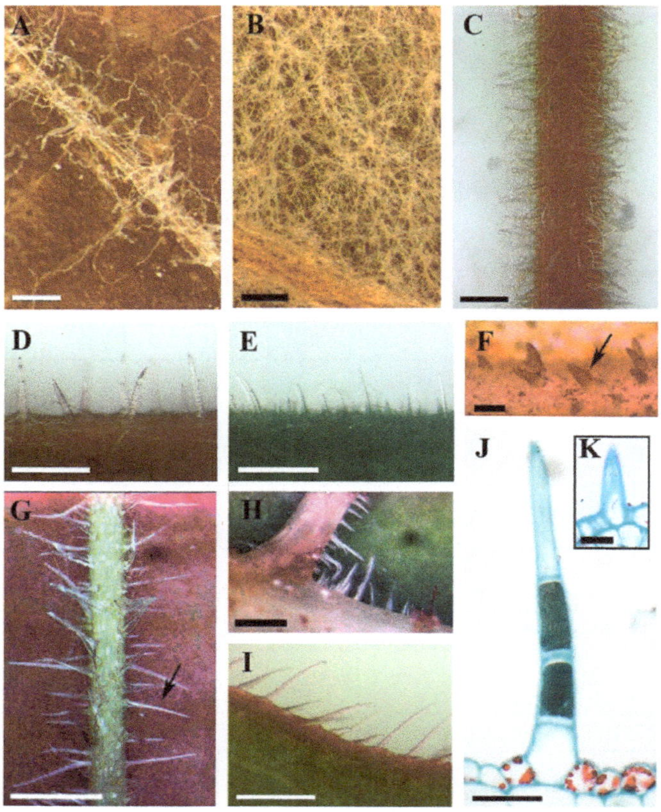

Figure 1: Distributions and structure of ribbon and simple trichomes. (A,B) Ribbon trichomes on different plant organs. Ribbon trichomes on adaxial leaf surfaces of *V. bryoniifolia* (A) and on abaxial leaf surfaces of *V. adenoclada* (B). (C–I) Simple trichomes on different plant organs. Simple trichomes on branchlets of *V. hancockii* (C), petioles of *V. pseudoreticulata* (D), adaxial leaf surfaces of *V. retordii* (E), adaxial leaf veins (F) and abaxial leaf veins of *V. hancockii* (G), abaxial leaf vein axils of *V. shenxiensis* (H) and leaf margins of *V. piasezkii* (I). Arrow in (F) shows a simple unicellular trichome and in (G) shows a simple uniseriate multicellular trichome. (J) Photomicrograph of microtome section of a simple trichome with uniseriate cells of *V. hancockii*. (K) A simple unicellular trichome of *V. davidii*. Scale bars: (A,B) = 0.25 mm; (C,E) = 1 mm; (D,G–I) = 0.5 mm; (F) = 100 μm; (J) = 50 μm; (K) = 20 μm.

Ribbon trichomes are flattened, slender, greatly elongated, twisted and unicellular (Figures 1A,B, 2A,B). Ribbon trichomes can be as long as several millimeters. In total, 32 of the 34 species sampled bear the ribbon trichomes (Table 1). Twenty-three of the 34 species sampled have ribbon trichomes on current year›s branchlets. On mature leaves, we observed ribbon trichomes on the petioles of 21 of the 34 species sampled. In addition, on the adaxial leaf surfaces of some species (e.g., *V. adenoclada* and *V. bryoniifolia*) only the leaf veins bear sparse ribbon trichomes (Figure 1A; Table 1). On the abaxial leaf surfaces, ribbon trichomes are found in 26 species, of which 11 species have non-deciduous dense arachnoid indumentum composed of this trichome type (Figure 1B; Table 1; *V. adenoclada*, *V. bellula* var. *pubigera*, *V. heyneana*, *V. heyneana* subsp. *ficifolia*, *V. hui*, *V. lanceolatifoliosa*, *V. retordii*, *V. sinocinerea*, *V. labrusca*, *V. mustangensis*, and *V. shutterworthii*).

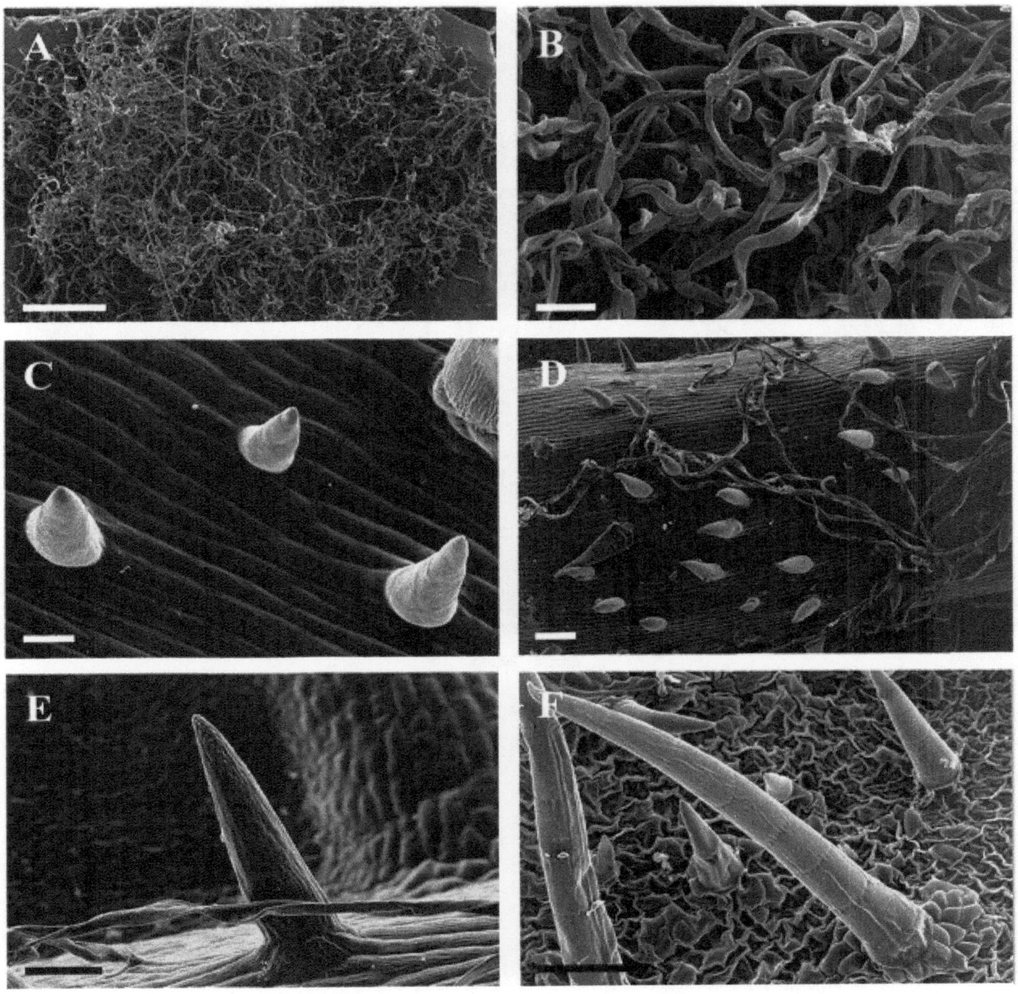

Figure 2: SEM micrographs of ribbon and simple trichomes. (A,B) Ribbon trichomes on abaxial leaf surfaces of a young leaf in *V. wilsoniae*. (C–F) Simple trihcomes: (C) young simple trichomes on veins of *V. romanetii*; (D,E) simple trichomes on veins of young leaves in *V. davidii*; (F) simple trichomes on adaxial leaf surfaces of *V. retordii*. Scale bars: (A) = 500 μm; (B,E) = 50 μm; (C) = 20 μm; (D,F) =100 μm.

Simple trichomes are unbranched, uniseriate, or unicellular (Figures 1C–J, 2C–F), and originate from the epidermis (Figure 2C). Simple trichomes are found in 30 of the 34 *Vitis* species examined, but their distribution on different plant organs varies (Table 1). On mature leaves, only eight species have simple trichomes on the current year's branchlets (Figure 1C), and 22 species have this trichome type on petioles (Figure 1D), while 28 species bear simple trichomes on adaxial leaf surfaces (Figures 1E,F). Simple trichomes are present on the veins of abaxial leaf surfaces in 30 species (Figure 1G), and 11 species (*V. amurensis*, *V. davidii*, *V. flexuosa*, *V. hancockii*, *V. piasezkii* var. pagnucii, *V. pseudoreticulata*, *V. shenxiensis*, *V. wilsoniae*, *V. riparia*, *V. rupestris*, and *V. rotundifolia*) bear simple trichomes on the abaxial leaf vein axils (Figure 1H). On leaf margins, 20 species bear simple trichomes (Figure 1I). We observed dramatic length differences in simple trichomes ranging from 50–200 μm in length on veins of adaxial and/or abaxial leaf surfaces in *V. adenoclada*, *V. davidii*, *V. hancockii*, *V. heyneana*, and *V. wilsoniae* (Figures 1F, 2D,E), while some of simple trichomes are up to ca. 750–900 μm long in *V. romanetii* and *V. retordii* (Figure 2F).

Almost all species examined (30/34) show an indumentum composed of ribbon trichomes on young leaves (Figures 3A–E), which may become sparse or glabrescent on mature leaves (Table 1). Ribbon trichomes are only occasionally observed on the young leaves of some species (*V. chunganensis* and *V. chungii*) (Table 1). Thirty species sampled were covered with simple trichomes on young leaves (Figure 3F; Table 1) and some species sampled are nearly glabrous (Figure 3G). Twenty-eight species sampled bear both ribbon and simple trichome types (Figure 3A, Table 1).

Figure 3: Young leaves with ribbon and/or simple trichomes. (A–E) Dense ribbon trichomes on abaxial leaf surfaces of *V. romanetii*(A), *V. lanceolatifoliosa* (B) and *V. retordii* (C) and adaxial leaf surfaces of *V. wuhanensis* (D) and *V. wilsoniae* (E). (F) Dense simple trichomes on abaxial leaf surfaces of *V. hancockii*. (G) *V. chungii* nearly glabrous on the abaxial leaf surfaces. Scale bars: (A–E,G) = 3 mm; (F) = 4 mm.

The non-glandular trichome density, especially the ribbon trichome density, changed on leaves of some species of *Vitis* (e.g., *V. amurensis* and *V. wilsoniae*) during development of leaves. The ribbon trichomes become lost with development of the leaves in some species of *Vitis*.

Glandular Trichomes and Prickles

Glandular trichomes are observed in six species (*V. romanetii*, *V. adenoclada*, *V. chunganensis*, *V. labrusca*, *V. rotundifolia*, and *V. shenxiensis*), but prickles are only observed in *V. davidii*. Glandular trichomes vary in shape, size, density, distribution and color among *Vitis* species in this study (Figures 4A–J, 5A–I). All of them are multicellular, non-vascularized and composed of both epidermis and subjacent layers (Figures 4B,E,G,J). The glandular trichomes of these species are composed of apparent glandular heads with multiple layers of epidermal cell and stalks with one layer of epidermal cell (Figure 4J). Stomata are frequently observed on the upper region of glandular heads in most *Vitis*species sampled except in *V. chunganensis* (Figures 5C,F,H,I).

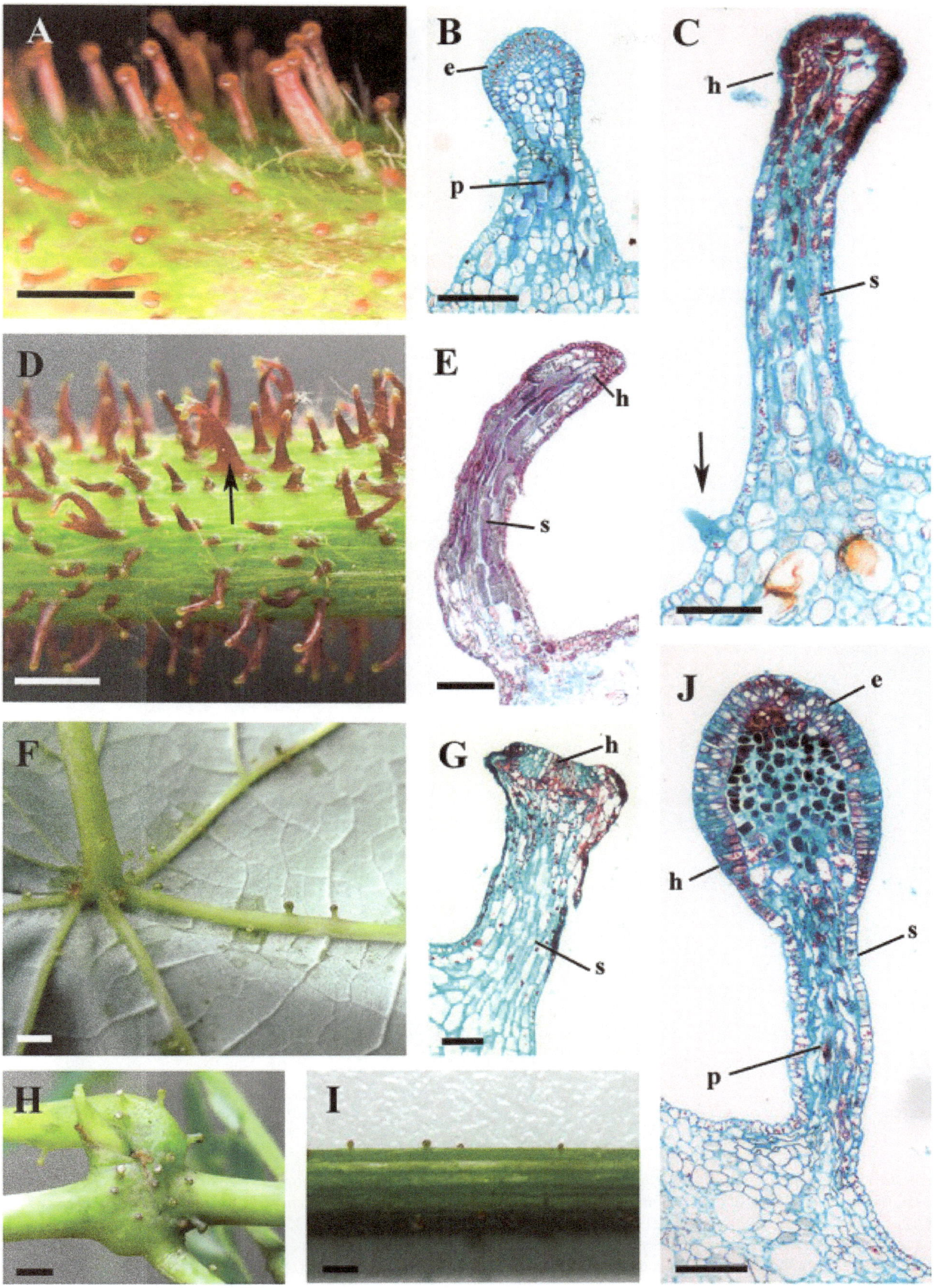

Figure 4: Structures and distributions of glandular trichomes in *Vitis*. (A–C) Glandular trichomes of *V. romanetii*: (A) glandular trichomes on stems; (B,C) structures of glandular trichomes and simple trichomes on the epidermis of stalks of glandular trichomes (arrow). (D,E) Glandular trichomes on the stems of *V. adenoclada*: glandular trichomes on stems and some of these trichomes form complexes (arrow) (D). (F–H) Glandular trichomes of *V. chunganensis*: glandular trichomes on basal parts of veins of abaxial leaf surfaces (F) and nodes (H). (I,J) Glandular trichomes on stems of *V. shenxiensis*. Abbreviations: e, epidermal cells; p, parenchymatic cells; h, head; s, stalk. Scale bars: (A,D,F,H,I) = 1 mm; (B,C,E,G,J) = 100 μm.

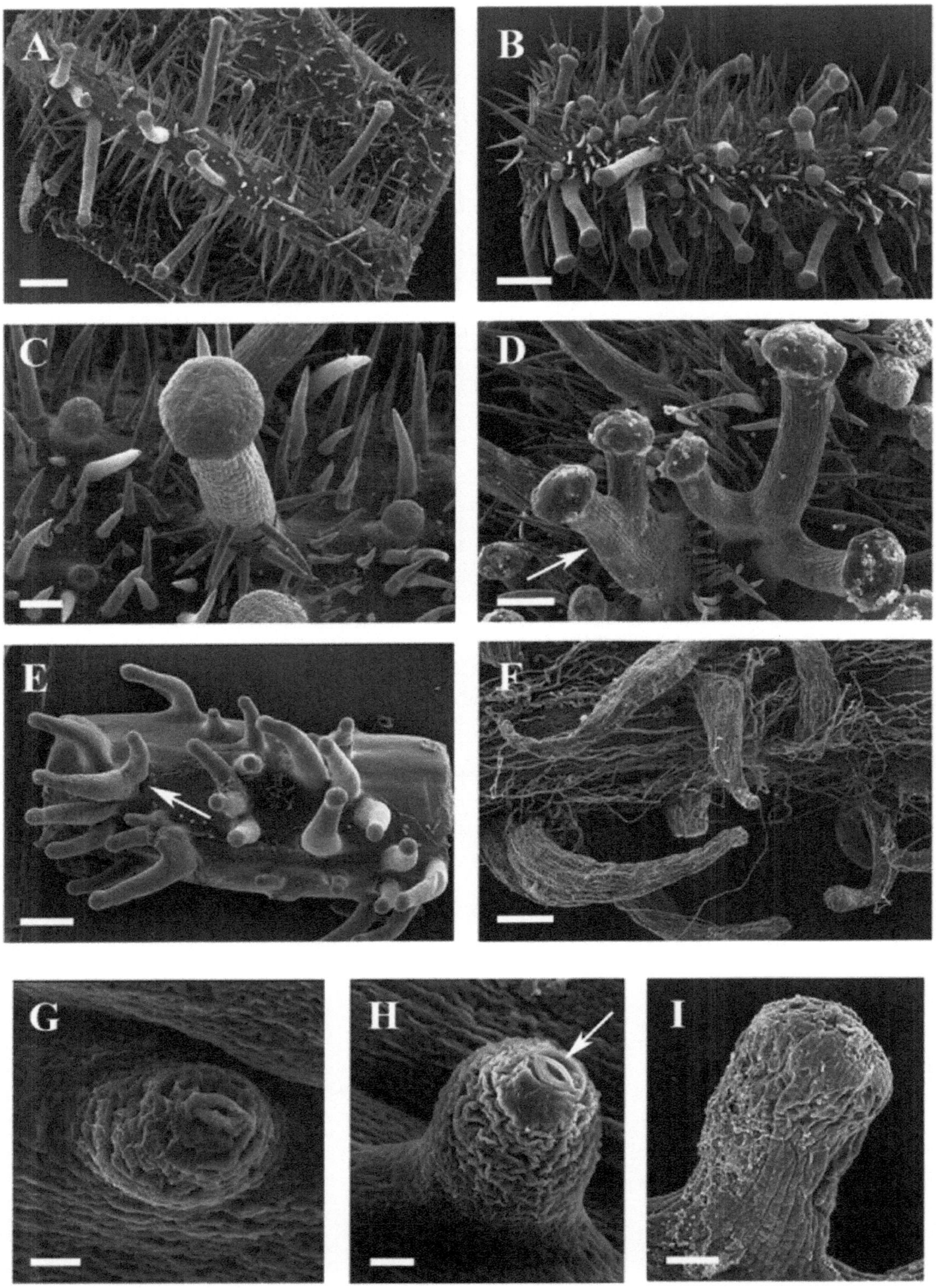

Figure 5: SEM micrographs of glandular trichomes of *Vitis*. (A,B) Glandular trichomes and simple trichomes on abaxial leaf surfaces (A) and petioles (B) of *V. romanetii*. (C) Simple trichomes on the epidermis of stalks of glandular trichomes of *V. romanetii*. (D) Trichome complexes of glandular trichomes (arrow) of *V. romanetii*. (E) Complexes (arrow) of prickles on petioles of *V. davidii*. (F) Glandular trichomes with hooked stalks on stems of *V. adenoclada*. (G,H) Emerging cell masses (G)and a glandular head without stalk and an obvious stoma (arrow) on the top of glandular head (H) in *V. shenxiensis*. (I) A glandular trichome on nodes of *V. chunganensis*. Scale bars: (A,B,E) = 500 μm; (C) = 100 μm; (D,F) = 200 μm; (G,H) =20 μm; (I) = 50 μm.

Developmentally, we can recognize three stages of glandular trichome growth (Figures 6A–E, 7A–F). Initially, cell masses emerge from the epidermis where stomata may develop with rapid proliferation and elongation of upper region epidermal cells (Figures (Figures 5G,5G, 7A,B), forming a sessile glandular head (Figures 5H, 6A–C, 7C). Secondly, rapid proliferation and expansion of epidermal cells and underlying parenchymatous cells cause a stalk to develop (Figures 6D, 7D,E). Finally, glandular trichomes fully mature (Figures 6E, 7F) and proceed to the senescence period in which they appear shrunken (Figure 5I).

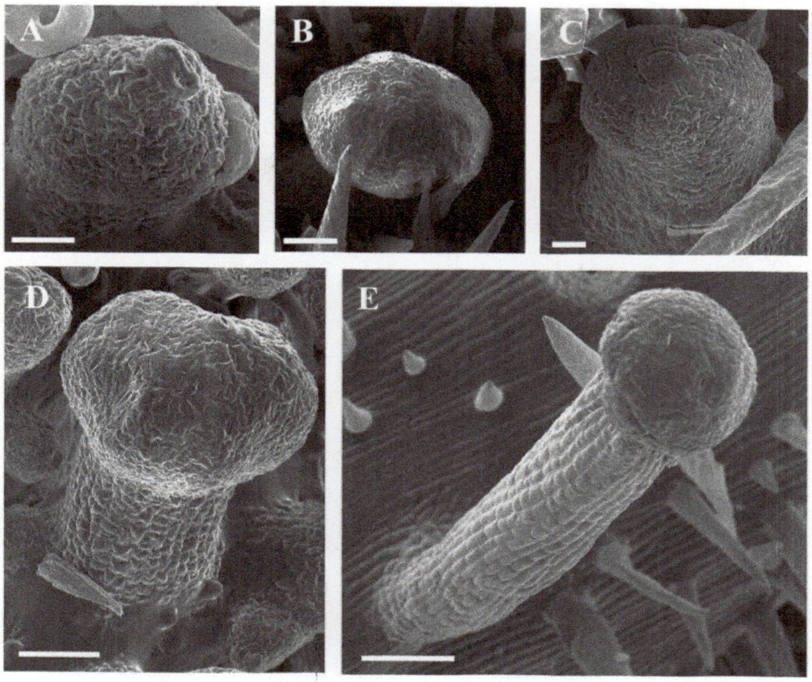

Figure 6: SEM micrographs of development of glandular trichomes of *V. romanetii*. (A–C) Glandular heads without apparent stalks.(D) Young glandular trichome with short stalk. (E) Mature glandular trichome with long stalk. Scale bars: (A) = 20 μm; (B) = 50 μm; (C) = 20 μm; (D) = 50 μm; (E) = 100 μm.

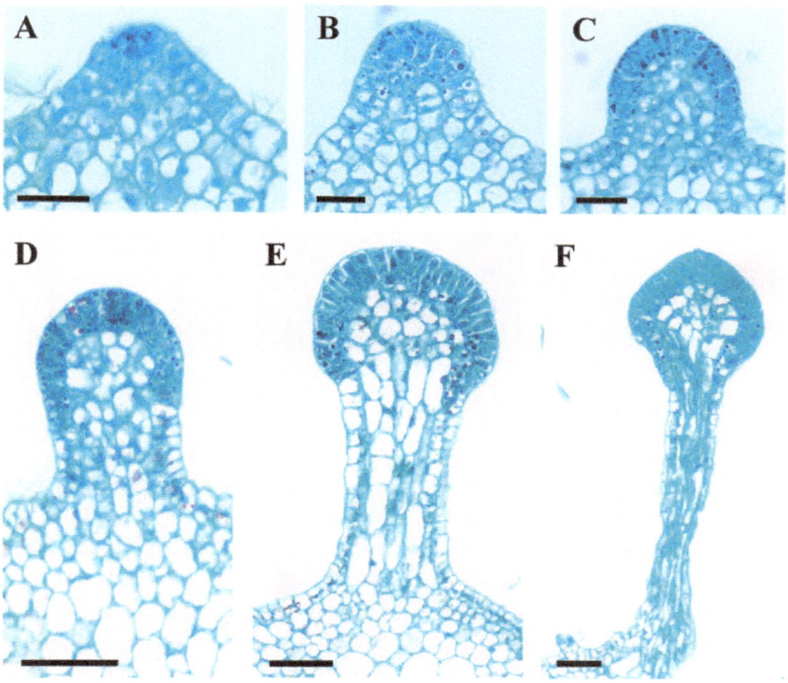

Figure 7: Development of glandular trichomes of *V. romanetii*. (A) Emerging cell masses. (B) Original gland head with rapid proliferation and elongation of upper region epidermal cells. (C) Emerging glandular head without stalk. (D) Young glandular trichome with short stalk. (E) Elongation of stalk of young glandular trichome. (F) Mature glandular trichome. Scale bars:(A–C) = 50 μm; (D–F) = 100 μm.

The predominant distributions of dense glandular trichomes of *V. romanetii* are on branchlets, petioles and veins of abaxial leaf surfaces alongside dense and long simple trichomes (Figures 4A–C, 5A–D). The glandular trichomes are pink/yellow, erect and ca. 1200 μm long. Diameters of their glandular heads are ca. 200 μm. They emerge and continue to grow with the extension of stalks until maturity. Interestingly, adjacent glandular trichomes may produce trichome complexes which may be formed after fusion or branching, meanwhile, this similar structure is also observed in *V. adenoclada* and *V. davidii* (Figures 4D, 5D, **E**). Simultaneously, many simple trichomes are observed on the epidermis of the stalks of glandular trichomes (Figures 4C, 5C, D).

The glandular trichomes of *V. romanetii* are mucilaginous and emit volatile substance based on our observation. Many insects are often glued to their glands (Figures 8A–C).

Figure 8: Insects were glued to the glandular trichomes of *V. romanetii*(A–C). Arrows indicate stuck insects. Scale bars: (A–C) = 3 mm.

Glandular trichomes of *V. adenoclada* have wine-red and hooked stalks and faint-yellow glandular heads (Figures 4D, E, 5F). They are commonly observed on branchlets and petioles. The height of these mature trichomes is up to 1100 µm and the diameter of the glandular heads is ca. 80–100 µm.

The glandular trichomes of *V. chunganensis* are present on basal veins of abaxial leaf surfaces, petioles and at along stem nodes (Figures 4F,H) and have short stalks and inverted trapezoid heads (Figures 4G, 5I). The height of these trichomes is only 80–550 µm and the diameter of the glandular heads is 100–350 µm.

Glandular trichomes of *V. shenxiensis* bear similar structures to those of *V. romanetii*, but are less dense on branchlets (Figures 4I,J). Furthermore, these trichomes are ca. 650 µm long and have large glandular heads with diameters up to 250 µm.

The prickles of *V. davidii* are similar to those of other glandular trichomes in *Vitis*. They are also multicellular, non-vascularized and composed of both epidermis and subjacent layers. The prickles have stalks with one layer of epidermal cells and apparent glandular heads with multiple layers of epidermal cells on which stomata are frequently observed. Similar to early developmental stages of glandular trichomes, initially, the cell masses emerge from the epidermis with notable stomata at ca. 200 µm in diameter (Figure 9A); then the glandular heads grow without stalks (Figure 9B); and finally, the stalks of prickles form with rapid proliferation and expansion of epidermal and parenchymatous cells (Figures 9C,D, 10A). However, the prickles increase in both height and basal width during development (Figure 9E). They become conical and glaucous (Figure 10B) and lignified at maturity (Figure 10C). Some glandular heads of prickles shrink with lignification (Figures 9F,G). Finally, they become hard and prickly (positive result for histochemical lignin estimation) (Figures 9H, 10D).

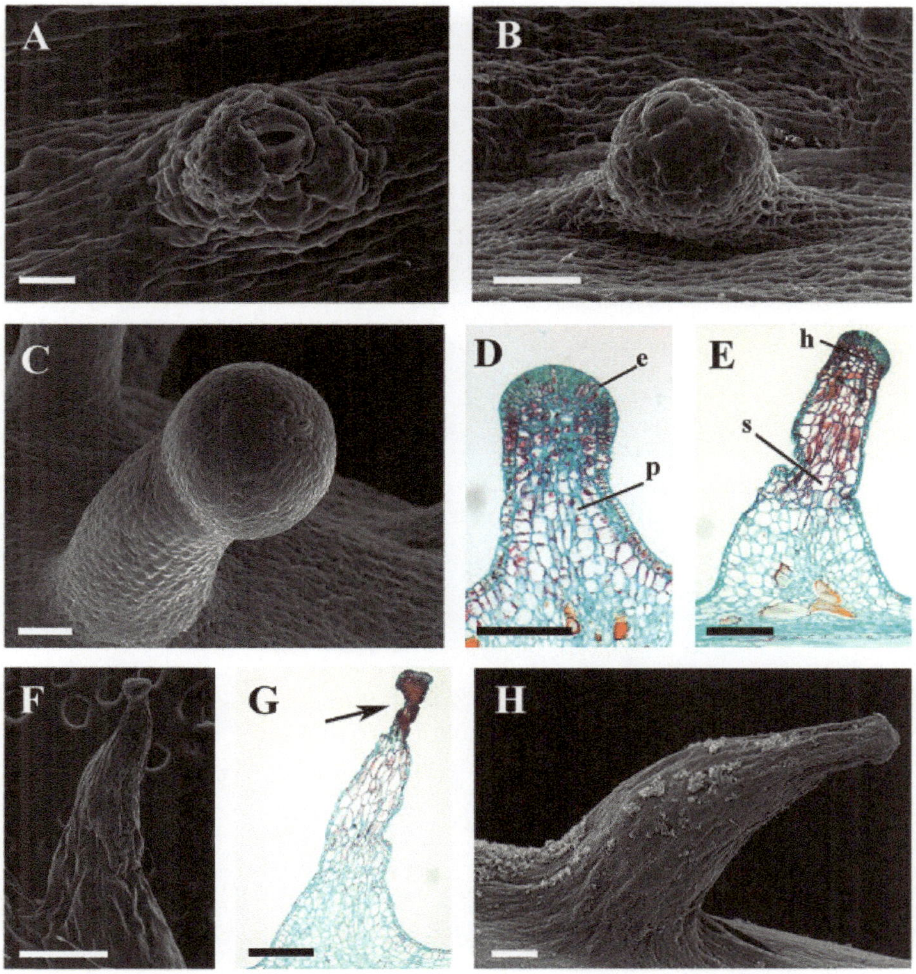

Figure 9: Development and structure of prickles of *V. davidii*. (A) Emerging cell masses with a notable stoma and (B) a glandular head without stalk. (C) Young prickle with stalk. (D) Structure of the young prickle. (E) Mature prickle. (F) Shrinking prickle. (G) Senescent prickle, arrow indicating the shrinking top of gland. (H) The lignifying prickle. Abbreviations: e, epidermal cells; p, parenchymatic cells; h, head; s, stalk. Scale bars: (A) = 20 μm; (B,C) = 50 μm; (D) = 100 μm; (E,H) = 200 μm; (F) = 500 μm; (G) =250 μm.

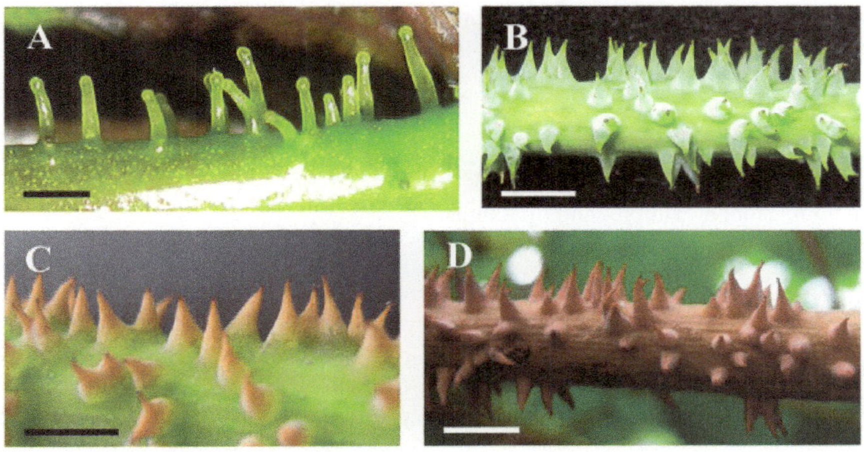

Figure 10: Development and morphology of prickles of *V. davidii*. (A)Young prickles on petiole. (B) Mature prickles on stem. (C) The withering and lignifying terminal part of prickles. (D) Fully lignifying prickles. Scale bars: (A) = 1 mm; (B–D) = 5 mm.

Ultrastructure

The glandular heads of *V. davidii*, *V. adenoclada*, *V. chunganensis*, and *V. romanetii* showed similar ultrastructure. There are numerous osmiophilic granular and irregular masses found in the dense cytoplasm and vacuoles in the epidermal cells of glandular heads of all species sampled (Figures 11A–E). The intercellular spaces of the epidermal cells contain granular and/or thready material in *V. davidii*, *V. adenoclada*, and *V. chunganensis* (Figures 11A–D).

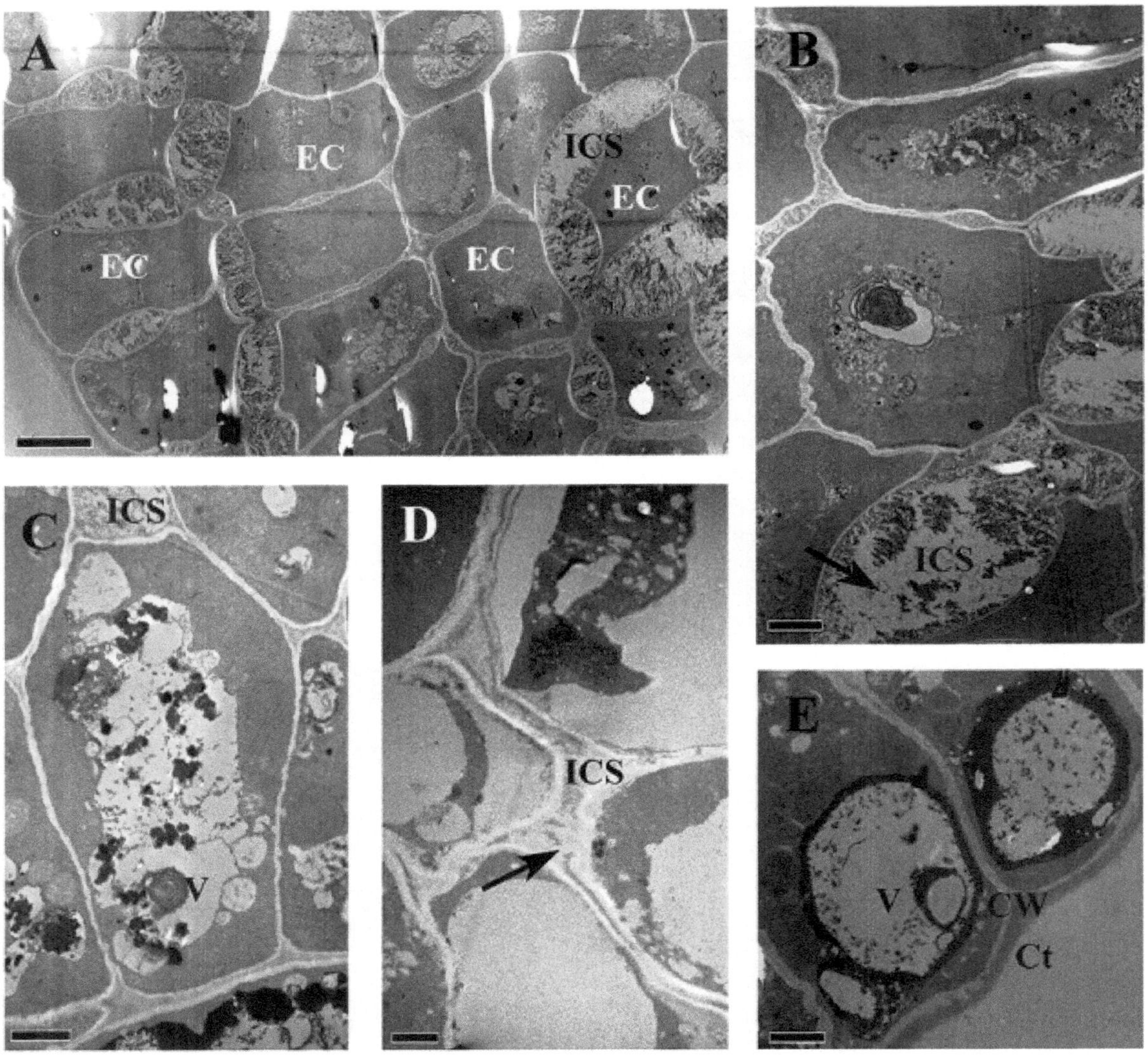

Figure 11: TEM images of the glandular trichomes and young prickles. (A) Epidermal cells of gland head in *V. davidii*, showing containing of dense thready and granular material in vacuoles and intercellular spaces. (B) Detail of the epidermal cells of *V. davidii*. Arrow indicate dense granular material. (C) Numerous irregular masses of osmiophilic material accumulating in vacuoles of epidermal cells of gland head in *V. adenoclada*. (D) Dense thready material (arrow) in the intercellular spaces of*V. chunganensis*. (E) Vacuoles are filled with osmiophilic material. Abbreviations: CW, cell wall; Ct, cuticle; EC, epidermal cells; ICS, intercellular spaces; V, vacuole. Scale bars: (A) = 5 μm; (B,C,E) = 2 μm; (D) = 1 μm.

DISCUSSION

The trichome types include ribbon, simple and glandular trichomes in representatives of *Vitis* on different vegetative plant organs. These trichomes may play different roles in plant physiology and ecology with variable morphological, mechanical and phytochemical characteristics (Kortekamp and Zyprian, 1999; Wagner et al., 2004).

Ribbon Trichomes

Ribbon trichomes are flattened and twisted with variable densities on the different vegetative organs among species of *Vitis* (Figures 1A,B; Table 1). The young leaves of most *Vitis* species are covered by ribbon trichomes, which may serve as a mechanical barrier against harsh spring temperature, herbivores and pathogens (Kortekamp and Zyprian, 1999; Werker, 2000) and restrict insect activities on the leaf surfaces (Wagner et al., 2004). They may also play a photo-protective role in decreasing light stress (Liakopoulos et al., 2006). Dense ribbon trichomes (Figures 3A–E) may be a significant defense mechanism against abiotic and biotic stresses for young plant organs of *Vitis*. These trichomes may also help decrease water loss and facilitate acclimation to xeric environments (Wagner et al.,2004). During development of these young leaves, in some *Vitis* species sampled (e.g., *V. flexuosa*, *V. rupestris*, *V. rotundifolia*, and *V. shenxiensis*), the ribbon trichomes are deciduous, becoming glabrescent at maturity (Table 1). This phenomenon may be due to the promotion of photosynthetic competence and establishment of phytochemical defense mechanism of leaves when they mature, at which point they no longer need the protection of ribbon trichomes.

Ribbon trichomes are frequently observed in both subgenera of *Vitis* (subgenus *Muscadinia* and subgenus *Vitis*) distributed in North America, East Asia, and Europe (Moore, 1991; Chen et al., 2007; our observations). Interestingly, ribbon trichomes also occur in other genera of the *Ampelocissus–Vitis*clade (e.g., *Ampelocissus* and *Pterisanthes*) (Liu et al., 2013, 2016; Wen et al., 2013) in Vitaceae according to the previous studies

(Latiff, 1982; Jackes, 1984; Ickert-Bond et al., 2015) and our observations, while, as far as we know they are not observed in any other genera of Vitaceae (e.g.,*Ampelopsis, Rhoicissus, Tetrastigma, Cayratia, Cissus, Parthenocissus*, and *Yua*). Our results support that *Vitis, Ampelocissus* and *Pterisanthes* have similar morphological characteristics, confirming the hypothesis that *Vitis* is most closely related to *Ampelocissus* and *Pterisanthes* based on the molecular evidence (Liu et al., 2013, 2016; Wen et al., 2013) and inflorescence morphology (Ickert-Bond et al.,2015). We infer that the ribbon trichome may be a morphological synapomorphy for the*Ampelocissus–Vitis* clade, which should be tested in conjunction with a fully resolved phylogeny of the*Ampelocissus–Vitis* clade in the future.

Simple Trichomes

In most *Vitis* species, simple trichomes differ in size, density, cell number and growth positon, and they are all unbranched, uniseriate or unicellular (Figures 1J,K). The development of uniseriate trichomes in the majority of flowering plants is achieved by initial enlargement and a series of periclinal divisions of an epidermal cell (Werker, 2000; Yang and Ye, 2013).

A dense indumentum composed of simple trichomes usually acts as a filter protecting plant tissues against damage of ultraviolet-B radiation and as a deterrent to insect activity, including oviposition and feeding in many species (Levin, 1973; Karabourniotis et al., 1995; Liakoura et al., 1997; Manetas,2003; Yan et al., 2012). Simple trichomes in wine grape leaves (*V. vinifera* L.) are also considered as a defense against photo-damage during leaf development by modifying and filtering both ultraviolet and visible lights (Karabourniotis et al., 1999; Liakopoulos et al., 2006). Kortekamp and Zyprian (1999) demonstrated that the indumentum of grape leaves with both simple and ribbon trichomes acts as a protective barrier and could restrict the development of fungi and prevent the fungal hyphae from reaching the leaf cuticle.

However, the density of leaf trichomes in grapevine cultivars has no influence on the activity and oviposition of leafhoppers, on the contrary, the density of simple trichomes on the leaf veins is negatively correlated with the parasitism rate by *Anagrus* spp., which is a biological control agent to parasitize the eggs of leafhoppers (Pavan and Picotti, 2009). Short simple trichomes did not affect the oviposition of the leafhoppers, but instead reduced the parasitism rate of *Anagrus* spp., while the long simple trichomes could influence the oviposition of the leafhopper on different herbaceous plants (Pavan and Picotti, 2009). We thus infer that different sizes of insects are influenced by the different height of simple trichomes in *Vitis*, with the short simple trichomes (e.g. in *V. davidii* and *V. vinifera*) impeding activities of smaller insects, and the long simple trichomes (e.g., in *V. retordii* and *V. romanetii*) may protect against bigger insects.

Glandular Trichomes and Prickles

Glandular trichomes show great variation in morphology and cell structure (Werker, 2000). They may have uni- or multiserate stalks and uni- or multicellular glandular heads. They can be derived from the epidermis (e.g., capitate glandular trichomes) or from both the epidermis and subjacent layers. Some types of glandular trichomes, which are solely derived from the epidermis are found in many species and are particularly prominent in Asteraceae, Cucurbitaceae, Cannabaceae, Lamiaceae, and Solanaceae (Akers et al., 1978; Werker et al., 1985; Nielsen et al., 1991; Ascensão and Pais, 1998; Kolb and Müller, 2004; Tissier, 2012; Celep et al., 2014).

However, all glandular trichomes and the prickles of *Vitis* which were examined histologically in our study are derived from both the epidermis and subjacent layers, which are similar to the glandular trichomes on the leaves of *Drosera* (Fahn, 1979) and those on *Passiflora foetida* stipules (Durkee et al.,1984), as well as the food bodies, which are globose structures attached to the plant by a short peduncle for promoting mutualism between plants (e.g., in Vitaceae) and ants (Buono et al., 2008; Paiva et al.,2009).

Glandular trichomes can secrete terpenes, phenolics, alkaloids, lipophilic or other substances deterring or poisoning herbivores and pathogens (Levin, 1973; Werker, 2000; Tissier, 2012) or secrete resin to protect the developing tissues from cold temperatures (Lapinjoki et al., 1991). Secreted mucilage from digestive glands of *Drosera* acts as a sticky trap for insects (Outenreath and Dauwaldert, 1986).

The glandular trichomes of *V. romanetii* can secret mucilage and emit volatile substance so that many insects are often glued to glands based on our observation (Figures 8A–C). This trichome type may act as a defense mechanism against insects. The composition of mucilage and volatile substance will be explored in future. However, the glandular trichomes of other species lacked apparent secretions.

Akers et al. (1978) reported that accumulation of the secretions of glandular trichomes of *Nicotiana tabacum* can be observed both within extraplasmic and intercellular spaces. These extracellular spaces for the preservation of the secretions were observed in other glandular trichomes as well (Vermeer and Peterson, 1979; Ascensão et al., 1997; Gravano et al., 1998; Turner et al., 2000). In the secretory cells of glandular trichomes of *Passiflora foetida*, dense masses of granular or fibrillar material accumulate in vacuoles, intercellular spaces and beneath the cuticle (Durkee et al., 1984). Similarly, the ultrastructure of the glandular trichomes and the young prickle of *Vitis* show that there are numerous osmiophilic granular, thread-like and irregular masses in the cytoplasm, vacuoles and/or the intercellular spaces in the epidermal cells of glandular heads. These results suggest that the glandular trichomes and the young prickle might excrete substances and the intercellular spaces may be the places of temporary storage of these secretions in the *Vitis* species examined. There is always one stoma at the apex of glandular trichome (except *V. chunganensis*) based on our observations.

Classification of glandular trichomes is based on morphological and ultrastructural differences (Werker, 2000). In *Vitis* the prickles are only present on stems and petioles of *V. davidii*. Considering the morphology, ultrastructural and development of glandular trichomes of other species in *Vitis*, the prickles of *V. davidii* and the glandular trichomes have similar morphological and ultrastructural characteristics and share similar early developmental stages. Our studies reveal that the prickles of *V. davidii* are homologous to glandular trichomes. The prickles undergo early developmental stages like other glandular trichomes (Figures 9A–D, 10A) and then increase in both height and basal width (Figures 9E–G, 10B), lignification (Figure 10C) and harden (Figures 9H, 10D). Our findings are consistent with the developmental process of prickles on raspberries and roses (Kellogg et al., 2011). Furthermore, trichomes as the precursors to prickles have been mentioned in many other species (Delbrouck, 1875; Leelavathi and Ramayya, 1983). The prickles of *V. davidii* may facilitate climbing and deter large herbivores from feeding on this species. Since birds are important seed dispersers of grapes (Hardie and Obrien, 1988), the prickles may protect fruits from frugivores as well.

Glandular trichomes and the prickles are here observed on Asian *Vitis* species, and they do not usually occur on species from North America and Europe. Moore and Wen (2016) suggest that only *V. labrusca* and *V. rotundifolia* occasionally show few glandular trichomes (Table 1). In addition, glandular trichomes are present on the stems of *Ampelocissus martini* and *A. arachnoidea* and on the pedicel of *A. xizangensis* (Latiff, 1982, 2001; Chen et al., 2007). Conical prickles are also present on the stems of *A. aculeata* and *A. acetosa* in Papua New Guinea and Northern Australia (Jackes, 1984; Latiff, 2001). Further study is needed to elucidate the origin of the glandular trichomes and prickles of*Vitis* and *Ampelocissus*.

Taxonomic Utility of Trichomes in *Vitis*

The trichome types and density of *Vitis* are influenced by different regional environment and growth stages based on our observation (Table 1). For example, not all *V. chunganensis* which were collected from different localities have glandular trichomes. Some plants of *V. piasezkii* only have ribbon trichomes while the others bear simple trichomes along the current years' branchlets. The ribbon trichomes are dense to nearly glabrous on leaves of *V. wilsoniae* during the growing season. The ribbon trichomes with variable density are present on different leaves of *V. balansana*.

Ribbon and simple trichomes are known from most species of *Vitis*, except for some nearly glabrous species (Table 1). Nineteen and three species sampled have dense ribbon and simple trichomes on young leaves respectively, which are useful taxonomic characters. Meanwhile, young leaves of many species often differ in their types and density of trichomes compared to mature leaves on some vegetative organs (Table 1), due to the loss of trichomes during development of leaves. So it is important for identification to examine mature leaves. For example, in *V. wilsoniae*, the young leaves bear very dense ribbon trichomes (Figure 3E), while, there are hardly any ribbon trichomes on the mature leaves (only a few of trichomes on the veins of abaxial leaf surfaces were observed).

Eleven species sampled in this study show a non-deciduous dense arachnoid indumentum composed of ribbon trichomes on the abaxial leaf surfaces, which distinguish these species from any of the other *Vitis* species examined. *Vitis retordii* shows a dense indumentum composed of long simple trichomes on adaxial leaf surfaces, which can easily distinguish this species from any other *Vitis* species examined. Simple trichomes on abaxial leaf vein axils or leaf margins are important taxonomic characters for some species (e.g., *V. flexuosa*, *V. piasezkii*, *V. piasezkii* var. *pagnucii*, and *V. shenxiensis*).

Glandular trichomes and prickles are only documented in seven species and these species can be further distinguished by the morphological differences of the glandular trichomes. The prickles on stems and petioles of *V. davidii* are considered an unique character that distinguishes this species from any other *Vitis* species (Chen et al., 2007). The glandular trichomes of *V. romanetii* are dense, long, pink/yellow and columnar on branchlets, petioles and veins of abaxial leaf surfaces, while those of *V. adenoclada* are dense, long, wine-red and hooked on stems and petioles. In comparison the glandular trichomes of *V. chunganensis* are less dense, short, green and have inverted trapezoid heads on basal veins of abaxial leaf surfaces, petioles and along stem nodes. The distribution of the glandular trichomes of *V. shenxiensis* is very sparse, and the glandular trichomes are short and the glandular heads are oval in *V. shenxiensis*. Details of the distribution of the three types of trichomes in different species in this study (Table 1) can contribute to increase our understanding of *Vitis* systematics and taxonomy.

ACKNOWLEDGMENTS

We thank Prof. C.H. Liu and Dr. J.F. Jiang (Zhengzhou Fruit Research Institute, Chinese Academy of Agricultural Science) who gave us some materials of *Vitis* and we would like to thank three master candidates (L.J. Duan, L. Dong and, L.L. Gui) in Huazhong Agricultural University for collecting additional samples.

REFERENCES

1. Akers C. P., Weybrew J. A., Long R. C. (1978). Ultrastructure of glandular trichomes of leaves of *Nicotiana tabacum* L., cv. Xanthi. Am. J. Bot. 65, 282–292. 10.2307/2442269
2. Aradhya M., Koehmstedt A., Prins B. H., Dangl G. S., Stover E. (2008). Genetic structure, differentiation, and phylogeny of the genus *Vitis*: implications for genetic conservation. Acta Hort. 799, 43–49. 10.17660/ActaHortic.2008.799.4
3. Ascensão L., Marques N., Pais M. S. (1997). Peltate glandular trichomes of *Leonotis leonurus*leaves: ultrastructure and histochemical characterization of secretions. Int. J. Plant Sci. 158, 249–258. 10.1086/297436
4. Ascensão L., Pais M. S. (1998). The leaf capitate trichomes of *Leonotis leonurus*: histochemistry, ultrastructure and secretion. Ann. Bot. 81, 263–271. 10.1006/anbo.1997.0550
5. Brizicky G. K. (1965). The genera of Vitaceae in the southeastern United States. J. Arnold Arbor. 46, 48–67.
6. Brown M. V., Moore J. N., McNew R. W., Fenn P. (1999). Inheritance of downy mildew resistance in table grapes. J. Amer. Soc. Hort. Sci. 124, 262–267.
7. Buono R. A., De Oliveira A. Í. D. B., Paiva E. A. S. (2008). Anatomy, ultrastructure and chemical composition of food bodies of *Hovenia dulcis* (Rhamnaceae). Ann. Bot. 101, 1341–1348. 10.1093/aob/mcn052
8. Celep F., Kahraman A., Atalay Z., Doğan M. (2014). Morphology, anatomy, palynology, mericarp and trichome micromorphology of the rediscovered Turkish endemic *Salvia quezelii*(Lamiaceae) and their taxonomic implications. Plant Syst. Evol. 300, 1945–1958. 10.1007/

s00606-014-1020-1

9. Chen Z. D., Ren H., Wen J. (2007). Vitaceae, in Flora of China, Vol. 12, eds Wu C. Y., Hong D. Y., Raven P. H., editors. (Beijing: Science Press, and St. Louis: Missouri Botanical Garden Press;), 173–222.

10. Cheng D. W., Jiang J. F., Fan X. C., Zhang Y., Zhang G. H., Liu C. H. (2013). Diversity analysis of Chinese wild grape species. J. Plant Genet. Resour. 14, 996–1012.

11. Delbrouck C. (1875). Die Pflanzen-Stacheln, in Botanische Abhandlungen, Vol. 2, ed. Hanstein J., editor. (Bonn: Adolph Marcus;), 1–119.

12. Durkee L. T., Baird C. W., Cohen P. F. (1984). Light and electron microscopy of the resin glands of *Passiflora foetida* (Passifloraceae). Am. J. Bot. 71, 596–602. 10.2307/2443335

13. Fahn A. (1979). Secretory Tissues in Plants. London: Academic Press.

14. Galet P. (1988). Cépages et Vignobles de France. Tome 1. Lesvignes Américaines. Montpellier: Pierre Galet.

15. Gerrath J., Posluszny U., Melville L. (2015). Taming the Wild Grape. Botany and Horticulture in the Vitaceae. Heidelberg: Springer.

16. Gravano E., Tani C., Bennici A., Gucci R. (1998). The ultrastructure of glandular trichomes of*Phillyrea latifolia* L.(Oleaceae) leaves. Ann. Bot. 81, 327–335. 10.1006/anbo.1997.0562

17. Hardie W. J., Obrien T. P. (1988). Considerations of the biological significance of some volatile constituents of grape (*Vitis* spp.). Aust. J. Bot. 36, 107–117. 10.1071/BT9880107

18. He Y. H., Li C. L., Cao Y. L. (1994). Comparative anatomy of vegetative organs in the genus*Vitis* L. and its systematic significance. Acta Phytotaxon. Sin. 32, 154–164.

19. Ickert-Bond S. M., Gerrath J. M., Posluszny U., Wen J. (2015). Inflorescence development in the*Vitis–Ampelocissus* clade of Vitaceae: the unusual lamellate inflorescence of *Pterisanthes*. Bot. J. Linn. Soc. 179, 725–741. 10.1111/boj.12348

20. Jackes B. R. (1984). Revision of the Australian Vitaceae, 1. *Ampelocissus* Planchon.Austrobaileya 2, 81–86.

21. Johansen D. A. (1940). Plant Microtechnique. New York, NY: McGraw Book Co.

22. Karabourniotis G., Bornman J. F., Liakoura V. (1999). Different leaf surface characteristics of three grape cultivars affect leaf optical properties as measured with fibre optics: possible implication in stress tolerance. Aust. J. Plant Physiol. 26, 47–53. 10.1071/PP98052

23. Karabourniotis G., Kotsabassidis D., Manetas Y. (1995). Trichome density and its protective potential against ultraviolet-B radiation damage during leaf development. Can. J. Bot. 73, 376–383. 10.1139/b95-039

24. Kellogg A. A., Branaman T. J., Jones N. M., Little C. Z., Swanson J. D. (2011). Morphological studies of developing *Rubus* prickles suggest that they are modified glandular trichomes. Botany89, 217–226. 10.1139/b11-008

25. Kolb D., Müller M. (2004). Light, conventional and environmental scanning electron microscopy of the trichomes of *Cucurbita pepo* subsp. pepo var. styriaca and histochemistry of glandular secretory products. Ann. Bot. 94, 515–526. 10.1093/aob/mch180[PubMed]

26. Kortekamp A., Zyprian E. (1999). Leaf hairs as a basic protective barrier against downy mildew of grape. J. Phytopathol. 147, 453–459. 10.1111/j.1439-0434.1999.tb03850.x

27. Lapinjoki S. E., Elo H. A., Taipale H. T. (1991). Development and structure of resin glands on tissues of *Betula pendula* Roth. during growth. New Phytol. 117, 219–223. 10.1111/j.1469-8137.1991.tb04902.x

28. Latiff A. (1982). Studies in Malesian Vitaceae, I–IV. Fed. Mus. J. 27, 46–93.

29. Latiff A. (2001). Diversity of the Vitaceae in the Malay Archipelago. Malayan Nat. J. 55, 29–42.

30. Leelavathi P., Ramayya N. (1983). Structure, distribution and classification of plant trichomes in relation to taxonomy III. Papilionoideae. Proc. Plant Sci. 92, 421–441.

31. Levin D. A. (1973). The role of trichomes in plant defense. Q. Rev. Biol. 48, 3–15. 10.1086/407484

32. Liakopoulos G., Stavrianakou S., Karabourniotis G. (2006). Trichome layers versus dehaired lamina of *Olea europaea* leaves: differences in flavonoid distribution, UV-absorbing capacity, and wax yield. Environ. Exp. Bot. 55, 294–304. 10.1016/j.envexpbot.2004.11.008

33. Liakoura V., Stefanou M., Manetas Y., Cholevas C., Karabourniotis G. (1997). Trichome density and its UV-B protective potential are affected by shading and leaf position on the canopy.Environ. Exp. Bot. 38, 223–229. 10.1016/S0098-8472(97)00005-1

34. Liu X. Q., Ickert-Bond S. M., Chen L. Q., Wen J. (2013). Molecular phylogeny of Cissus L. of Vitaceae (the grape family) and evolution of its pantropical intercontinental disjunctions. Mol. Phylogenet. Evol. 66, 43–53. 10.1016/j.ympev.2012.09.003

35. Liu X. Q., Ickert-Bond S. M., Nie Z. L., Zhou Z., Chen L. Q., Wen J. (2016). Phylogeny of the*Ampelocissus–Vitis* clade in Vitaceae supports the New World origin of the grape genus. Mol. Phylogenet. Evol. 95, 217–228. 10.1016/j.ympev.2015.10.013

36. Manetas Y. (2003). The importance of being hairy: the adverse effects of hair removal on stem photosynthesis of *Verbascum speciosum* are due to solar U-VB radiation. New Phytol. 158, 503–508. 10.1046/j.1469-8137.2003.00768.x

37. Miller A. J., Matasci N., Schwaninger H., Aradhya M. K., Prins B., Zhong G. Y., et al. . (2013).*Vitis* phylogenomics: hybridization intensities from a SNP array outperform genotype calls.PLoS ONE 8:e78680. 10.1371/journal.pone.0078680

38. Moore M. O. (1991). Classification and systematics of eastern North American *Vitis* L. (Vitaceae) north of Mexico. SIDA 14, 339–367.

39. Moore M. O., Wen J. (2016). Vitaceae, in Flora of North America, North of Mexico, Vol. 12, eds Flora of North

America Editorial Committee, editor. (New York, NY; Oxford: Oxford University Press;).

40. Mullins M. G., Bouquet A., Williams L. E. (1992). Biology of the Grapevine. Cambridge: Cambridge University Press.

41. Nielsen M. T., Akers C. P., Järlfors U. E., Wagner G. J., Berger S. (1991). Comparative ultrastructural features of secreting and nonsecreting glandular trichomes of two genotypes of*Nicotiana tabacum* L. Bot. Gaz. 152, 13–22. 10.1086/337858

42. Outenreath R. L., Dauwaldert M. (1986). Ultrastructural and radioautographic studies of the digestive gland cells of *Drosera capensis* II. Changes induced by stimulation. J. Ultrastruct. Mol. Struct. Res. 95, 164–174. 10.1016/0889-1605(86)90038-8

43. Paiva E. A. S., Buono R. A., Lombardi J. A. (2009). Food bodies in *Cissus verticillata*(Vitaceae): ontogenesis, structure and functional aspects. Ann. Bot. 103, 517–524. 10.1093/aob/mcn237

44. Pavan F., Picotti P. (2009). Influence of grapevine cultivars on the leafhopper *Empoasca vitis* and its egg parasitoids. Biocontrol 54, 55–63. 10.1007/s10526-008-9151-3

45. Payne W. W. (1978). A glossary of plant hair terminology. Brittonia 30, 239–255. 10.2307/2806659

46. Péros J. P., Berger G., Portemont A., Boursiquot J. M., Lacombe T. (2011). Genetic variation and biogeography of the disjunct *Vitis* subg. *Vitis* (Vitaceae). J. Biogeogr. 38, 471–486. 10.1111/j.1365-2699.2010.02410.x

47. Pillonel C., Mulder M. M., Boon J. J., Forster B., Binder A. (1991). Involvement of cinnamyl-alcohol dehydrogenase in the control of lignin formation in *Sorghum bicolor* L. Moench. Planta185, 538–544. 10.1007/bf00202964

48. Robards A. W. (1978). An introduction to techniques for scanning electron microscopy of plant cells, in Electron Microscopy and Cytochemistry of Plant Cells, ed Hall J. L., editor. (New York, NY: Elsevier;), 343–403.

49. Roland J. C., Vian B. (1991). General preparation and staining of thin sections, in Electron Microscopy of Plant Cells, eds Hall J. L., Hawes C., editors. (London: Academic Press;), 1–66.

50. Soejima A., Wen J. (2006). Phylogenetic analysis of the grape family (Vitaceae) based on three chloroplast markers. Am. J. Bot. 93, 278–287. 10.3732/ajb.93.2.278

51. Staudt G. (1997). Evaluation of resistance to grapevine powdery mildew (*Uncinula necator*[Schw.] Burr., anamorph *Oidium tuckeri* Berk.) in accessions of *Vitis* species. Vitis 36, 151–154.

52. Staudt G., Kassemeyer H. H. (1995). Evaluation of downy mildew resistance in various accessions of wild *Vitis* species. Vitis 34, 225–228.

53. Tissier A. (2012). Glandular trichomes: what comes after expressed sequence tags? Plant J. 70, 51–68. 10.1111/j.1365-313X.2012.04913.x

54. Tröndle D., Schröder S., Kassemeyer H. H., Kiefer C., Koch M. A., Nick P. (2010). Molecular phylogeny of the genus *Vitis* (Vitaceae) based on plastid markers. Am. J. Bot. 97, 1168–1178. 10.3732/ajb.0900218

55. Turner G. W., Gershenzon J., Croteau R. B. (2000). Development of peltate glandular trichomes of peppermint. Plant Physiol. 124, 665–680. 10.1104/pp.124.2.665

56. Uphof J. C. T. (1962). Plant hairs, in Encyclopedia of Plant Anatomy, Vol. 5, eds Zimmermann W., Ozenda P. G., editors. (Berlin: Gebrüder Borntrager;), 1–206.

57. Vermeer J., Peterson R. L. (1979). Glandular trichomes on the inflorescence of *Chrysanthemum morifolium* cv. Dramatic (Compositae). II. Ultrastructure and histochemistry. Can. J. Bot. 57, 714–729. 10.1139/b79-091

58. Wagner G. J., Wang E., Shepherd R. W. (2004). New approaches for studying and exploiting an old protuberance, the plant trichome. Ann. Bot. 93, 3–11. 10.1093/aob/mch011

59. Wan Y., Schwaninger H. R., Baldo A. M., Labate J. A., Zhong G. Y., Simon C. J. (2013). A phylogenetic analysis of the grape genus (*Vitis* L.) reveals broad reticulation and concurrent diversification during Neogene and Quaternary climate change. BMC Evol. Biol. 13:141. 10.1186/1471-2148-13-141

60. Wan Y., Schwaninger H., He P., Wang Y. (2007). Comparison of resistance to powdery mildew and downy mildew in Chinese wild grapes. Vitis 46, 132–136.

61. Wang Y., Liu Y., He P., Chen J., Lamikanra O., Lu J. (1995). Evaluation of foliar resistance to*Uncinula necator* in Chinese wild *Vitis* species. Vitis 34, 159–164.

62. Wen J. (2007). Vitaceae, in The families and Genera of Vascular Plants, Vol. 9, ed. Kubitzki K., editor. (Berlin: Springer;), 466–478.

63. Wen J., Xiong Z., Nie Z. L., Mao L., Zhu Y., Kan X. Z., et al. . (2013). Transcriptome sequences resolve deep relationships of the grape family. PLoS ONE 8:e74394. 10.1371/journal.pone.0074394

64. Werker E. (2000). Trichome diversity and development. Adv. Bot. Res. 31, 1–35. 10.1016/S0065-2296(00)31005-9

65. Werker E., Ravid U., Putievsky E. (1985). Structure of glandular hairs and identification of the main components of their secreted material in some species of the Labiatae. Israel J. Bot. 34, 31–45.

66. Yan A., Pan J. B., An L. Z., Gan Y. B., Feng H. Y. (2012). The responses of trichome mutants to enhanced ultraviolet-B radiation in *Arabidopsis thaliana*. J. Photochem. Photobiol. B 113, 29–35. 10.1016/j.jphotobiol.2012.04.011

67. Yang C. X., Ye Z. B. (2013). Trichomes as models for studying plant cell differentiation. Cell. Mol. Life Sci. 70, 1937–1948. 10.1007/s00018-012-1147-6

68. Zecca G., Abbott J. R., Sun W. B., Spada A., Sala F., Grassi F. (2012). The timing and the mode of evolution of wild grapes (*Vitis*). Mol. Phylogenet. Evol. 62, 736–747. 10.1016/j.ympev.2011.11.015

69. Zhang J., Wu X., Niu R., Liu Y., Liu N., Xu W., et al. (2012). Cold-resistance evaluation in 25 wild grape species. Vitis 51, 153–160.

CHAPTER 3

BIOMECHANICS AND FUNCTIONAL MORPHOLOGY OF A CLIMBING MONOCOT

Linnea Hesse[1,2,3], Sarah T. Wagner[1], and Christoph Neinhuis[1]

[1]Institut für Botanik, Technische Universität Dresden, D-01062 Dresden, Germany

[2]Institut für Spezielle Botanik, Johannes Gutenberg—Universität Mainz, D-55112 Mainz, Germany

[3]Present address: Plant Biomechanics Group Freiburg, University of Freiburg, Schänzlestrasse 1, D-79104 Freiburg im Breisgau, Germany

ABSTRACT

Plants with a climbing growth habit possess unique biomechanical properties arising from adaptations to changing loading conditions connected with close attachment to mechanical supports. In monocot climbers, mechanical adaptation is restricted by the absence of a bifacial vascular cambium. Flagellaria indica was used to investigate the mechanical properties and adaptations of a monocot climber that, uniquely, attaches to the surrounding vegetation via leaf tendrils. Biomechanical methods such as three-point bending and torsion tests were used together with anatomical studies on tissue development, modification and distribution. In general, the torsional modulus was lower than the bending modulus; hence, torsional stiffness was less than flexural stiffness. Basal parts of mature stems showed the greatest stiffness while that of more apical stem segments levelled off. Mechanical properties were modulated via tissue maturation processes mainly affecting the peripheral region of the stem. Peripheral vascular bundles showed a reduction in the amount of conducting tissue while the proportion and density of the bundle sheath increased. Furthermore, adjacent bundle sheaths merged resulting in a dense ring of fibrous tissue. Although F. indica lacks secondary cambial growth, the climbing habit is facilitated by a complex interaction of tissue maturation and attachment.

INTRODUCTION

One potential cause of species richness in angiosperms is a high diversity in growth form that enables colonization of a wide range of contrasting habitats (Ricklefs and Renner 1994; Bresinsky *et al.* 2008; Wagner *et al.* 2012). The development of secondary growth via cambial activity and the linked ability to change structural and mechanical properties during ontogeny are considered to be important novel properties promoting this diversity. In climbing plants, developmental changes in stem biomechanics become apparent in the transition from their juvenile self-supporting stage, which enables plants to reach and attach to host supports, to their non-self-supporting climbing stage, which tolerates mechanical stresses (Hoffmann *et al.* 2003; Rowe *et al.* 2004; Isnard and Silk 2009; Chen *et al.* 2014). During the initial self-supporting phase, stem stiffness is often achieved by a ring of primary fibres in the peripheral regions of the stem or by the early production of a dense secondary xylem (Isnard and Silk 2009). With the onset of the climbing phase, however, cambial activity often leads to a highly compliant 'lianoid' stem. In addition, cambial variants such as irregular, supernumerary, unilateral, concentric, multiple and lobed cambia can lead to anatomically anomalous structures in lianas (Carlquist 1988; Rowe *et al.* 2004; Isnard and Silk

Citation: Linnea Hesse, Sarah T. Wagner, and Christoph Neinhuis (2015). Biomechanics and functional morphology of a climbing monocot, Oxford Journals, Science & Mathematics, AoB Plants published online March 13, 2016, Volume 8, DOI: 10.1093/aobpla/plw005,

2009), which contribute to their particular biomechanical properties. Thus, it is mainly cambial activity that contributes to the anatomical adaptations and modifications leading to the specific mechanical and architectural properties of climbing plants.

Therefore, the question arises as to how monocots that have lost the secondary vascular cambium realize the specific mechanical requirements of the climbing growth. These plants are limited in modulating material properties by secondarily produced tissues and thereby in their ability to generate different growth habits. This is because their growth and development are confined to primary-type growth. Nevertheless, monocots have evolved tall self-supporting tree-like forms and even climbers (Rowe *et al.* 2004, 2006; Isnard *et al.* 2005; Isnard and Silk 2009).

Compared with biomechanical studies in other angiosperm taxa, monocots are poorly investigated with respect to the biomechanics of growth forms. Studies have mainly focussed on climbing palms such as *Calamus* (rattans) and *Desmoncus*, which cling to the surrounding vegetation via specialized spines or hooks along their elongated leaf rachis (Isnard *et al.* 2005;Isnard and Silk 2009; Isnard and Rowe 2008*b*; Rowe and Isnard 2009). These studies provide valuable results on the mechanisms that monocots have developed to realize a non-self-supporting growth habit that depends on a mechanically supporting host.

Flagellaria indica promises further insights into biomechanical traits of monocotyledonous climbers and their mechanical adaptation due to a different attachment mechanism compared with *Desmoncus* and *Calamus*. *Flagellaria indica* climbs via coiled tendrils terminating the leaves and assuring a close contact to supports (Tomlinson and Posluszny 1977; Takhtadzhi-an 1997; Kubitzki *et al.* 1998).

Flagellaria is the sole genus of the family Flagellariaceae and consists of five species, which are distributed throughout the tropical and subtropical regions of the Old World (southeast Asia, tropical Africa, northern Australia and the Pacific Islands) (Kubitzki *et al.* 1998; Rabenantoandro *et al.* 2007; Wepfer and Linder 2014). *Flagellaria* species occur mainly in coastal regions, in moist lowland forests and mangroves, often along forest margins and in secondary forests (Larsen 1972; Linder 1987;Kubitzki *et al.* 1998). The solid, cane-like, aerial stems, which originate from sympodial rhizomes, can reach a length of 20 m and lack axillary buds (Kubitzki *et al.* 1998; Wepfer and Linder 2014). Dichotomizing apices have previously drawn attention to this species (Tomlinson and Posluszny 1977). Leaves are distichous with a closed leaf sheath and differentiate into coiled tendrils, which enable attachment to surrounding vegetation.

This study addresses (i) the overall biomechanical architecture of *F. indica*, (ii) the functional traits of *F. indica* by discussing the results of biomechanical measurements via bending and torsion tests in relation to anatomical observations and (iii) the mechanical adaptation of *F. indica* and other climbing monocots, despite the lack of cambial growth.

METHODS

Sampling

All investigations were carried out on one mature, intensely sprouting individual plant cultivated in a greenhouse of the Botanic Garden at Dresden, Germany (Fig. 1B). Entire aerial stems were cut for biomechanical and anatomical investigation, and were kept under moist conditions prior to measurements (maximum 24 h) to prevent dehydration and to maintain natural material properties. The longest aerial stems investigated had a length of 15 m, comparable with the stem length (15–20 m) recorded at natural sides (Larsen 1972; Kubitzki *et al.* 1998; Wepfer and Linder 2014). Branches overgrew the surrounding vegetation reaching the greenhouse roof top (5 m). Thus, growth habit resembles natural observations (Fig. 1), even though long aerial shoots were cut back by the garden staff to reduce their extensive growth. As most plant species show variation in their growth habit, also under natural conditions, the cultivated individual might also differ slightly in size, in growth rate and biomechanical traits compared with traits of other individuals in greenhouses and in the natural habitat. However, each individual, whether cultivated or growing in the wild, must cope with habitat-specific challenges, by adjustments to growth habit and architecture.

Figure 1. Growth habit and morphology of *F. indica*. (A) The morphology and growth habit of an individual observed in its natural habitat. (B) Morphology and growth habit of the cultivated plant. Basal self-supporting region with vertical growth; distal, climbing regions are partly fixed underneath the roof of the greenhouse. (C) The leaf apex, curled to a tendril. The adaxial side of the tendril is thickened. (D) Attachment via leaf tendril to an adjacent host support.

Biomechanics

Stem segments of five aerial shoots (AS 1–5) were measured first in bending and subsequently in torsion. Aerial shoots 1–3 represented young and relatively short stems (max. 4 m), while AS 4 and 5 are older and have already reached the greenhouse roof top (max. stem length 15 m). Their apices were no longer active. The position of each tested sample was recorded to note its distance from the base (i.e. from basal to the most aerial part) and/or apex of the stem. Only relatively straight stem segments were used.

Samples were assigned to three developmental stages, based on comparable characteristics and distance from either the apex or base of the stem. Stage G1 (n= 6) includes most apical stem sections located in 0–1 m distance to an active apex (only AS 1–3). Stage G2 (n= 19) consists of samples located 4–13 m from the base (only AS 4 and 5). Stage G3 (n= 9) contains old, mostly basal stem sections with a maximal distance to the base of 1 m (only AS 2–5). Some additional segments of the transition zones between the defined developmental stages were also included in the measurements.

To avoid bias to measurements caused by shear forces, a suitable span-to-depth ratio of >20 in bending as well as torsion tests was maintained for all samples (Vincent 1992; Beismann *et al*. 2000; Gallenmüller *et al*. 2001). Here, span describes the distance between the two support points of the bending device or the two chucks of the torsion device, respectively; depth representing the average diameter of the sample based on three measurements along the stem segment. The diameter considers the leaf sheath, and thus, measurements of stem biomechanics reflect the entire axis.

The structural Young's (structural bending modulus) and structural torsional modulus were used to compare the developmental stages in terms of their biomechanical properties. They were calculated from flexural and torsional stiffness determined in experiments described below. The term 'structural' emphasizes that, unlike according to the definition of Young's modulus and the modulus of torsion, not a single material but a composite is measured. As plant stems and many other biological materials have a complex structure, both moduli are necessarily interpreted as spatially averaged values across the entire heterogeneous plant tissues (Rowe and Speck 1998; Isnard *et al*. 2005; Rowe *et al*. 2006).

Three-point-bending measurements were conducted using a Zwick/Roell BZ 2.5/TS1S bench-top mechanical testing device (Zwick/Roell, Ulm, Germany), following the protocol of Wagner *et al*. (2012). Structural Young's modulus (E) (referred to as bending modulus in the following) of each stem segment was determined by dividing the flexural stiffness (EI) by the second moment of area (I).

Tests in torsion were carried out using the Zwick/Roell testing machine with a set-up consisting of two chucks, one fixed (Fig. 2A) and one rotatable (Fig. 2B). The moment of force was applied by moving the cross head upward (5 mm s^{-1}; Fig. 2C). This movement was transmitted to the rotatable chuck by a Kevlar cord (Grebenstein Kevlarvorfach, 15 kg lifting capacity; Fig. 2D), fixed to the force transducer at the cross head (Fig. 2E) and to a transmission disk (Fig. 2F) at the rotatable chuck. Kevlar chord was chosen as it has a low strain of rupture (2.99 %), a high tensile strength (F_{max} = 150 kg m s^{-2}) and little internal strain, which allows measurements of both stiff and soft materials.

Figure 2. Experimental set-up for torsional loading of plant material. Tests in torsion were carried out using the Zwick/Roell testing machine. Each sample (G) is fixed to two chucks of which one is fixed (A) and the other one is rotatable (B). The moment of force is applied by moving the cross head upward (C). This movement is transmitted to the rotatable chuck by a Kevlar cord (D), which is fixed to the force transducer at the cross head (E) and to a transmission disk (F) at the rotatable chuck.

Each sample (Fig. 2G) was fixed to the chucks and twisted by the applied moment of force. The translation of the cross head, moving a distance x, was converted to an angle of torsion (φ) with the formula $\varphi = x \times r^{-1}$[rad], where r is the radius of the transmission disk.

The modulus of torsion was calculated using the length of a sample L (the distance between the two chucks), the torsional constant J, the angle of torsion φ and the applied moment of force T (Niklas 1991; Etnier 2003;Vogel 2007): $G = (F \times x^{-1}) \times ((r^2 \times L) \times J^{-1})$, in which $F \times x^{-1}$ is the slope of the linear regression of the force–deflection curve and r is the radius of the transmission disk.

The torsional constant J takes the geometry of the sample into account and thus the influence of its cross-sectional shape under torsion. The formula $J = (\varpi \times 16)^{-1} \times ((a^3 \times b^3) \times (a^2 + b^2))^{-1}$ for elliptic cross-sections incorporates natural variations of stem shape, with a being the major and b the minor diameter of the stem.

Torsional stiffness GJ is analogous to the flexural stiffness EI, the product of the material property G of the sample and its geometry factor J. It represents the resistance of a stem fragment to twisting loads (Vogel 2007).

The ratio of flexural stiffness EI and torsional stiffness GJ gives a dimensionless value that allows comparison of different developmental stages while considering geometrical differences. Two stem fragments can have the same EI/GJ ratio though showing differences in either material or geometrical properties or even both (Niklas 1992; Etnier 2003). Furthermore, the EI/GJ ratio allows comparing different species and provides information about the degree of anisotropy (Vogel 1992, 2007). An isotropic material would have an EI/GJ ratio of 1. A high EI/GJ ratio indicates a strong directional dependence of mechanical traits and thereby a strong anisotropy. Determining the anisotropy of the tested samples will, therefore, provide a better understanding of how the material properties are directionally dependant.

Due to the lack of secondary cambial growth, stems of *F. indica* are relatively homogenous in diameter from base to apex compared with dicotyledons. Thus, stem geometrical traits of *F. indica* cannot be as easily correlated with stem developmental traits, as it has been done in studies on non-monocots (Gallenmüller *et al.* 2001; Isnard *et al.* 2003*a*, *b*; Wagner*et al.* 2012, 2014). Instead, considering the base (here: the position where the plant emerges from the soil) and/or the apex as reference points gives insights into changes of mechanical properties in relations to developmental change.

Statistical Analysis

Material properties of the three developmental stages were statistically analysed using Kruskal–Wallis tests and pairwise Mann–Whitney *U*-tests (*post hoc* testing). The *P*-values were adjusted using Holm-correction. All statistical tests were carried out using the stats package of the freeware 'R' (R Development Core Team 2014). The boxplots were created using the ggplot2 package (Wickham 2009).

Anatomy

Anatomical investigations were carried out subsequent to the biomechanical measurements to correlate tissue distribution and differentiation with biomechanical properties. Samples were preserved in 70% ethanol, and transverse and longitudinal sections of different developmental stages were prepared with a vibratome (Microtome Hyrax V50, Carl Zeiss MicroImaging GmbH, Jena, Germany). Amplitude, frequency and razor blade speed were adjusted for each sample. Stiff samples were cut with a standard sliding wood microtome or cut by hand with a razor blade. Fully developed, rather stiff stem regions were softened with ethylenediamine solution (6.8–10 %) for 4 days prior to sectioning (Kukachka 1977; Carlquist 1982; Angyalossy *et al.* 2010). All sections were stained with safranin/astra blue and photographed using the ProgRes® Capture Pro 2.7 Camera (Jenoptik) and the software Image-Pro Plus 7.0. Phloroglucinol–HCl was used to confirm the staining results by staining lignified tissues pink/red. Further anatomical investigations were conducted using a scanning electron microscope (Supra 40 VP, Zeiss) to gain better insights on tissue development and distribution. Samples were then photographed using the Software SmartSEM® control.

RESULTS

Growth Habit of *F. Indica*

Aerial shoots of *F. indica* show a self-supporting phase up to 2–3 m and grow vertically towards the greenhouse rooftop searching for supporting vegetation (Figs 1B and 3D). The basal regions of the stems are characterized by incompletely developed, quickly senescing and scarious leaves, consisting mainly of sheaths (Fig. 3A). They have only a rudimentary lamina and no climbing function. Leaves are fully developed at a height of 2.5 m with a well-developed lamina and leaf tendrils (Figs 1C and D and 3B and C) and extensions of the midrib, which enable attachment to supports.

Figure 3. The stem surface of three developmental stages G1–G3. The black arrows point towards the lamina of the leaf. (A) Old, basal stem surface (G3) with incompletely developed (rudimentary lamina without climbing function), quickly senescing and scarious leaves, consisting mainly of sheath. (B) Leaf tendrils begin to develop at a height of 2.5–3 m, which enables the attachment to supports (basal G2 stems). (C) After a height of 2.5 m, leaves are fully developed, becoming apparent by a well-developed lamina and leaf tendrils. (D) The leaf tips begin to coil very early in their development, which allows a first attachment to the surrounding vegetation (G1).

Leaf tips begin to coil very early in their development (Fig. 3D) and the coiled structures allow a first attachment to the surrounding vegetation. Once support is located, tendrils begin to coil further and thicken adaxially (S. Wagner, A. Rjosk and C. Neinhuis, unpubl. data). Hence, a strong and close contact to a support is achieved (Fig. 1D). A close contact to supporting hosts could only be detected after a height of 3 m or higher was accomplished. Attachment to adjacent plants allows *F. indica* to maintain a more-or-less upright growth without relying on its own stem as mechanical support (Fig. 1A and B).

The morphology and growth habit of the cultivated plant resemble that of individuals observed in their natural habitat (e.g. Linder 1987; Kubitzki *et al.* 1998; Wepfer and Linder 2014; C. Neinhuis, pers. comm.; Fig. 1A). Nevertheless, all data obtained from greenhouse material will be discussed with caution since it may differ somewhat from that growing in natural environments.

Biomechanical Properties in Bending and Torsion

In general, the modulus of torsion of a stem segment is lower than its bending modulus and, if *E* increases, so does *G* (Fig. 4A). This is also the case for the flexural and torsional stiffness (Fig. 4B). Each developmental stage is characterized by particular biomechanical properties.

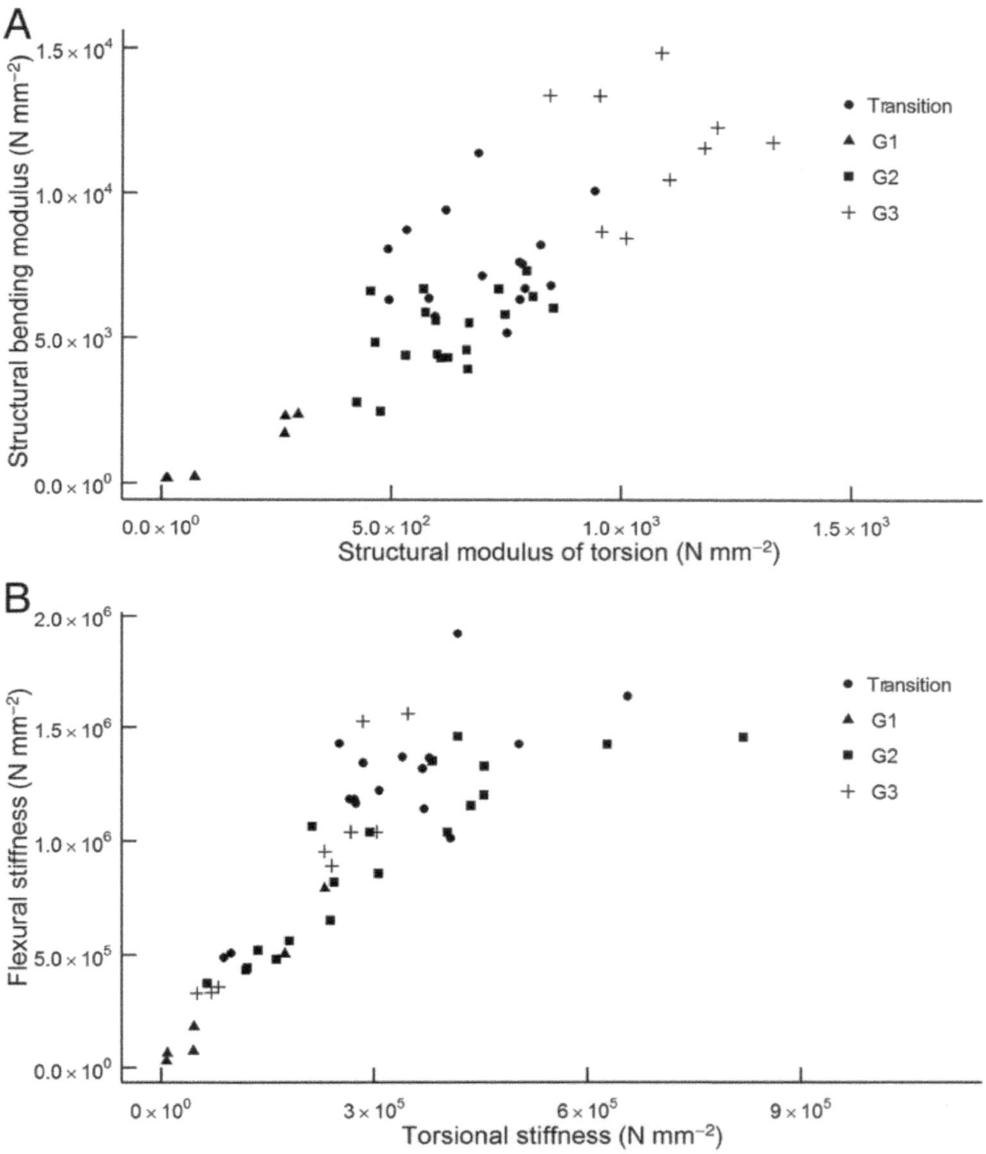

Figure 4. Correlation of structural bending modulus and structural modulus of torsion (A) and bending stiffness and torsional stiffness (B) of stem segments assigned to three developmental stages. Samples, which are not assigned to developmental stages, are transitional stem segments between the groups G1–G3. (A) Structural bending modulus and modulus of torsion show a linear relationship; if *E* increases, so does *G*. *E* is always higher than *G*(*E*> *G*). (B) Flexural stiffness always shows higher values compared with torsional stiffness (*EI*> *GJ*). If *EI* increases, so does*GJ*. Again, a linear relationship is apparent.

Stage G1

Young apical stem regions show the lowest values with moduli of torsion below 200 N mm^{-2} (155.6 ± 124.7 N mm^{-2}) and bending moduli below 2000 N mm^{-2} (1153.4 ± 991.1 N mm^{-2}) (G1 in Figs 4A and 5).

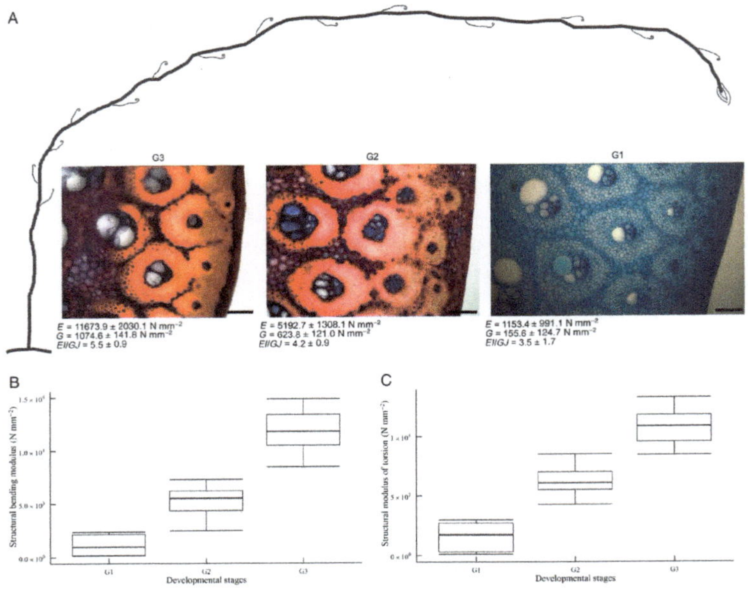

Figure 5. Changes of material properties during development. (A) Schematic representation of one aerial shoot. The anatomical changes during development are visualized for G1–G3 via small representative images of the peripheral region of the transverse stem section. The changes of material properties, structural bending modulus (*E*) and structural modulus of torsion (*G*), and changes of the *EI/GJ* ratio are also given with the mean values for each developmental stage G1–G3. (B) Boxplot showing the highly significant differences of the structural bending modulus of G1–G3. (C) Boxplot showing the highly significant differences of the structural modulus of torsion of G1–G3.

Stage G2

Apical stem regions are followed by stem segments with almost constant material properties. The bending modulus of G2 samples was maintained between 4000 and 6000 N mm^{-2} (5192.7 ± 1308.1 N mm^{-2}) (Figs 4A, 5and 6A), while the modulus of torsion always lay between 400 and 800 N mm^{-2} (623.8 ± 121.0 N mm^{-2}) (Figs 4A and 5 and 6B).

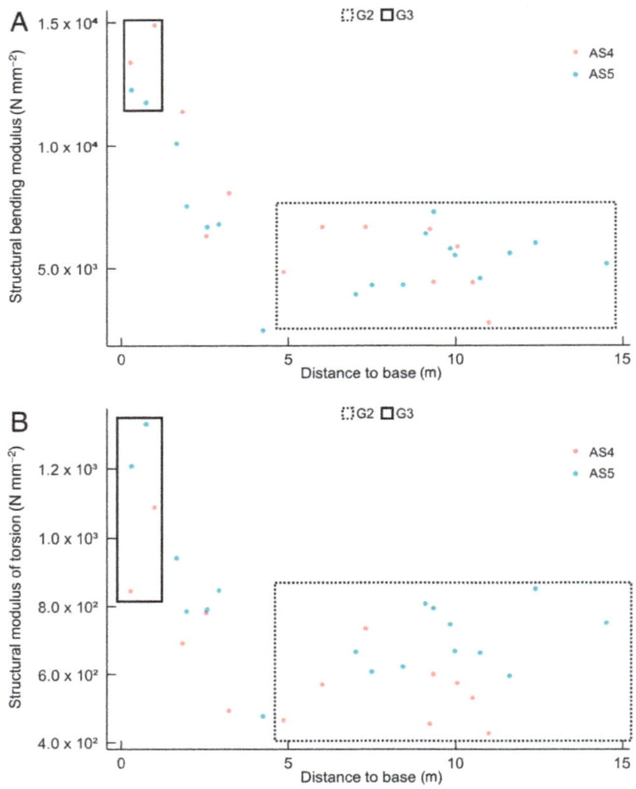

Figure 6. Changes of material properties, structural bending modulus (*E*) and structural modulus of torsion (*G*), along the stems of two aerial shoots (AS 4 and AS 5) of *F. indica*. Samples, which are assigned to the developmental stages G2 and G3, are highlighted by boxes with dashed or continuous lines. Due to spatial limitations in the greenhouse, the apical stages were cut back and, thus, cannot be plotted here. Old G3 segments close to the base of the shoots have highest material properties (A, structural bending modulus; B, structural modulus of torsion). Both *E* and *G* decrease with distance from the base and level off when the developmental stage G2 is reached.

Stage G3

Old, basal stem segments have the highest material properties (Figs 4⇑–6). The average bending modulus (11 673.9 ± 2030.1 N mm^{-2}) is more than twice the average of that measured at the G2 stage. The average of the modulus of torsion (1074.6 ± 141.8 N mm^{-2}) is almost twice as high as that of G2 segments (Fig. 5; Table 1).

Table 1. Biomechanical characteristics of developmental stages. Group G1 includes young, most apical stems (0–1 m from apex). G2 contains samples of 4–13 m distance to base. G3 consists of old, most basal stem fragments between 0 and 1 m distances to base. All moduli, as well as E/G and EI/GJ ratios, and diameter, are averaged and given with the corresponding standard deviation (mean ± SD).

Stage	*E* (N mm^{-2})	*G* (N mm^{-2})	*E/G*	*EI/GJ*	*d* (mm)	*n*
G1	1153.4 ± 991.1	155.6 ± 124.7	7.4 ± 4.1	3.5 ± 1.7	8.5 ± 1.0	6
G2	5192.7 ± 1308.1	623.8 ± 121.0	8.3 ± 2.2	4.2 ± 0.9	8.3 ± 1.6	19
G3	11 673.9 ± 2030.1	1074.6 ± 141.8	10.9 ± 2.6	5.5 ± 0.9	6.5 ± 1.1	9

Overall, the values of material properties decrease with increasing distance from the base (Fig. 5). The oldest and longest stems analyzed (AS 4–5), with a maximum length of 15 m (Fig. 6), are characterized by a continuous decrease of moduli from the base (G3) towards the transition zone and reach almost constant moduli in the stage G2 (Fig. 6). Statistical analyses indicate a significant increase of the modulus of torsion and the bending modulus from G1 to G3 (*E*: $\chi^2(6) = 226.16$, $P < 0.001$; *G*: $\chi^2(6) = 226.16$, $P < 0.001$; Table 2). Although an increase of the *EI/GJ* ratio is indicated by the mean values of the developmental stages (Table 1), these differences are not highly significant in the pairwise Mann–Whitney *U*-test ($P > 0.001$; Table 2). This could be due to the low sample size per group. However, the increase of the *EI/GJ* ratio indicates an increase in anisotropy from G1 to G3 (Table 1). In general, all samples are easier to twist than to bend (*E*>*G*); thus, the ratio of *EI* and *GJ* is always above 1 (Fig. 4B and Table 1).

Table 2. Comparison of mechanical properties of the three developmental stages G1 ($n = 6$), G2 ($n = 19$) and G3 ($n = 9$). The Kruskal–Wallis test indicates significant differences between the groups (*E*: $\chi^2(6) = 226.16$, $P < 0.001$; *G*: $\chi^2(6) = 226.16$, $P < 0.001$; *EI/GJ*: $\chi^2(6) = 226.16$, $P < 0.001$). Significant differences in *E* and *G* for G1–G3 could be identified using the pairwise Mann–Whitney *U*-test (*post hoc*). Only G2 and G3 show differences in the *EI/GJ* ratio with $P = 0.003$.

	E (N mm^{-2})			*G* (N mm^{-2})			*EI/GJ*		
	G1	G2	G3	G1	G2	G3	G1	G2	G3
G1	–	<0.001	<0.001	–	<0.001	<0.001	–	0.400	0.289
G2	<0.001	–	<0.001	<0.001	–	<0.001	0.400	–	0.003
G3	<0.001	<0.001	–	<0.001	<0.001	–	0.289	0.003	–

While the plot of *E* against *G* shows distinct areas for all three developmental stages (Fig. 4A), the pattern in the plot of *EI* against *GJ* is not as clear (Fig. 4B). This is mainly due to differences of the stem diameter of the samples. Although the stem diameter of one aerial shoot is quite homogenous, the stem diameter between aerial shoots or between the main stem of an aerial shoot and its branches can differ greatly (Table 1). This has an influence on the geometrical factors, second moment of area (*I*) and torsional constant (*J*), which both contribute to flexural (*EI*) and torsional rigidity (*GJ*).

Anatomy

Flagellaria indica shows the characteristic monocot stem organization, with scattered vascular bundles surrounded by parenchyma. Each vascular bundle consists of phloem and xylem, and is protected by a sclerenchymatous bundle sheath. Vascular bundles vary in terms of vessel number and size, degree of lignification, thickness of cell walls and thickness of bundle sheath (Fig. 7). In general, the number of vascular bundles and the density of the bundle sheath increase towards the periphery of the stem. Thus, the stem is differentiated into two regions with specific anatomical characteristics, i.e. the peripheral and the central regions (Fig. 8). Conductive tissues of the vascular bundles in the peripheral region are reduced, while the amount of sclerenchymatous bundle sheath fibres is comparably high (Fig. 7A and B). In the central region, vessels are much larger and the bundle sheath fibres are less dominant (Figs 7E and F and 9). Whereas the parenchyma forms a distinct part of the central stem, it is reduced towards the periphery and/or lignified. Leaf sheaths closely envelope the stems (Figs 3 and 10A). The thin cortex of the stem consists of four to five cell layers outside the peripheral region of dense vascular bundles (Fig. 11A–C). In addition to these overall anatomical features, each developmental stage possesses particular anatomical characteristics.

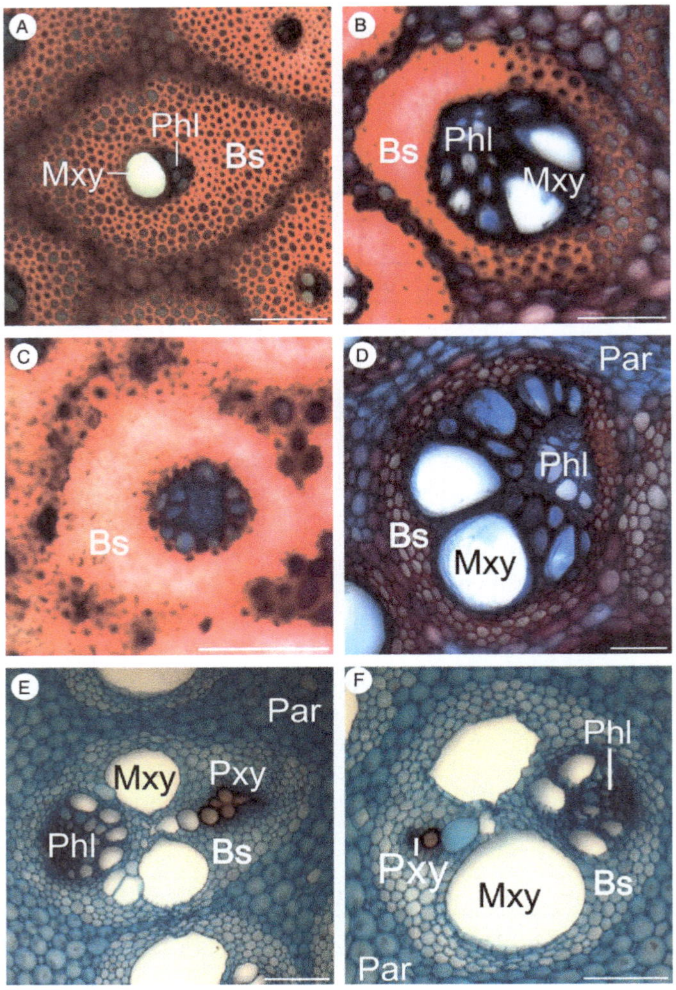

Figure 7. Variation of vascular bundles. (A and C) Vascular bundles of the peripheral region, with abundant and strongly lignified bundle sheaths, as well as reduced conductive tissues. (B and D) Vascular bundles at the transition zone of peripheral and central region. (B) shows higher bundle sheath density towards the periphery. (D) is characterized by a large number of conductive elements. (E and F) Vascular bundles of the central region of a young stem fragment differing in shapes from oval to circular, and having a large amount of non-lignified cells. Par, parenchyma; Phl, phloem; Pxy, protoxylem; Bs, bundle sheath; Mxy, metaxylem. Scale bars are 100 µm.

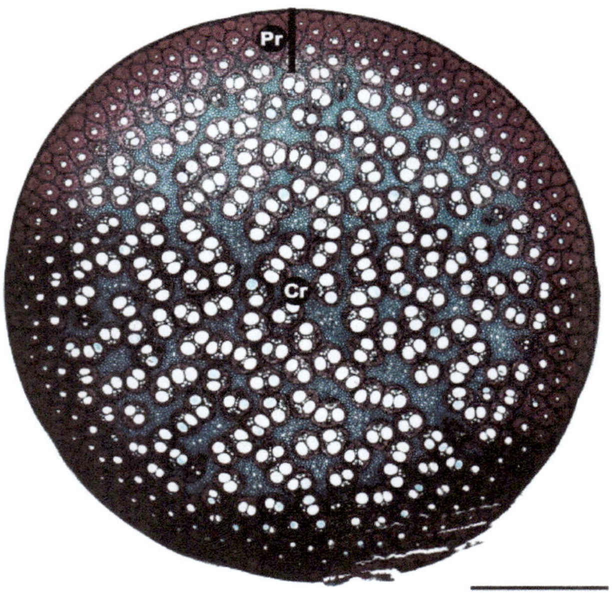

Figure 8. Transverse section of *F. indica*, without the leaf sheath. The stem is differentiated into two areas with specific anatomical characteristics. The region towards the periphery shows an increased number of vascular bundles and density of the bundle sheath. The central region shows a typical monocot stem anatomy, with scattered vascular bundles surrounded by parenchyma. Pr, peripheral region; Cr, central region. Scale bar is 2 mm.

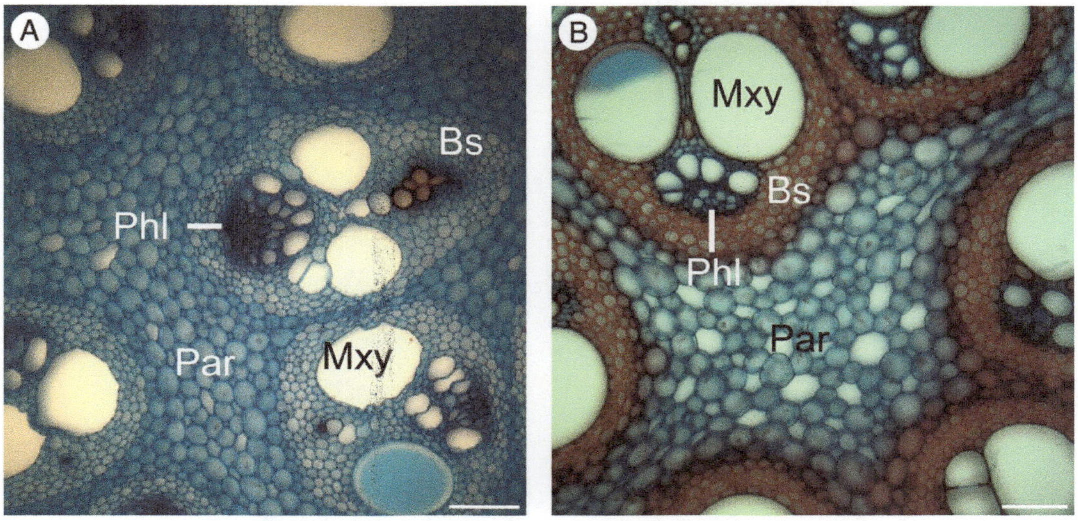

Figure 9. Transverse sections of two developmental stages from central stem regions of *F. indica*. Differences are not as pronounced as in the peripheral region. (A) Young apical stem sections show low degrees of lignification, with the exception of the protoxylem. (B) Differentiation processes of the central region involve lignification of the bundle sheath and the metaxylem. Par, parenchyma; Phl, phloem; Bs, bundle sheath; Mxy, metaxylem. Scale bars are 100 μm.

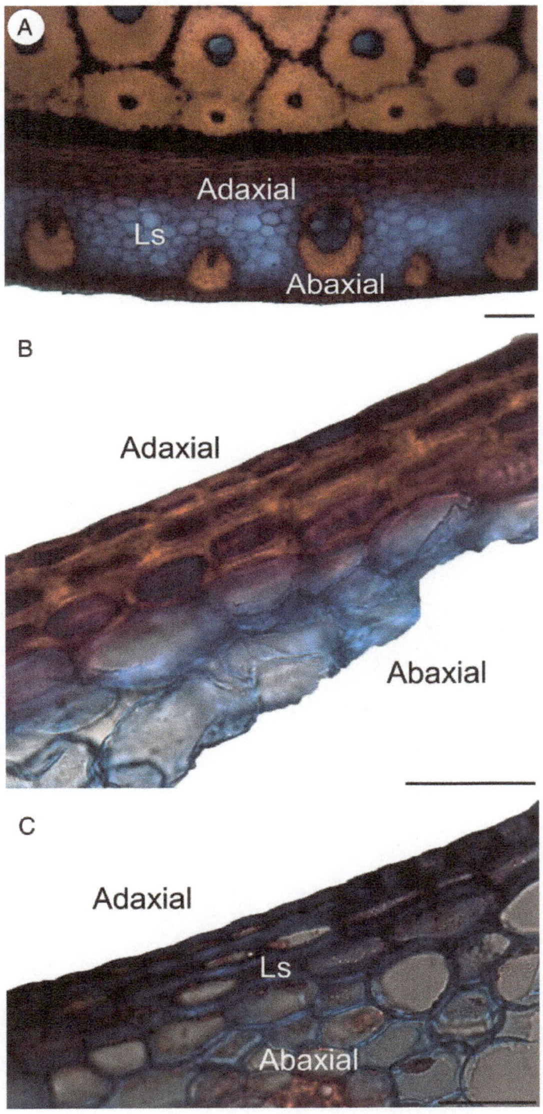

Figure 10. Developmental changes in the leaf sheath. (A) Close contact of leaf sheath and stem of developmental stage G2. (B) Lignification of leaf sheath tissue orientated towards the stem. (C) In younger stages, no lignification at the adaxial side of leaf sheaths is observed. Ls, leaf sheath. Scale bars are (A) 100 μm; (B and C) 50 μm.

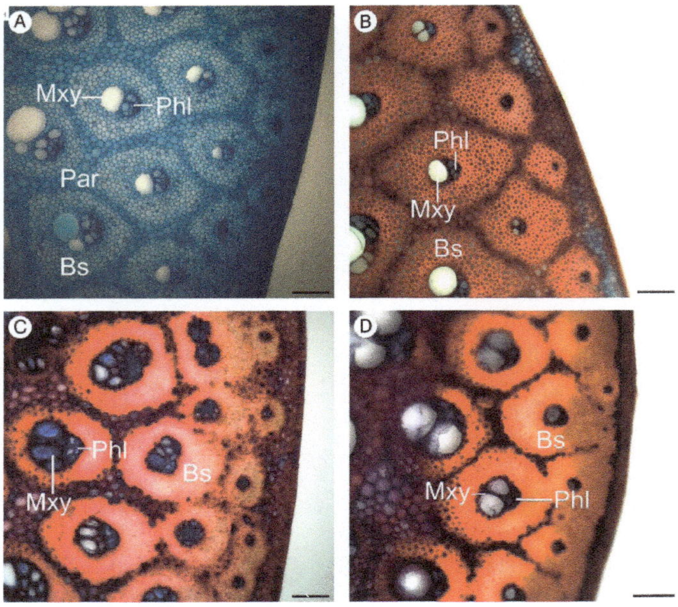

Figure 11. Details of developmental changes in transverse sections of the peripheral region of the stem, with increased maturity from A to D. (A) Young apical stem sections hardly show any lignification. Furthermore, vascular bundles are still separated by parenchyma. (B) Onset of lignification and increase of tissue density; bundle sheaths are still not fused. (C) Fusion of bundle sheaths and lignification of parenchyma increases. (D) In older, rigid and basal stems, maximal bundle sheath fusion and tissue density are observed. Par, parenchyma; Phl, phloem; Bs, bundle sheath; Mxy, metaxylem. Scale bars are 100 μm.

Stage G1

In young apical stems, the peripheral regions are not yet or only slightly lignified (Fig. 11A and B), while vascular bundles of the central zone show lignification of protoxylem cells (Fig. 9A). The bundle sheaths of peripheral vascular bundles are clearly separated and the cell walls of fibre cells are thin.

Stage G2

During further development, cell walls of the xylem and the bundle sheath of the central vascular bundles completely lignify while the parenchyma between the bundles remains largely non-lignified (Fig. 9). In the central region, no further distinct developmental differences between G1 and G2 were observed. However, in the peripheral region, tissue density and lignification increase during development (Fig. 11A–C), and cell walls of the fibres of peripheral bundle sheaths increase in thickness. Furthermore, the bundle sheaths of adjacent vascular bundles begin to fuse to a continuous layer of lignified cells in the peripheral region (Fig. 11C). This fusion is accomplished by lignification and cell wall thickening of parenchyma tissue between the vascular bundles (Fig. 12). The amount of fusion between adjacent vascular bundles is highest in the outer most ring of vascular bundles, which also shows the largest reduction of conductive tissue. Fusion affects three to four rings of vascular bundles. The pattern reveals a gradual decrease in the amount of fusion from periphery to centre (Fig. 11C).

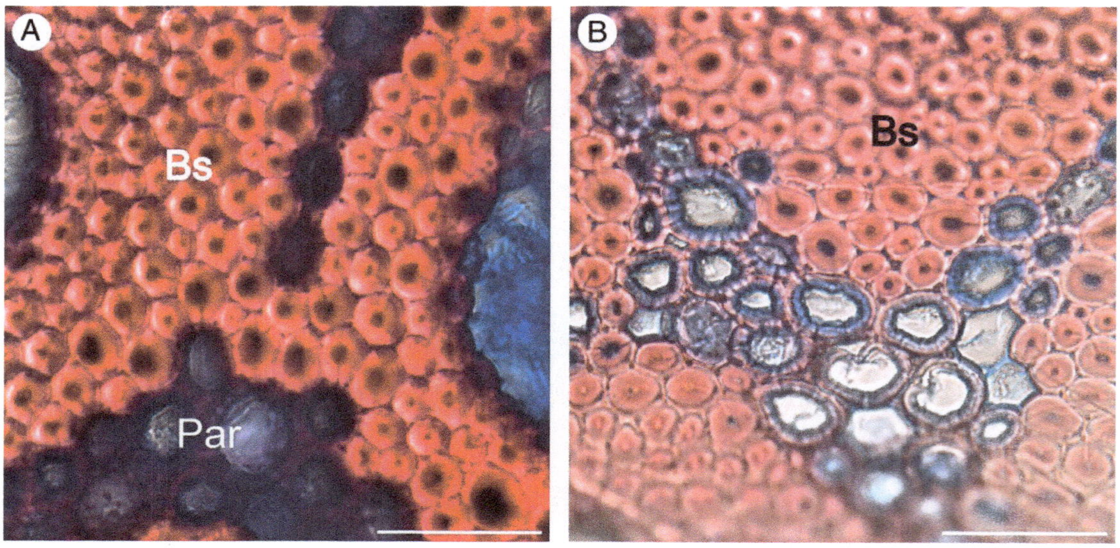

Figure 12. Fusion of bundle sheaths observed in the periphery of older stems (A) and lignification of parenchyma between bundle sheaths (B). Par, parenchyma; Bs, bundle sheath. Scale bars are 50 μm.

Stage G3

The largest amount of lignification and bundle fusion is reached in the most basal and rigid region of the stem and, therefore, increases with age. In addition, peripheral parenchyma cells show a notable elongation during maturity in addition to their lignification (Fig. 13). During development (from G1 to G3), the fibres of vascular bundle sheaths located between the peripheral and central region that are orientated towards the periphery increase in density while those directed towards the centre remain unchanged and are less prominently lignified (Fig. 11D). Changes in leaf sheath anatomy were also observed with respect to age-related lignification of the tissue orientated towards the stem (Fig. 10).

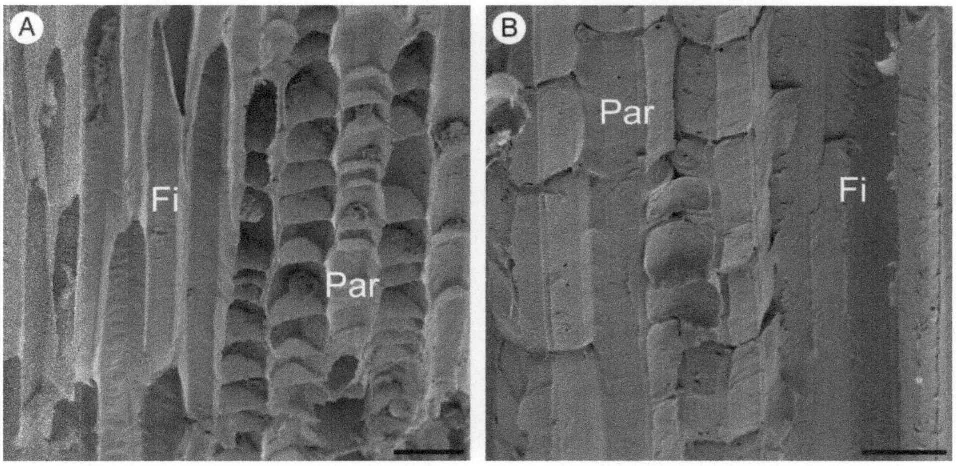

Figure 13. Longitudinal sections from the peripheral region of a stem of *F. indica*. (A) Scanning electron microscopic image of a young ontogenetic stage of the peripheral region. Between elongate fibre cells, small (~12 μm) isodiametric cells of parenchyma are observed. (B) Scanning electron microscopic image of a mature stage from the peripheral region. Between fibrous cells, elongated (~34 μm) parenchyma cells are visible. Par, parenchyma; Fi, fibres. Scale bars are 20 μm.

DISCUSSION

Changes in Growth Habit, Biomechanics and Stem Anatomy during Ontogeny

Plant stems are fibrous composite materials and their mechanical traits result from mechanical properties of both fibre and matrix, as well as fibre orientation within the stem (Hornbogen *et al.* 2011). In *F. indica*, this is, without exception, reflected in a higher bending modulus compared with the modulus of torsion correspondingly leading to high values of *EI*/*GJ*(*EI*/*GJ* > 1; Table 1). A high *EI*/*GJ* ratio indicates a high degree of anisotropy, suggesting a strong lignification and degree of fibre orientation. In general, stems with a high proportion of parenchyma and, therefore, a high amount of hydrostatic tissue show a higher resistance to torsional loads (Niklas 1992; Vogel 1992).

Apical stem sections (G1) have the lowest moduli measured (Table 1; Figs 4 and 5). Lignification and cell wall thickening, both being of high mechanical relevance (Gierlinger and Schwanninger 2006; Horvath *et al.* 2010; Schreiber *et al.* 2010), have not yet been initiated, even though tissue distribution and longitudinal orientation of fibres in the stem is determined. Thus, the contribution of mechanically relevant tissues (i.e. with lignified and thickened cell walls) to the material properties is relatively low, Lignification and thereby stiffness of young stems increase with distance from the apex. Samples of G1, thus, include different degrees of lignification. However, their cell walls have not thickened and lignified as prominently as it is the case in older stages (G2 and G3) and bundle sheaths are not fused by lignified parenchyma cells in the peripheral region. A rapid increase of rigidity in apical parts of shoots is essential for the survival and ecology of climbing plants. It allows potential supports to be located and gaps in the vegetation to be successfully spanned until attachment is assured (Rowe *et al.* 2006). The *EI*/*GJ* ratio of G1 segments is low (Table 1, G1) and attributable to the large amount of parenchyma (Niklas 1992; Vogel 1992).

During development (G2), lignification and cell wall thickening of mechanical relevant tissues, and thus their contribution to the material stiffness, is increased. As a result, the *EI*/*GJ* ratio increases (Table 1). The fusion of bundle sheaths due to parenchyma lignification in the peripheral region increases the amount of sclerenchymatous tissue in the stem (Fig. 12), raising stiffness further compared with younger stages. The resulting gradient in tissue density towards the periphery is known to increase rigidity and to be congruent with the gradient of tension during bending and torsional loads (Amada *et al.* 1997). Tension increases with distance to the neutral axis or neutral plane and is highest in the most peripheral fibres (Bürgel 2005; Dankert and Dankert 2006; Läpple 2008;Wagner *et al.* 2012). Thus, the high density of mechanically relevant tissue in the periphery has a large influence on the material properties of the stem. G2 is characterized by somewhat consistent material properties (*E*and *G*) along most of the aerial stem. This mechanical architecture facilitates the climbing growth habit by reducing the risk of mechanical failure, e.g. allowing stem position to be maintained if attachment to supports is partially lost. This is especially beneficial in low height forest communities, such as mangroves or at forest margins, where climbers such as *F. indica* have no need to reach tree canopies of 50 m or more above the ground. These stiff stem characteristics, which are also found in Stage G3, are mainly brought about through changes in the peripheral region of the stem. These include larger numbers of vascular bundles and greater tissue density towards the periphery.

In basal stem segments (G3), bundle sheath fusion is highly advanced (Fig. 11). The vascular bundles of the outermost ring fuse completely, resulting in a cylinder of dense, stiff fibres and parenchyma cells. The latter are lignified and elongated (Fig. 13). Furthermore, most of the adjacent layer of vascular bundles fuses, although not to the same degree as in the outermost ring. Prominent fusion of bundle sheath, lignification and elongation of parenchyma cells increase the rigidity of basal stem regions by contributing to the material properties (*E* and *G*). The high *EI*/*GJ* ratio of old, basal stem segments (G3; Table 1) most likely depends on the degree of vascular bundle fusion in the stem periphery (Figs 11 and 12), resulting in a cylinder of stiff lignified fibre tissue with high resistance to mechanical loads. Furthermore, the elongation of parenchyma cells (Fig. 13) contributes to the direction dependency of mechanical properties and thereby to the anisotropy of the stem.

When taking the low safety factors of the long and slender shoots of *F. indica* into account (Niklas 1998; Niklas and Spatz 1999), it is evident that the amount of self-supporting growth must be limited. However, when assessing the energetic costs for developing stiffness in basal self-supporting stem regions (G3), it is apparent that it is energetically more beneficial to reduce tissue maturation processes in climbing stem regions (G2) because these are mechanically supported by hosts via the clinging leaf tendrils.

Climbing Habit of *F. Indica* and Other Monocots

The general trend observed in *F. indica* broadly resembles that discovered for woody self-supporting trees and shrubs (Speck *et al.* 2003), with an overall decrease of stiffness from basal towards apical stem segments. In particular, the stems of Stage G2 reflect a semi-self-supporting stage with nearly constant material properties (Fig. 6). Thus, the biomechanical behaviour discovered for the climber *F. indica* differs from that observed for most climbers. Thus, while basal parts of the stem become more flexible with age in most climbers (e.g. Speck *et al.* 2003; Rowe *et al.* 2004; Wagner *et al.* 2014; Angyalossy *et al.* 2015), those of *F. indica*increase in stiffness reaching material properties comparable with those of the bamboo *Phyllostachys edulis* (Table 1). This is a tubular self-supporting member of the Poaceae with mean values of bending modulus ranging from 11 000 to 17 000 N mm^{-2} (Amada and Untao 2001). This demonstrates how little is known about climbing monocots outside of the palms and that further investigations of the growth habit of these plants are much needed. The few monocot climbers studied thus far suggest that the growth form diversity of monocots and in particular of monocot climbers is much higher than expected: *Desmoncus polyacanthos*, a climbing palm reaching stem diameters comparable with those of *F. indica*, possesses a less distinctive self-supporting phase and instead attaches rapidly to a support in dense vegetation, or discontinues apical growth by entering a latency period before growth continues (Isnard *et al.* 2005). The juvenile self-supporting phase of *Desmoncus orthacanthos* instead can, nevertheless, achieve an unsupported height of 2–3 m, but unlike in *F. indica*, this is effected via large basal stem diameter (up to 3 cm) and high stem stiffness, which changes little with distance from the base (Isnard *et al.* 2005).

In contrast to the climbing palms, which attach via cirri (modified leaf rachis) or flagella (modified inflorescences) (Isnard and Rowe 2008*b*;Isnard and Silk 2009; Rowe and Isnard 2009), *F. indica* climbs by means of leaf tendrils, which coil around mechanical hosts. While the diameter of potential supporting structures is limited, a strong and close contact to various kinds of support is achieved through tendrils thicken adaxially (S. Wagner, A. Rjosk and C. Neinhuis, unpubl. data). In contrast, the climber-support-connection of flagellate or cirri climbers like *Desmoncus* and*Calamus* is relatively loose and detachable (Isnard and Rowe 2008*b*; Rowe and Isnard 2009). While the latter can easily find new supporting structures after detachment, one main challenge for tendril climbers is the tight anchorage to the mechanical host, which can sway in the wind, expand or die back. Preliminary results have shown that the attachment of leaf tendrils of *F. indica* is relatively strong, facilitated by a high degree of lignification in the tendril and by friction between tendril and support (S. Wagner, A. Rjosk and C. Neinhuis, unpubl. data). If the stress acting on the tendril reaches a critical level, the tendril fails and separates the stem of *F. indica* from the host (S. Wagner, A. Rjosk and C. Neinhuis, unpubl. data). Thus, the integrity of the stem is preserved, and further attachment will occur via remaining leaf tendrils. The integrity of the stem is especially crucial for the survival of monocot climbers since damaged tissue, especially of the conducting vessels, cannot be repaired by cambial activity (Tomlinson and Zimmermann 2003).

Climbing Habit to Escape Constraints by Primary Growth

The climbing habit evolved several times independently within monocotyledons (e.g. *Bowiea volubilis*, *Bambusa moreheadiana*, *Calamus*, *Desmoncus* and *Flagellaria*). *Calamus* is assumed to be one of the most successful climbing genera and moreover comprises the longest known, singly rooted vascular plant species (Weinstock 1983; Uhl and Dransfield 1987; Isnard and Rowe 2008*a*). Since growth form diversity reaches its peak in the tropics, the respective climatic conditions might be a driving force towards the development of the climbing habit (Couvreur *et al.* 2014). In agreement, rattan palms as well as *Flagellaria* are distributed in the tropics and subtropics of the Old World (Weinstock 1983; Kubitzki *et al.* 1998; Rabenantoandro *et al.* 2007; Isnard and Rowe 2008*a*; Couvreur*et al.* 2014). Because climbing palms and *Flagellaria* are constitutively confined to primary growth and therefore unable to develop large diameter stems and height in order to escape shading by rainforest understorey, they have seemingly evolved alternative mechanical characteristics to survive the competition. These include specialized attachment structures, such as spiny structures of some climbing palms (Couvreur *et al.* 2014) or the tendrils of *F. indica*. Such attributes may have facilitated the evolution of several climbing monocot taxa. Climbing by means of a compliant stem is a limiting factor due to the anatomical constraints of monocot plants lacking secondary growth. Nevertheless, some monocot climbers, i.e. *Calamus*, have achieved high stem flexibility by loosing of the leaf sheath (Isnard and Rowe 2008*a*). The stem organization of monocots seems simple, but is, nevertheless, capable of developing contrasting growth forms by relatively simple modifications (Couvreur *et al.* 2014).

Although there are many monocot climbers, only a few species belong to the species-rich Poales (e.g. Tomlinson and Fisher 2000). The hypothesis that simple changes lead to a climbing habit in monocots is supported by the atypical climbing architecture of *F. indica*, which does not show the characteristic stem biomechanical pattern observed in non-monocot climbers. The modification towards the climbing habit in monocots might involve the development of attachment structures with appropriate mechanical characteristics of stems and leaf tendrils described here.

CONCLUSIONS

Despite being confined by primary growth, monocot climbers can develop into large-bodied plants. In *F. indica*, maturation processes lead to higher densities of mechanically relevant tissues in the periphery of the stem. Attachment via leaf tendrils enables the transition from self-supporting to climbing growth followed by further substantial growth height thereby gaining access to high light intensities. Furthermore, adaptations to altered mechanical requirements, such as collapse of or disconnection from the mechanical support, become possible. Even though the range of mechanical modification in *F. indica* is limited by the lack of secondary cambial growth, this has not prevented the development of specialized attachment structures. The appearance of such structures has probably underpinned the evolution of numerous large-bodied climbing monocot taxa.

ACKNOWLEDGEMENTS

We thank M. Horbens and A. Kempe for advice and guidance for biomechanical measurements, M. Günther for assistance with SEM and the staff of the botanic garden Dresden for logistical support. We further thank the reviewers very much for the time and effort they have put into the substantial improvement of our manuscript.

REFERENCES

1. Amada S, Untao S. 2001. Fracture properties of bamboo. Composites Part B: Engineering 32:451–459. doi:10.1016/S1359-8368(01)00022-1
2. Amada S, Ichikawa Y, Munekata T, Nagase Y, Shimizu H. 1997. Fiber texture and mechanical graded structure of bamboo. Composites Part B: Engineering 28:13–20. doi:10.1016/S1359-8368(96)00020-0
3. Angyalossy V, Barbosa ACF, Pace MR, Witovisk L. 2010. A new method to obtain good anatomical slides of heterogeneous plant parts. IAWA Journal 31:373–383.
4. Angyalossy V, Pace MR, Lima AC. 2015. Liana anatomy: a broad perspective on structural evolution of the vascular system. In: Schnitzer SA, Bongers F, Burnham RJ, Putz FE, eds. Ecology of Lianas. Chichester, UK: John Wiley & Sons, Ltd.
5. Beismann H, Wilhelmi H, Bailleres H, Spatz H-C, Bogenrieder A, Speck T. 2000. Brittleness of twig bases in the genus Salix: fracture mechanics and ecological relevance. Journal of Experimental Botany 51:617–633. doi:10.1093/jexbot/51.344.617
6. Bresinsky A, Körner C, Kadereit JW, Neuhaus G, Sonnewald U. 2008. Strasburger - Lehrbuch der Botanik, 36th edn. Heidelberg: Spektrum Akademischer Verlag.
7. Bürgel R. 2005. Festigkeitslehre und Werkstoffmechanik 1: Festigkeitslehre, 1st edn. Wiesbaden: Friedr. Vieweg & Sohn Verlag/GWV Fachverlage GmbH.
8. Carlquist S. 1982. The use of ethylenediamine in softening hard plant structures for paraffin sectioning. Biotechnic & Histochemistry 57:311–317. doi:10.3109/10520298209066729
9. Carlquist S. 1988. Comparative wood anatomy: systematic, ecological, and evolutionary aspects of dicotyledon wood. Berlin: Springer.
10. Chen Y-J, Bongers F, Zhang J-L, Liu J-Y, Cao K-F. 2014. Different biomechanical design and ecophysiological strategies in juveniles of two liana species with contrasting growth habit. American Journal of Botany 101:925–934. doi:10.3732/ajb.1300429
11. Couvreur TLP, Kissling DW, Condamine FL, Svenning JC, Rowe NP, Baker WJ. 2014. Global diversification of a tropical plant growth form: environmental correlates and historical contingencies in climbing palms. Frontiers in Genetics 5:452.
12. Dankert J, Dankert H. 2006. Technische Mechanik. Statik, Festigkeitslehre, Kinematik/Kinetik, 4th edn. Wiesbaden: BG Teubner Verlag/GWV Fachverlage GmbH.
13. Etnier SA. 2003. Twisting and bending of biological beams: distribution of biological beams in a stiffness mechanospace. The Biological Bulletin 205:36–46. doi:10.2307/1543443
14. Gallenmüller F, Müller U, Rowe N, Speck T. 2001. The growth form of Croton pullei (Euphorbiaceae): functional morphology and biomechanics of a Neotropical liana. Plant Biology 3:50–61. doi:10.1055/s-2001-11750
15. Gierlinger N, Schwanninger M. 2006. Chemical imaging of poplar wood cell walls by confocal Raman microscopy. Plant Physiology 140:1246–1254. doi:10.1104/pp.105.066993
16. Hoffmann B, Chabbert B, Monties B, Speck T. 2003. Mechanical, chemical and X-ray analysis of wood in the two tropical lianas Bauhinia guianensis and Condylocarpon guianense: variations during ontogeny. Planta 217:32–40.
17. Hornbogen E, Eggeler G, Werner E. 2011. Werkstoffe: Aufbau und Eigenschaften von Keramik-, Metall-, Polymer- und Verbundwerkstoffen, 10th edn. Berlin: Springer.
18. Horvath L, Peszlen I, Peralta P, Kasal B, Li L. 2010. Mechanical properties of genetically engineered young aspen with modified lignin content and/or structure. Wood and Fiber Science 42:310–317.
19. Isnard S, Rowe NP. 2008a. Mechanical role of the leaf sheath in rattans. New Phytologist 177:643–652. doi:10.1111/j.1469-8137.2007.02308.x
20. Isnard S, Rowe NP. 2008b. The climbing habit in palms: biomechanics of the cirrus and flagellum. American Journal of Botany 95:1538–1547. doi:10.3732/ajb.0700005

21. Isnard S, Silk WK. 2009. Moving with climbing plants from Charles Darwin's time into the 21st century. American Journal of Botany 96:1205–1221. doi:10.3732/ajb.0900045

22. Isnard S, Speck T, Rowe NP. 2003a. Growth habit and mechanical architecture of the sand dune-adapted climber Clematis flammula var. maritima L. Annals of Botany 91:407–417. doi:10.1093/aob/mcg044

23. Isnard S, Speck T, Rowe NP. 2003b. Mechanical architecture and development in Clematis: implications for canalised evolution of growth forms. New Phytologist 158:543–559. doi:10.1046/j.1469-8137.2003.00771.x

24. Isnard S, Speck T, Rowe NP. 2005. Biomechanics and development of the climbing habit in two species of the South American palm genus Desmoncus (Arecaceae). American Journal of Botany 92:1444–1456. doi:10.3732/ajb.92.9.1444

25. Kubitzki K, Huber H, Rudall PJ, Stevens PS, Stützel T. 1998. Flowering plants. Monocotyledons, Alismatanae and Commelinanae (except Gramineae). Berlin: Springer.

26. Kukachka BF. 1977. Sectioning refractory woods for anatomical studies. US For Serv Res Note FPL 0236: 1–9 (also Microscopia Acta 80: 301–307, 1978).

27. Läpple V. 2008. Einführung in die Festigkeitslehre: Lehr- und Übungsbuch, 2nd edn. Wiesbaden: Vieweg+Teubner Verlag.

28. Larsen K. 1972. Flagellariaceae; Hanguanaceae. Flora of Thailand 2:162–166.

29. Linder HP. 1987. The evolutionary history of the Poales/Restionales: a hypothesis. Kew Bulletin 42:297–318. doi:10.2307/4109686

30. Niklas KJ. 1991. The elastic moduli and mechanics of Populus tremuloides (Salicaceae) petioles in bending and torsion. American Journal of Botany 78:989–996. doi:10.2307/2445178

31. Niklas KJ. 1992. Plant biomechanics: an engineering approach to plant form and function, 1st edn. Chicago: University of Chicago Press.

32. Niklas KJ. 1998. A statistical approach to biological factors of safety: bending and shearing in Psilotum axes. Annals of Botany 82:177–187. doi:10.1006/anbo.1998.0669

33. Niklas KJ, Spatz H. 1999. Methods for calculating factors of safety for plant stems. Journal of Experimental Biology 202:3273–3280.

34. Rabenantoandro J, Randrihasipara L, Randriatafika F, Vincelette M, Rakoto J. 2007. Testing the propagation and growth of the liana Flagellaria indica, used to make lobster traps, and Bambusa multiplex as an alternative source. In: Ganzhorn JU, Goodman SM, Vincelette M, eds. Biodiversity, ecology and conservation of littoral forest ecosystems in southeastern Madagascar, Tolagnaro (Fort Dauphin). Washington, DC: Smithsonian Institution, 363–367.

35. R Development Core Team. 2014. R: a language and environment for statistical computing. Vienna, Austria: R Foundation for Statistical Computing. http://www.R-project.org/ (25 September 2015).

36. Ricklefs RE, Renner SS. 1994. Species richness within families of flowering plants. Evolution 48:1619–1636. doi:10.2307/2410252

37. Rowe NP, Isnard S. 2009. Biomechanics of climbing palms and how they climb. Plant Signaling & Behavior 4:875–877. doi:10.4161/psb.4.9.9426

38. Rowe NP, Speck T. 1998. Biomechanics of plant growth forms: the trouble with fossil plants. Palaeobotany and Palynology 102:43–62. doi:10.1016/S0034-6667(98)00013-X

39. Rowe NP, Isnard S, Speck T. 2004. Diversity of mechanical architectures in climbing plants: an evolutionary perspective. Journal of Plant Growth Regulation 23:108–128. doi:10.1007/s00344-004-0044-0

40. Rowe NP, Isnard S, Gallenmüller F, Speck T. 2006. Diversity of mechanical architectures in climbing plants: an ecological perspective. In: Herrel A, Speck T, Rowe NP, eds. Ecology and biomechanics—a mechanical approach to the ecology of animals and plants. Boca Raton, FL: CRC Press, 35–59.

41. Schreiber N, Gierlinger N, Pütz N, Fratzl P, Neinhuis C, Burgert I. 2010. G-fibres in storage roots of Trifolium pratense (Fabaceae): tensile stress generators for contraction. The Plant Journal 61:854–861. doi:10.1111/j.1365-313X.2009.04115.x

42. Speck T, Rowe N, Civeyrel L, Classen-Bockhoff R, Neinhuis C, Spatz C. 2003. The potential of plant biomechanics in functional biology and systematics. In: Stuessy TF, Mayer V, Hörandl E, eds. Deep morphology: toward a renaissance of morphology in plant systematics. Königstein: Koeltz Scientific Books, 241–271.

43. Takhtadzhi�an AL. 1997. Diversity and classification of flowering plants. New York: Columbia University Press.

44. Tomlinson PB, Fisher JB. 2000. Stem vasculature in climbing monocotyledons: a comparative approach. In: Wilson KL, Morrison DA, eds. Monocots: systematics and evolution, Melbourne: CSIRO Publishing, 89–96.

45. Tomlinson PB, Posluszny U. 1977. Features of dichotomizing apices in Flagellaria indica (Monocotyledones). American Journal of Botany 64:1057–1065. doi:10.2307/2442161

46. Tomlinson PB, Zimmermann MH. 2003. Stem vascular architecture in the American climbing palm Desmoncus (Arecaceae-Arecoideae-Bactridinae). Botanical Journal of the Linnean Society 142:243–254. doi:10.1046/j.1095-8339.2003.00174.x

47. Uhl N, Dransfield J. 1987. Genera Palmarum: a classification of palms based on the work of Harold E. Moore, Jr, 1st edn. Lawrence: Allen Press.

48. Vincent JFV. 1992. Biomechanics—materials: a practical approach. Oxford: Oxford University Press.

49. Vogel S. 1992. Twist-to-bend ratios and cross-sectional shapes of petioles and stems. Journal of Experimental Botany 43:1527–1532. doi:10.1093/jxb/43.11.1527

50. Vogel S. 2007. Living in a physical world XI. To twist or bend when stressed. Journal of Biosciences 32:643–655. doi:10.1007/s12038-007-0064-6

51. Wagner ST, Isnard S, Rowe NP, Samain M-S, Neinhuis C, Wanke S. 2012. Escaping the lianoid habit: evolution of shrub-like growth forms in Aristolochia subgenus Isotrema (Aristolochiaceae). American Journal of Botany 99:1609–1629. doi:10.3732/ajb.1200244

52. Wagner ST, Hesse L, Isnard S, Samain M-S, Bolin J, Maass E, Neinhuis C, Rowe NP, Wanke S. 2014. Major trends in stem anatomy and growth forms in the perianth-bearing Piperales, with special focus on Aristolochia. Annals of Botany 113:1139–1154. doi:10.1093/aob/mcu044

53. Weinstock JA. 1983. Rattan: ecological balance in a Borneo rainforest swidden. Economic Botany 37:58–68. doi:10.1007/BF02859305

54. Wepfer PH, Linder HP. 2014. The taxonomy of Flagellaria (Flagellariaceae). Australian Systematic Botany 27:159–179. doi:10.1071/SB13048

55. Wickham H. 2009. ggplot2: elegant graphics for data analysis. New York: Springer.

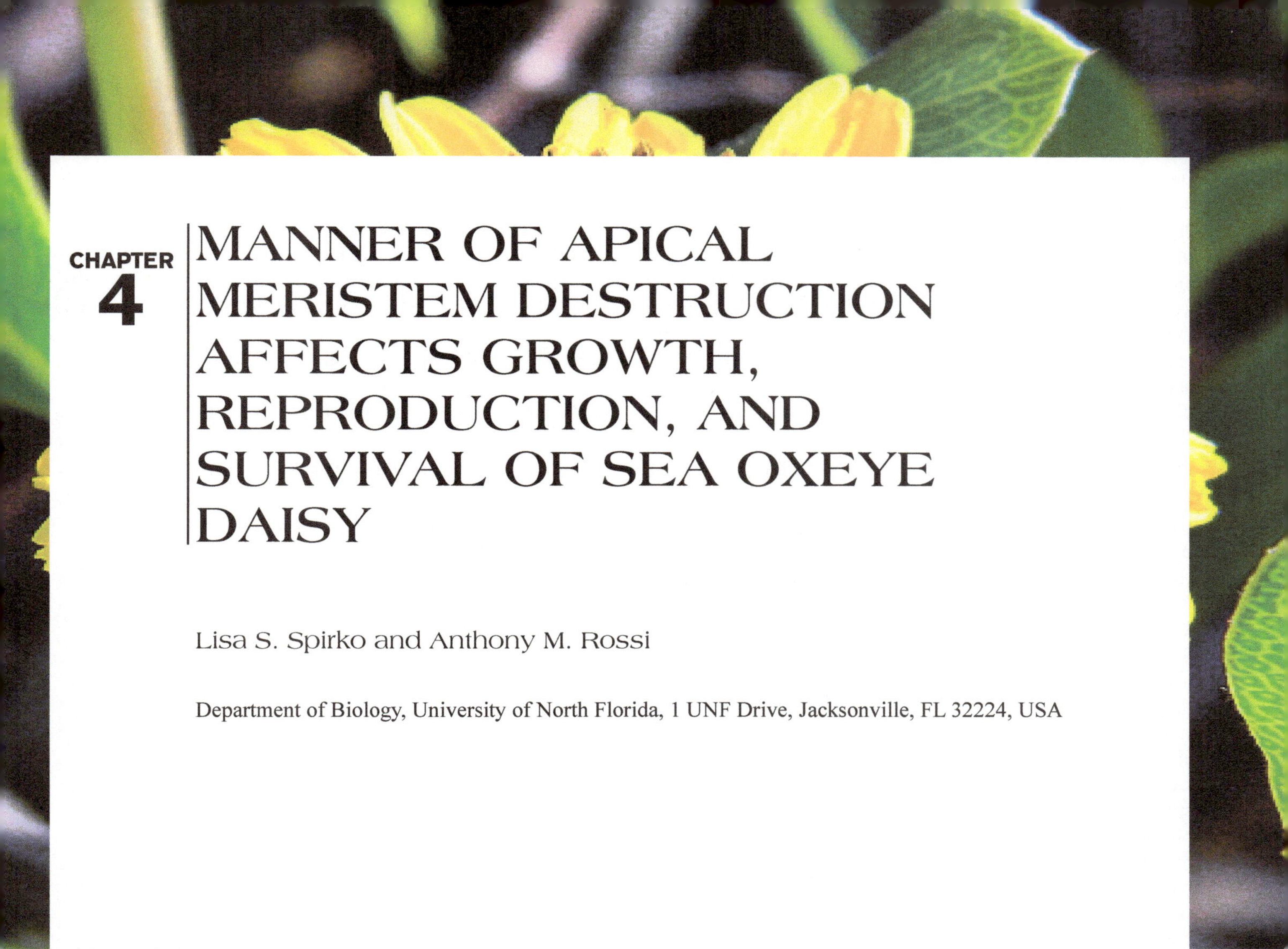

CHAPTER 4

MANNER OF APICAL MERISTEM DESTRUCTION AFFECTS GROWTH, REPRODUCTION, AND SURVIVAL OF SEA OXEYE DAISY

Lisa S. Spirko and Anthony M. Rossi

Department of Biology, University of North Florida, 1 UNF Drive, Jacksonville, FL 32224, USA

ABSTRACT

Although herbivory may result in plant death, the mode and timing of damage may produce variable wounding responses in the attacked plant. In this study, effects of different types of apical meristem damage on growth and performance of sea oxeye daisy Borrichia frutescens (L.) DC were compared. Damage involved either clipping or galling of the apical meristem by the gall midge Asphondylia borrichiae Rossi and Strong. Apical dominance was relatively weak before flowering and stronger in short ramets that were shaded by taller neighbors later in the season. Clipped ramets delayed sprouting new stems, and galled ramets sprouted new stems quickly compared to intact ramets, but final stem counts were similar across treatments. Clipping significantly delayed flowering, reduced seed head ripening time, and resulted in fewer seed heads and seeds. Galling did not significantly impact reproduction compared to intact ramets. Nitrogen supplementation significantly increased stem count and seed count and lengthened seed ripening time. Borrichia frutescensres ponds differently to clipping versus galling by A. borrichiae and better tolerates galling in terms of nonreproductive performance and survival. Galling from A. borrichiae likely prevents Borrichia frutescens from flowering, which suggests resource regulation of meristems by the midge.

INTRODUCTION

Plant-animal interactions such as herbivory and parasitism can vary greatly in their effect on plants, ranging from death and decreased plant performance and fitness at one end of the spectrum [1, 2] to resistance (e.g., chemical or physical defenses) and tolerance or compensatory regrowth at the other end [3–12]. Many of these relationships have existed for long periods, and coevolution, or the extent to which species are closely adapted (i.e., coadapted) to each other, has long been understood as the driver by which mechanisms of resistance [13,14] and tolerance [9, 14, 15] develop.

Tolerance arises from the complex interactions of numerous genetic, physiological, and morphological traits, including plant architecture [5, 8], meristem allocation [7, 8, 10, 11, 15], nutrient uptake [15], photosynthetic rates [8, 15], and phenological timing [3, 15–17], under influence of environmental factors [10, 11, 15]. For example, plant species that hold axillary buds dormant may regain lost tissue if the apical meristem, or actively growing stem tip, is damaged because it disrupts apical dominance, or hormonal suppression of axillary buds by the apical meristem, and allows new growth [3–7, 9–11]. In some cases, overcompensation, or regrowth that exceeds the amount of tissue lost and even results in improved fecundity, has been documented [3, 18–21]. The

Citation: Lisa S. Spirko and Anthony M. Rossi, "Manner of Apical Meristem Destruction Affects Growth, Reproduction, and Survival of Sea Oxeye Daisy," Journal of Botany, vol. 2015, Article ID 480891, 11 pages, 2015. doi:10.1155/2015/480891

degree of tolerance or compensation is often subject to such environmental factors as the timing of herbivory or parasitism (i.e., the season in relation to the plant's phenological state) and nutrient availability. For instance, Ipomopsis aggregata significantly increased its flowering stems after browsing early in the growing season by ungulates [3]. In another study [16], Gentianella campestris overcompensated in terms of fruits and seeds, but only when simulated herbivory (clipping) occurred during a narrow window in Sweden's growing season. Benner [17] also found overcompensation in transplants of Thlaspi arvense occurred if nutrient supplementation was provided early in the growing season.

Apical meristem damage also may occur as a result of cecidogenesis (gall-making), which involves a unique, intimate relationship between the gall-inducer, typically an insect or fungus, and its host plant, and results in a tumor-like overgrowth of plant tissue [22–28]. Gall-inducers tend to be specialists and highly specific to their particular host species, genus, or family and to the organ they attack, suggesting tight coevolutionary relationships in these systems [22, 23, 28, 29]. Gagné [28] reports that gall midges (Cecidomyiidae) are usually monophagous (one host plant species) or oligophagous (a few closely related species). In a review of the literature spanning 196 neotropical species of gall midges, Carneiro et al. [30] found that 92.4% were monophagous and 5.6% were oligophagous. Additionally, 90.8% of the cecidomyiid species reviewed were found to induce galls on only one plant organ (leaves, flowers, or stems), further supporting the claim that they are host specialists. Galling of the apical meristem can disrupt apical dominance and initiate axillary shoots [27,31, 32]. Thus, since host plants likely coadapted with their associated gall inducers, it is reasonable that plants attacked by apical gall inducers may demonstrate tolerance or compensation in response to the damage.

Galls of the midge Asphondylia borrichiae on the apical meristem of sea oxeye daisy (Borrichia frutescens) are associated with stem bifurcation below the gall [33]. Other types of naturally occurring apical meristem damage, such as chewing by grasshoppers (e.g., Orchelimum fidicinium), are seen rarely in the field (Anthony M. Rossi, personal observation).

The overall goal of this research was to examine the compensatory response of B. frutescens to different types of apical meristem damage in young single-stemmed ramets. Damage to the meristem was either sudden (i.e., clipping to simulate chewing) or gradual via galling, which slowly destroyed the apical meristem. Because the plant's ability to compensate depends largely on having a bank of dormant buds in reserve [3–5, 10, 11, 34], the strength of apical dominance in B. frutescens was investigated in relation to its response to meristem damage. A second experiment investigated how B. frutescens responds to clipping, galling by A. borrichiae, and nutritional differences in terms of growth and reproduction (i.e., flower head and seed production).

MATERIALS AND METHODS

Study System

Borrichia frutescens, commonly known as sea oxeye daisy or bushy seaside tansy, is an herbaceous, halophytic, perennial shrub found in salt marshes along the U.S. Atlantic and Gulf of Mexico coasts [35]. It propagates primarily through extensive rhizomes [36–38] but also produces small, monoecious flower heads consisting of 15–30 petal-like ray florets and 20–75 disc florets per flower head [35–39]. Each floret produces an achene, or dry seed-bearing fruit, and each achene contains a single seed. Peak flowering is between May and July [38], and flowering rates, or the number of flower heads per ramet, range from 3% to over 20% (although these rates can be much higher), depending on factors such as season or access to tidal inundations of nutrients [40, 41]. Under greenhouse conditions, seed viability of 90% was reported by Biber et al. [39].

Borrichia frutescens is the primary host plant of the gall midge Asphondylia borrichiae [33, 41–43]. Adult A. borrichiae are small (<3 mm long), fragile, weak-flying gall midges that live for only 2-3 days, during which their primary activity is breeding [2, 44–46]. Female A. borrichiae collect the conidia of a fungal symbiont and deposit the conidia with her eggs inside the apical meristematic tissue of the host plant during oviposition, which induces a gall to form [42, 45, 47, 48]. Larvae feed on the fungal symbiont, which draws nutrients from plant tissues, and they pupate within the gall [47, 48]. With occasional exceptions, female midges appear to avoid ovipositing in flower heads [40, 42], and the presence of a gall typically prevents the apical meristems from flowering [33, 42, 45]. Galls persist on the host plant during spring and summer for about seven to ten weeks and senesce after midge eclosion, typically resulting in the death of the galled apical meristem [33, 45]. A previous study at the site in 2005 found moderate correlation between stem count and gall count (Anthony M. Rossi, unpublished data). Aside from galling by A. borrichiae, the sap feeding leafhopper Pissonotus quadripustulatus and the aphid Dactynotus sp. are the only other herbivores regularly associated with B. frutescens in Florida [49–51].

Field Site

This study was conducted at Timucuan Ecological and Historic Preserve in Jacksonville, Florida, USA (N30°22′45″, W81°28′49″). This site features an intertidal back-marsh community bordered by hammock and consisting primarily of Borrichia frutescens. Previous studies at this site [43, 47] have found galling on B. frutescens to be very high (approximately 30–50% of apical meristems) and a strong positive correlation between stem number and gall count (Anthony M. Rossi, unpublished data).

First Bud Break Experiment

To investigate the strength of apical dominance in B. frutescens and to test its response to various types of apical meristem damage, 25 single-stemmed B. frutescens ramets (i.e., individual plants) were randomly assigned to a stem treatment in May 2013: (1) galled, (2) clipped, or (3) intact. Ramets in the galled group began the experiment with a single stem, but also had a single, live gall (i.e., green, no emergence holes). Clipping was performed by removing the apical meristem and the terminal pair of leaves with scissors (removing approximately 1 cm of the stem tip). Intact ramets were protected from galling by a mesh bag placed over the apical meristem and secured with a twist tie. Two-way ANOVA on light readings (Basic Quantum Meter, model BQM-SUN,

Apogee Instruments Inc., Logan, Utah, USA) taken inside and outside the mesh bags in sunny and shady conditions revealed no effect of the mesh bags or their interaction with light conditions on light penetration (inside versus outside mesh: $F_{1,39} = 0.004$, $P = 0.949$; sunny versus shady: $F_{1,39} = 3277.478$, $P < 0.001$; mesh × light: $F_{1,39} = 0.022$, $P = 0.882$). Ramets were monitored for bud break biweekly and, after eight weeks, the node number bearing the bud to break first was recorded (0 = apical meristem or clipped stem tip, 1 = first node below apical meristem, etc.) and its distance measured from the apical meristem or clipped stem tip (nearest 0.1 cm). Differences in the number of surviving ramets that broke bud at eight weeks were tested using a χ^2 test. One-way ANOVA was used to test the number of days and distance to first bud break. Variances of time data were homogeneous, but due to small sample sizes of surviving ramets, normality was inferred based on a linear probability plot of residuals. Distance data were square-root transformed to achieve normality and homogeneity of variances. After a significant main effect, group means were analyzed with Tukey's HSD test to examine planned pairwise comparisons between treatments. The ratio of ramets with different frequencies of nodes bearing bud break was tested using χ^2 test. Due to low sample sizes, frequencies for nodes two and more distal from the apical meristem were pooled, and the χ^2 test compared nodes 0, 1, and 2+ across treatments.

Stem Condition and Nutritional Status Experiment

To test the effects of apical meristem damage on regrowth, seed set, quality, and viability and any interaction with plant nutritional status (i.e., nitrogen levels), a 3 × 2 fully factorial experiment involving three stem treatments (galled, clipped, and intact) and two nutrient treatments (fertilized and not fertilized) was initiated in January, about 11 weeks before the flowering season started. For this experiment, approximately 300 single-stemmed Borrichia frutescens were selected haphazardly and assigned randomly to one of six groups: (1) galled and fertilized; (2) galled and not fertilized; (3) clipped and fertilized; (4) clipped and not fertilized; (5) stem intact and fertilized; and (6) stem intact and not fertilized (control). Each of the previous treatments began the experiment with 50 ramets (i.e., individual plants). Ramets in the galled groups began the experiment with a single stem and gall. Clipped ramets were clipped once at the beginning of the experiment. Intact ramets were left unprotected from galling. However, many intact ramets were expected to be galled during the experiment [43, 47] and may have severely reduced the sample size of the control group. Therefore, in addition to the first control group of 50 ramets, a second control group of 20 additional intact, unfertilized ramets was selected haphazardly and protected from galling by a mesh bag. After experiment completion, all statistical analyses were first conducted on the bagged and unbagged control groups only and, finding no significant differences between them, data were pooled into a single control group for all remaining analyses.

Ramets in the three fertilized treatment groups received a small amount (ca. 2 g) of supplemental nitrogen in the form of blood meal (N-P-K: 12-0-0; Miracle-Gro Organic Choice Blood Meal, Scotts Miracle-Gro Products, Marysville, Ohio, USA) every four weeks. The blood meal was applied by carefully forming a hole approximately 3-4 cm deep, or as allowed by the soil compacted with oyster shells, near the base of the ramet using a screwdriver, pouring the blood meal into the hole, and covering it with substrate. Effects of the hole were controlled in the nonfertilized groups by creating a hole and then filling it in.

Total stem counts (i.e., main stem and axillary stems) and the presence of flower head buds on ramets were assessed every two weeks from January through June 2013. Stems were counted if they were alive and at least one centimeter long. Flower head buds were tagged at first appearance through the end of June and were monitored biweekly until seed head harvesting, which concluded in early September, with the initiation date of each phenological event (i.e., budding, blooming, and ripening) and harvesting recorded. After each flower head finished blooming (i.e., all florets had senesced and/or abscised), a mesh bag was placed over them and seed heads were harvested when they were brown and dry and the stalk was senescent. Each seed head was carefully broken apart and each "seed" (achene, or dry seed-bearing fruit) separated from its bract with forceps. Seeds that were abnormally developed (shriveled, malformed) were removed. Normal seeds were counted and weighed to the nearest 10^{-4} mg.

Germination success of the normally developed seeds produced was determined in September after all seed heads had been harvested and processed; abnormally developed seeds were excluded from the germination phase of the experiment. Twenty-five seeds or fewer, if less than 25 seeds developed, were randomly selected from each seed head and soaked for 48–60 hours in tap water [39]. All seeds for each seed head were planted together in the cell of a horticultural flat or a seedling pot in a 1 : 1 blend of Ace Horticultural Grade Vermiculite (A. H. Hoffman, Inc., Landisville, Pennsylvania, USA) and Organic Seed Starting Jiffy-Mix (Ferry-Morse Seed Company, Fulton, Kentucky, USA), the latter which contained 48–52% sphagnum peat moss, 48–52% vermiculite, lime, and an organic wetting agent. No fertilizers were included in the original manufactured products or added to the final planting mixture. Seeds were very lightly covered with planting medium, placed in light shade in a greenhouse (≈25–30°C, ≈60% relative humidity), and lightly watered under timed misters for four minutes every four hours for 12 weeks. Because unequal numbers of seed heads were produced in each treatment, a Latin square layout was not possible. Treatments were randomized across rows and columns as possible.

Except where noted, all statistical analyses were conducted on ramets surviving to the end of the experiment and on intact ramets remaining intact (ungalled) until the end of the experiment. Mean (±SD) overall survivorship across all treatments at the end of the experiment was s 76.6 ± 0.9% and ranged from a low of 74.2% for clipped ramets to a high of 78.8% for intact ramets. These differences were not significantly different across treatments (stem: $\chi^2 = 0.629$, $df = 2$, and $P = 0.730$; fertilizer: $\chi^2 = 0.357$, $df = 1$, and $P = 0.550$); therefore statistical analyses of counts could be performed with the assumption that the number of surviving ramets was similar across treatments. Stem counts at the end of the experiment were found to follow a Poisson distribution because ramets began the experiment with a single stem and the experiment was conducted over a relatively short period of time (25 weeks). Thus, stem counts were not parametric and were tested across stem treatments and fertilization treatments using Kruskal-Wallis tests. Chi-square (χ^2) tests were used to test the number of ramets that produced flower head buds and the total number of seed heads produced across treatments. The number of days to first flowering, or the number of days between the start of the experiment and the day the first flower head bud on each flowering ramet was recorded, was tested across stem treatments and fertilization treatments using Kruskal-Wallis tests. Kruskal-Wallis was used in this instance because assumptions of ANOVA could not be met for some groups and data (and associated residuals) could not be transformed to achieve normality and homogeneity of variances for all groups. Initial height was statistically different among the stem treatments; galled ramets (mean ± SE: 33.0 ± 1.9 cm) were between 41 and 50% taller than clipped (23.4 ± 2.6 cm) and intact ramets (22.0 ± 1.7 cm), respectively (Kruskal-Wallis: $\chi^2 = 23.794$, $df = 2$, and $P < 0.001$). This difference is likely because apical meristems above the canopy are more likely to be visible to ovipositing females; thus, two-way analysis of covariance (ANCOVA) was used on seed count, seed mass, and the number of days budding, blooming, and ripening (phenological timing) with initial height as the covariate. ANCOVA included bootstrapping (1,000 bootstrap samples) to improve confidence limits due to small sample size for some groups. For phenological timing, although variances were homogeneous, sample sizes for flowering ramets in the clipped group were small ($n = 7$); thus, these results should be viewed with caution. Seed germination rates were tested using chi-square (χ^2) tests. Significant ANCOVA results were analyzed using contrasts with automatic Sidak correction. Significant ANOVA results among stem treatments were analyzed using Tukey's HSD tests. Significant Kruskal-Wallis results among stem treatments were analyzed using Mann-Whitney U with Bonferroni correction ($\alpha = 0.017$). All statistical tests were conducted using PASW Statistics (currently known as SPSS Statistics), version 18.0.0 (IBM, Armonk, NY).

RESULTS

First Bud Break Experiment

Stem treatment had a highly significant effect on the number of days to first bud break ($F_{2,24} = 45.568$, $P < 0.001$). Both galled and clipped surviving ramets broke bud within the first 40 days (mean ± SE: 36.4 ± 6.3 and 33.6 ± 4.3, resp.) and were statistically similar, while intact surviving ramets were significantly slower and broke bud about 80 days later (113.8 ± 1.8) (Figure 1(a)). Stem treatment also had a highly significant effect on the number of surviving ramets that broke bud by eight weeks ($\chi^2 = 46.000$, $df = 2$, and $P < 0.001$) (Figure 1(b)). Slightly more clipped ramets than galled ramets broke bud in the first five weeks. By eight weeks, 100% of surviving galled and clipped ramets had broken bud, while no intact ramets had broken bud by this time. Intact ramets did not begin to break bud until sixteen weeks.

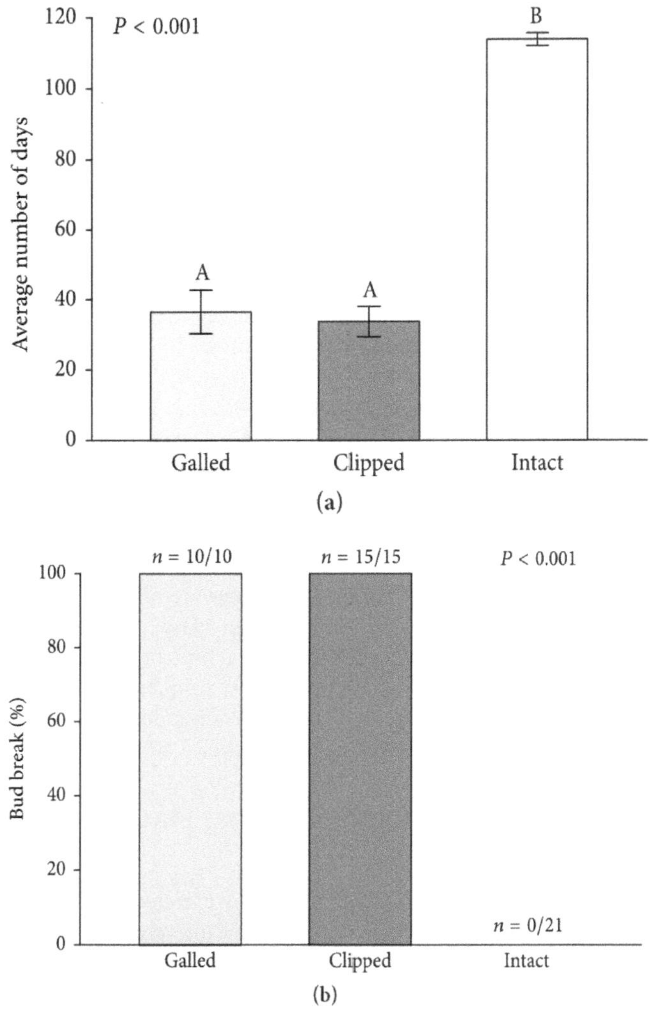

Figure 1: In the first bud break experiment, (a) average number of days to first bud break with SEM error bars across stem treatments ($F_{2,24} = 45.568$, $P < 0.001$) and (b) percent bud break occurring by eight weeks across stem treatments ($\chi^2 = 46.000$, $df = 2$, and $P < 0.001$). Different letters in (a) indicate significant differences among treatments (Tukey's post hoc with Bonferroni correction to $\alpha = 0.017$).

Stem treatment had a highly significant effect on the distance (cm) from the stem terminal to the first bud break ($F_{2,24} = 12.474$, $P < 0.001$). Clipped ramets broke bud significantly closer to the terminal, with an average distance of 1.3 ± 0.2 cm from the terminal, compared to intact ramets, which broke bud an average distance of 6.6 ± 2.0 cm (Figure 2(a)). Galled ramets broke bud an average distance of 3.0 ± 0.7 cm from the terminal and were statistically similar to both the clipped and intact groups. Stem treatment also had a highly significant effect on the node bearing first bud break ($\chi^2 = 14.534$, $df = 4$, and $P = 0.006$) as counted from the stem terminal (node 0) (Figure 2(b)). Clipped ramets first broke bud at an average node of 0.8 ± 0.2. Approximately 25% of clipped ramets broke bud directly at the stem terminal and were the only group to do so. Galled ramets broke bud at an average node of 1.9 ± 0.3, with most bud breaks occurring at node 1 and the remaining breaks being split evenly between nodes 2 and 3. Intact ramets broke bud at an average node of 4.5 ± 1.2, and never broke bud closer to the terminal than node 3. One ramet broke bud first at node 8.

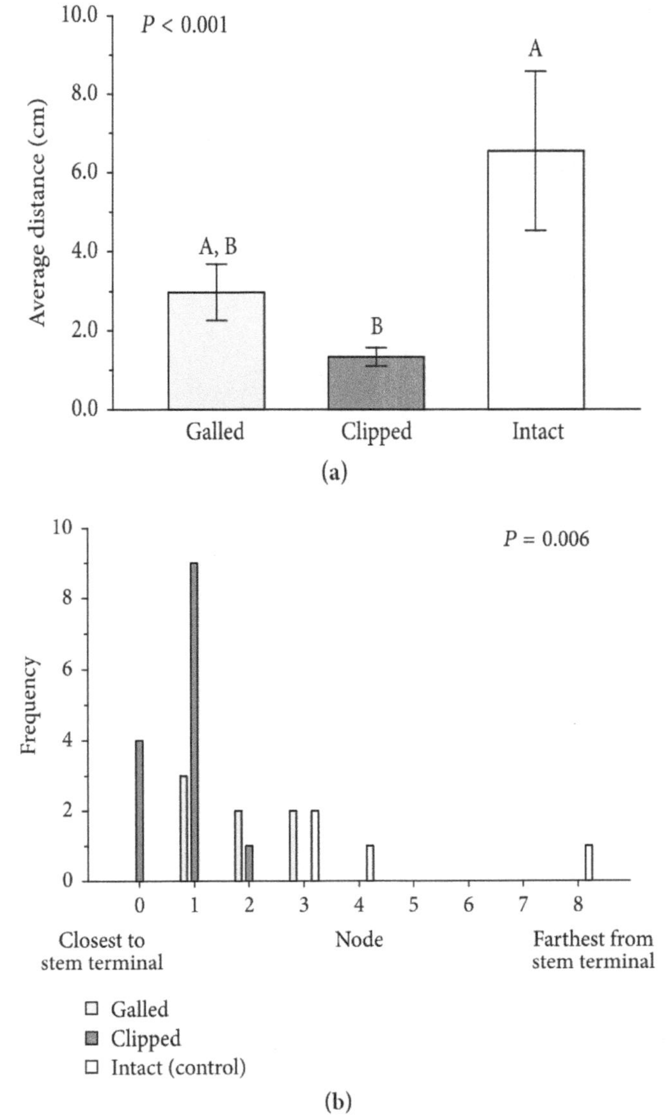

Figure 2: In first bud break experiment, (a) average distance to first bud break in centimeters, as measured from the stem terminal to the node bearing the first lateral meristem to break bud, with SEM error bars ($F_{2,24} = 12.474$, $P < 0.001$) and (b) frequencies of first bud breaks occurring at different nodes (0 = stem terminal, 1 = first node below terminal, etc.) ($\chi^2 = 14.534$, $df = 4$, and $P = 0.006$). In (a), data were square-root transformed to meet ANOVA assumptions but are presented untransformed for clarity. Different letters indicate significant differences among treatments (Tukey's post hoc with Bonferroni correction to $\alpha = 0.017$).

Stem Condition and Nutritional Status Experiment

Stem treatment had no effect on the total number of stems at the end of the experiment (week 25) ($\chi^2 = 0.986$, $df = 2$, and $P = 0.611$) with galled ramets having an average of 4.1 ± 0.6 (SE) stems, compared to 4.1 ± 0.5 stems for clipped ramets and 4.6 ± 0.5 stems for intact ramets (Figure 3(a)). Galled ramets had dramatic increases in stem count within the first month, while clipped ramets appeared to be delayed in sprouting new stems. Stem count increased modestly in intact ramets until week 15, after which average stem count in that group was comparable to that in galled and clipped plants. Approximately 27% of intact/unfertilized (control) ramets broke bud by week 3 and 44% by week 11, when flower buds began to appear on the ramets. Fertilization treatment had a highly significant effect on the number of stems at the end of the experiment ($\chi^2 = 11.973$, $df = 1$, and $P = 0.001$) with fertilized ramets having an average maximum of 4.9±0.4 stems, compared to 3.6 ± 0.5 stems for unfertilized ramets (Figure 3(b)).

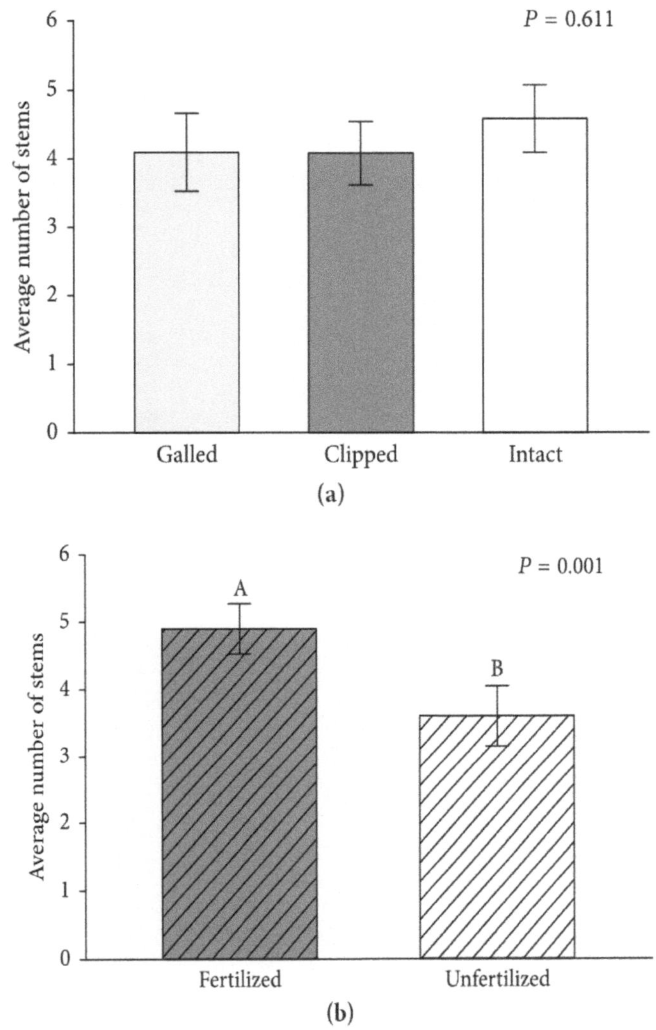

Figure 3: In stem-nutrition experiment, average number of stems at week 25 with SEM error bars across (a) stem treatments ($\chi^2 = 0.986$, $df = 2$, and $P = 0.611$) and (b) fertilization treatments ($\chi^2 = 11.973$, $df = 1$, and $P = 0.001$). Different letters indicate significant differences among treatments.

Stem treatment had a highly significant effect on the average number of days before first flowering occurred in flowering ramets ($\chi^2 = 19.747$, $df = 2$, and $P < 0.001$) with clipped ramets flowering significantly later than either galled and intact ramets (Figure 4(a)). With Bonferroni correction ($\alpha = 0.017$), galled and intact ramets were not significantly different. Clipped ramets flowered at an average of 136.2 ± 4.7 days after the beginning of the experiment, three weeks later than galled ramets (115.9 ± 3.2 days) and a full month later than intact ramets (106.2 ± 2.8 days). Fertilization treatment had no effect on the number of days to first flowering ($\chi^2 = 1.323$, $df = 1$, and $P = 0.250$), with fertilized ramets flowering an average of about five days earlier (111.6 ± 2.8 days) than unfertilized ramets (116.6±4.0 days) (Figure 4(b)).

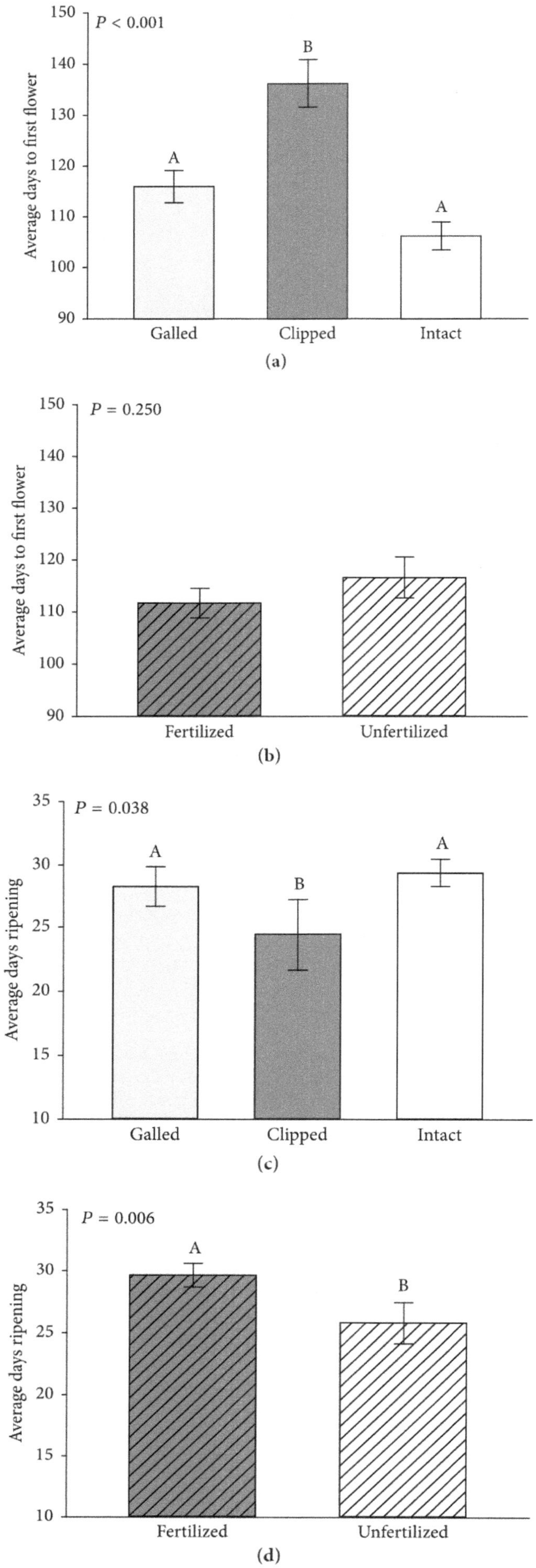

Figure 4: In stem-nutrition experiment, average number of days until appearance of first flower bud in flowering ramets only with SEM error bars across (a) stem treatments ($\chi^2 = 19.747$, $df = 2$, and $P < 0.001$) and (b) fertilization treatments ($\chi^2 = 1.323$, $df = 1$, and $P = 0.250$) and average number of days seed heads ripened with SEM error bars across (c) stem treatments and (d) fertilization treatments (stem treatment: $F_{2,47} = 3.566$, $P = 0.038$; fertilization treatment: $F_{1,47} = 8.281$, $P = 0.006$; stem × fertilization: $F_{2,47} = 1.401$, $P = 0.258$; covariate: $F_{1,47} = 0.013$, $P = 0.911$). Different letters indicate significant differences among treatments (Mann-Whitney for Kruskal-Wallis and Tukey's for ANCOVA for stem treatments with Bonferroni correction to $\alpha = 0.017$).

Stem treatment had no effect on the number of days in flower bud stage ($F_{2,46} = 0.298$, $P = 0.744$) with flower heads on galled ramets spending an average of 24.5 ± 2.0 days as flower buds, clipped ramets an average of 22.0 ± 2.8 days, and intact ramets an average of 23.5 ± 1.4 days. Fertilization treatment also had no effect on the length of flower bud stage ($F_{1,46} = 0.004$, $P = 0.950$) with flower heads on fertilized ramets spending an average of 23.6 ± 1.4 days as flower buds and on unfertilized ramets an average of 23.7 ± 1.7 days. No significant interaction between stem and fertilization treatments was revealed ($F_{2,46} = 0.057$, $P = 0.945$). The covariate, initial height, also had no significant influence on the length of flower bud stage ($F_{1,46} = 0.004$, $P = 0.949$).

Stem treatment had no effect on the number of days in the blooming stage ($F_{2,46} = 1.859$, $P = 0.169$) with flower heads on galled ramets blooming for an average of 25.5±1.4 days, on clipped ramets for an average of 21.3 ± 2.7 days, and on intact ramets for an average of 26.5±1.5 days. Fertilization treatment also had no effect on the length of blooming stage ($F_{1,46} = 0.848$, $P = 0.363$) with flower heads on fertilized ramets blooming for an average of 25.8±1.3 days and on unfertilized ramets for an average of 24.6 ± 1.5 days. No significant interaction between stem and fertilization treatments was revealed ($F_{2,46} = 0.267$, $P = 0.767$). Again, the covariate, initial height, also had no significant influence on blooming time ($F_{1,46} = 0.261$, $P = 0.612$).

However, stem treatment had a significant effect on the number of days seed heads spent ripening ($F_{2,47} = 3.566$, $P = 0.038$) with those on clipped ramets ripening significantly earlier than those on galled and intact ramets, which were statistically similar to each other (Figure 4(c)). Planned contrasts indicated that clipping had a significantly negative effect on ripening time compared to the galled group ($P = 0.006$, 95% CI [1.800, 10.291]) and the intact group ($P = 0.011$, 95% CI [−12.616, −1.719]). Again, the galled group was intermediate between the control and clipped groups, but it was not significantly different from the intact group ($P = 0.419$, 95% CI [−6.035, 2.563]). Seed heads on clipped ramets ripened for an average of 24.4 ± 2.8 days, or about 4-5 fewer days than galled (28.3 ± 1.6 days) and intact ramets (29.4 ± 1.1 days). Fertilization treatment also had a highly significant effect on the length of the ripening stage ($F_{1,47} = 8.281$, $P = 0.006$) with seed heads on fertilized ramets ripening for an average of 29.6 ± 1.0 days, or nearly four days longer than unfertilized ramets (25.8 ± 1.7 days) (Figure 4(d)). No significant interaction between stem and fertilization treatments was revealed ($F_{2,47} = 1.401$, $P = 0.258$). The covariate, initial height, also had no significant influence on ripening time ($F_{1,47} = 0.013$, $P = 0.911$).

Stem treatment had a highly significant effect on the number of ramets producing flower buds (χ^2 = 19.229, df = 2, and P < 0.001), with intact ramets producing twice as many total flower buds (40) as galled ramets (19) and nearly four times as many flower buds as clipped ramets (11) (Figure 5(a)). Fertilization treatment had no effect on the number of ramets producing flower buds (χ^2 = 0.700, df = 1, and P = 0.403), with fertilized ramets producing 39 flower buds and unfertilized ramets producing 31 flower buds (Figure 5(b)). Stem treatment also had a highly significant effect on the total number of seed heads produced (χ^2 = 9.234, df = 2, and P = 0.010) with intact ramets producing a total of 24 seed heads compared to 16 for galled ramets and 7 for clipped ramets (Figure 5(c)). Although suggestive, fertilization treatment had no effect on the number of seed heads produced (χ^2 = 3.064, df = 1, and P = 0.080) with fertilized ramets producing 30 seed heads and unfertilized ramets producing 17 seed heads (Figure 5(d)).

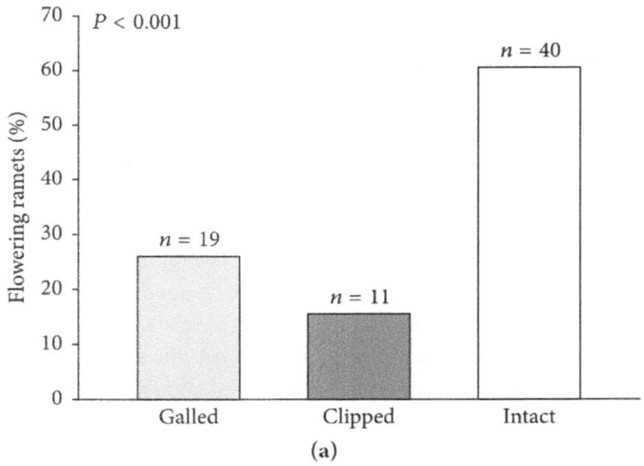

(a)

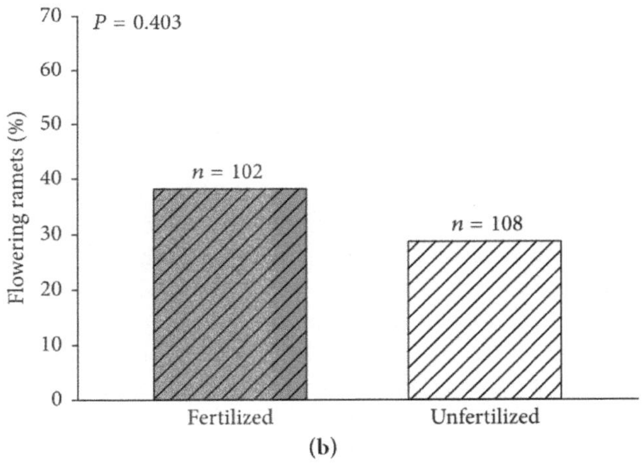

(b)

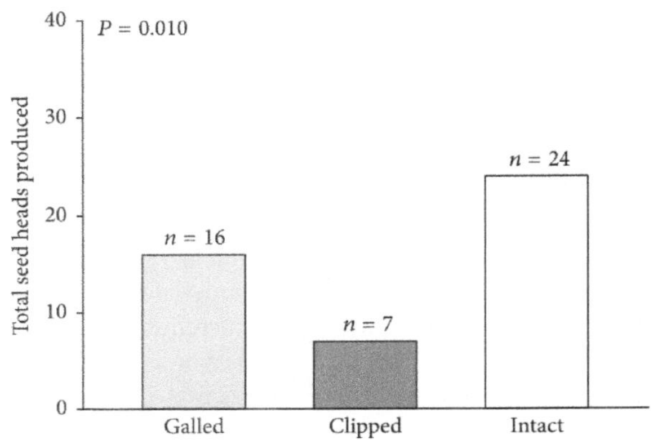

(c)

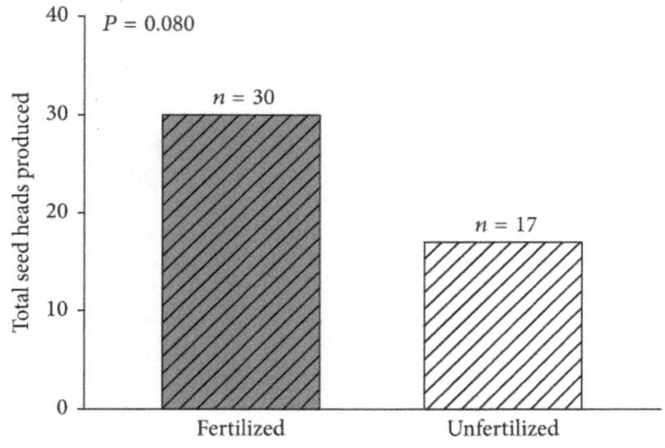

(d)

Figure 5: In stem-nutrition experiment, total number of ramets producing flower buds across (a) stem treatments (χ^2 = 19.229, df = 2, and P < 0.001) and (b) fertilization treatments (χ^2 = 0.700, df = 1, and P = 0.403) and total number of seed heads produced across (c) stem treatments (χ^2 = 9.234, df = 2, and P = 0.010) and (d) fertilization treatments (χ^2 = 3.064, df = 1, and P = 0.080).

Both stem treatment and fertilization treatment had highly significant effects on normal versus abnormal seed development (stem: χ_2 = 13.499, df = 2, and P < 0.01; fertilization: χ_2 = 60.314, df = 1, and P < 0.001). Stem treatment also had a significant effect on the average number of apparently normally developed seeds per seed head ($F_{2,47}$ = 4.370, P = 0.019) (Figure 6(a)). Planned contrasts indicated that clipping had a significantly negative effect on seed count compared to the galled group (P = 0.020, 95% CI [6.039, 66.450]) and the intact group (P = 0.005, 95% CI [−95.368, −17.834]). The galled group was not significantly different from the intact group (P = 0.005, 95% CI [−95.368, −17.834]) but was lower, as expected. Average seed count per seed head was significantly lower for clipped ramets at 58.1 ± 14.0 per seed head, compared to 83.9 ± 11.4 seeds per head for galled ramets and 100.7 ± 8.6 seeds per head for intact ramets. Fertilization treatment also had a significant effect on seed count ($F_{1,47}$ = 5.881, P = 0.020) with average seed count per seed head being significantly higher in fertilized ramets (95.2 ± 8.5) compared to unfertilized ramets (77.1 ± 9.6) (Figure 6(b)). No significant interaction between stem and fertilization treatments was revealed ($F_{2,47}$ = 1.223, P = 0.305). The covariate, initial height, also had no significant influence on seed count ($F_{1,47}$ = 2.941, P = 0.094).

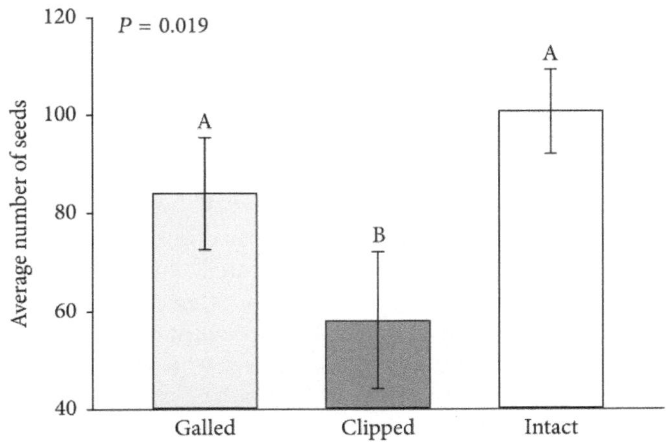

(a)

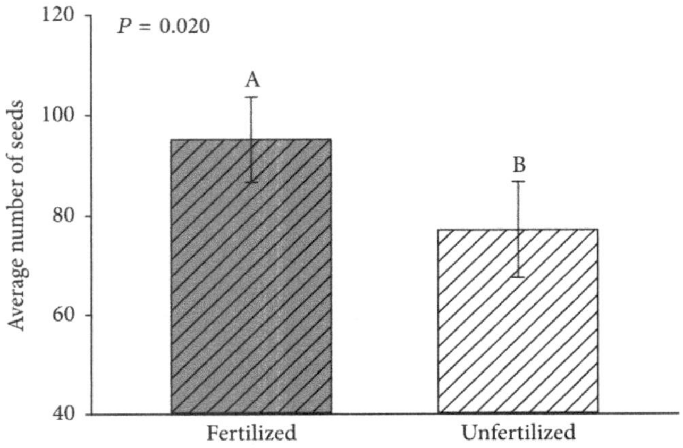

(b)

Figure 6: Average seed count per seed head with SEM error bars across s (a) stem treatments and (b) fertilization treatments in the stemnutrition experiment (stem treatment: $F_{2,47} = 4.370$, $P = 0.019$; fertilization treatment: $F_{1,47} = 5.881$, $P = 0.020$; stem × fertilization: $F_{2,47} = 1.223$, $P = 0.305$; covariate: $F_{1,47} = 2.941$, $P = 0.094$). Different letters indicate significant differences among treatments (Tukey's with Bonferroni correction to $\alpha = 0.017$ for stem treatments).

Stem treatment had no effect on seed mass of normally developed seeds ($F_{2,47} = 0.856$, $P = 0.433$) with average mass per seed being 0.9 ± 0.1 mg for galled ramets, compared to 0.8 ± 0.1 mg for clipped ramets and 0.8 ± 0.0 mg for intact ramets. Although fertilization produced slightly larger seeds, the effect on seed mass was not significant ($F_{1,47} = 0.362$, $P = 0.551$) with average mass per seed being 0.9 ± 0.0 mg for fertilized ramets, compared to 0.8 ± 0.1 mg for unfertilized ramets. No significant interaction between stem and fertilization treatments was found ($F_{2,47} = 0.060$, $P = 0.942$). The covariate, initial height, also had no significant influence on seed mass ($F_{1,47} = 1.595$, $P = 0.214$). Seed germinations were low across all treatments, with only 9.2% of all sowed seeds germinating. Stem treatment had no effect on germination ($\chi^2 = 1.582$, $df = 2$, and $P = 0.453$) with 9% germination success in the galled group, compared to 7% in the clipped group and 10% in the intact group. Curiously, fertilization treatment had a highly significant effect on seed germination success ($\chi^2 = 8.418$, $df = 1$, and $P = 0.004$) with 7% germination success in the fertilized group, compared to 13% in the unfertilized group.

DISCUSSION

The first bud break experiment investigated the strength of apical dominance and the effects on different types of stem damage on it. The distance at which intact ramets broke bud and the extended amount of time it took for them to break bud suggest that apical dominance is strong in Borrichia frutescens. However, stem count at the end of the stem-nutrition experiment (25 weeks) was similar across all treatment groups, and at least some ramets in all treatments broke bud shortly after the experiment began. This apparent contradiction is likely due to differences in the strength of apical dominance at different times of the season; the stem-nutrition experiment was started in January, before flowering, while the first bud break experiment was started in May, after flowering began. Changes in photoperiod in spring [52–54] and flowering [4, 55] are both known to trigger a weakening in apical dominance in some plant species, possibly to maximize the number of meristems available for flower head production. Additionally, foliage in January was sparse and became denser by May, at which point single-stemmed (i.e., young) B. frutescens used in the bud break experiment were notably shorter and more shaded than their neighbors. Light competition has been known to strengthen apical dominance and reduce branching [4, 53, 56]. The finding that apical dominance in intact ramets weakens before flowering is important because overcompensation can occur only when apical dominance is strong and the plant has a bank of dormant buds in reserve for recovery after damage [3–5, 10, 11, 34]. In order for reproductive benefits to be realized, strong apical dominance must be maintained before and during flowering so that its disruption leads to significantly more flower heads and seeds compared to intact ramets. Clearly, that was not the case in this study.

Borrichia frutescens is a long-lived perennial and even under nonstressful conditions (e.g., readily available water, soil nutrients, etc.), damage may extend phenological stages (e.g., growth, flowering, etc.). For instance, if the flowering stage is delayed, it may affect the fitness of the plant by altering pollination opportunities [57]. In the stem-nutrition experiment, clipping was detrimental to reproduction in terms of significant delays in flowering and shorter seed head ripening times. Delays in flowering suggest that florets on clipped ramets may have missed the window of opportunity for cross-pollinating with other florets in the population, while shortened ripening time may have resulted in poor seed development. Such phenological abnormalities can be expected to affect seed production, and indeed, clipped ramets produced significantly fewer seed heads and normally developed seeds per seed head.

Clipping may have been detrimental because removing the apical meristem also involved removing the terminal pair of leaves associated with it. Since single-stemmed ramets were sparsely foliated in January, removing even a single pair of leaves may have drastically reduced the plants' photosynthetic area and caused a carbon limitation. Such a limitation may delay regrowth, which was evidenced by the one-month lag in stem count in clipped ramets. Since regrowth can redirect nutrients away from reproduction when it precedes flowering in the population [3, 16, 58], the subsequent one-month delay in flowering in clipped ramets is not surprising. Additionally, naturally occurring apical meristem damage similar to clipping, such as chewing by grasshoppers (e.g., Orchelimum fidicinium), is seen rarely in the field (Anthony M. Rossi, personal observation), suggesting that this type of damage probably does not exert strong selective pressure and that B. frutescens is poorly adapted to it. A high risk of chewing would need to exist for a beneficial or neutral response to it to be adaptive [16, 19, 52].

Both clipping and galling had detrimental effects on B. frutescens, but galling had less impact. Borrichia frutescens has had a long-term association with galling by Asphondylia borrichiae [43], which may have exerted strong selective pressure over time to which B. frutescens is better adapted and not overly burdened. Indeed, galled ramets were slightly, but not significantly, different from intact ramets for several of the performance variables assessed in the current study. However, galled ramets produced 50% fewer flower buds than intact ramets, suggesting that galling may present a form of resource regulation, which has been identified in other galling systems [31, 32]. Meristems damaged by cecidogenesis or other types of herbivory are essentially removed from the pool of meristems available for the plant's growth and reproduction [5, 7, 8, 10, 11]. Additionally, although leaves at the galled terminal as well as the gall itself can remain green for some time, galls can significantly redirect nutrients from plant functions to feed the gall-inducing organism [5, 22, 23, 26,59], making any photosynthesis by those tissues likely beneficial only to the gall-inducer and not the plant itself. Thus, the reallocation of meristems and nutrients from plant growth and reproduction to A. borrichiaegall development may ensure a steady supply of available (nonflowering) meristems for future generations of gall midges by maintaining apical meristems of B. frutescens in a nonflowering state.

Borrichia frutescens occurs primarily in saline habitats and, as Fernandes and Ribeiro [57] point out, stress and nutrient availability are likely to affect a plant's ability to recover from damage. The fate of regrown shoots (e.g., survival, growth rate, and galling rate) was not assessed in the current study, but they may have performed poorly compared to undamaged primary shoots. However, being a tidal environment, water is not likely to be limiting. Additionally, only three variables increased significantly under nitrogen supplementation: final stem count, number of days for seed head ripening, and the number of normal seeds per seed head. These results suggest that nitrogen also is probably not a limiting nutrient at this site, and the small amount used in this experiment may have been insignificant compared to the twice-daily tidal influx of nutrients. A previous fertilization study reported vigorous regrowth and other beneficial responses of Borrichia frutescens, but that study utilized relatively high concentrations of highly soluble inorganic ammonium nitrate [40]. Thus, stem number and seed count may be phenotypically plastic characteristics that respond more readily to changes in nutrient level and other environmental conditions compared to the other response variables tested in this experiment. Indeed, previous studies suggest that nutrient supplementation via roots may override apical dominance when it is weak and result in the growth of new stems [4, 11, 54, 55]. Nutrient supplementation can also increase the production of normally developed seeds by allowing plants to allocate more nutrients to reproduction [58]. An increase in the number of days for seed head ripening may also enhance seed development and decrease the chances of seed abortion by allowing fertilized ova more time to acquire adequate nutrient and energy stores. Finally, while the count of normally developed seeds varied significantly across both stem treatments and fertilizer treatments in this experiment, seed mass, an important measure of seed quality and energy reserves [3, 60, 61], did not. This result supports the observation that seed count in many plant species tends to vary in response to environmental changes, while seed mass remains relatively constant [60, 61].

CONCLUSION

Clearly, the manner of apical meristem damage had an effect on the plant's compensatory response. Borrichia frutescens appears to be poorly adapted to damage that abruptly removes the apical meristem (e.g., chewing damage), especially when it occurs early in the season. However, even clipped plants, which exhibited more adverse effects than galled ones, largely recovered vegetative growth by the end of the study. Results from this study suggest that B. frutescens is better adapted to galling by A. borrichiae in terms of regrowth and reproduction compared to chewing damage. Moreover, we suggest possible resource regulation of apical meristems by the gall-inducer.

ACKNOWLEDGMENTS

The authors thank Dr. D. Moon (University of North Florida) and Dr. J. Smith (University of Florida) for reviewing this paper; M. Masdea, N. Millan, and R. Fuller for assisting with field research; and J. Evans (Sanibel-Captiva Conservation Foundation Native Plant Nursery) for information about Borrichia frutescensseed germination. In addition, we would like to thank the anonymous reviewers for their comments, which improved the paper. This work was conducted under National Park Service permit No. TIMU-2013-SCI-0001, study No. TIMU-00014. It was funded in part by a Timucuan Preserve Research Grant awarded by the Timucuan Ecological and Historic Preserve (U.S. NPS) and Timucuan Trail Parks Foundation and a University of North Florida Travel Award from the UNF Biology Department.

REFERENCES

1. D. A. Jameson, "Responses of individual plants to harvesting," The Botanical Review, vol. 29, no. 4, pp. 532–594, 1963.
2. D. R. Strong, J. H. Lawton, and R. Southwood, Insects on Plants: Community Patterns and Mechanisms, Harvard University Press, Cambridge, Mass, USA, 1984.
3. K. N. Paige and T. G. Whitham, "Overcompensation in response to mammalian herbivory: the advantage of being eaten," American Naturalist, vol. 129, no. 3, pp. 407–416, 1987.
4. L. W. Aarssen, "Hypotheses for the evolution of apical dominance in plants: implications for the interpretation of overcompensation," Oikos, vol. 74, no. 1, pp. 149–156, 1995.
5. R. J. Marquis, "Plant architecture, sectoriality and plant tolerance to herbivores," Vegetatio, vol. 127, no. 1, pp. 85–97, 1996.
6. A.-P. Huhta, T. Lennartsson, J. Tuomi, P. Rautio, and K. Laine, "Tolerance of Gentianella campestris in relation to damage intensity: an interplay between apical dominance and herbivory," Evolutionary Ecology, vol. 14, no. 4–6, pp. 373–392, 2000.
7. K. Lehtilä, "Modelling compensatory regrowth with bud dormancy and gradual activation of buds,"Evolutionary Ecology, vol. 14, no. 4–6, pp. 315–330, 2000.
8. K. A. Stowe, R. J. Marquis, C. G. Hochwender, and E. L. Simms, "The evolutionary ecology of tolerance to consumer damage," Annual Review of Ecology and Systematics, vol. 31, pp. 565–595, 2000.
9. J. Fornoni, J. Núñez-Farfán, and P. L. Valverde, "Evolutionary ecology of tolerance to herbivory: advances and perspectives," Comments on Theoretical Biology, vol. 8, no. 6, pp. 643–663, 2003.
10. M. J. Wise and W. G. Abrahamson, "Beyond the compensatory continuum: environmental resource levels and plant tolerance of herbivory," Oikos, vol. 109, no. 3, pp. 417–428, 2005.

11. M. J. Wise and W. G. Abrahamson, "Applying the limiting resource model to plant tolerance of apical meristem damage," American Naturalist, vol. 172, no. 5, pp. 635–647, 2008.

12. S. Strauss and A. Zangerl, "Plant-insect interactions in terrestrial ecosystems," in Plant-Animal Interactions: An Evolutionary Approach, C. Herrera and O. Pellmyr, Eds., pp. 77–106, Wiley-Blackwell, Hoboken, NJ, USA, 2009.

13. P. R. Ehrlich and P. H. Raven, "Butterflies and plants: a study in coevolution," Evolution, vol. 18, no. 4, pp. 586–608, 1964.

14. A. Kessler and I. T. Baldwin, "Plant responses to insect herbivory: the emerging molecular analysis,"Annual Review of Plant Biology, vol. 53, pp. 299–328, 2002.

15. J. P. Rosenthal and P. M. Kotanen, "Terrestrial plant tolerance to herbivory," Trends in Ecology and Evolution, vol. 9, no. 4, pp. 145–148, 1994.

16. T. Lennartsson, P. Nilsson, and J. Tuomi, "Induction of overcompensation in the field gentian,Gentianella campestris," Ecology, vol. 79, no. 3, pp. 1061–1072, 1998.

17. B. L. Benner, "Effects of apex removal and nutrient supplementation on branching and seed production inThlaspi arvense (Brassicaceae)," American Journal of Botany, vol. 75, no. 5, pp. 645–561, 1988.

18. D. W. Inouye, "The consequences of herbivory: a mixed blessing for Jurinea mollis (Asteraceae)," Oikos, vol. 39, no. 2, pp. 269–272, 1982.

19. T. G. Whitham, J. Maschinski, K. C. Larson, and K. N. Paige, "Plant responses to herbivory: the continuum from negative to positive and underlying physiological mechanisms," in Plant-Animal Interactions: Evolutionary Ecology in Tropical and Temperate Regions, P. W. Price, T. M. Lewinsohn, G. W. Fernandes, and W. W. Benson, Eds., pp. 227–256, John Wiley & Sons, New York, NY, USA, 1991.

20. R. Karban and S. Y. Strauss, "Effects of herbivores on growth and reproduction of their perennial host,Erigeron glaucus," Ecology, vol. 74, no. 1, pp. 39–46, 1993.

21. P. Olejniczak, "Overcompensation in response to simulated herbivory in the perennial herb Sedum maximum," Plant Ecology, vol. 212, no. 11, pp. 1927–1935, 2011.

22. A. E. Weis, R. Walton, and C. L. Crego, "Reactive plant tissue sites and the population biology of gall makers," Annual Review of Entomology, vol. 33, pp. 467–486, 1988.

23. F. Dreger-Jauffret and J. D. Shorthouse, "Diversity of gall-inducing insects and their galls," in Biology of Insect-Induced Galls, J. D. Shorthouse and O. Rohfritsch, Eds., pp. 8–33, Oxford University Press, New York, NY, USA, 1992.

24. M. S. Mani, "Introduction to cecidology," in Biology of Insect-Induced Galls, J. D. Shorthouse and O. Rohfritsch, Eds., pp. 3–7, Oxford University Press, New York, NY, USA, 1992.

25. O. Rohfritsch, "Patterns in gall development," in Biology of Insect-Induced Galls, J. D. Shorthouse and O. Rohfritsch, Eds., pp. 60–86, Oxford University Press, New York, NY, USA, 1992.

26. A. Raman, C. W. Schaefer, and T. M. Withers, "Galls and gall-inducing arthropods: an overview of their biology, ecology, and evolution," in Biology, Ecology, and Evolution of Gall-Inducing Arthropods, A. Raman, C. W. Schaefer, and T. M. Withers, Eds., pp. 1–34, Science Publishers, Enfield, NH, USA, 2005.

27. J. Yukawa and O. Rohfritsch, "Biology and ecology of gall-inducing Cecidomyiidae (Diptera)," in Biology, Ecology, and Evolution of Gall-Inducing Arthropods, A. Raman, C. W. Schaefer, and T. M. Withers, Eds., pp. 273–304, Science Publishers, Enfield, NH, USA, 2005.

28. R. J. Gagné, The Plant-Feeding Gall Midges of North America, Cornell University Press, Ithaca, NY, USA, 1989.

29. A. M. Rossi, "Speciation processes among insects," in Encyclopedia of Entomology, J. L. Capinera, Ed., pp. 2060–2062, Kluwer Academic Publishers, London, UK, 2004.

30. M. A. A. Carneiro, C. S. A. Branco, C. E. D. Braga et al., "Are gall midge species (Diptera, Cecidomyiidae) host-plant specialists?" Revista Brasileira de Entomologia, vol. 53, no. 3, pp. 365–378, 2009.

31. T. P. Craig, P. W. Price, and J. K. Itami, "Resource regulation by a stem-galling sawfly on the arroyo willow," Ecology, vol. 67, no. 2, pp. 419–425, 1986.

32. H. Roininen, P. W. Price, and J. Tahvanainen, "Field test of resource regulation by the bud-galling sawfly,Euura mucronata, on Salix cinerea," Holarctic Ecology, vol. 11, no. 2, pp. 136–139, 1988.

33. A. M. Rossi and D. R. Strong, "A new species of Asphondylia (Diptera: Cecidomyiidae) on Borrichia(Asteraceae) from Florida," Proceedings of the Entomological Society of Washington, vol. 92, no. 4, pp. 732–735, 1990.

34. M. M. Espírito-Santo, F. D. S. Neves, F. R. Andrade-Neto, and G. W. Fernandes, "Plant architecture and meristem dynamics as the mechanisms determining the diversity of gall-inducing insects," Oecologia, vol. 153, no. 2, pp. 353–364, 2007.

35. J. M. Carlton, A Guide to Common Florida Salt Marsh and Mangrove Vegetation, Florida Department of Natural Resources, Florida Marsh Research Publications, St. Petersburg, Fla, USA, 1975.

36. A. Cronquist, Vascular Flora of the Southeastern United States: Vol. 1, Asteraceae, University of North Carolina Press, Chapel Hill, NC, USA, 1980.

37. A. E. Antlfinger, "The genetic basis of microdifferentiation in natural and experimental populations ofBorrichia frutescens in relation to salinity," Evolution, vol. 35, no. 6, pp. 1056–1068, 1981.

38. W. H. Duncan and M. B. Duncan, Seaside Plants of the Gulf and Atlantic States, Smithsonian Institution Press, Washington, DC, USA, 1987.

39. P. Biber, J. D. Caldwell, S. R. Caldwell, and M. Marenberg, Sea-Oxeye Daisy, Borrichia Frutescens, Propagation Guide, Center for Plant Restoration

and Coastal Plant Research, University of Southern Mississippi, Ocean Springs, Miss, USA, 2013.

40. A. M. Rossi, P. D. Stiling, D. R. Strong, and D. M. Johnson, "Does gall diameter affect the parasitism rate of Asphondylia borrichiae (Diptera: Cecidomyiidae)?" Ecological Entomology, vol. 17, no. 2, pp. 149–154, 1992.

41. A. M. Rossi and P. D. Stiling, "Intraspecific variation in growth rate, size, and parasitism of galls induced by Asphondylia borrichiae (Diptera: Cecidomyiidae) on three host species," Annals of the Entomological Society of America, vol. 88, no. 1, pp. 39–44, 1995.

42. A. M. Rossi, P. Stiling, M. V. Cattell, and T. I. Bowdish, "Evidence for host-associated races in a gall-forming midge: trade-offs in potential fecundity," Ecological Entomology, vol. 24, no. 1, pp. 95–102, 1999.

43. K. Stokes, P. Stiling, M. Gilg, and A. M. Rossi, "The gall midge Asphondylia borrichiae Rossi & Strong (Diptera: Cecidomyiidae): an indigenous example of host-associated sympatric genetic divergence,"Environmental Entomology, vol. 41, no. 5, pp. 1246–1254, 2012.

44. J. C. Roskam, "Evolution of the gall-inducing guild," in Biology of Insect-Induced Galls, J. D. Shorthouse and O. Rohfritsch, Eds., pp. 34–49, Oxford University Press, New York, NY, USA, 1992.

45. P. Stiling, A. M. Rossi, D. R. Strong, and D. M. Johnson, "Life history and parasites of Asphondylia borrichiae (Diptera: Cecidomyiidae), a gall maker on Borrichia frutescens," Florida Entomologist, vol. 75, no. 1, pp. 130–137, 1992.

46. J. Yukawa, "Synchronization of gallers with host plant phenology," Population Ecology, vol. 42, no. 2, pp. 105–113, 2000.

47. A. M. Rossi, M. Murray, K. Hughes, M. Kotowski, D. C. Moon, and P. Stiling, "Non-random distribution among a guild of parasitoids: implications for community structure and host survival," Ecological Entomology, vol. 31, no. 6, pp. 557–563, 2006.

48. D. Te Strake, A. H. Keagy, and P. D. Stiling, "Fungi associated with Borrichia frutescens (Asteraceae): insect galls and endophytes," SIDA, Contributions to Botany, vol. 22, no. 1, pp. 755–763, 2006.

49. P. Stiling and A. M. Rossi, "Coastal insect herbivore communities are affected more by local environmental conditions than by plant genotype," Ecological Entomology, vol. 20, no. 2, pp. 184–190, 1995.

50. P. Stiling, A. M. Rossi, M. Cattell, and T. I. Bowdish, "Weak competition between coastal insect herbivores," Florida Entomologist, vol. 82, no. 4, pp. 599–608, 1999.

51. D. C. Moon and P. Stiling, "The influence of species identity and herbivore feeding mode on top-down and bottom-up effects in a salt marsh system," Oecologia, vol. 133, no. 2, pp. 243–253, 2002.

52. E. Van Der Meijden, "Herbivory as a trigger for growth," Functional Ecology, vol. 4, no. 4, pp. 597–598, 1990.

53. K. A. Franklin, "Shade avoidance," New Phytologist, vol. 179, no. 4, pp. 930–944, 2008.

54. M. A. Domagalska and O. Leyser, "Signal integration in the control of shoot branching," Nature Reviews Molecular Cell Biology, vol. 12, no. 4, pp. 211–221, 2011.

55. M. G. Cline, "Apical dominance," The Botanical Review, vol. 57, no. 4, pp. 318–358, 1991.

56. H. Smith and G. C. Whitelam, "The shade avoidance syndrome: multiple responses mediated by multiple phytochromes," Plant, Cell and Environment, vol. 20, no. 6, pp. 840–844, 1997.

57. G. W. Fernandes and S. P. Ribeiro, "Plant response to herbivory: two examples from the neotropics,"Ecotropicos, vol. 3, no. 2, pp. 77–86, 1990.

58. A. G. Stephenson, "Flower and fruit abortion: proximate causes and ultimate functions," Annual Review of Ecology and Systematics, vol. 12, pp. 253–279, 1981.

59. O. Rohfritsch, "Plants, gall midges, and fungi: a three-component system," Entomologia Experimentalis et Applicata, vol. 128, no. 1, pp. 208–216, 2008.

60. J. L. Harper, Population Biology of Plants, Academic Press, London, UK, 1977.

61. B. A. Schaal, "Reproductive capacity and seed size in Lupinus texensis," American Journal of Botany, vol. 67, no. 5, pp. 703–709, 1980.

CHAPTER 5

IMPACT OF GENOTYPE, PLANT GROWTH REGULATORS AND ACTIVATED CHARCOAL ON EMBRYOGENESIS INDUCTION IN MICROSPORE CULTURE OF PEPPER (CAPSICUM ANNUUM L.)

Yan Cheng[a], Rong-li Ma[a], Yan-sheng Jiao[a], Ning Qiao[a], and Ting-ting Li[b]

[a]Institute of Vegetable Research, Shanxi Academy of Agricultural Sciences, China

[b]Horticulture College, Shanxi Agricultural University, China

ABSTRACT

The production of double haploids through androgenesis is used by breeders to produce homozygous lines in a single generation. Androgenesis can be achieved by isolated microspore culture, which, however, allows the production of embryogenesis with a very low efficiency. In order to improve the overall embryogenesis in pepper, we study the differences of microspore embryogenesis in different genotypes of pepper, and also document the effect of growth regulators in pretreatment media, and activated charcoal (AC) on embryogenesis induction. Fifty different pepper genotypes were evaluated, and the swollen rate of microspores from different genotypes varied from 3.11% to 29.56% with the mean value of 13.13%. Microspores from genotype '36' had the highest swollen rate, and the lowest swollen rate of microspores was observed in genotype '26'. It was concluded from the statistical results of L_9 (3^3) orthogonal test that changes in the level of BA influenced the swollen rate of microspores more significantly, and the combination of 0 mg·l^{-1} 6-benzyladenine (BA), 0.2 mg·l^{-1} α-naphthaleneacetic acid (NAA) and 0.5 mg·l^{-1} kinetinin (Kin) was best. AC at a concentration of 0.05% could act as a promoter of embryogenesis in the microspore culture of different pepper genotypes, while the more significant effect was observed with the low responsive genotypes.

INTRODUCTION

Pepper (*Capsicum annuum* L.) is one of the most variable genera among horticultural species, and its fruits range from tiny hot chillies to long cayenne chillies (Derera et al., 2005). And pepper is a widely grown crop all over the world, which can be used as a vegetable, an ingredient of sauces, a coloring and pungent agent of foods, or in pharmaceutical applications. At present, people mainly adopt the method of continuous selfing to purify parent in heterosis breeding. Therefore, it usually needs 5–6 generations or even more generations of continuous selfing and selection to cultivate the excellent inbred line from heterozygous breeding materials. The major advantage of doubled haploids in plant breeding is the immediate achievement of complete homozygosity. Desired genotypes are thus fixed in one generation, reducing time and cost for cultivar or inbred development. Among the different technologies to produce doubled haploids, microspore embryogenesis is by far the most common.

For some vegetable species, such as *Brassica rapa* (Cao et al., 1994), *Brassica oleracea* (Dias, 1999), and *Raphanus sativus* (Lichter, 1989), establishing isolated microspore cultures has been readily well set up. In pepper, Dumas de

Citation: Yan Cheng, Rong-li Ma, Yan-sheng Jiao, Ning Qiao, Ting-ting Li, Impact of genotype, plant growth regulators and activated charcoal on embryogenesis induction in microspore culture of pepper (Capsicum annuum L.), South African Journal of Botany, Volume 88, September 2013, Pages 306-309, ISSN 0254-6299, http://dx.doi.org/10.1016/j.sajb.2013.08.012.

Vaulx et al. (1981)was the first to publish a reproducible in vitro anther culture method which is still the basis of any other modified techniques used in all over the world. Now, private seed companies and certain biotechnology laboratories in Europe show significant practical activity on production of doubled haploid pepper lines via in vitro anther-culture. Supena et al. (2006) published the first microspore culture-derived haploids in Indonesian hot pepper by a shed-microspore culture protocol. Later, Kim et al. (2008) reported on establishing haploids from the hot pepper 'Milyang-jare' using microspore culture, and optimized plating density of microspores (8×10^4–10×10^4) as well as source and amount of added carbon (9% w/v sucrose) in the medium. Recently, isolated microspore cultures of Hungarian and Spanish spice pepper genotypes were improved by using the wheat ovary co-culture method (Lantos et al., 2009). Moreover, experiments conducted with sweet pepper aimed at improving the efficiency of microspore cultures (Gémes Juhász et al., 2009, Seguí-Simarro et al., 2011 and Lantos et al., 2012). Whereas, the genotype is still the most important and often limiting factor in the pepper androgenic reaction.

In those pioneer experiments, androgenesis was induced in growth regulator-free medium (Kim et al., 2008) as well as in media supplemented with 2.5 µM zeatin and 5 µM indole-3-acetic acid (IAA) (Supena and Custers, 2011 and Supena et al., 2006), or in the induction media of 0.5 $mg{\cdot}l^{-1}$ 2,4-dichlorophenoxyacetic acid (2,4-D) and 0.5 $mg{\cdot}l^{-1}$kinetinin (Kin) (Lantos et al., 2009). A clear understanding of the effects of growth regulators in microspore culture of pepper requires further studies.

Activated charcoal (AC) could absorb substances, including phenols and abscisic acid inhibitory to microspore embryogenesis from the environment of culture vessel and is often added to tissue culture media. In *B. oleracea* Dias (1999) observed that AC seems to have a stimulatory effect on microspore embryogenesis of the different accessions and, in contrast to other factors, seems not to be genotype dependent. In shed-microspore culture of Indonesian hot pepper 'Tombak', Supena et al. (2006) described a method using a double layer culture medium containing AC which was needed, as embryo formation almost completely failed without its addition.

The objective of the present research was to study the differences on microspore embryogenesis in different genotypes of pepper, and also document the effect of growth regulators in pretreatment media, and AC on embryogenesis induction of pepper.

MATERIAL AND METHODS

Plant Material

Fifty pepper genotypes (obtained from Chinese seed companies or Institute of Vegetable Research, Shanxi Academy of Agricultural Sciences, China) were used. Seedlings were cultivated in cold-bed in March, and eighteen plants of each genotype were planted in the field in May 2011. Plants were grown with two main stems, and young fruits were removed and plants were severely pruned to stimulate the development of new young side shoots for flower bud production.

Donor Bud Collection and Pretreatment of Anthers

Flower buds of the desired size (petals equal or slightly longer than sepals and anthers with a faint purple tip), which contained more than 50% of microspores in the late unicellular stage, with amounts of mid unicellular microspores and of bicellular pollen not exceeding 20% and 50% respectively, were harvested from young inflorescences in the morning.

After being rinsed under tap water for 30 min, flower buds were sprayed with 70% ethanol, then surface-sterilized in a 0.1% (w/v) mercuric chloride containing 0.05% (v/v) Tween 20 and periodically agitated for 12 min before three rinses in sterile distilled water for 5 min each. Anthers isolated from one bud were placed on one Petri dish (60 mm × 15 mm), with their inner part facing the sucrose-starvation medium containing 0.37 M mannitol, 10 mM $CaCl_2$, 1 mM $MgSO_4{\cdot}7H_2O$, 1 mM KNO_3, 200 µM KH_2PO_4, 1 µM KI, 100 nM $CuSO_4{\cdot}5H_2O$, and 0.5% agar (pH 5.8). For the first seven days anthers were incubated in the darkness at the temperature of 33 °C.

Isolation Protocol and Culture of Microspores

For both isolation and culture of the microspores, a filter sterilized Nitsch and Nitsch (1967) medium, as modified by Lichter (1981) and by Takahata and Keller (1991) with 9% sucrose (NLN-9), at pH 5.8 was used. After heat treatment for seven days, the swollen anthers were chosen, mixed with NLN-9 medium and immediately extrusioned with glass rod to isolate microspores. Microspore suspension was obtained by filtration through 150 µm nylon screens. This suspension was centrifuged 3 times at 500 rpm for 5 min and resuspended in NLN-9 medium. The microspore density was adjusted to 8×10^4–10×10^4 per ml. Isolated microspores were cultured in 60 mm Petri dishes containing 2 ml microspore suspension.

The cultures were incubated at 28 °C with 80% air humidity and maintained in the dark for three weeks, and then transferred to a gyratory shaker agitating at 60 rpm in darkness at 28 °C.

Testing of the Embryogenic Potential of Different Genotypes

Only the swollen microspores had the possibility to develop into embryoid-like structures (ELSs), so in the first part of our experiments the swollen rate of microspores (mean fraction of the number of swollen microspores and the number of total microspores in five fields of microscope) for each genotype was analyzed by isolating microspores of 6 swollen anthers in 1 ml NLN-9 medium after anther pretreatment.

Testing of the Effects of Growth Regulators and Activated Charcoal

In order to find the optimum ratio of BA (0 $mg{\cdot}l^{-1}$, 0.2 $mg{\cdot}l^{-1}$, 0.4 $mg{\cdot}l^{-1}$), NAA (0 $mg{\cdot}l^{-1}$, 0.1 $mg{\cdot}l^{-1}$, 0.2 $mg{\cdot}l^{-1}$) and Kin (0 $mg{\cdot}l^{-1}$, 0.2 $mg{\cdot}l^{-1}$, 0.5 $mg{\cdot}l^{-1}$) in pretreatment medium, each factor was set three levels according to L_9 (3^3) orthogonal design and the total were 9 level combinations.

Activated charcoal suspension preparation and addition of a 0.1 ml drop to the culture dishes in charcoal treatments were made following Dias' (1999) procedure: autoclaved suspension of 1 g of activated charcoal, 0.5 g of agarose and 100 ml of bi-distilled water. It is important to associate the charcoal with agarose because suspended charcoal without agarose adheres to the microspores and hampers embryogenesis.

Data Collection and Statistical Analysis

At least ten dishes of each treatment were made with three replications, giving a total of thirty dishes per single treatment. Statistical analyses were performed using SPSS 17.0 for Windows software, and means were compared according to pairwise *t* tests. The number of microspore-derived ELSs was visually counted after six weeks of culture.

RESULTS AND DISCUSSION

Embryogenic Potential of Different Genotypes

After 7 d anther pretreatments in sucrose-starvation medium, significant differences on the swollen rate of microspores were recorded among the fifty genotypes in this experiment (Fig. 1). The swollen rate of microspores from different genotypes varied from 3.11% to 29.56% with the mean value of 13.13%. Microspores from genotype '36' had the highest swollen rate. While, the least swollen rate was observed in genotype '26', which were in accordance with the results of subsequent culture of ELSs (data not shown). Thus, genotype '36' was selected as an experimental material to study the effects of growth regulators.

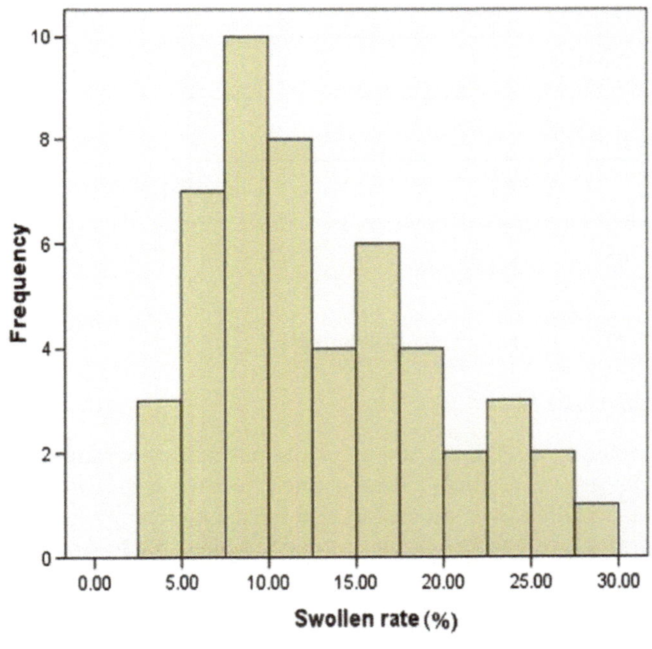

Figure 1. Frequency distribution of the swollen rate of microspores in fifty different genotypes of pepper.

Based on the present results, there was demonstrated a visible heterogeneity in the effectiveness of androgenesis in respective plants of the genotypes researched. Some genotypes naturally respond better than others to pre-treatment applied, and the high and low responding genotypes have different requirements for optimal pre-treatment conditions in the same species. Supena et al. (2006) used ten varieties of Indonesian hot pepper and the quantity and quality of embryos obtained depended on the variety used for the same pretreatment. Similarly, Nowaczyk et al. (2009) showed that high temperature (35°C) of anther pretreatment, which is used frequently in pepper and which has improved the embryogenesis and regeneration of green plants, proved ineffective in certain recalcitrant pepper genotypes. In our work, although genotype '36' and '26' do belong to the same type, they have different fruit character. So it is not surprising to see the huge difference of microspore embryogenesis between them.

Whereas, the possibility of applying different types of inductive protocols allows for the choice of the most convenient for each variety. As an example, the variety 'Piquillo' shows a null response to the method described by Dumas de Vaulx (Mityko et al., 1995), and a positive response to the biphasic method described by Dolcet-Sanjuan et al. (1997). Therefore, before starting a breeding program based on doubled haploid (DH) production it is advisable to assess the response of each variety to the different protocols available.

Effects of Different Combination of Plant Growth Regulators on Embryogenesis

Different combinations of BA, NAA and Kin were tested in genotype "36" which had the highest swollen rate of microspores. Data of the swollen rate of microspores were collected to analyze the combined effect of the three growth regulators in the pretreatment medium. The statistical results of L_9 (3^3) orthogonal test were listed inTable 1 and Table 2. It was concluded from square sum of growth regulator that the influences on the swollen rate of microspores were listed in the following order BA > NAA > Kin (Table 2). The influences of BA and NAA on the swollen rate of microspores were at 1% or 5% significant level, while the influence of Kin was not at significant level. Average swollen rate of microspores could reflect the effects of different combination of plant growth regulators on embryogenesis at different levels of the same factor. Range of average swollen rate of microspores could reflect the effects of changes in the level of various factors on embryogenesis. Changes in the level of BA influenced the swollen rate of microspores more significantly, and the combination of plant growth regulators $A_1B_3C_3$ was best (Table 2). However, the terms and conditions of the test did not appear in the 9 combinations of the orthogonal table. So additional combination of 0 mg·l^{-1} BA, 0.2 mg·l^{-1} NAA and 0.5 mg·l^{-1} Kin was tested, and the highest swollen rate of microspores at 38.47% was got.

Table 1. Variance of plant growth regulator on the swollen rate of microspore in genotype "36".

Growth regulator	F value	P value
BA	32.852**	< 0.01
NAA	17.189*	< 0.05
Kin	2.721	

*means square sum of grows regulator significant at 0.05.

**means square sum of grows regulator significant at 0.01.

Table 2. Effect of different concentration of plant growth regulators on the swollen rate of microspore in genotype "36".

Code	Error term	BA (A, mg·l⁻¹)	NAA (B, mg·l⁻¹)	Kin (C, mg·l⁻¹)	Swollen rate of microspores (%)
1	1	0	0	0	29.56
2	1	0.2	0.1	0.2	22.72
3	1	0.4	0.2	0.5	26.33
4	2	0	0.1	0.5	29.17
5	2	0.2	0.2	0	28.83
6	2	0.4	0	0.2	18.25
7	3	0	0.2	0.2	34.16
8	3	0.2	0	0.5	29.94
9	3	0.4	0.1	0	18.75
K_1	78.61	92.89	77.75	77.14	
K_2	76.25	81.49	70.64	75.13	
K_3	82.85	63.33	89.32	85.44	
$\overline{K}_1$		30.96	25.92	25.71	
$\overline{K}_2$		27.16	23.55	25.04	
$\overline{K}_3$		21.11	29.77	28.48	
R		9.85	6.22	3.44	

K_1, K_2 and K_3 are the sums of swollen rate of microspores for each factors and levels respectively.

$\overline{K}_1$, $\overline{K}_2$, and $\overline{K}_3$ are the average swollen rates of microspores for each factors and levels respectively.

R is the range of average swollen rate of microspores at different levels of the same factor.

In embryogenesis of microspore, an appropriate concentration and ratio of auxin and cytokinin plays an important role in early embryogenesis processes. In the present study, when exogenous growth regulators were omitted from the induction media, the basic media can not only ensure the survival and physiological activities of cultures, but also induce more embryogenesis than some combination of growth regulator (Table 2). The different androgenic responsiveness is due to genetic predetermination of the genotype and the ratio between endogenous and exogenous plant growth regulators (Irikova et al., 2011). Only with proper addition of plant growth regulator, it can induce the proper change of start-up of cell division and callus growth.

Effect of Activated Charcoal on Embryogenesis

ELS yields were significantly increased in all of the genotypes by the addition of a 0.1 ml drop of AC to the microspore culture media (Table 3). In no instances did the addition of AC have any detrimental effect on ELSs yield. Therefore AC at a concentration of 0.05% seems to act as a promoter of embryogenesis in the microspore culture of different pepper genotypes. Although the more significant effect was observed with the low responsive genotypes '04' and '46', it was observed that even the high responsive '36' responded positively to the addition of AC. The magnitude of the response to the addition of AC varied with the different genotypes. The best results were obtained with '46' with increases in ELSs yield of 208.3%. For all the other accessions, the increasing efficiency of AC on ELSs yield varied from 38.5% in '36' to 171.2% in '04'.

Table 3. Effect of activated charcoal (AC) on embryogenesis of microspore culture in different genotypes of pepper.

Genotype	Treatments	No. of ELSs per dish
04	NLN-9	0.73 ± 0.1 b [y]
	NLN-9 + AC	1.98 ± 0.2 a
10	NLN-9	4.83 ± 0.7 b
	NLN-9 + AC	7.97 ± 1.1 a
33	NLN-9	1.60 ± 0.3 b
	NLN-9 + AC	4.13 ± 0.4 a
36	NLN-9	7.20 ± 0.8 b
	NLN-9 + AC	9.97 ± 1.1 a
40	NLN-9	1.33 ± 0.2 b
	NLN-9 + AC	3.20 ± 0.5 a
46	NLN-9	0.12 ± 0.06 b
	NLN-9 + AC	0.37 ± 0.1 a

ELSs = embryoid-like structures.

[y]Values represent means ± SEM. The means followed by the same letter do not significantly differ at $p = 0.05$ according to pairwise t-tests.

The present results are in accordance with the observations of Supena et al. (2006) who saw a significant increase in the in vitro androgenesis of pepper on 1% AC medium no matter the low and high responsive microspore culture genotypes.

The explanation for the promotion of embryogenesis by AC is not yet well clarified. AC possesses strong adsorptive properties and is usually used in chemistry to absorb both gases and dissolved solids. When added to tissue culture media, AC is commonly thought to remove growth inhibitory substances exuded by tissues or present in the ingredients of the medium, but it is now clear that promontory substances can also be absorbed and made unavailable to the plant. The effects of increasing concentrations of charcoal on the total embryo yield and the yield of normal-looking embryos did not run in parallel, and 2% charcoal impaired embryonic shoot development (Supena et al., 2006).

SUPPLEMENTAL MATERIAL

Fruit characteristics of fifty pepper genotypes

Genotype	Types (hot or bell pepper)	Weight of fruit (g)	Length of fruit (mm)	Width of fruit (mm)	Pericarp thickness (mm)	No. of fruit per plant	Yield of fruit per plant (g)
01	Hot	11	95	19	3.0	20	220
02	Hot	54	235	55	2.0	12	648
03	Bell	74	80	67	5.0	11	814
04	Bell	150	86	80	7.0	10	1500
05	Bell	223	104	104	5.0	5	1115
06	Bell	350	110	93	9.0	4	1400
07	Bell	163	140	90	4.0	6	978
08	Bell	235	140	95	6.0	5	1175
09	Bell	145	95	75	7.0	7	1015
10	Bell	91	90	70	4.0	7	637
11	Bell	129	100	51	5.0	6	774
12	Hot	29	185	21	2.1	31	899
13	Hot	38	200	25	2.3	29	1102
14	Hot	42	230	26	2.5	33	1386
15	Hot	41	170	34	3.4	30	1230
16	Hot	26	230	29	2.0	12	312
17	Hot	37	220	25	2.3	13	481
18	Hot	42	220	30	3.2	32	1344

19	Hot	29	200	28	2.1	28	812
20	Hot	98	240	40	2.9	8	784
21	Hot	62	220	37	2.2	15	930
22	Hot	9	70	13	1.5	54	486
23	Hot	26	120	15	2.6	25	650
24	Hot	54	170	28	3.1	16	864
25	Hot	2.9	45	11	0.9	226	655
26	Hot	3.1	80	10	1.0	125	388
27	Hot	2.3	75	9	1.0	182	419
28	Hot	2.5	84	13	1.5	40	100
29	Hot	2.6	57	9	1.2	63	164
30	Hot	67	220	55	3.0	6	402
31	Hot	89	215	54	4.0	19	1691
32	Hot	87	201	51	4.1	23	2001
33	Hot	78	162	32	3.1	14	1092
34	Hot	68	78	46	2.2	13	884
35	Hot	42	158	37	2.3	7	294
36	Hot	58	155	50	3.1	5	290
37	Hot	39	164	43	2.0	9	351
38	Hot	32	140	29	2.0	9	288
39	Hot	13	95	26	2.0	20	260
40	Hot	85	153	46	4.0	13	1105
41	Hot	66	195	40	4.0	19	1254
42	Hot	17	110	15	2.1	25	425
43	Hot	7	181	11	0.7	161	1127
44	Hot	12	180	12	1.2	60	720
45	Hot	13	235	19	0.8	25	325
46	Hot	9	250	11	1.0	33	297
47	Hot	11	185	13	1.0	32	352
48	Hot	11	200	15	0.8	72	792
49	Hot	33	33	56	5.1	7	231
50	Hot	3	20	18	1.0	17	51

ACKNOWLEDGMENTS

Financial support in the form of the research grant (ybsjj1103) from Shanxi Academy of Agricultural Sciences of China and Basic Research Project of Shanxi Province (2013021026-1) are gratefully acknowledged.

REFERENCES

1. Cao, M.Q., Li, Y., Liu, F., Doré, C., 1994. Embryogenesis and plant regeneration of pakchoi (Brassica rapa L. ssp. chinensis) via in vitro isolated microspore culture. Plant Cell Reports 13, 447–450.
2. Derera, N.F., Nagy, N., Hoxha, A., 2005. Condiment paprika research in Australia. Journal of Business Chemistry 2, 4–18.
3. Dias, J.S., 1999. Effect of activated charcoal on Brassica oleracea microspore culture embryogenesis. Euphytica 108, 65–69.
4. Dolcet-Sanjuan, R., Claveira, E., Huerta, A., 1997. Androgenesis in Capsicum annuum L. — effects of carbohydrate and carbon dioxide enrichment. Journal of the American Society for Horticultural Science 122, 468–475.
5. Dumas de Vaulx, R., Chambonnet, D., Pochard, E., 1981. Culture in vitro d'anthéres de piment (Capsicum annuum L.): amelioration des taux d'obtention de plantes chez différents génotypes par des traitements á +35 °C. Agronomie 1, 859–864.
6. Gémes Juhász, A., Kristóf, Z., Vági, P., Lantos, C., Pauk, J., 2009. In vitro anther and isolated microspore culture as tools in sweet and spice pepper breeding. Acta Horticulturae 829, 61–64.
7. Irikova, T., Grozeva, S., Rodeva, V., 2011. Anther culture in pepper (Capsicum annuum L.) in vitro. Acta Physiologiae Plantarum 33, 1559–1570.
8. Kim, M., Jang, I.C., Kim, J.A., Park, E.J., Yoon, M., Lee, Y., 2008. Embryogenesis and plant regeneration of hot pepper (Capsicum annuum L.) through isolated microspore culture. Plant Cell Reports 27, 425–434.
9. Lantos, C., Juhász, A.G., Somogyi, G., Ötvös, K., Vági, P., Mihály, R., Kristóf, Z., Somogyi, N., Pauk, J., 2009. Improvement of isolated microspore culture of pepper (Capsicum annuum L.) via co-culture with ovary of pepper or wheat. Plant Cell, Tissue and Organ Culture 97, 285–293.
10. Lantos, C., Juhász, A.G., Somogyi, G., Vági, P., Mihály, R., Kristóf, Z., Pauk, J., 2012. Androgenesis induction in microspore culture of sweet pepper (Capsicum annuum L.). Plant Biotechnology Reports 6, 123–132.
11. Lichter, R., 1981. Anther culture of Brassica napus in a liquid culture media. Z Pflanzenphysiology 103, 229–237.
12. Lichter, R., 1989. Efficient yield of embryoids by culture of isolated microspores of different Brassicaceae species. Plant Breeding 103, 119–123.
13. Mityko, J., Andrasfalvy, A., Csillery, G., Fari, M., 1995. Anther culture response in different genotypes and F1 hybrids of pepper (Capsicum annuum L.). Plant Breeding 114, 78–80.
14. Nitsch, C., Nitsch, J.P., 1967. The induction of flowering in vitro in stem segments of Plumbago indica L. I. The production of vegetative buds. Planta 72, 355–370.
15. Nowaczyk, P., Olszewska, D., Kisiala, A., 2009. Individual reaction of Capsicum F2 hybrid genotypes in anther cultures. Euphytica 168, 225–233.
16. Seguí-Simarro, J.M., Corral-Martínez, P., Parra-Vega, V., González-García, B., 2011. Androgenesis in recalcitrant solanaceous crops. Plant Cell Reports 30, 765–778.
17. Supena, E.D.J., Custers, J.B.M., 2011. Refinement of shed-microspore culture protocol to increase normal embryos production in hot pepper (Capsicum annuum L.). Scientia Horticulturae 130, 769–774.
18. Supena, E.D.J., Suharsono, S., Jacobsen, E., Custers, J.B.M., 2006. Successful development of a shed-microspore culture protocol for doubled haploid production in Indonesian hot peppers (Capsicum annuum L.). Plant Cell Reports 25, 1–10.
19. Takahata, Y., Keller, W.A., 1991. High frequency embryogenesis and plant regeneration in isolated microspore culture of Brassica oleracea. Plant Science 74, 235–242

CHAPTER 6

MAXIMAL STOMATAL CONDUCTANCE TO WATER AND PLASTICITY IN STOMATAL TRAITS DIFFER BETWEEN NATIVE AND INVASIVE INTRODUCED LINEAGES OF PHRAGMITES AUSTRALIS IN NORTH AMERICA

V. Douhovnikoff[1], S. H. Taylor[1], E. L. G. Hazelton[2], C. M. Smith[1], and J. O'Brien[1]

[1]Bowdoin College, 6500 College Station, Brunswick, ME 04011, USA

[2]Department of Watershed Sciences, Ecology Center, Utah State University, Logan, UT 84322, USA

ABSTRACT

The fitness costs of reproduction by clonal growth can include a limited ability to adapt to environmental and temporal heterogeneity. Paradoxically, some facultatively clonal species are not only able to survive, but colonize, thrive and expand in heterogeneous environments. This is likely due to the capacity for acclimation (sensu stricto) that compensates for the fitness costs and complements the ecological advantages of clonality. Introduced Phragmites australis demonstrates great phenotypic plasticity in response to temperature, nutrient availability, geographic gradient, water depths, habitat fertility, atmospheric CO_2, interspecific competition and intraspecific competition for light. However, no in situ comparative subspecies studies have explored the difference in plasticity between the non-invasive native lineage and the highly invasive introduced lineage. Clonality of the native and introduced lineages makes it possible to control for genetic variation, making P. australis a unique system for the comparative study of plasticity. Using previously identified clonal genotypes, we investigated differences in their phenotypic plasticity through measurements of the lengths and densities of stomata on both the abaxial (lower) and adaxial (upper) surfaces of leaves, and synthesized these measurements to estimate impacts on maximum stomatal conductance to water (g_{wmax}). Results demonstrated that at three marsh sites, invasive lineages have consistently greater g_{wmax} than their native congeners, as a result of greater stomatal densities and smaller stomata. Our analysis also suggests that phenotypic plasticity, determined as within-genotype variation ing_{wmax}, of the invasive lineage is similar to, or exceeds, that shown by the native lineage.

INTRODUCTION

The capacity for clonal growth is often given as an explanation for the invasive character of many introduced species (Thompson *et al.* 1995). Clonal growth affords species a capacity for reproduction despite small initial population sizes. It also offers competitive advantages such as the ability to nurse new ramets (sprouts), share resources between ramets and avoid the costly risks involved in sexual reproduction. However, the fitness costs of reproduction by clonal growth can include a limited ability to adapt to environmental and temporal heterogeneity (Alpert and Simms 2002). Recombination of genetic material and associated natural selection are not available for the rapid innovation and trial of new genotypes in clones, suggesting that the range of habitats invaded by clonal lineages should be more limited than that inhabited by competitors exhibiting more frequent sexual reproduction. Paradoxically, some facultatively clonal species are not only able to survive, but colonize, thrive and expand in heterogeneous environments. What factors underlie the success of particularly invasive

Citation: Douhovnikoff V, Taylor SH, Hazelton ELG, Smith CM, and O'Brien J (2016). Maximal stomatal conductance to water and plasticity in stomatal traits differ between native and invasive introduced lineages of Phragmitesaustralis in North America. AoB PLANTS, 8:plw006. doi: 10.1093/aobpla/plw006 clonal lineages? We hypothesize that these lineages are able to compete with, and ultimately outcompete, species with more diverse gene pools, greater rates of recombination or longer history of local adaptation, through the process of acclimation (*sensu stricto*) and a potentially greater range of phenotypic plasticity, which compensates for the fitness costs and complements the ecological advantages of clonality.

Phragmites australis is a large stature clonal grass that is found in a wide range of wetland and marsh-like ecosystsems and occurs on every continent but Antarctica. In North America, several lineages have been recognized, while two are most prevalent: *P. australis* (Trin. × Steud.) is an invasive lineage (introduced), and *P. australis* subspecies *americanus*(Saltonstall, PM Peterson and Soreng) is a native lineage (native) (Saltonstall *et al.* 2004). Both the native and introduced lineages have the capacity for extensive clonal growth (Douhovnikoff and Hazelton 2014). However, the introduced lineage is expanding its range and outcompeting many native species across a broad range of local conditions and wetland types throughout North America.

Introduced *P. australis* demonstrates great phenotypic plasticity in response to temperature and nutrient availability (Eller and Brix 2012), geographic gradient (Bastlova *et al.* 2004), water depths (Vretare *et al.* 2001), habitat fertility (Clevering 1999), atmospheric CO_2 (Mozdzer and Megonigal 2012), interspecific competition (Bellavance and Brisson 2010) and intraspecific competition for light (Bellavance and Brisson 2010). However, the majority of prior work focussed on common garden studies with the European ancestral lineage, and not plants collected in North America. Further, no *in situ* comparative lineage studies have explored the difference in plasticity between the invasive introduced and non-invasive native lineages (reviewed in Mozdzer *et al.* 2013). Despite the obvious comparative potential, such closely related groups have rarely been examined with respect to the ecology of invasion. Among 93 comparative studies of plasticity in invasive plants identified by Palacio-López and Gianoli (2011), the closest shared taxon was at the genus level.

In addition to closely related (conspecific) lineages, the clonality of the native and introduced lineages make *P. australis* a unique system for the comparative study of phenotypic plasticity. Phenotype is a result of genetic (G) × environmental (E) interactions (Via and Lande 1985). Clonal plants are powerful model systems as they control for genetics (G). Assuming moderate mutation rates and developmental differences among compared groups, observed variation would largely be explained by plastic responses to environmental (E) conditions. Naturally occurring replicates (ramets) of a given genotype (genet) make it possible to measure and compare the reaction norms within and between genotypes permitting a better understanding of the role plasticity plays in plant ecology (Douhovnikoff and Dodd 2015) from the ramet to the lineage scale (Gianoli and Valladares 2012).

The size and spacing of stomata on leaves are simple measurements that provide a strong framework within which to explore phenotypic plasticity linked with physiological performance (Hetherington and Woodward 2003). Stomata permit and regulate gas exchange between the inner plant and the atmosphere, facilitating the exchange of gases necessary for photosynthesis and transpiration. In moving air, stomatal conductance is the principal control over leaf gas exchange with direct consequences for both leaf metabolism and energy balance (Schulze *et al.* 1994). Stomatal morphometrics provide an accurate representation of the capacity for leaf gas exchange through the calculation of maximal conductance (g_{max}, Dow*et al.* 2014), which incorporates the influences of stomatal pore area and pore depth (Brown and Escombe 1900). The multi-dimensional framework for the assessment of stomatal variation provided by g_{max} has been used to demonstrate both heritable variation and environmental plasticity (Franks *et al.* 2009; Fanourakis *et al.* 2015). Differences in stomatal morphometrics have previously been identified for *P. australis* lineages (Hansen *et al.* 2007; Saltonstall *et al.* 2007). Differences in plasticity of stomatal morphology could further permit a single genotype to acclimate to a range of conditions, making it a strong competitor in heterogeneous environments such as tidal wetlands.

Introduced *Phragmites* produces biomass more quickly, metabolizes carbon and nitrogen more quickly, and it is suspected that the introduced lineage has a photosynthetic advantage over its native conspecific (Mozdzer *et al.* 2013). Using previously identified clonal genotypes (Douhovnikoff and Hazelton 2014), we took advantage of the g_{max}framework to investigate variation in stomatal conductance and its dependence on stomatal morphometrics within and between *P. australis*lineages, stands and genets. We quantified maximum stomatal conductance to water, g_{wmax}, and its plasticity, through measurements of the lengths and densities of stomata on the abaxial (lower) and adaxial (upper) surfaces of leaves. We tested the hypotheses that (i) there are genetic effects on g_{wmax} differentiating native and introduced *P. australis*lineages and genotypes and (ii) variation in g_{wmax} in response to local site conditions is greater in clones of introduced *P. australis*, indicating greater physiological plasticity that may contribute to the invasive character of this lineage.

METHODS

Site Description

Three marshes in Southern Maine were systematically surveyed for stand scale *P. australis* clonal structure, which was mapped on a 5 × 5 m grid (Douhovnikoff and Hazelton 2014). Marsh sites were Libby (70.310W, 43.563N), Spurwink (70.250W, 43.589N) and the more distant Webhannet (70.585W, 43.286N). Maximum and minimum marsh-to-marsh distances were 43.2 and 5.6 km, respectively. The marshes are back barrier dune systems, and are well suited for comparisons of lineages among stands within the respective marshes; both native and introduced *P. australis* were present, in proximity to each other, at all sites. In the case of the Libby marsh, the introduced and native stands abut each other and overlap in some areas (E. L. G. Hazelton, pers. obs.). The most developed of these sites is the Webhannet marsh, the Spurwink marsh abuts agricultural land and the Libby marsh occupies a watershed with relatively little development or agriculture.

Sample Collection and DNA Extraction

Samples were collected in the summer of 2011. The most apical fully expanded leaves were collected from the nearest stem to each sample grid point. Earlier work had determined that the 5 × 5 m sampling grid was ideal for the efficient mapping of genotypic diversity at the sites (Douhovnikoff and Hazelton 2014). Lineages were differentiated by morphological characteristics (Swearingen and Saltonstall 2010), and microsatellite markers (Saltonstall 2003) were used to establish clonal identities (detailed methods in Douhovnikoff and Hazelton 2014).

Stomatal Morphometrics and g_{wmax}

Leaf material was stored at −20 °C prior to analysis. Epidermal impressions were made using clear nail polish (ethyl acetate) applied directly to the leaf surface, and were mounted on slides. Preliminary measurements indicated that stomatal traits varied systematically along the length of leaves, so middle-adaxial and middle-abaxial leaf surfaces were sampled for consistency. Slides were viewed on Olympus BX-51 microscopes and stomatal morphometrics were determined from images captured at ×400 total magnification using QCapture software (QImaging). ImageJ software (Abramoff *et al.* 2004) was used to count the total number of stomata and measure the lengths of five randomly chosen stomata within a standardized 200 × 200 μm area within each image.

Maximum stomatal conductance to water vapour (mol m^{-2} s^{-1}) was calculated using the formula of Brown and Escombe (1900, see alsoWeyers and Meidner 1990; Franks and Farquhar 2006) parameterized for grass stomata (Taylor *et al.* 2012). Briefly, g_{wmax} for each leaf is the sum of maximum conductance values for leaf surfaces ($g_{wmax,}i$, where *i* is abaxial or adaxial), calculated as:

$$g_{wmax,i} = \frac{d}{v} \times D \times \frac{a_{max}}{l + (\pi/2)\sqrt{a_{max}/\pi}}$$

The diffusivity of water in air (d, $m^2\,s^{-1}$, at 25 °C), the molar volume of air (v, $m^3\,mol^{-1}$, at 25 °C) and ϖ are physical and geometric constants. Stomatal density (D, m^{-2}) and stomatal length (L, m) were determined from our measurements and used to derive (i) stomatal size (S, m^2), as $0.25L^2$ (stomatal width = $0.25L$, Taylor *et al.* (2012)); (ii) depth of stomatal pores (l, m), as $0.125L$ (equal to guard cell width, Franks *et al.* 2009) and (iii) the maximum stomatal pore area (a_{max}, m^2), as $0.4S$ (an empirical relationship for grass stomata determined by Taylor *et al.* 2012). Calculations were made using R Language and Environment (version 3.1.3,R Development Core Team 2015).

Statistical Analysis

We $\log_e$ transformed g_{wmax} prior to statistical analysis. We employed standard approaches for an unbalanced nested 2 × 2 analysis of variance, using the R Language and Environment (version 3.1.3, R Development Core Team 2015), as follows. We performed a Type III conditioning procedure (Fox 2008), initially testing for interactions between the two putative fixed effects, site and lineage, holding the clones as random effects. We detected no significant interactions in the complete data set, though we did find weak but statistically significant interactions when several highly variable clones were excluded from the data. We inferred the significant effects using the complete data set, employing a Type II procedure to ensure full power to determine effects (Langsrud 2003): all factors (site, lineage and clone) exhibited effects with P-values $<10^{-16}$. We also employed a more advanced model selection machinery available to Bayesian approaches to calculate the Bayes factors across a wide variety of possible analytic frameworks (Rouder *et al.* 2012), garnering additional support for our choice of analysis. For clones with $N > 11$ ramets, robust estimates of within-clone spatial variation, mean and standard deviation (SD) in $\log_e(g_{wmax})$ were made using a permutation test that preserved the variation intrinsic to the data accounting for the variable number of ramets within each clone. This test proceeds by generating two distributions of statistics, a null distribution reflecting the correlation expected under no spatial effect but accounting for unevenness in the underlying spatial distribution of ramets and a corresponding distribution reflecting the correlation observed within the data. The first was generated by randomly permuting which g_{wmax} values associate with which (x,y) position pair for a given ramet, and repeating 10 000 times; for each permutation, a subset of size 10 was taken and a simple Spearman (rank order) correlation was calculated between the pairwise distance between ramets and the difference in their g_{wmax} values. The latter distribution was generated to represent the observed data by sampling 10 000 size 10 subsets and again calculating the Spearman correlation. A P-value was calculated by finding the fraction of replicates in the observed distribution that were more extreme than all values in the null distribution. While similar in concept to a Mantel test, this permutation approach is significantly more conservative in its P-value calculation while still sensitive to even mild (correlation values of 0.1) levels of spatial structure. To ensure that the results were independent of coordinate frame, the test was repeated having rotated the axes by 45°.

RESULTS

Site, Lineage and Clone as Factors Influencing g_{wmax}

Our model of $\log_e(g_{wmax})$ identified significant additive effects of site, lineage and clone (clones having been identified as unique to each site, i.e. completely nested; F values 48.06, 495.70 and 4.50 with df = 2, 1 and 68, respectively, $P < 10^{-16}$ for all). At the three sites, *P. australis* showed greater mean $\log_e(g_{wmax})$ at Webhannet (2.28) and Libby (2.26) than at Spurwink (1.93). When grand means for the native and introduced lineages were compared, $\log_e(g_{wmax})$ of the introduced lineage was 21 % greater than the native lineage (Fig. 1A), equivalent to an increase of 54 % when back-transformed to the original scale (mean (2.5–97.5 % quantile): native, 7.5 (4.5–12.1) mol m^{-2} s^{-1}; introduced, 11.5 (6.7–18.1) mol m^{-2} s^{-1}). This substantial difference between the lineages was relatively consistent across the three sites (16–31 % increase on $\log_e$ scale depending on site; Fig. 1B). When clones were treated as independent of their classification by site and lineage, and when lineage was excluded from consideration, among-clone variation explained the majority of variance in $\log_e(g_{wmax})$ (55 %). Differences among clones were, however, strongly structured by contrasts between native and introduced lineages and sites (Fig. 1C).

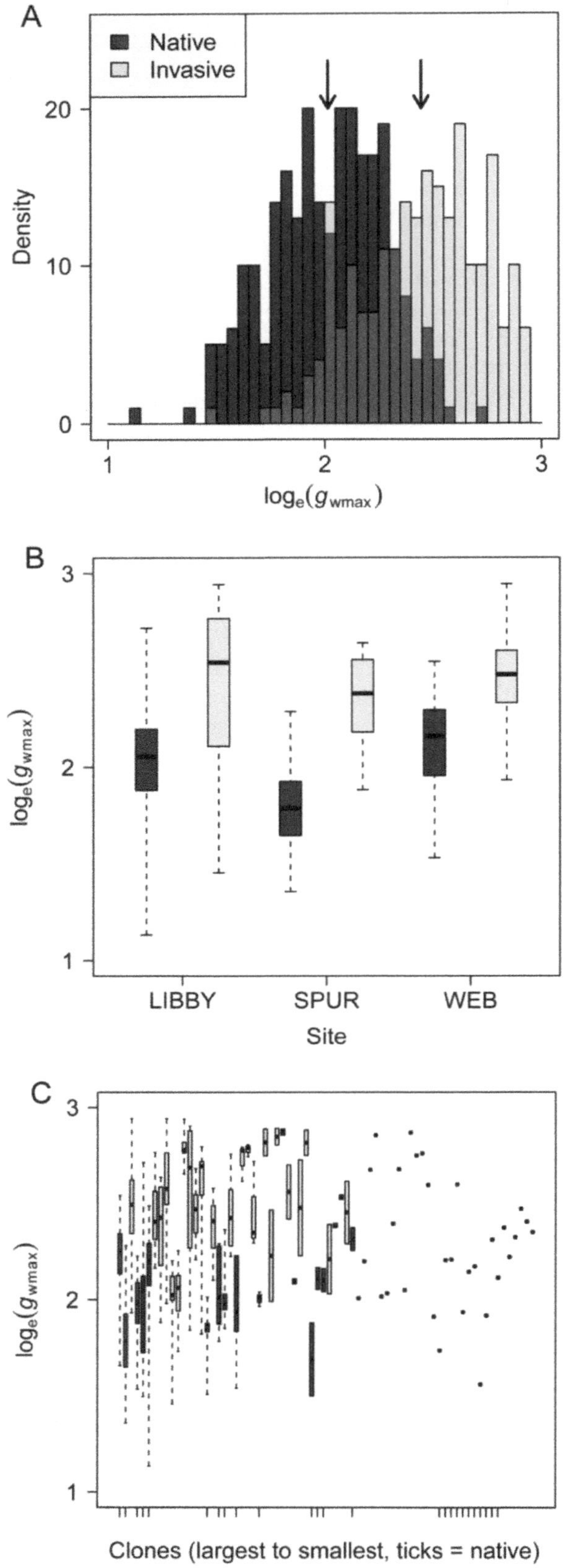

Figure 1. (A) Native and invasive lineages of *P. australis* show significantly different g_{wmax} determined on the basis of stomatal morphometrics. (B) Differences in g_{wmax} between native and invasive lineages of *P. australis* are consistent between marsh sites in Maine, and are substantially greater than differences in g_{wmax} between sites. (C) When comparing unique clones of *P. australis* across three marsh sites in Maine, g_{wmax} differentiates clones belonging to native and invasive lineages.

Plasticity (Within-clone Variation) in g_{wmax}

Using our entire data set, plasticity in $\log_e(g_{wmax})$, determined as the SD of $\log_e(g_{wmax})$ conditioned for clone identity (Fig. 2), was greater within the introduced lineage at the Libby (SD: introduced, 0.36; native, 0.27) and Spurwink (SD: introduced, 0.22; native, 0.18) marshes. At the Webhannet marsh, the opposite was true (Fig. 2), but the lineages were also more similar (SD: introduced, 0.20; native, 0.22).

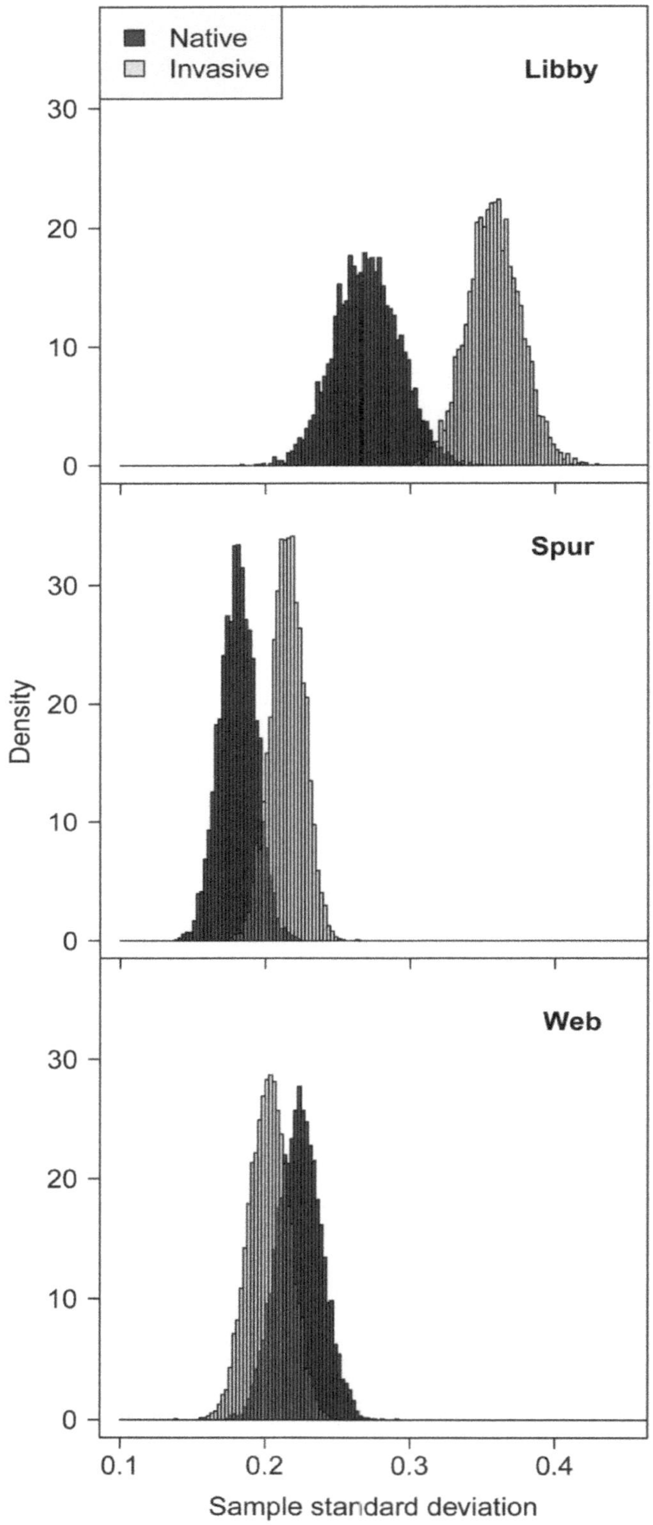

Figure 2. Plasticity (SD) in g_{wmax} within the invasive lineage of *P. australis* is similar to, or exceeds, plasticity measured within the native lineage at Libby, Spurwink and Webhannet marshes, in Maine.

Our investigation of both spatial variation and phenotypic variation in $\log_e(g_{wmax})$ within the 10 clones having $N > 11$ ramets found no evidence for significant within-clone spatial structure (permutation test null distribution construction described in Methods with 9999 degrees of freedom, $P > 0.291$). The test used does not rule out spatial autocorrelation as a determinant of finer-scale patterns. Distributions of SDs for $\log_e(g_{wmax})$ within large clones at the Libby site, in particular, were multimodal (Fig. 3). The permutation distributions shown in Fig. 3 were realized for each clone by holding the number of ramets to 10 and resampling from the full collection of observed values with replacement: for each clone, 1000 resamplings were made, with the sample mean and sample SD calculated for each sample. This analysis indicates that within these large, genetically homogeneous clones, subsets of ramets showed uniquely identifiable levels of plasticity, perhaps linked by epi-genotype.

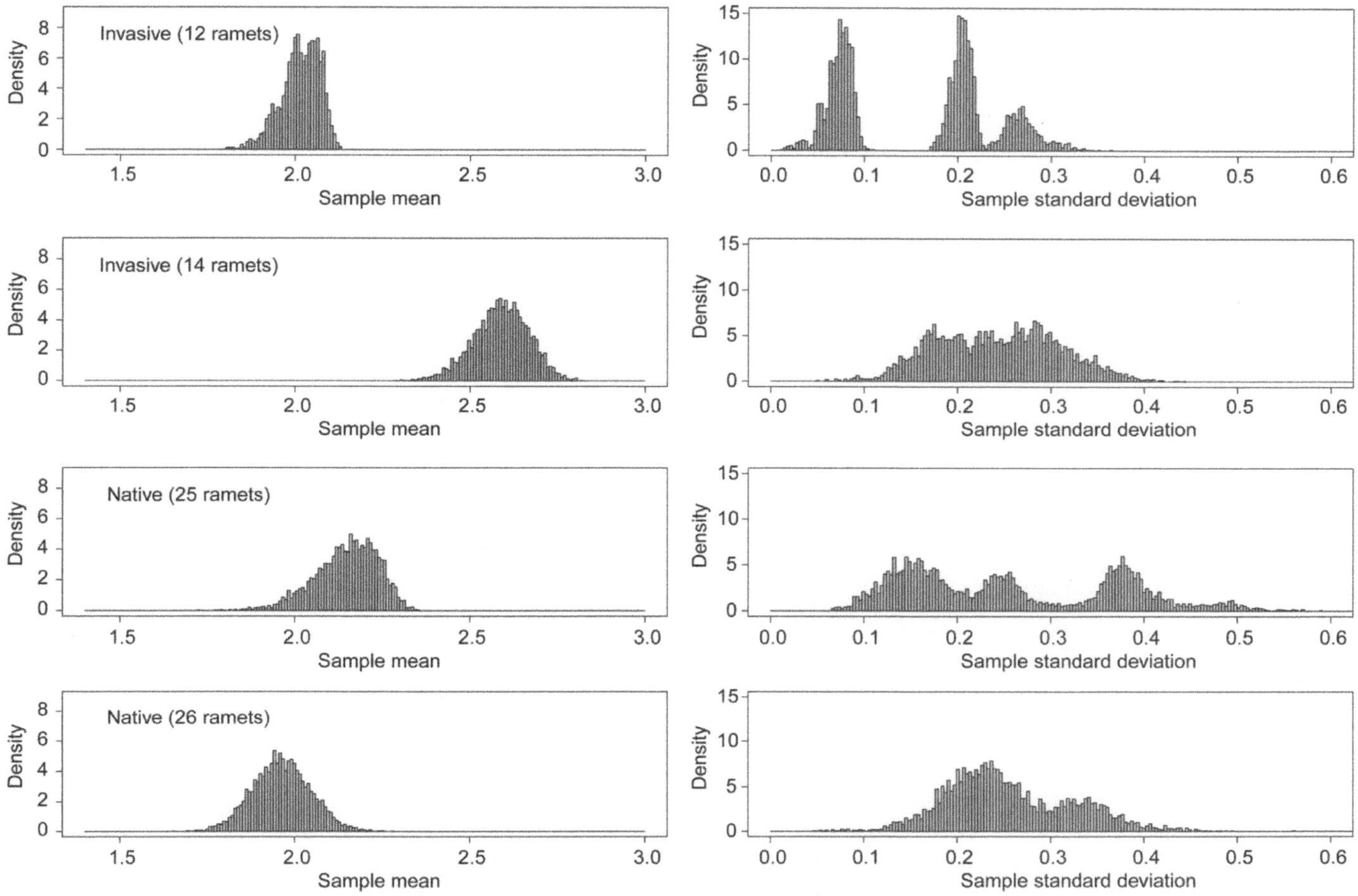

Figure 3. Permutation analysis demonstrates that unimodal distributions for means of g_{wmax} within clones of *P. australis* (left column) are linked with multimodal distributions for SDs (right column); clusters of ramets within each genet show unique levels of variability. Results shown are for invasive (BI24, BI51) and native (BN11, BN8) clones at the Libby marsh.

Lineage Differences in Stomatal Morphometrics Underpinning g_{wmax}

The consistently greater g_{wmax} of introduced lineages of *Phragmites* was a result of increases in both adaxial and abaxial g_{wmax} (Fig. 4A). Size (*S*)–density (*D*) plots indicated that differences in *S* and *D* between the lineages were broadly consistent with a size–density trade-off: the introduced lineage had relatively smaller and more abundant stomata than the native lineage (Fig. 4B and C). Shifts in *S* and *D* among native ramets resulted in conservation of g_{wmax} (data for native ramets fall along g_{wmax} isoclines in Fig. 4B and C). Among ramets of the introduced lineage, variation in g_{wmax} arose from variation in *D* that was not matched by shifts in *S* (Fig. 4B and C).

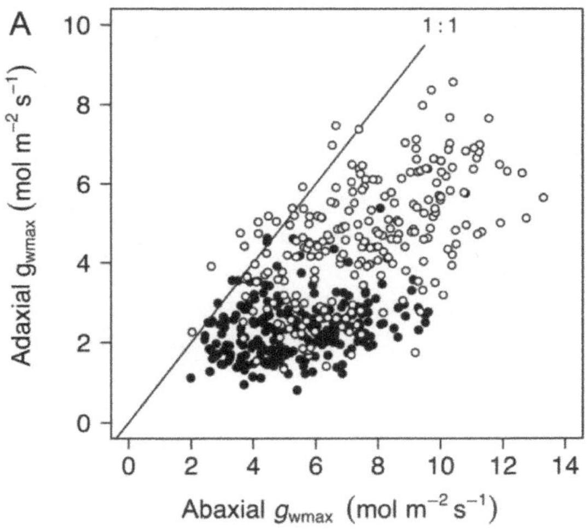

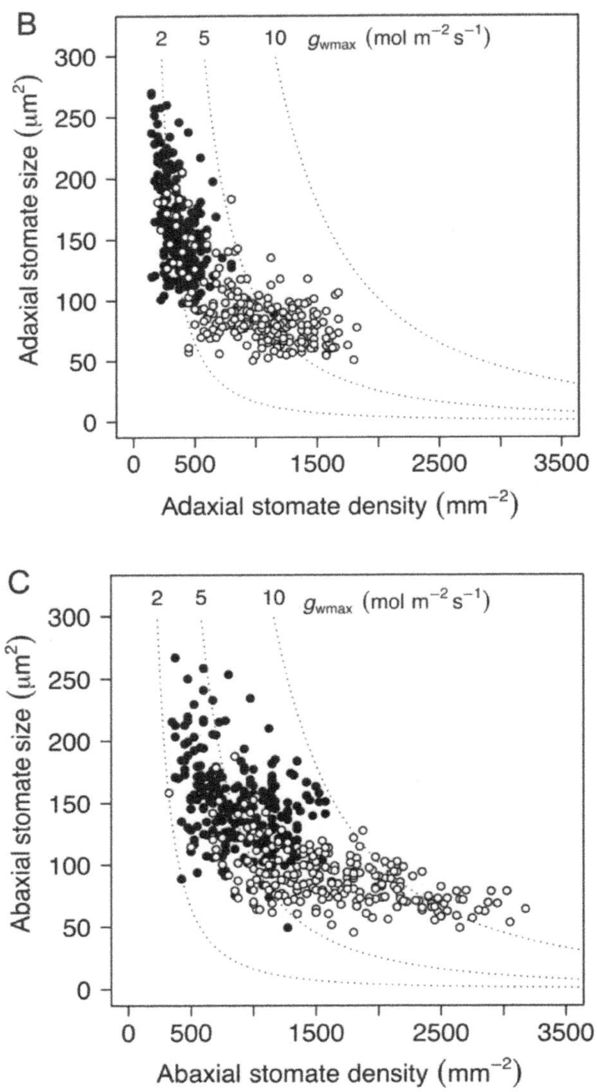

Figure 4. Components of g_{wmax} for lineages of *P. australis* native (filled symbols), and invasive (open symbols), to North America. (A) Leaf g_{wmax} is the sum of g_{wmax} for the adaxial and abaxial leaf surfaces; higher leaf g_{wmax} among invasive lineages is a result of increases in both adaxial and abaxial g_{wmax}. Stomate size shows a negative relationship with stomate density on both the adaxial (B) and abaxial (C) leaf surfaces: higher g_{wmax} on abaxial surfaces are linked with greater stomate densities, and the higher stomate densities among invasive *P. australis* are linked with reduced stomate size compared with the native lineage.

DISCUSSION

Previous demonstrations that g_{wmax} is reliably linked with gas exchange performance (Dow *et al.* 2014) and demonstrates both heritable variation and environmental plasticity (Franks *et al.* 2009; Fanourakis *et al.* 2015) suggested that simple measurements of the size and spacing of stomata on leaves would provide a strong framework within which to explore phenotypic plasticity in *P. australis*. Our results confirm this expectation; we were able to characterize plasticity in stomatal morphometrics that contributed to differences in g_{wmax} between native and invasive lineages. We found that at three marsh sites separated by as much as 43 km, introduced lineages have consistently greater g_{wmax} than their native congeners. Thus, g_{wmax} can be added to an already extensive list of functional traits that distinguish these genetic variants (stem densities, heights, above ground biomass, leaf area, leaf nitrogen and chlorophyll content, rates of photosynthesis, relative growth rates (RGR) and carbon fixation; reviewed in Mozdzer *et al.* 2013). Our analysis also indicates that plasticity of the introduced lineage, determined as within-genotype variation in g_{wmax}, is similar to or exceeds that shown by the native lineage. These results provide insights that scale up from stomatal morphometrics to community dynamics.

Phenotypic Variation in Stomatal Morphometrics

We observed inverse relationships between stomatal size and density, as have been commonly reported in the literature for multiple taxa (Kawamitsu *et al.* 1996; Hetherington and Woodward 2003; Franks *et al.* 2009). The derivation of g_{wmax} based on the work of Brown and Escombe (1900) suggests that a trade-off between stomate size and density will be broadly linked with conservation of g_{wmax}; decreases in stomatal size without a compensatory increase in density should result in decreases in g_{wmax} (the relative effect of decreased stomatal size on g_{wmax} is smaller when stomata are large because while pore resistance is increased by declines in pore area, parallel decreases in pore depth act to decrease pore resistance; see discussion by Franks *et al.* 2009). We interpret our results as pointing to size–density trade-offs linked with conservation of g_{wmax} among leaves from native *P. australis*. Meanwhile, plasticity in g_{wmax} among ramets of introduced *P. australis* was linked with greater plasticity in densities of stomata and was sometimes greater than for native clones.

Smaller stomata, as observed for the introduced lineage of *P. australis*, may improve water use efficiency. They are expected to be capable of opening and closing more rapidly (Aasamaa *et al.* 2001; Drake *et al.* 2013); in combination with lower resistance offered by shorter diffusion paths through smaller pores, rapid adjustment should lead to tighter linkage between stomatal responses and the need to regulate transpiration (Knapp 1993). In the case of *P. australis*, improvements in stomatal feedback could allow introduced lineage access to more exposed ground with less reliable water supply, contributing to their observed capacity to reduce soil moisture levels (by accretion, Rooth *et al.* 2003; by transpiration, Windham 2001; by Venturi Effect ventilation, Armstrong and Armstrong 1991). Detailed physiological work assessing the components of leaf gas exchange and hydraulics will be necessary to fully resolve whether differences in water use efficiency are mechanistically linked with stomatal morphometrics in these *Phragmites* lineages.

The g_{wmax} values we determined for *P. australis* in Maine, particularly the introduced lineage, were very high (Table 1). They exceeded measurements made by one of the authors in a previous pot-based greenhouse study (Taylor *et al.* 2012). A broad survey of other grass species (Kawamitsu *et al.* 1996) indicates that stomatal morphometrics of cultivated rice (*Oryza sativa*) were most similar to *P. australis*, but g_{wmax} values for *P. australis* were higher. This is despite the expectation that a hydrophytic habit and selection for high productivity in rice would be expected to have maximized g_{wmax}. The g_{wmax} values we determined are underpinned by similar stomatal morphometrics to those demonstrated in a previous study that addressed the potential for ploidy level variation of stomatal traits in field collected samples across north-eastern North America (Saltonstall *et al.* 2007; Table 1). Indeed, the stomatal traits reported by Saltonstall *et al.* (2007) suggest even more extreme values forg_{wmax} than in our sample (Table 1). Although our study is limited to three marshes in Maine, our results parallel those from a broader set of populations support differences in mean g_{wmax} between native and introduced lineages as a general feature of *P. australis*, at least across its north-eastern North American range. Comparison of our measurements, those made by Saltonstall *et al.* (2007), and material of a European origin (Table 1, Taylor *et al.* 2012) also suggests a strong conservation of between-lineage differences in stomatal size while density is more variable (Table 1): plastic responses of g_{wmax} in *P. australis* may depend strongly on variation in density of stomata.

Table 1. *Phragmites australis* as a species show exceptionally high g_{wmax}. [1] Saltonstall *et al.* (2007), [2] Taylor *et al.* (2012) and [3] Kawamitsu *et al.* (1996). [1]Saltonstall *et al.* did not determine lengths of adaxial stomata for most populations, there being no significant difference between surfaces in a subset. [2]Most extreme among 41 cultivars.

Species/lineage	Mean length of stomata (μm)		Mean density of stomata (mm^{-2})		g_{wmax} ($mol\ m^{-2}\ s^{-1}$) predicted from mean values	Location, data source
	Adaxial	Abaxial	Adaxial	Abaxial		
P. australis native	25	24	394	924	7.94	Maine, this study
P. australis invasive	19	19	1002	1635	12.42	
P. australis native	25[1]	25	804	1147	12.09	Northeast USA and Canada [1]
P. australis invasive	19[1]	19	1726	2167	18.34	
P. australis	19	19	419	510	4.38	Glasshouse, UK [2]
Oryza sativa cv. Raikei[2]	18	18	646	844	6.65	Japan [3]

Phragmites australis is a water-loving species characteristic of marshes and wetlands. Reliable availability of water can relax selection against increases in transpiration (Dudley 1996), allowing for improved net carbon gain or nutrient acquisition (Donovan *et al.* 2007). In hot environments, increased transpiration can improve photosynthetic efficiency and leaf survival by helping to decrease leaf temperatures (Lu *et al.* 1998). In the cool climate of New England, it seems likely that the principal advantage of high stomatal conductances would be to decrease resistance to CO_2diffusion into leaves and improve net carbon gain, consistent with observations that the introduced lineage shows greater productivity, responsiveness to carbon enrichment (Mozdzer and Megonigal 2012) and higher RGR, the latter being a proposed factor driving invasion (Mozdzer*et al.* 2013). More broadly, high rates of productivity and the capacity for local habitat modification, e.g. by drying, are traits common to many invasive plants (Cuddington and Hastings 2004); our demonstration thatg_{wmax} values for introduced *Phragmites* stands exceed those for native stands fits with reports of local drying effects linked the introduced lineage, mediated by both evapotranspiration and sediment accretion (Rooth *et al.* 2003). Summarizing, advantages under a variety of field conditions could arise from increases in transpiration linked with higherg_{wmax} that would provide for increased conductance to CO_2 and reduction in leaf temperature, or improved water use efficiency linked with decreases in stomatal size.

Community Dynamics

High levels of plasticity in stomatal traits support the description of introduced *P. australis* as a 'Jack-and-master' of change (Mozdzer and Megonigal 2012; Mozdzer *et al.* 2013). Plasticity in stomatal morphology would be expected to permit a single genotype to acclimate to a range of conditions and make it a strong competitor in a heterogeneous environment. Marsh systems susceptible to Phragmites invasion are starkly heterogeneous in many factors, for example sharp gradients from waterline to bank in salinity, aeration, nutrient availability and water depth (reviewed in Engloner 2009). Comparing North American lineages,Holdredge *et al.* (2010) described a cline ranging from lower elevation associated with waterlogged soils up to higher elevation characterized by high levels of interspecific competition. A single clonal genotype of *P. australis* might span multiple microhabitat transitions in this setting. Genotypes with a plastic localized response at the scale of the ramet could minimize the risks, costs or genetic resources associated with adaptation through sexual reproduction while best optimizing potential opportunities for resource sharing and economies of scale inherent in integrated clonality.

Indeed, 'Theory predicts that plasticity in ... morphologies of plants can transmit heterogeneity from the environment to the population or community' (Callaway *et al.* 2003). Thus, we can predict that significant variation should be identifiable from the among-lineage down to the among-ramet scales dependent upon local conditions. The lack of spatial structure to our data suggests that drivers of heterogeneity in stands of *P. australis* operate at a scale smaller than the 5 × 5 m scale measured here.

Plasticity is important for both lineages (Mozdzer and Megonigal 2012) and worth comparison against other non-clonal species. However, the lower levels of native plasticity suggest that there may be a cost involved. Net fitness, which synthesizes survival, growth and fecundity, does not necessarily benefit from plasticity (Palacio-López and Gianoli 2011;Pichancourt and Van Klinken 2012). In some circumstances, plasticity can be disadvantageous, for example, when there are costs of inappropriate specialized phenotypes, when environmental cues are unreliable, when the environment is not variable or when the plastic response lags too far behind environmental change (Vretare *et al.* 2001; Callaway *et al.* 2003). Thus, narrower plasticity in the native lineage may constrain optimal microhabitat range or reflect the more homogeneous sites it occupies.

A frequent assertion in invasive plant literature is that phenotypic plasticity is common in invasive species, making possible a broader ecological niche through the expression of site-specific advantageous phenotypes (Richards *et al.* 2006; Davidson *et al.* 2011). Previous work has shown that invading populations have the potential for rapid adaptive evolution (Dlugosch and Parker 2008), which would select for a 'general purpose genotype' (Moroney *et al.* 2013). There is some evidence that introduced *P. australis* may be less plastic in its native range (Rolletscheke*t al.* 1999) warranting further study of reaction norms in common gardens (e.g. Křiváčková-Suchá *et al.* 2007; Achenbach *et al.* 2012).

After within-genotype variation (plasticity), genetic variation (diversity) was the most important contributor to heterogeneity in phenotypes in this study, with relatively little variation being explained by among-site comparisons. Limited variation among sites may result from an emphasis on clonal reproduction, with limited sexual reproduction, natural selection and genetic drift. Initial models of *P. australis* establishment focussed on the transport of vegetative propagules and would lead to low genet richness at a given site (Bart *et al.* 2006); however, recent research indicates a greater role for sexual reproduction (McCormick *et al.* 2010) with clonal growth clearly important on a local scale (Kettenring and Mock 2012; Douhovnikoff and Hazelton 2014). Instead local genetic diversity can remain relatively high due to long lifespans and mechanisms such as remnant regional dynamics (Douhovnikoff and Hazelton 2014).

CONCLUSIONS

Plasticity in the introduced lineage of *P. australis* is similar to or exceeds that of native stands, both in our results and other reports (Mozdzer and Megonigal 2012; Mozdzer *et al.* 2013). This suggests that capacity for greater plasticity may be a major driver in the introduced lineage's invasiveness. Nonetheless, native *P. australis* does demonstrate considerable plasticity, which may underpin observations of long-term resistance to invasion, resilience and site consolidation. For example, the native lineage is well adapted to both low nutrient environments and exploitation of increasing nitrogen (sensu Hazelton *et al.* 2010). In contrast, the invader consistently outperforms the native in biomass production, nitrogen assimilation and various aspects of carbon metabolism (Mozdzer *et al.* 2013). These differences in physiological traits and trait plasticity may be indicators of different life-history strategies underpinning the ecological success and evolutionary maintenance of the two *P. australis* lineages in North America.

REFERENCES

1. Aasamaa K, Sõber A, Rahi M. 2001. Leaf anatomical characteristics associated with shoot hydraulic conductance, stomatal conductance and stomatal sensitivity to changes of leaf water status in temperate deciduous trees. Australian Journal of Plant Physiology 28:765–774.
2. Abramoff MD, Magalhaes PJ, Ram SJ. 2004. Image processing with ImageJ. Biophotonics International 11:36–42.
3. Achenbach L, Lambertini C, Brix H. 2012. Phenotypic traits of Phragmites australis clones are not related to ploidy level and distribution range. AoB PLANTS 2012: pls017; doi:10.1093/aobpla/pls017.
4. Alpert P, Simms EL. 2002. The relative advantages of plasticity and fixity in different environments: when is it good for a plant to adjust? Evolutionary Ecology 16:285–297. doi:10.1023/A:1019684612767
5. Armstrong J, Armstrong W. 1991. A convective through-flow of gases in Phragmites australis (Cav.) Trin. ex Steud. Aquatic Botany 39:75–88. doi:10.1016/0304-3770(91)90023-X
6. Bastlova D, C˘íźková H, Bastl M, Kvet J. 2004. Growth of Lythrum salicaria and Phragmites australis plants originating from a wide geographical area: response to nutrient and water supply. Global Ecology and Biogeography 13:259–271. doi:10.1111/j.1466-822X.2004.00089.x
7. Bart D, Burdick D, Chambers R, Hartman JM. 2006. Human facilitation of Phragmites australis invasions in tidal marshes: A review and synthesis. Wetlands Ecology and Management 14:53–65. doi:10.1016/j.aquabot.2010.04.003
8. Bellavance M-E, Brisson J. 2010. Spatial dynamics and morphological plasticity of common reed (Phragmites australis) and cattails (Typha sp.) in freshwater marshes and roadside ditches. Aquatic Botany 93:129–134. doi:10.1016/j.aquabot.2010.04.003
9. Brown HT, Escombe F. 1900. Static diffusion of gases and liquids in relation to the assimilation of carbon and translocation in plants. Philosophical Transactions of the Royal Society B: Biological Sciences 193:223–291. doi:10.1098/rstb.1900.0014
10. Callaway RM, Pennings SC, Richards CL. 2003. Phenotypic plasticity and interactions among plants. Ecology 84:1115–1128. doi:10.1890/0012-9658(2003)084[1115:PPAIAP]2.0.CO;2
11. Clevering OA. 1999. The effects of litter on growth and plasticity of Phragmites australis clones originating from infertile, fertile or eutrophicated habitats. Aquatic Botany 64:35–50. doi:10.1016/S0304-3770(99)00009-1
12. Cuddington K, Hastings A. 2004. Invasive engineers. Ecological Modelling 178:335–347. doi:10.1016/j.ecolmodel.2004.03.010
13. Davidson AM, Jennions M, Nicotra AB. 2011. Do invasive species show higher phenotypic plasticity than native species and, if so, is it adaptive? A meta-analysis. Ecology Letters 14:419–431. doi:10.1111/j.1461-0248.2011.01596.x
14. Dlugosch KM, Parker IM. 2008. Founding events in species invasions: genetic variation, adaptive evolution, and the role of multiple introductions. Molecular Ecology 17:431–449. doi:10.1111/j.1365-294X.2007.03538.x
15. Donovan LA, Dudley SA, Rosenthal DM, Ludwig F. 2007. Phenotypic selection on leaf water use efficiency

and related ecophysiological traits for natural populations of desert sunflowers. Oecologia 152:13–25. doi:10.1007/s00442-006-0627-5

16. Douhovnikoff V, Dodd RS. 2015. Epigenetics: a potential mechanism for clonal plant success. Plant Ecology 216:227–233. doi:10.1007/s11258-014-0430-z
17. Douhovnikoff V, Hazelton ELG. 2014. Clonal growth: invasion or stability? A comparative study of clonal architecture and diversity in native and introduced lineages of Phragmites australis (Poaceae). American Journal of Botany 101:1577–1584. doi:10.3732/ajb.1400177
18. Dow GJ, Bergmann DC, Berry JA. 2014. An integrated model of stomatal development and leaf physiology. New Phytologist 201:1218–1226. doi:10.1111/nph.12608
19. Drake PL, Froend RH, Franks PJ. 2013. Smaller, faster stomata: scaling of stomatal size, rate of response, and stomatal conductance. Journal of Experimental Botany 64:495–505. doi:10.1093/jxb/ers347
20. Dudley SA. 1996. Differing selection on plant physiological traits in response to environmental water availability: a test of adaptive hypotheses. Evolution 50:92–102. doi:10.2307/2410783
21. Eller F, Brix H. 2012. Different genotypes of Phragmites australis show distinct phenotypic plasticity in response to nutrient availability and temperature. Aquatic Botany 103:89–97. doi:10.1016/j.aquabot.2012.07.001
22. Engloner AI. 2009. Structure, growth dynamics and biomass of reed (Phragmites australis)—a review. Flora - Morphology, Distribution, Functional Ecology of Plants 204:331–346. doi:10.1016/j.flora.2008.05.001
23. Fanourakis D, Giday H, Milla R, Pieruschka R, Kjaer KH, Bolger M, Vasilevski A, Nunes-Nesi A, Fiorani F, Ottosen C-O. 2015. Pore size regulates operating stomatal conductance, while stomatal densities drive the partitioning of conductance between leaf sides. Annals of Botany 115:555–565.
24. Fox J. 2008. Applied regression analysis and generalized linear models, 2nd edn. New York, NY: Sage.
25. Franks PJ, Farquhar GD. 2006. The mechanical diversity of stomata and its significance in gas-exchange control. Plant Physiology 143:78–87. doi:10.1104/pp.106.089367
26. Franks PJ, Drake PL, Beerling DJ. 2009. Plasticity in maximum stomatal conductance constrained by negative correlation between stomatal size and density: an analysis using Eucalyptus globulus. Plant, Cell and Environment 32:1737–1748. doi:10.1111/j.1365-3040.2009.002031.x
27. Gianoli E, Valladares F. 2012. Studying phenotypic plasticity: the advantages of a broad approach. Biological Journal of the Linnean Society 105:1–7.
28. Hansen DL, Lambertini C, Jampeetong A, Brix H. 2007. Clone-specific differences in Phragmites australis: effects of ploidy level and geographic origin. Aquatic Botany 86:269–279. doi:10.1016/j.aquabot.2006.11.005
29. Hazelton ELG, Knight TJ, Theodose TA. 2010. Glutamine synthetase partitioning in native and introduced salt marsh grasses. Marine Ecology Progress Series 414:57–64. doi:10.3354/meps08704
30. Hetherington AM, Woodward FI. 2003. The role of stomata in sensing and driving environmental change. Nature 424:901–908. doi:10.1038/nature01843
31. Holdredge C, Bertness MD, Von Wettberg E, Silliman BR. 2010. Nutrient enrichment enhances hidden differences in phenotype to drive a cryptic plant invasion. Oikos 119:1776–1784. doi:10.1111/j.1600-0706.2010.18647.x
32. Kawamitsu Y, Agata W, Hiyane S, Murayama S, Nose A, Shinjyo C. 1996. Relation between leaf gas exchange rate and stomata I. Stomatal frequency and guard cell length in C3 and C4 grass species. Japanese Journal of Crop Science 65:626–633.
33. Kettenring KM, Mock KE. 2012. Genetic diversity, reproductive mode, and dispersal differ between the cryptic invader, Phragmites australis, and its native conspecific. Biological Invasions 14:2489–2504. doi:10.1007/s10530-012-0246-5
34. Knapp AK. 1993. Gas exchange dynamics in C3 and C4 grasses: consequence of differences in stomatal conductance. Ecology 74:113–123. doi:10.2307/1939506
35. Křiváčková-Suchá O, Vávřová P, Čížková H, Čurn V, Kubátová B. 2007. Phenotypic and genotypic variation of Phragmites australis: a comparative study of clones originating from two populations of different age. Aquatic Botany 86:361–368. doi:10.1016/j.aquabot.2007.01.001
36. Langsrud O. 2003. ANOVA for unbalanced data: use Type II instead of Type III sums of squares. Statistics and Computing 13:163–167. doi:10.1023/A:1023260610025
37. Lu Z, Percy RG, Qualset CO, Zeiger E. 1998. Stomatal conductance predicts yields in irrigated Pima cotton and bread wheat grown at high temperatures. Journal of Experimental Botany 49:453–460. doi:10.1093/jxb/49.Special_Issue.453
38. Mccormick MK, Kettenring KM, Baron HM, Whigham DF. 2010. Spread of invasive Phragmites australis in estuaries with differing degrees of development: genetic patterns, Allee effects and interpretation. Journal of Ecology 98:1369–1378. doi:10.1111/j.1365-2745.2010.01712.x
39. Moroney JR, Rundel PW, Sork VL. 2013. Phenotypic plasticity and differentiation in fitness-related traits in invasive populations of the Mediterranean forb Centaurea melitensis (Asteraceae). American Journal of Botany 100:2040–2051. doi:10.3732/ajb.1200543
40. Mozdzer TJ, Megonigal JP. 2012. Jack-and-master trait responses to elevated CO2 and N: a comparison of native and introduced Phragmites australis. PLoS ONE 7:e42794. doi:10.1371/journal.pone.0042794
41. Mozdzer TJ, Brisson J, Hazelton EL. 2013. Physiological ecology and functional traits of North American native and Eurasian introduced Phragmites australis lineages. AoB PLANTS 5: plt048; doi:10.1093/aobpla/plt048.
42. Palacio-López K, Gianoli E. 2011. Invasive plants do not display greater phenotypic plasticity than their native or non-invasive counterparts: a meta-analysis. Oikos 120:1393–1401. doi:10.1111/j.1600-0706.2010.19114.x

43. Pichancourt J-B, Van Klinken RD. 2012. Phenotypic plasticity influences the size, shape and dynamics of the geographic distribution of an invasive plant. PLoS ONE 7:e32323. doi:10.1371/journal.pone.0032323

44. R Development Core Team. 2015. R: a language and environment for statistical computing. Vienna, Austria: R Foundation for Statistical Computing. http://www.R-project.org/ (March 2015).

45. Richards CL, Bossdorf O, Muth NZ, Gurevitch J, Pigliucci M. 2006. Jack of all trades, master of some? On the role of phenotypic plasticity in plant invasions. Ecology Letters 9:981–993. doi:10.1111/j.1461-0248.2006.00950.x

46. Rolletschek H, Rolletschek A, Kühl H, Kohl JG. 1999. Clone specific differences in a Phragmites australis stand: II. Seasonal development of morphological and physiological characteristics at the natural site and after transplantation. Aquatic Botany 64:247–260. doi:10.1016/S0304-3770(99)00054-6

47. Rooth JE, Stevenson JC, Cornwell JC. 2003. Increased sediment accretion rates following invasion by Phragmites australis: the role of litter. Estuaries 26:475–483. doi:10.1007/BF02823724

48. Rouder JN, Morey RD, Speckman PL, Province JM. 2012. Default Bayes factors for ANOVA designs. Journal of Mathematical Psychology 56:356–374. doi:10.1016/j.jmp.2012.08.001

49. Saltonstall K. 2003. Microsatellite variation within and among North American lineages of Phragmites australis. Molecular Ecology 12:1689–1702.

50. Saltonstall K, Peterson PM, Soreng RJ. 2004. Recognition of Phragmites australis subsp. americanus (Poaceae: Arundinoideae) in North America: evidence from morphological and genetic analysis. SIDA, Contributions to Botany 21:683–692.

51. Saltonstall K, Glennon K, Burnett A, Hunter RB, Hunter KL. 2007. Comparison of morphological variation indicative of ploidy level in Phragmites Australis (Poaceae) from Eastern North America. Rhodora 109:415–429. doi:10.3119/0035-4902(2007)109[415:COMVIO]2.0.CO;2

52. Schulze E, Kelliher FM, Körner C, Lloyd J, Leuning R. 1994. Relationships among maximum stomatal conductance, ecosystem surface conductance, carbon assimilation rate, and plant nitrogen nutrition: a global ecology scaling exercise. Annual Review of Ecology and Systematics 25:629–662. doi:10.1146/annurev.es.25.110194.003213

53. Swearingen J, Saltonstall K. 2010. Phragmites field guide: distinguishing native and exotic forms of common reed (Phragmites australis) in the United States. Plant Conservation Alliance, Weeds Gone Wild. http://www.nps.gov/plants/alien/fact/pdf/phau1-powerpoint.pdf.

54. Taylor SH, Franks PJ, Hulme SP, Spriggs E, Christin P-A, Edwards EJ, Woodward FI, Osborne CP. 2012. Photosynthetic pathway and ecological adaptation explain stomatal trait diversity amongst grasses. New Phytologist 193:387–396. doi:10.1111/j.1469-8137.2011.03935.x

55. Thompson K, Hodgson JG, Rich TC. 1995. Native and alien invasive plants: more of the same? Ecography 18:390–402. doi:10.1111/j.1600-0587.1995.tb00142.x

56. Via S, Lande R. 1985. Genotype-environment interaction and the evolution of phenotypic plasticity. Evolution 39:505–522. doi:10.2307/2408649

57. Vretare V, Weisner SEB, Strand JA, Granéli W. 2001. Phenotypic plasticity in Phragmites australis as a functional response to water depth. Aquatic Botany 69:127–145. doi:10.1016/S0304-3770(01)00134-6

58. Weyers JDB, Meidner H. 1990. Methods in stomatal research. Harlow: Longman Scientific and Technical.

59. Windham L. 2001. Comparison of biomass production and decomposition between Phragmites australis (common reed) and Spartina patens (salt hay grass) in brackish tidal marshes of New Jersey, USA. Wetlands 21:179–188. doi:10.1672/0277-5212(2001)021[0179:COBPAD]2.0.CO;2

CHAPTER 7

RHIZOSPHERE ACID AND ALKALINE PHOSPHATASE ACTIVITY AS A MARKER OF P NUTRITION IN NODULATED CYCLOPIA AND ASPALATHUSSPECIES IN THE CAPE FYNBOS OF SOUTH AFRICA

S.T. Maseko[a] and F.D. Dakora[b]

[a]Department of Crop Sciences, Tshwane University of Technology, Private Bag X680, Pretoria 0001, South Africa

[b]Chemistry Department, Tshwane University of Technology, Private Bag X680, Pretoria 0001, South Africa

ABSTRACT

Cyclopia and Aspalathus species are important economic legumes in the Cape fynbos of South Africa, as they are used for making Honeybush and Rooibos tea, and for trade in the cut wild flower industry. The aim of this study was to assess acid and alkaline phosphatase activity in the rhizosphere of Cyclopia genistoides, Cyclopia subternata, Aspalathus caledonensis and Aspalathus aspalathoides as an indicator of P supply and P nutrition in the nutrient-poor soils of the Cape fynbos. Whether at Kokrivier or Kanetberg, the P enzyme activities were much higher in the rhizospheres of the legumesC. genistoides, C. subternata, A. caledonensis, and A. aspalathoides compared to those of the non-legumes Leucadendron strictum, Elegia thyrsoidea and Mimetes cucullatus, or bulk soil. As a result, plant-available P concentration in the rhizosphere, as well as shoot P levels closely mirrored acid and alkaline phosphatase activity in the rhizosphere of each plant species. Relative to younger plants, older Cyclopia species exhibited, much greater acid and alkaline phosphatase activity in the rhizosphere and this again resulted in much higher plant-available rhizosphere P. C. subternata plants developed from cuttings at Kanetberg showed greater rhizosphere acid and alkaline phosphatase activity than seedlings and bulk soil. As a result, the concentration of plant available-P and organic P were much higher in the rhizosphere of cuttings than seedlings, leading to greater shoot P in cuttings than seedlings. Taken together, these data suggest that rhizosphere P enzyme activity can be used as a good indicator of P supply and P nutrition in Cyclopia cuttings and seedlings, but less so in Aspalathus species in the Cape fynbos. The enhanced P nutrition in plants from cuttings probably accounts for the higher tea yields obtained by farmers when they use cuttings instead of seedlings in their plantations.

INTRODUCTION

Phosphorus is the second most important mineral nutrient for plant growth after N. Although total phosphate is abundant in many soils, it is largely unavailable for root uptake (Lambers et al., 2008), because, about 20–80% of soil P occurs in organic forms (Richardson et al., 2009). Soil P easily forms insoluble complexes with cations, and/or incorporated into organic matter by microbes (Tran et al., 2010 and Vance et al., 2003). Even where P is applied as fertilizer, more than 80% becomes unavailable for plant uptake through adsorption, precipitation, or microbial immobilization (Holford, 1997 and Mózner et al., 2012). Thus crop yield on the 30 − 40% of the world's arable land is limited by P availability (Kochian et al., 2004, Lambers et al., 2006 and Vance et al., 2003). In low P soils, both plants and microbes have developed various strategies for enhancing

Citation: S.T. Maseko, F.D. Dakora, Rhizosphere acid and alkaline phosphatase activity as a marker of P nutrition in nodulated Cyclopia and Aspalathus species in the Cape fynbos of South Africa, South African Journal of Botany, Volume 89, November 2013, Pages 289-295, ISSN 0254-6299, http://dx.doi.org/10.1016/j.sajb.2013.06.023.

P supply. These include acidification of the rhizosphere, exudation of organic acids (Dakora and Phillips, 2002), and secretion of extracellular phosphatases (Kanu and Dakora, 2009, Kanu and Dakora, 2012, Kanu et al., 2007, Makoi et al., 2010 and Zhao et al., 2010). Acid phosphatases are enzymes of plant origin (Tarafdar et al., 2001), while alkaline phosphatases are secreted by bacteria, fungi and earthworms (Hebrien and Neal, 1990) and function catalytically above pH 7. These enzymes catalyze the cleavage of mineral P from organic phosphate esters, in acidic and alkaline soils that are low in P (Nannipieri et al., 2011), thus making P more available in these soils. Organic P from soil organic matter is therefore the main substrate for P supply by phosphatases in natural and agricultural ecosystems.

Soils of the Cape fynbos are characterized by low organic matter, low pH (pH 3–5), and low nutrient levels (Dakora, 2012), especially plant-available N (0.001–0.002% N, or 1 to 2 mg N g^{-1}soil) and P (0.00004 to 0.00037% P, or 0.4–3.7 µg P g^{-1}; Cramer, 2010). Furthermore, soils of the fynbos biome contain between 58–77% organic P and 2–88% inorganic P, which is Fe-bound and unavailable to plants (Straker, 1996). Despite these nutrient constraints, the fynbos is rich in many indigenous legume genera and species. For example, the genus *Aspalathus* consists of 281 species that are endemic to the Cape Floristic Region (Boatwright and Cupido, 2011), while *Cyclopia* has 24 species that uniquely occur only in the Cape fynbos (Du Toit et al., 1998). N limitation in legumes of the fynbos is overcome by high levels of N_2 fixation. For example, *Aspalathus linearis* and*Cyclopia* species can contribute over 100 kg N ha^{-1} annually to the fynbos biome (Moufhe and Dakora, 1999 and Spriggs and Dakora, 2008). However, little is known about P nutrition in fynbos legumes. Different mechanisms have been proposed for enhanced P supply and mineral uptake by fynbos species. Mycorrhizal infection of fynbos legumes has been suggested to enhance P supply by the symbiosis (Harrison, 1999), just as formation of cluster roots has been regarded as a way of increasing P nutrition in fynbos legumes (Kanu et al., 2013, Lambers and Shane, 2007 and Lamont, 1982). We have, in fact, shown that cluster root formation by *A. linearis* in the Cape fynbos does not only improve P nutrition, but also increases the assimilation of other nutrients such as Ca, K, Mg, Na, Al, Fe, Mn, Zn and Cu (data not shown), clearly indicating that cluster root formation is a major mechanism for nutrient acquisition in the nutrient-poor fynbos.

So far, little information exists on the mechanisms of P supply to native legumes of the fynbos. Yet some of them have economic importance. The 24 *Cyclopia* species, for example, are a source of Honeybush tea, a herbal beverage that is emerging as a major income earner that contributes about R12 million to the South African economy (DAFF, Department of Agriculture, Fisheries and Forestry, 2012). Similarly, some of the 281*Aspalathus* species and most *Leucadendron* species are exported as cut wild flowers (Criley, 2000 and Reinten et al., 2011), while *A. linearis* produces Rooibos tea, another herbal beverage that contributes about R500 million to the South African economy (Joubert and De Beer, 2011 and http://www.busrep.co.za). *A. linearis* is among the many species of the genus *Aspalathus* that is perceived to be suitable for phytoremediation (Kanu et al., 2013). There is therefore a need to understand the mechanisms underlying P supply to native species in a very low P environment. The aim of this study was to measure rhizosphere acid and alkaline phosphatase activity as an indicator of P supply and availability. This was achieved by measuring acid and alkaline phosphatase activity in bulk and rhizosphere soils and comparing that with P concentrations in both systems. Finally, a measure of increased P availability in the rhizosphere was used to track P levels in plant shoots. For comparison, non-legume plant species were sampled from each site as the legumes and included in this study. The test plant species used in this study are shown in Table 1, together with their collection sites, uses and nutrition-related biological traits. The data showed a relationship between acid and alkaline phosphatase in the rhizosphere and P availability in the rhizosphere, which was ultimately linked to P concentration in shoots.

Table 1. Plant species studied, their sites of collection, families, tribes, uses and nutrition-related traits.

Plant species	Family	Tribe	**Plant age**	**Site**	**Uses**	**Nutrition-related traits**	**References**
*Cyclopia genistoides*and *Cyclopia subternata*	Fabaceae	Podalyrieae	10 & 2 8 & 5	Koksrivier Kanetberg	Honeybush tea, fruit juice mixtures, cosmetic industry, food flavourant, food dye	N_2-fixing, mycorrhizal, cluster root-forming	Spriggs et al. (2003)Spriggs (2004) Power et al (2010) DAFF, Department of Agriculture, Fisheries and Forestry (2012) Van Wyk (2011)
Aspalathus caledonensis	Fabaceae	Crotalarieae	2	Koksrivier	Harvested as fresh and cut dried flowers	N_2-fixing, mycorrhizal cluster root-forming	Allsopp and Stock (1994) Maseko (2013) Maseko and Dakora (2013)
Aspalathus aspalathoides	Fabaceae	Crotalarieae	2	Koksrivier	Harvested as fresh and cut dried flowers	N_2-fixing, mycorrhizal, cluster root-forming	Allsopp and Stock (1994) Maseko (2013) Maseko and Dakora (2013)
Leucadendron strictum	Proteaceae	Senecioneae	8	Kanetberg	Harvested as fresh and cut dried flowers	Non-N_2-fixing, mycorrhizal, cluster root-forming	Criley (2000) Lambers and Shane (2007) Reinten et al (2011) Maseko and Dakora (2013)

Mimetes cucullatus	Proteaceae	Corotocina	2	Koksrivier	Harvested as fresh and cut dried flowers	Non-N_2-fixing	Reinten et al (2011)
Gnidia anomala	Thymelaeceae	Gnidieae	10	Koksrivier	Medicinal, thatching, wood, clothing, dyeing leather	Non-N_2-fixing	Bhandurge et al (2013)
Tetraria bromoides	Cyperaceae	Schoeneae	2	Koksrivier	Grazing, erosion control, water purification	Non-N_2-fixing	Lambers and Shane (2007) http://www.inarchive.com
Elegia thyrsoidea	Restionaceae	Restioneae	5	Kanetberg	Thatching, horticultural industry	Non-N_2-fixing	Kubitzki and Huber (1998)
Erica multiflora	Ericaceae	Ericeae	5	Kanetberg	Grazing, medicinal	Non-N_2-fixing, mycorrhizal	Straker (1996) Rogosic et al (2006) Harnafi et al (2007)

MATERIALS AND METHODS

Collection and Processing of Rhizosphere and Bulk Soils

Rhizosphere soil (defined as soil rich in roots, and/or soil adhering to the roots and influenced by root activity) was collected together with bulk soils from 0 to 30 cm depth at Koksrivier near Gansbaai, and Kanetberg near Riversdale, in the Western Cape Province of South Africa. A minimum of 5 and maximum of 10 soil samples were cored from each farm as rhizosphere or non-rhizosphere bulk soils. The rhizosphere soil was cored at a distance of 1 to 2 mm from the plant roots, while bulk non-rhizosphere soil (free from roots) was collected in between rows of *Cyclopia* and *Aspalathus* plants. The samples were transferred into pre-labelled plastic bags and transported to the laboratory, where they were stored at − 4 °C until bioassay of enzyme activity. Subsamples of soil were however air-dried, sieved (2.0 mm), and used for measuring soil P concentration. Rhizosphere soils of non-legumes were similarly collected and processed for comparison with those of legumes.

Plant Harvest and Processing

Young shoots of 5 to 10 replicate legume plants (10 shoots per plant) (from which rhizosphere soils were collected) were harvested at crown level, oven-dried (60 °C), weighed, and milled to a fine powder (0.85 mm) for analysis of P. Adjacent non-legume plants were similarly harvested and processed for P analysis for comparison with legumes.

Determination of pH and Plant-available P in Rhizosphere and Bulk Soils

To measure pH, 50 mL 0.01 M $CaCl_2$ was added to 10 g of soil sample which had been air-dried and sieved (2 mm). The mixture was shaken for 30 min, allowed to settle, and the pH measured using a pH meter.

Extractable P in each of the 5 to 10 replicate soils was determined by the citric acid method, using the modification of Du Plessis and Burger (1964). A 20 g of the air-dried soil sample was extracted in 200 mL of 1% (w/v) citric acid, heated to 80 °C shaken for 2 min at 10-min intervals over 1-h period and filtered. A 50 mL aliquot was heated to dryness on a water bath, digested with 5 mL of concentrated HCl and HNO_3, evaporated to dryness on water bath, and 5 mL of concentrated HNO_3 and 20 mL of deionized water added. The mixture was then heated to dissolve the dry residue, and the sample filtered. Measurement of P was done directly by aspiration on a calibrated simultaneous inductively coupled plasma-mass spectrometer (IRIS/AP HR DUO Thermo Electron Corporation, Franklin, Massachusetts, USA).

Measurement of P in Plant Shoots

Measurement of P in plant shoots was determined by ashing 1.0 g of finely ground sample (0.85 mm) in a porcelain crucible at 500 °C overnight. The ash was dissolved in 5 mL of 6 M HCl and placed in an oven at 50 °C for 30 min, followed by addition of 35 mL deionised water. The mixture was filtered through Whatman #1 filter paper, and the concentration of P in plant extracts determined using inductively coupled plasma mass spectrometry (Giron, 1973).

Assay of Acid and Alkaline Phosphatase Activity in Bulk and Rhizosphere Soils

Bioassay of phosphatase activity was done as described by Tabatabai (1994). About 1 g of moist soil sample (stored at − 4 °C) was weighed in duplicates into polypropylene vials for each of the 5 to 10 replicate soil samples. Acid or alkaline phosphatase activity was determined by adding 4 mL of a pH buffer (pH 6.5 for acid and pH 11 for alkaline phosphatases), and 1 mL of 0.1 M disodium phenylphosphate as a substrate. The mixture in polypropylene vials was incubated at 37 °C for 1 h. At the end of incubation, 1 mL of 0.5 M $CaCl_2$ and 4 mL of 0.5 M NaOH were added to stop the reaction, the mixture was filtered through Whatman #2 filter paper and the supernant analyzed using a UV–VIS spectrophotometer at 420 nm. Absorbance of filtrates was compared with *p*-nitrophenol standards. For each assay, a control was included to account for non-enzymatic substrate hydrolysis.

Statistical Analysis

Statistical analyses were carried out using a STATISTICA 2 analytical software program 2007 (StatSoft Inc, 2005). A 1-Way ANOVA was performed to compare the means of phosphatase enzyme activity as well as P concentration in soil and shoots of each species. Where significant differences were found, the Duncan Multiple Range Test was used to separate treatment means at $p \leq 0.05$.

RESULTS

Soil pH

Although the rhizosphere pH of each species was generally lower than that of the corresponding bulk soil, these were not statistically significant. The mean values of rhizosphere and bulk soil were respectively 4.32 and 4.48, 4.12 and 4.48, and 4.12 and 4.20 for *Cyclopia genistoides*, *Aspalathus* species and *Cyclopia subternata*. The rhizosphere pH of the test species were also similar in magnitude.

Soil P, Shoot P and Acid/Alkaline Phosphatase Activity in Rhizosphere and Bulk Soils from Koksrivier and Kanetberg

Acid and alkaline phosphatase activity in rhizosphere soil of *C. genistoides* established in 1997 and 2005 at Koksrivier were measured and found to be significantly ($p \leq 0.05$) higher than that of bulk soil (Table 2). A comparison of phosphatase activity between years showed that *C. genistoides* planted in 1997 had higher rhizosphere acid and alkaline phosphatase activity than those established in 2005 (Table 2). P concentration was measured in both bulk and rhizosphere soil of *C. genistoides* to determine whether the differences in acid and alkaline phosphatase activity affected P availability. Plant-available inorganic P concentration was much greater in the rhizosphere soil of *C. genistoides* than bulk soil (Table 2). Organic P levels however were greater in non-rhizosphere bulk soil compared to rhizosphere of *C. genistoides* established in 1997 (Table 2). The greater enzyme activity in the rhizosphere of *C. genistoides* from the 1997 plantation resulted in equally greater rhizosphere inorganic P concentration compared to*C. genistoides* in 2005 plantation (Table 2). A comparison of shoot P concentration between *C. genistoides* grown in 1997 and a non-legume, *Gnidia anomala* showed no differences, younger shoots of the legume however accumulated greater P concentration compared to another non-legume, *Tetraria bromoides* (Table 2). A similar assay of enzymes was done for *C. subternata* planted at Kanetberg in 1999 and 2002. The data showed that acid phosphatase activity was again higher in the rhizosphere soils of *C. subternata* compared to bulk soil (Table 3). Rhizosphere soil of *C. subternata*established in 1999 showed significantly greater acid and alkaline phosphatase activity to that grown in 2002 (Table 3). Measurement of inorganic and organic P levels in both bulk and rhizosphere soil revealed significantly greater P levels in rhizosphere soil than bulk soil (Table 3). Determination of P concentration in shoots of *C. subternata* plants showed higher P levels when compared to the two non-legumes, *Leucadendron strictum*and *Elegia thyrsoidea* (Table 3).

Table 2. P nutrition in two fynbos species at Koksrivier. Acid (APase) and alkaline (Alk) phosphatase activity and plant-available inorganic phosphorus (Pi) and organic phosphorus (Po) associated with rhizosphere and non-rhizosphere soil of *Cyclopia genistoides* established in 1997 and 2005 at Koksrivier farm. Values (Mean ± SE) followed by dissimilar letters in a column for year of planting and in a row for enzyme activity for soil or plant P are significantly different at $p \leq 0.05$. Lower case letters compare means of phosphorus level in a column and upper case letters compare means in a row. 1997 (non-legume: *Gnidia anomala*) and 2005 (non-legume: *Tetraria bromoides*).

Plant age	Soil APase activity		Soil Alk activity		Pi		Po		Shoot P	
	Rhizosphere	Bulk	Rhizosphere	Bulk	Rhizosphere	Bulk	Rhizosphere	Bulk	Legume	Non-legume
	μgρ-nitrophenol g^{-1} F wt soil h^{-1}				**mg.kg^{-1}**		**g.kg^{-1}**		**mg.g^{-1}**	
10	826 ± 37.3aA	379 ± 17.9aB	168 ± 9.9aC	75 ± 10.8aD	10 ± 2.0aA	3.5 ± 1.8aB	372 ± 34.4aB	782 ± 25.6aA	0.57 ± 0.0aA	0.625 ± 0.1aA
2	586 ± 35.4bA	263 ± 90.5aB	93 ± 4.9bC	64 ± 28.3aD	8 ± 0.9bA	3.7 ± 1.8aB	499 ± 74.5aA	604 ± 58.6bA	0.88 ± 0.0aA	0.575 ± 0.1aB
F-Statistics	21.86**	1.58^{ns}	46.56***	0.13^{ns}	4.83*	0.13^{ns}	2.41^{ns}	6.47*	0.00^{ns}	0.00^{ns}

Table 3. P nutrition in two fynbos species at Koksrivier. Acid (APase) and alkaline (Alk) phosphatase activity and plant-available inorganic phosphorus (Pi) and organic phosphorus (Po) associated with rhizosphere and non-rhizosphere soil of *Cyclopia subternata* established in 1997 and 2005 at Koksrivier farm. Values (Mean ± SE) followed by dissimilar letters in a column for the year of planting and in a row for enzyme activity for soil or plant P are significantly different at $p \leq 0.05$. Lower case letters compare means of phosphorus level in a column and upper case letters compare means in a row.1999 (non-legume:*Leucadendron strictum*) and 2002 (non-legume: *Elegia thyrsoidea*).

Plant age	Soil APase activity		Soil Alk activity		Pi		Po		Shoot P	
	Rhizosphere	Bulk	Rhizosphere	Bulk	Rhizosphere	Bulk	Rhizosphere	Bulk	Legume	Non-legume
	μg ρ-**nitrophenol g^{-1} F wt soil h^{-1}**				**mg.kg^{-1}**		**g.kg^{-1}**		**mg.g^{-1}**	
8	839 ± 50.3aA	644 ± 51.9aB	266 ± 15.9aC	242 ± 9.7aC	20.4 ± 1.8aA	14.8 ± 1.4aB	2064 ± 92.8aA	1706 ± 121.0aB	1.1 ± 0.0aA	0.75 ± 0.0aB
5	726 ± 48.8bA	561 ± 34.4aB	193 ± 21.4bC	138 ± 37.1bC	24.0 ± 2.7aA	8.8 ± 2.1bB	2114 ± 164.3aA	1636 ± 152.4aB	1.1 ± 0.0aA	0.62 ± 0.2aB
F-Statistics	8.85***	1.77^{ns}	7.47**	7.39**	1.24^{ns}	5.92*	0.07^{ns}	0.13^{ns}	0.31^{ns}	0.41^{ns}

Acid Phosphatase Activity, Soil P and Shoot P of *C. Genistoides* Compared with *Aspalathus Caledonensis, A. Aspalathoides* and the Non-legume *Mimetes Cucullatus* at Koksrivier

The marked differences in acid phosphatase between bulk and rhizosphere soil of *C. genistoides* and *C. subternata* led to further field tests that compared the enzyme activity of *C. genistoides* with two *Aspalathus* species and the non-legume *Mimetes cucullatus* at Koksrivier. The data generally showed much higher acid phosphatase activity in the legumes relative to non-legume. *C. genistoides* had the highest enzyme activity, followed by *A. caledonensis*, *A. aspalathoides*, and *Mimetes cucullatus* (Fig. 1A). For each species, rhizosphere plant-available inorganic P concentration closely mirrored rhizosphere enzyme activity, except for *A. caledonensis* which had high enzyme activity but low rhizosphere inorganic P level (Fig. 1B). In a trend almost opposite that of the inorganic P, the non-legume, *Mimetes cucullatus* showed greater rhizosphere organic P (Fig. 1C). To ascertain plant response to the species differences in rhizosphere P availability, shoot P concentration was measured and found to closely mirror rhizosphere inorganic P concentration of *C. genistoides*, *A. aspalathoides* and *M. cucullatus*, but not*A. caledonensis* (Fig. 1D).

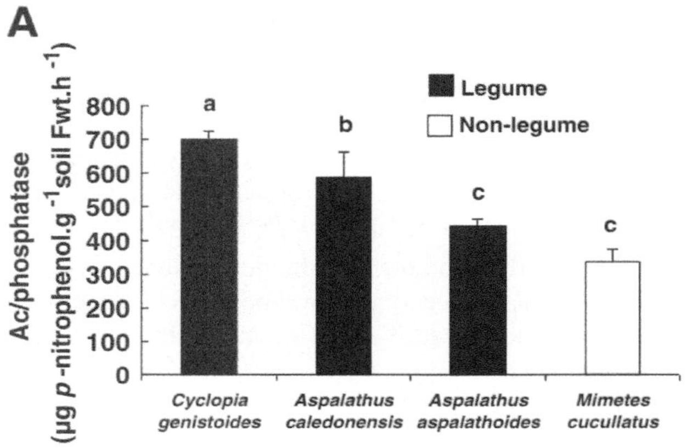

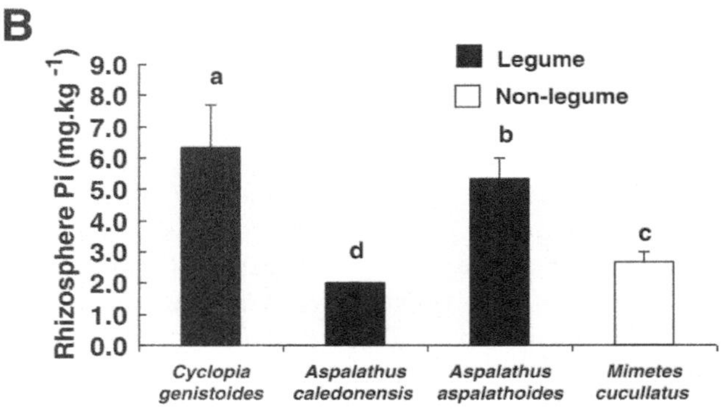

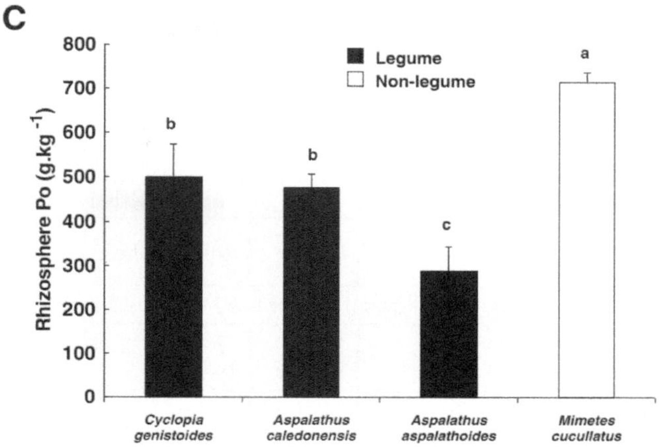

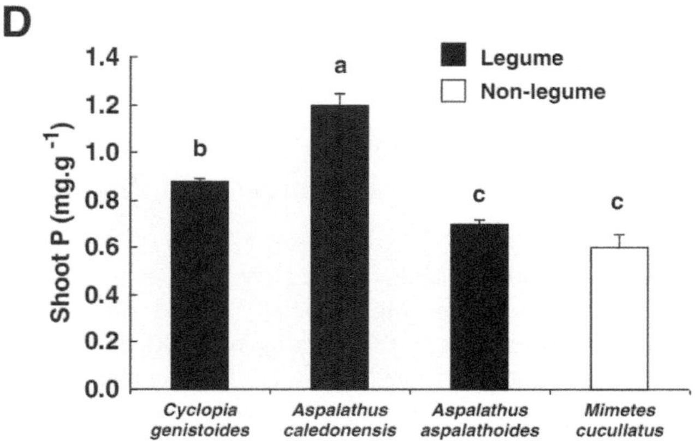

Figure 1. Phosphorus nutrition in four fynbos species: A) Acid phosphatase activity associated with rhizosphere soils of *Cyclopia genistoides*, *Aspalathus caledonensis*, *Aspalathus aspalathoides*, and *Mimetes cucullatus*sampled from Koksrivier farm; B) Pi levels in the rhizosphere soils of the four fynbos species, C) Po percentage in rhizosphere of the test species and D) P in shoots of *Cyclopia genistoides*, *Aspalathus caledonensis*, *Aspalathus aspalathoides*, and *Mimetes cucullatus*. Bars followed by dissimilar letters are significantly different at $p \leq 0.05$. Vertical lines on bars represent SE (n = 5).

A Comparison of Acid/Alkaline Phosphatase Activity, Soil P, and Shoot P in the Rhizosphere of Cyclopia Cuttings vs. Seedlings

Two adjacent *C. subternata* plantations developed from seedlings and cuttings at Kanetberg were assessed for phosphatase enzyme activity in bulk and rhizosphere soil. Rhizosphere soil from both cuttings and seedlings showed greater enzyme activity than bulk soil. However, the acid and alkaline phosphatase activity was higher in cuttings than seedlings (Table 4). Determination of both inorganic and organic P showed both planting material to possess greater P level in rhizosphere than non-rhizosphere bulk soil (Table 4). However, the cuttings showed much greater plant-available inorganic and organic P concentration in rhizosphere soil than seedlings (Table 4). The marked differences in P availability and accumulation in the rhizosphere resulted in statistically greater shoot P levels in cuttings than seedlings, legume shoot P was also statistically greater than those of the non-legumes *Erica multiflora* and *E. thyrsoidea* (Table 4).

Table 4. P nutrition in two fynbos species at Koksrivier. Acid (APase) and alkaline (Alk) phosphatase activity and plant-available inorganic phosphorus (Pi) and organic phosphorus (Po) associated with rhizosphere and non-rhizosphere soil of *Cyclopia subternata* planted using cuttings and seeds in 2002 at Kanetberg farm. Values (Mean ± SE) followed by dissimilar letters in a column for year of planting and in a row for enzyme activity for soil or plant P are significantly different at $p \leq 0.05$. Lower case letters compare means of phosphorus level in a column and upper case letters compare means in a row. Cuttings (non-legume:*Elegia thyrsoidea*) and seedlings (non-legume: *Erica multiflora*).

Treatment	Soil APase activity		Soil Alk activity		Pi		Po		Shoot P	
	Rhizosphere	Bulk	Rhizosphere	Bulk	Rhizosphere	Bulk	Rhizosphere	Bulk	Legume	Non-legume
	μg ρ-nitrophenol g^{-1} F wt soil h^{-1}				mg kg^{-1}		g kg^{-1}		mg g^{-1}	
Cuttings	726.1 ± 48.8aA	561.2 ± 34.4aB	229.5 ± 19.9aC	102.5 ± 15.3aC	24.0 ± 2.7aA	8.8 ± 2.1aB	2114 ± 164.3aA	1636 ± 152.4aB	1.1 ± 0.0aA	0.63 ± 0.2aB
Seedlings	528.2 ± 46.4bA	318.8 ± 51.6bB	136.6 ± 20.0bC	42.7 ± 1.5bD	13.2 ± 1.1bA	7.8 ± 2.5aB	1642 ± 839.3bA	1154 ± 100.1bB	1.0 ± 0.0bA	0.68 ± 0.1aB
F-Statistics	8.63*	15.30**	10.85**	15.09**	13.99**	0.09^{ns}	6.55*	6.99*	6.60**	0.06^{ns}

DISCUSSION

Can Acid/Alkaline Phosphatase Activity in the Rhizosphere Serve as an Indicator of P Supply to Cyclopia Species?

The Cape fynbos is characterized by low concentrations of mineral nutrients in soil, especially P and N. While N requirement in legumes is often met by symbiotic N_2 fixation, P supply is enhanced through a range of mechanisms which include development of mycorrhizal symbiosis, formation of cluster roots, alteration in root architecture, production of root exudates, as well as induction of phosphatase enzyme activity (Maseko and Dakora, 2013). Some nutrition-related biological traits of the test species used in this study are listed in Table 1. The aim of this part of the study was to assess acid and alkaline phosphatase activity as an indicator of P supply in the rhizosphere of *C. genistoides* from a plantation at Koksrivier. The data revealed a significant relationship between acid/alkaline phosphatase activity in the rhizosphere of *C. genistoides* and P concentration in the soil around the roots. Whether planted in 1997 or 2005, both acid and alkaline phosphatase activities in the rhizosphere of *C. genistoides* showed higher activity compared to bulk soils (Table 2). There was also increased P enzyme activity in the rhizosphere of older *C. genistoides* plants compared to their younger counterparts (Table 2). These higher P enzyme activities resulted in markedly greater plant-available inorganic P concentrations in the rhizosphere compared to bulk soil, and in the rhizosphere of older plants compared to younger ones (Table 2). The greater acid/alkaline phosphatase activity also resulted in changes in soil organic P. As the activity of soil phosphatases increased (Table 2), there was a marked and significant decrease in rhizosphere organic P compared to the bulk soil, and a decrease in rhizosphere organic P of older plants compared to their younger counterparts (Table 2). Similar findings have been reported for a number of agricultural species (Asmar et al., 1995, Chen et al., 2002, Radersma and Grierson, 2004 and Tarafdar and Jungk, 1987). Our data therefore suggest that the increase in acid/alkaline phosphatase activity in the rhizosphere of *Cyclopia* resulted in greater inorganic P supply from organic sources relative to bulk soil, where there was low P enzyme activity, and therefore low P concentration. The greater acid phosphatase activity in the rhizosphere of older *Cyclopia*plants also resulted in increased Pi supply in the rhizosphere of older compared to younger plants (Table 2). However, the increased plant-available P in the rhizosphere was not reflected in tissue P concentrations, perhaps suggesting that plants adapted to low-nutrient soils such as fynbos biome, inherently take up small quantities of nutrients even with high nutrient supply (Witkowski, 1991).

To confirm that rhizosphere P enzyme activity is a proxy for P supply in the rhizosphere, a similar study was conducted at Kanetberg using *C. subternata* planted in 1999 and 2002. The data again revealed markedly high acid phosphatase enzyme activity in the rhizosphere than bulk soil, and in the rhizosphere of older than younger *C. subternata*plants. As a result, organic and inorganic P levels were lower in magnitude in the rhizosphere of older plants, but not statistically (Table 3). There was also much greater organic and inorganic and P concentrations in the rhizosphere relative to bulk soil (Table 3).

It is important to note that the non-legume *T. bromoides* recorded lower P concentration in shoots at Koksrivier in 2005 compared to nodulated *C. genistoides* (Table 2). Similarly, the non-legumes *L. strictum* and *E. thyrsoidea* also showed much lower P levels in shoots at Kanetberg in 1999 and 2002, respectively, relative to the N_2-fixing *C. subternata* (Table 3). While these high P concentrations in shoots of *Cyclopia* species could be attributed to P enzyme activity in the rhizosphere, it is well documented that N_2-fixing legumes generally accumulate greater mineral nutrients in their organs than cereals (Broadley et al., 2003, Fageria, 2004 and Pedersen et al., 2002). While the mechanism remains unknown, the greater P concentration in *Cyclopia* shoots over the non-legumes *T. bromoides*, *L. strictum* and *E. thyrsoidea* could well be further evidence that N_2-fixing shrub/tree legumes also accumulate more minerals than their non-fixing counterparts. Whatever the case, the greater P assimilation by *C. genistoides*, *C. subternata*, *A. caledonensis* and *A. aspalathoides* over non-legumes at Koksrivier and Kanetberg could suggest that, for these legume species endemic to the low-P fynbos environment, rhizosphere acid/alkaline phosphatase activity is potentially a good measure of P availability.

To further assess if the increase in P supply observed here enhances P nutrition in other fynbos legumes, rhizosphere soils of two *Aspalathus* species growing in the same field at Koksrivier as *C. genistoides* were sampled together with the non-legume *Mimetes cucullatus* and *C. genistoides* as control. The rhizosphere soils were assayed for acid/alkaline phosphatase activity. The data revealed marked differences between and among the four test species, with *C. genistoides* exhibiting significantly higher acid phosphatase activity, followed by *A. caledonensis*, *A. aspalathoides* and *M. cucullatus* (Fig. 1A). Although the rhizosphere soil of *Aspalathus caledonensis* had the second highest P enzyme activity, rhizosphere Pi was lowest for that species (Fig. 1B). However, of the other test plants, the rhizosphere concentration of plant-available Pi was higher in*C. genistoides* and *A. aspalathoides*, and least in the non-legume, *M. cucullatus* (Fig. 1B). Because of the low acid phosphatase activity in the rhizosphere of the non-legume *M. cucullatus*, the level of organic P substrate was much greater compared to *C. genistoides*, *A. caledonensis* and *A. aspalathoides* (Fig. 1C). However, measuring P in shoots revealed significantly greater concentrations in *A. caledonensis* (Fig. 1D) [the species with the lowest rhizosphere Pi], suggesting that the low concentration in the *A. caledonensis* rhizosphere was due to rapid uptake by roots, thus creating a depletion in rhizosphere inorganic P (Fig. 1B). Such variation in P assimilation has been reported for other plant species and genotypes (Betrand et al., 1999, Chen et al., 2002, Hedley et al., 1994, Shi et al., 2004, Zhao et al., 2010, Zoysa et al., 1997 and Zoysa et al., 1999), and could be inherently related to a greater P demand by *A. caledonensis* compared to the other co-occurring fynbos species such as *A. aspalathoides* and *C. genistoides*. As found by Silberbush et al. (1981), the intensity of acid phosphatase enzyme exudation is strongly influenced by the P demand of the plant species.

Is the Expression of Acid/Alkaline Phosphatase Activity Different in the Rhizosphere of Cuttings and Seedlings of Cyclopia Species?

The Honeybush tea farmers in the Cape of South Africa use both seeds and cuttings as planting material in their tea plantations with greater preference for cuttings. From interviews with farmers regarding the choice of Honeybush tea planting material,Gleason (2004) found that farmers have observed a huge variation in tea yield during the second and third year of tea production from seeds. Moreover, the majority of *Cyclopia*plants grown from seeds generally exhibited retarded growth. Furthermore, less than 15% of *C. subternata* plants established from seed were productive by the fifth year of cultivation. By contrast, *C. subternata* plants established from cuttings generally produced twice the dry matter yield of seedlings of the same age.

The aim of this part of the study was to assess whether P nutrition is a factor in the variation of tea yield on farmers' fields. As described before, this was done using bioassays of P enzyme activity in rhizosphere soils, as well as P in shoots of tea plants. The data revealed higher acid and alkaline phosphatase activity in the rhizosphere soil of cuttings than seedlings (Table 4). As a result, there was markedly greater organic and inorganic P concentrations in the rhizosphere soil of cuttings relative to seedlings (Table 4). Possibly due to increased uptake of this available P, shoot P concentrations were also greater in *C. subternata* plants established from cuttings than seedlings. As observed at Koksrivier and Kanetberg, *Cyclopia* grown from seedlings also showed greater P levels in shoots compared to the non-legumes *E. thyrsoidea* and *Erica multiflora*, a finding consistent with data for N_2-fixing grain legumes and non-fixing cereals (Broadley et al., 2003, Fageria, 2004 and Pedersen et al., 2002). At Koksrivier, the shoot P levels of nodulated *A. caledonensis*, *A. aspalathoides* and *C. genistoides*were also greater than those of the non-legume *M. cucullatus* (Fig. 1D), suggesting that even with tree/shrub species, legumes accumulate more nutrient elements than non-legumes.

Using an agricultural legume, Makoi et al (2010) reported significant correlations between plant-available P and acid phosphatase activity in the rhizosphere of cowpea grown in fynbos soils. In this study, we also found a significantly positive correlation between acid phosphatase activity and organic P levels in the rhizosphere of the test legumes (Table 5). This could be interpreted to mean that the more the organic-P substrate in the rhizosphere, the greater the enzyme activity, and hence P availability in the rhizosphere for root uptake. However, with accumulation of inorganic P in the rhizosphere, acid phosphatase activity can get suppressed (Nannipieri et al., 2011 and Treseder and Vitousek, 2001).

Table 5. Correlation analysis of acid phosphatase activity against organic P (Po) and inorganic P (Pi) in rhizosphere soils of *C. genistoides*, *C. subternata*, *Aspalathus caledonensis* and *A. aspalathoides*. Mean values are significantly different at $^*P \leq 0.05$, $^{**}P \leq 0.01$, $^{***}P \leq 0.001$.

Species name/Year	Significance	
	Po	Pi
C. genistoides (1997)	0.9995***	– 0.8995*
C. genistoides (2005)	0.9141*	– 0.8138*
C. subternata (1999)	0.9884***	– 0.9399**
C. subternata (2002)	0.9100*	– 0.8902*
A. aspalathoides	0.9612*	– 0.9940***
Cuttings	0.9100*	– 0.8724*

In conclusion, our results suggest that acid/alkaline phosphatase activity in the rhizosphere of *Cyclopia* and *Aspalathus* species growing in the low-P soils of the Cape fynbos can be used as a predictor of P supply and P assimilation in the rhizosphere of those legumes. We, however, acknowledge that other mechanisms such as cluster root formation and mycorrhizal infection could have contributed to P uptake by the test legumes (Table 1), as cluster root-forming *A. linearis* has been found to accumulate P in addition to other essential nutrient elements in the nutrient-poor Cape fynbos.

ACKNOWLEDGEMENTS

The South African Research Chair in Agrochemurgy and Plant Symbioses, the National Research Foundation, and the Tshwane University of Technology are acknowledged for supporting the research of FDD, and for providing a bursary to STM. We are grateful to the farmers (Mr Fritz Joubert and Mrs Rica Joubert at Koksrivier, and Mr Matie Taljaard and Mrs Erica Taljaard at Kanetberg) for allowing us to collect plant and soil samples from their commercial Honeybush tea plantations.

REFERENCES

1. Allsopp, N., Stock, W.D., 1994. VA mycorrhizal infection in relation to edaphic characteristics and disturbance regime in three lowland plant communities in the south-

western Cape, South Africa. Journal of Ecology 82, 271–279.

2. Asmar, F., Gahoonia, T.S., Nielsen, N.E., 1995. Barley genotypes differ in activity of soluble extracellular phosphatase and depletion of organic phosphorus in the rhizosphere soil. Plant and Soil 172, 117–122.
3. Betrand, I., Hinsinger, P., Jaillard, B., Arvieu, J.C., 1999. Dynamics of phosphorus in the rhizosphere of maize and rape grown on synthetic, phosphate calcite and goethite. Plant and Soil 211, 111–119.
4. Bhandurge, P., N, R., Ganapaty, S., Pattanshetti, S., 2013. The Gnidia genus: a review. Asian Journal of Biomedical & Pharmaceutical Sciences 3, 1–31.
5. Boatwright, J.S., Cupido, C.N., 2011. Aspalathus crewiana (Crotalarieae, Fabaceae), a new species from the Western Cape Province of South Africa. Nordic Journal of Botany 29, 513–517.
6. Broadley, M.R., Bowen, H.C., Cotterill, H.L., Hammond, J.P., Meacham, M.C., Mead, A., White, P.J., 2003. Variation in the shoot calcium content of angiosperms. Journal of Experimental Botany 54, 1431–1446.
7. Chen, C.R., Condron, L.M., Davis, M.R., Sherlock, R.R., 2002. Phosphorus dynamics in the rhizosphere of perennial ryegrass (Lolium perenne L.) and radiata pine (Pinus radiata D. Don.). Soil Biology and Biochemistry 34, 487–499.
8. Cramer, M.D., 2010. Phosphate as a limiting resource: introduction. Plant and Soil 334, 1–10.
9. Criley, R.A., 2000. Leucospermum: botany and horticulture. In: Janick, J. (Ed.), Horticultural Reviews. John Wiley, Oxford, pp. 28–88.
10. DAFF, Department of Agriculture, Fisheries and Forestry, 2012. Report, Honeybush tea, Honeybush tea industry profile. http://www.nda.agric.za (15/09/2012).
11. Dakora, F.D., 2012. Root-nodule bacteria isolated from native Amphithalea ericifolia and four Aspalathus species from the acidic soils of the South African fynbos are tolerant to very low pH. African Journal of Biotechnology 11, 3766–3772.
12. Dakora, F.D., Phillips, D.A., 2002. Root exudates as mediators as mineral acquisition in low-nutrient environments. Plant and Soil 245, 35–47.
13. Du Plessis, S.F., Burger, R.D.T., 1964. A comparison of chemical extraction methods for the evaluation of phosphate availability of top soils. South African Journal of Agricultural Science 8, 11–13.
14. Du Toit, J., Joubert, E., Britz, T.J., 1998. Honeybush tea — a rediscovered indigenous South African herbal tea. Journal of Sustainable Agriculture 12, 67–84.
15. Fageria, N.K., 2004. Dry matter yield and shoot nutrient concentration of upland rice, common bean, corn, and soybean grown in rotation on an Oxisol. Communications in Soil Science and Plant Analysis 35, 961–964.
16. Giron, H.C., 1973. Comparison between dry ashing and wet digestion in the preparation of plant materials for atomic absorption analysis. Atomic Absorption Newsletters 12, 28–29.
17. Gleason, G., 2004. Cuttings are key to commercial Cape honeybush production. Farmers' Weekly No. 94029, 30 July. 40–41.
18. Harnafi, H., Bouanani, N.H., Serghini, C.H., Ghalim, N., Amrani, S., 2007. The hypolipidaemic activity of aqueous Erica multiflora flowers extract in Triton WR—1339 induced hyperlipidaemic rats: a comparison with fenofibrate. Journal of Ethnopharmacology 3, 156–160.
19. Harrison, M.J., 1999. Molecular and cellular aspects of the arbuscular mycorrhizal symbiosis. Plant Physiology and Plant Molecular Biology 50, 361–389.
20. Hebrien, S.A., Neal, J.L., 1990. Soil pH and phosphatase activity. Communications in Soil Science and Plant Analysis 21, 439–456.
21. Hedley, M.J., Kirk, G.J., Santos, M.B., 1994. Phosphorus efficiency and the forms of soil phosphorus utilized by upland rice cultivars. Plant and Soil 158, 53–62.
22. Holford, I.C.R., 1997. Soil phosphorus: its measurement, and its uptake by plants. Australian Journal of Soil Research 35, 227–239. http://www.busrep.co.za/business-news/25/08/10 (23 March 2011). http://www.inarchive.com (12 June 2013). StatSoft Inc, 2005. STATISTICA (data analysis software System), version 7.1. www. statsoft.com.
23. Joubert, E., De Beer, D., 2011. Rooibos (Aspalathus linearis) beyond the farm gate: from herbal tea to potential phytopharmaceutical. South African Journal of Botany 77, 869–886.
24. Kanu, S.A., Dakora, F.D., 2009. Thin-layer chromatographic analysis of lumichrome, riboflavin and indole acetic acid in cell-free culture filtrate of Psoralea nodule bacteria grown at different pH, salinity and temperature regimes. Symbiosis 48, 173–181.
25. Kanu, S.A., Dakora, F.D., 2012. Effect of N and P nutrition on extracellular secretion of lumichrome, riboflavin and indole acetic acid by N2-fixing bacteria and endophytes isolated from Psoralea nodules. Symbiosis 57, 15–22.
26. Kanu, S., Matiru, V.N., Dakora, F.D., 2007. Strain and species differences in rhizobial secretion of lumichrome and riboflavin, measured using thin-layer chromatography. Symbiosis 43, 37–43.
27. Kanu, S.A., Okonkwo, J.O., Dakora, F.D., 2013. Aspalathus linearis (Rooibos tea) as potential phytoremediation agent: a review on tolerance mechanisms for aluminium uptake. Environmental Reviews 21, 85–92.
28. Kochian, L.V., Hoekenga, O.A., Pineros, M.A., 2004. How do crop plants tolerate acid soils? Mechanism of aluminium tolerance and phosphorus efficiency. Annual Review of Plant Biology 55, 459–493.
29. Kubitzki, K., Huber, H., 1998. Flowering plants, monocotyledons: Lilianae (except Orchidaceae). In: Kubitzki, K. (Ed.), The Families and Genera of Vascular Plants. , vol. 3. Springer, Berlin, Heidelberg.
30. Lambers, H., Shane, M.W., 2007. Role of root clusters in phosphorus acquisition and increasing biological diversity in agriculture. In: Spiertz, J.H.J., Struik, P.C., Van Laar, H.H. (Eds.), Scale and Complexity in Plant

Systems Research: Gene-PlantCrop Relations. Springer, Dordrecht, The Netherlands, pp. 237–250.

31. Lambers, H., Shane, M.W., Cramer, M.D., Pearse, S.J., Veneklaas, E.J., 2006. Root structure functioning for efficient acquisition of phosphorus: matching morphological and physiological traits. Annals of Botany 98, 693–713.
32. Lambers, H., Raven, J.A., Shaver, G.R., Smith, S.E., 2008. Plant nutrient-acquisition strategies change with soil age. Trends in Ecology & Evolution 23, 95–103.
33. Lamont, B., 1982. Mechanisms for enhancing nutrient uptake in plants, with particular reference to Mediterranean South Africa and Western Australia. The Botanical Review 48, 597–689.
34. Makoi, J.H.J.R., Chimphango, S.B.M., Dakora, F.D., 2010. Elevated levels of acid and alkaline phosphatase activity in roots and rhizosphere of cowpea (Vigna unguiculata L. Walp.) genotypes grown in mixed culture and at different densities with sorghum (Sorghum bicolor L.). Crop & Pasture Science 61, 279–286.
35. Maseko, S.T., 2013. N2 fixation, P nutrition, C and mineral accumulation, and δ13C of nodulated natural stands of legumes in the Cape fynbos, South Africa. D-Tech. Agric. (Crops Sciences) Thesis Tshwane University of Technology.
36. Maseko, S.T., Dakora, F.D., 2013. Plant enzymes, root exudates, cluster roots and mycorrhizal symbiosis are the drivers of P nutrition in native legumes growing in P deficient soil of the Cape fynbos in South Africa. Journal of Agricultural Science and Technology A 3, 331–340.
37. Moufhe, M.L., Dakora, F.D., 1999. Nitrogen nutrition in nodulated field plants of the shrub tea legume Aspalathus linearis assessed using 15N natural abundance. Plant and Soil 209, 181–186.
38. Mózner, Z., Tabi, A., Csutora, M., 2012. Modifying the yield factor based on more effi- cient use of fertilizer-the environmental impacts of intensive agricultural practices. Ecological Indicators 16, 58–66.
39. Nannipieri, P., Giagnoni, L., Landi, L., Renetla, G., 2011. Role of phosphatase enzymes in soil. In: Bünemann, E.K., Oberson, A., Frossard, E. (Eds.), Phosphorus in action, soil biology 26. Springer, Berlin, pp. 215–243.
40. Pedersen, G.A., Brink, G.E., Fairbrother, T.E., 2002. Nutrient uptake in plant parts of sixteen forages fertilized with poultry litter. Agronomy Journal 94, 895–904.
41. Power, S.C., Cramer, M.D., Verboom, G.A., Chimphango, S.B.M., 2010. Does phosphate acquisition constrain legume persistence in the fynbos of the Cape Floristic Region? Plant and Soil 334, 33–46.
42. Radersma, S., Grierson, P.F., 2004. Phosphorus mobilization in agroforestry: Organic anions, phosphatase activity and phosphorus fractions in the rhizosphere. Plant and Soil 259, 209–219.
43. Reinten, E.Y., Coetzee, J.H., Van Wyk, B.-E., 2011. The potential of South African indigenous plants for the international cut flower trade. South African Journal of Botany 77, 934–946.
44. Richardson, A.E., Barea, J.M., Mcneill, A.M., Prigent-Combaret, C., 2009. Acquisition of phosphorus and nitrogen in the rhizosphere and plant growth promotion by microorganisms. Plant and Soil 321, 305–339.
45. Rogosic, J., Pfister, J.A., Provenza, F.D., Grbesa, D., 2006. Sheep and goat preference for and nutritional value of Mediterranean maquis shrubs. Small Ruminant Research 64, 169–179.
46. Shi, W.M., Wang, X.C., Yan, W.D., 2004. Distribution patterns of available P and K in rape rhizosphere in relation to genotypic difference. Plant and Soil 261, 11–16.
47. Silberbush, M., Shomer-Ilan, A., Waisel, Y., 1981. Root surface phosphatase activity in ecotypes of Aegilops peregrine. Physiologia Plantarum 53, 501–504.
48. Spriggs, A.C., 2004. Symbiotic N2 fixation in Cyclopia Vent. spp. (honeybush): Towards sustainable cultivation in the Western Cape of South Africa. Doctoral degree Thesis University of Cape Town.
49. Spriggs, A.C., Dakora, F.D., 2008. Field assessment of symbiotic N2 fixation in wild and cultivated Cyclopia species in the South African fynbos by 15N natural abundance. Tree Physiology 29, 239–247.
50. Spriggs, A.C., Stock, W.D., Dakora, F.D., 2003. Influence of mycorrhizal associations on foliar δ15N values of legume and non-legume shrubs and trees in the fynbos of South Africa: implications for estimating N2 fixation using the 15N natural abundance method. Plant and Soil 255, 495–502.
51. Straker, C.J., 1996. Ericoid mycorrhiza: ecological and host specificity. Mycorrhiza 6, 215–225.
52. Tabatabai, M.A., 1994. Soil enzymes. In: Weaver, R.W., Angle, S., Bottomley, P., Bezdicek, D., Smith, S., Tabatabai, A., Wollum, A. (Eds.), Methods of soil analysis. Part 2. Microbiological and biochemical properties. Soil Science Society of America, Madison, pp. 775–833.
53. Tarafdar, J.C., Jungk, A., 1987. Phosphatase activity in the rhizosphere and its relation to the depletion of soil organic phosphorus. Biology and Fertility of Soils 3, 199–204.
54. Tarafdar, J.C., Yadav, R.S., Meena, S.C., 2001. Comparative efficiency of acid phosphatase originated from plant and fungal sources. Journal of Plant Nutrition and Soil Science 164, 279–282.
55. Tran, H.T., Hurley, B.A., Plaxton, W.C., 2010. Feeding hungry plants: the role of purple acid phosphatases in phosphate nutrition. Plant Science 179, 14–27.
56. Treseder, K.K., Vitousek, P.M., 2001. Effects of soil nutrient availability on investment in acquisition of N and P in Hawaiian rain forests. Ecology 82, 946–954.
57. Van Wyk, B.-E., 2011. The potential of South African plants in the development of new food and beverage products. South African Journal of Botany 77, 857–868.
58. Vance, C.P., Uhde-Stone, C., Allan, D.L., 2003. Phosphorus acquisition and use: critical adaptations by plants for securing a non renewable resource. The New Phytologist 157, 423–447.

59. Witkowski, E.T.F., 1991. Growth and competition between seedlings of Protea repens (L.) L. and the alien invasive, Acacia saligna (Labill.) Wendl. in relation to nutrient availability. Functional Ecology 5, 101–110.

60. Zhao, Q., Zeng, D.H., Fan, Z.P., 2010. Nitrogen and phosphorus transformations in the rhizospheres of three tree species in a nutrient-poor sandy soil. Applied Soil Ecology 46, 341–346.

61. Zoysa, A.K.N., Loganathan, P., Hedley, M.J., 1997. A technique for studying rhizosphere processes in tree crops: soil phosphorus depletion around camellia (Camellia japonica L.) roots. Plant and Soil 190, 253–265.

62. Zoysa, A.K.N., Loganathan, P., Hedley, M.J., 1999. Phosphorus utilization efficiency and depletion of phosphate fractions in the rhizosphere of three tea (Camellia sinensis L.) clones. Nutrient Cycling in Agroecosystems 53, 189–201.

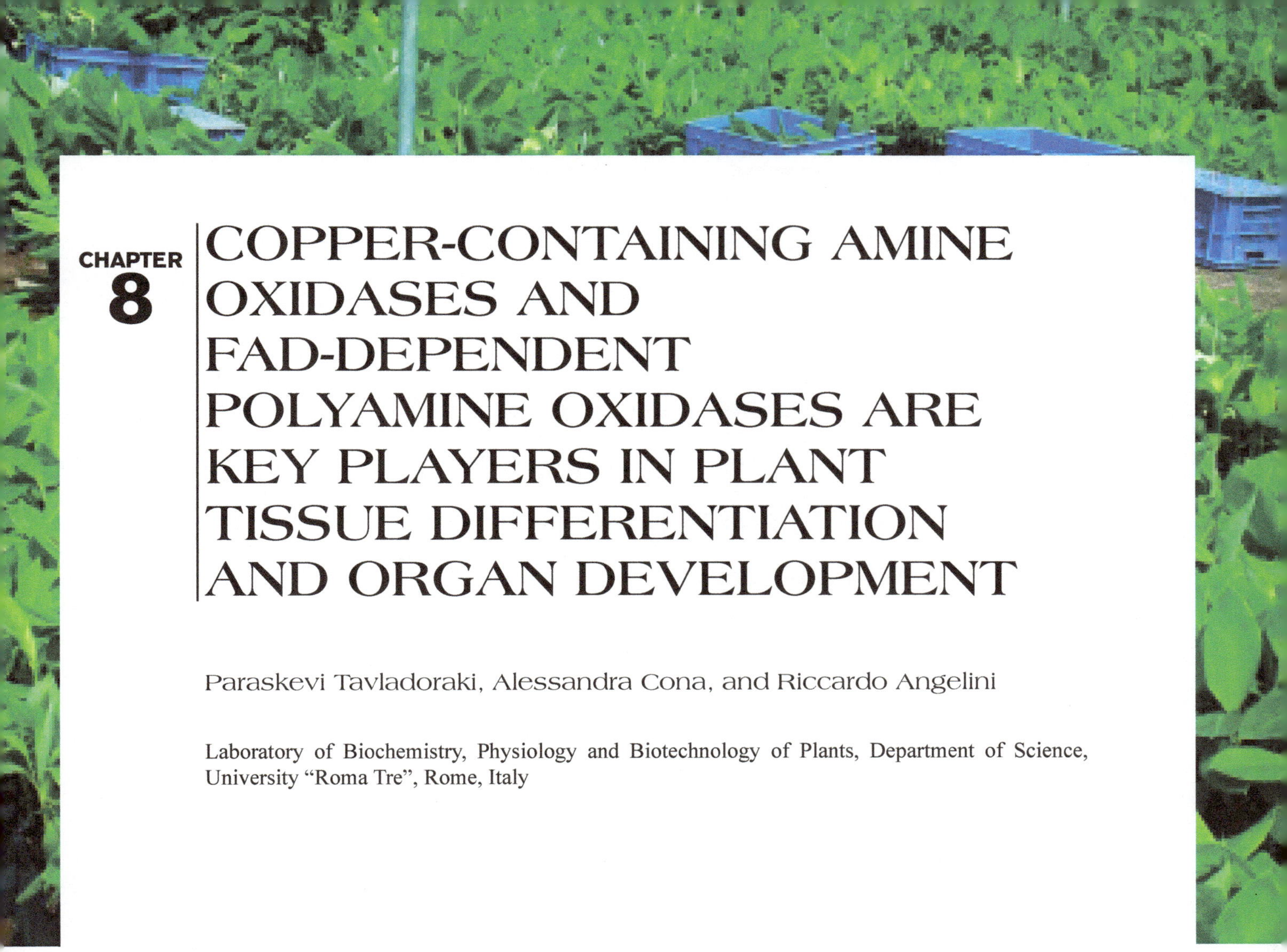

CHAPTER 8

COPPER-CONTAINING AMINE OXIDASES AND FAD-DEPENDENT POLYAMINE OXIDASES ARE KEY PLAYERS IN PLANT TISSUE DIFFERENTIATION AND ORGAN DEVELOPMENT

Paraskevi Tavladoraki, Alessandra Cona, and Riccardo Angelini

Laboratory of Biochemistry, Physiology and Biotechnology of Plants, Department of Science, University "Roma Tre", Rome, Italy

Plant polyamines are catabolized by two classes of amine oxidases, the copper amine oxidases (CuAOs) and the flavin adenine dinucleotide (FAD)-dependent polyamine oxidases (PAOs). These enzymes differ to each other in substrate specificity, catalytic mechanism and subcellular localization. CuAOs and PAOs contribute to several physiological processes both through the control of polyamine homeostasis and as sources of biologically-active reaction products. CuAOs and PAOs have been found at high level in the cell-wall of several species belonging to Fabaceae and Poaceae families, respectively, especially in tissues fated to undertake extensive wall loosening/stiffening events and/or in cells undergoing programmed cell death (PCD). Apoplastic CuAOs and PAOs have been shown to play a key role as a source of H_2O_2 in light- or developmentally-regulated differentiation events, thus influencing cell-wall architecture and maturation as well as PCD. Moreover, growing evidence suggests a key role of intracellular CuAOs and PAOs in several facets of plant development. Here, we discuss recent advances in understanding the contribution of different CuAOs/PAOs, as well as their cross-talk with different intracellular and apoplastic metabolic pathways, in tissue differentiation and organ development.

Citation: Tavladoraki P, Cona A and Angelini R (2016) Copper-Containing Amine Oxidases and FAD-Dependent Polyamine Oxidases Are Key Players in Plant Tissue Differentiation and Organ Development. Front. Plant Sci. 7:824. doi: 10.3389/fpls.2016.00824

COPPER-CONTAINING AMINE OXIDASES AND FAD-DEPENDENT POLYAMINE OXIDASES: A COMPLEX NETWORK

In plants, the polyamines (PAs) putrescine (Put), cadaverine (Cad), spermidine (Spd), spermine (Spm), and thermospermine (Therm-Spm) are involved in several physiological processes, such as cell proliferation, differentiation and defense responses (Takahashi and Kakehi, 2010; Marina et al., 2013; Jiménez-Bremont et al., 2014; Tiburcio et al., 2014; Pál et al., 2015; Strohm et al., 2015; Yoshimoto et al., 2016). PAs are oxidized by a heterogeneous class of enzymes which includes copper-containing amine oxidases (CuAOs) and FAD-dependent polyamine oxidases (PAOs) (Cona et al., 2006a; Angelini et al., 2010; Tavladoraki et al., 2012). CuAOs oxidize mainly Put and Cad, and less efficiently Spd and Spm at the primary amino groups, producing ammonia, H_2O_2 and an aminoaldehyde and are thus considered involved in PA terminal catabolism. In *Arabidopsis thaliana* 10 *CuAO* genes are present, among which only eight encode for putative functional CuAOs [*AtCuAOα1* (*At1g31670*); *AtCuAOα2* (*At1g31690*); *AtCuAOα3* (*At1g31710*; previously *AtCuAO2*); *AtCuAOβ* (*At4g14940*; prev. *ATAO1* or *AtAO1*); *AtCuAOγ1* (*At1g62810*; prev. *AtCuAO1*);*AtCuAOγ2* (*At3g43670*); *AtCuAOδ* (At4g12290, prev. *AtCuAOδ2*); *AtCuAOζ* (*At2g42490*; prev. *AtCuAO3* or *AtCuAO1*)][1](Figure 1). The remaining two genes *AtCuAOε1* (*At4g12270*; prev. *AtCuAOε*) and *AtCuAOε2* (*At4g12280*; prev.*AtCuAOδ1*) are consecutive fragments of a copy of *AtCuAOδ* gene. Phylogenetic analysis evidenced that plant

CuAOs form three clades (I-III), clade I being composed of three subgroups (Ia-Ic) and clade II of two (IIa and IIb; Figure 1A). Furthermore, genomic sequence analysis demonstrated that the Arabidopsis CuAOs of clades I and II, but not of clade III, present a similar gene structure to each other with three introns at conserved positions (Figure 1B). This suggests that *AtCuAOα1* to *AtCuAOδ* are recent derivatives of a common ancestor.

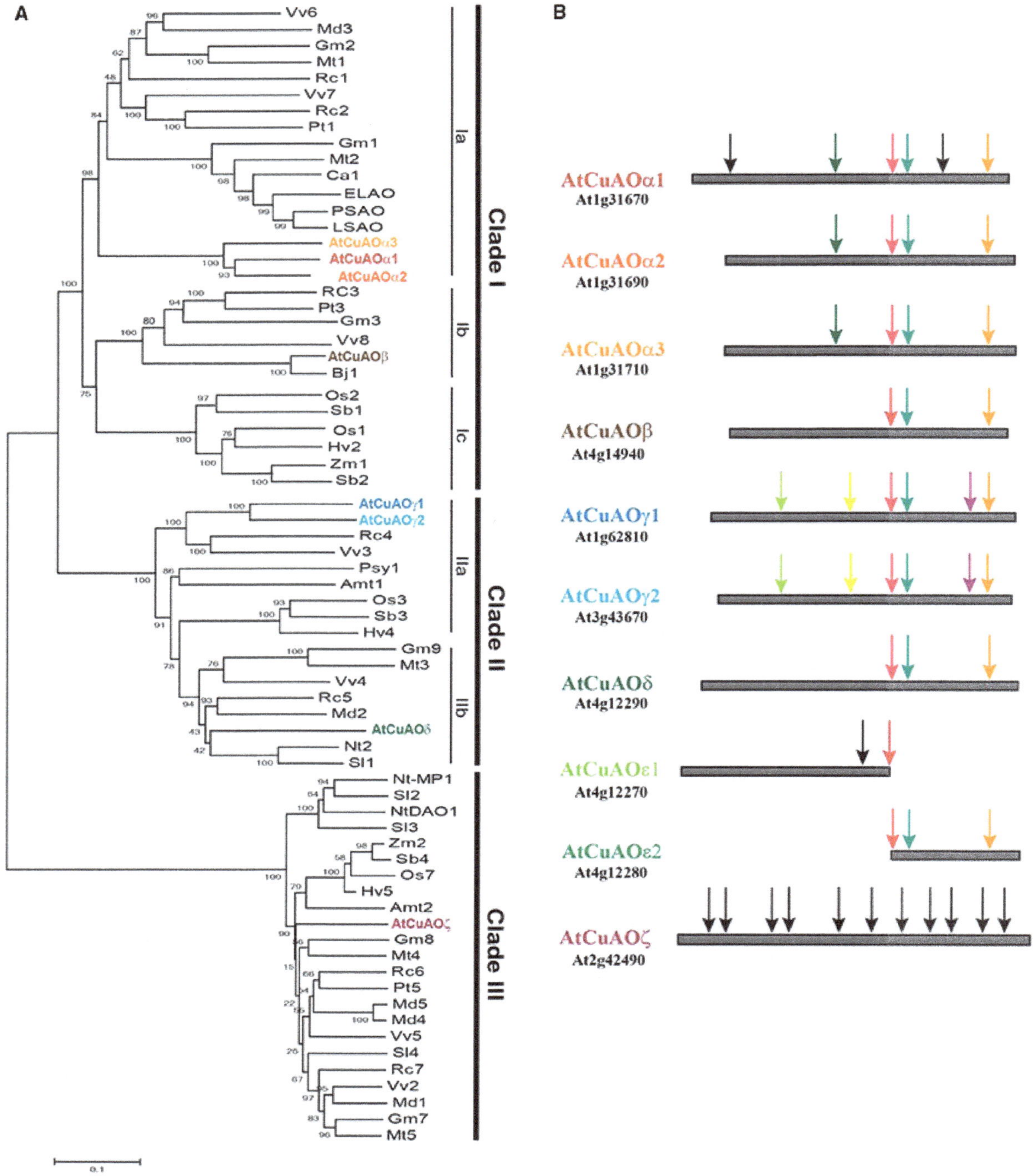

Figure 1. Sequence analysis of plant CuAOs. (A) Phylogenetic analysis of CuAOs from selected spermatophytes. Plant CuAOs form three principal supported clades (I-III). Clade I consists of three groups (groups a and b consisting of CuAOs from dicots and group c from monocots), the reciprocal relationship of which is not well resolved. For simplicity reasons, for each distinct group only a representative CuAO from each plant species was considered. Amino acid sequences were aligned with ClustalW (McWilliam et al., 2013) and phylogenetic analysis was performed using MEGA5 (Tamura et al., 2011) software with the neighbor-joining algorithm. Bootstrap values obtained with 1000 replicates are indicated at the nodes. (B)Genomic sequence analysis of Arabidopsis *CuAOs* (*AtCuAOs*). Black arrows show not conserved intron positions, while arrows of the same color indicate conserved intron positions. All *AtCuAOs* of clades I and II have three introns at conserved positions which suggests a recent common ancestor. However, based on the presence of additional introns, some of them placed at positions conserved among the members of the same group, but not among those of different groups, an independent evolution of the *AtCuAOs* from the different groups can be suggested. *AtCuAOζ* of clade III appears evolutionarily distant from *AtCuAOs* of clade I and II. Amt, *Amborella trichopoda*; Bj, *Brassica juncea*; Ca, *Cicer arietinum*; ELAO, CuAO from latex of*Euphorbia characias*; Gm, *Glycine max*; Hv, *Hordeum vulgare*; LSAO, CuAO from seedlings of *Lens culinaris*; Md, *Malus domestica*; Mt,*Medicago truncatula*; Nt, *Nicotiana tabacum*; Os, *Oryza sativa*; PSAO, CuAO from seedlings of *Pisum sativum*; Psy, *Pinus sylvestris*; Pt, *Populus trichocarpa*; Rc, *Ricinus communis*; Sb, *Sorghum bicolor*; Sl, *Solanum lycopersicum*; Vv, *Vitis vinifera*; Zm, *Zea mays*.

AtCuAOβ (clade Ic), AtCuAOγ1 (clade IIa) as well as *Pisum sativum, Lens culinaris* and *Euphorbia characias* CuAOs (PSAO, LSAO, and ELAO, respectively; clade Ia) are localized in the apoplast (Rossi et al., 1992; Tipping and McPherson, 1995; Møller and McPherson, 1998; Padiglia et al., 2002; Boudart et al., 2005; Planas-Portell et al., 2013), whereas AtCuAOζ, *Malus domestica* CuAO1 (MdAO1) and the other members of clade III in peroxisomes (Planas-Portell et al., 2013; Naconsie et al., 2014; Qu et al., 2014; Zarei et al., 2015a; Table 1). Peroxisomal localization was also shown for AtCuAOα3 (Planas-Portell et al., 2013) despite the apparent lack of canonical signal for peroxisomal localization and the fact that it is clustered together with the extracellular PSAO and LSAO. For MdAO2, which is clustered together with AtCuAOδ in clade IIb, both intracellular and apoplastic localization was shown (Zarei et al., 2015a).

Table 1. Characteristics and functions of plant CuAOs and PAOs.

		Localization	Substrate Preference	Function	References
COPPER AMINE OXIDASES					
Clade Ia	AtCuAOα1[a]	-	-	-	-
	AtCuAOα2	-	-	-	-
	AtCuAOα3	Peroxisomes	Put, Spd	-	Planas-Portell et al., 2013
	PSAO	Apoplast	Put, Spd, Spm	-	Tipping and McPherson, 1995; Moschou et al., 2012
	ELAO	Apoplast	Put, Benzylamine, Tyramine	-	Pintus et al., 2013
	LSAO	Apoplast	Put, Spd, Spm, Tryptamine	-	Rossi et al., 1992; Medda et al., 1997; Tavladoraki et al., 2012
	CaCuAO	Apoplast	Put	Wound healing, defense response	Rea et al., 2002
Clade Ib	AtCuAOβ	Apoplast	Put, Spd	Vascular development	Møller and McPherson, 1998; Ghuge et al., 2015a
Clade IIa	AtCuAOγ1	Apoplast	Put, Spd	PA- and ABA-mediated NO production	Wimalasekera et al., 2011; Planas-Portell et al., 2013
	AtCuAOγ2	-	-	-	-
Clade IIb	AtCuAOδ	-	-	-	-
	MdAO2	Apoplast	2-Phenylethylamine, Tyramine, Ethanolamine, Ethylamine, Tryptamine	Fruit flavor, flower fragance	Zarei et al., 2015a
Clade III	AtCuAOζ		Put, Spd, *N*-methyl-Put, Cad	ABA-induced stomatal closure	Planas-Portell et al., 2013; Naconsie et al., 2014; Qu et al., 2014
	NtDAO1	Peroxisomes	Put, *N*-methyl-Put, Cad	-	Naconsie et al., 2014
	MdAO1		Dap, Put, Cad	-	Zarei et al., 2015a
	Nt-MPO1		*N*-methyl-Put, *N*-methyl-Dap, Put, Dap, Cad	Alkaloid synthesis	Katoh et al., 2007; Naconsie et al., 2014
POLYAMINE OXIDASES					
Clade I	AtPAO1	Cytosol	Nor-Spm, Therm-Spm, Spm	Stress response	Tavladoraki et al., 2006; Takahashi et al., 2010; Sagor et al., 2016
	GhPAO1		Spm	Defense response, differentiation of embryogenic callus	Cheng et al., 2015; Mo et al., 2015
Clade II	ZmPAO1	Apoplast	Spd, Spm	Cell wall differentiation	Cona et al., 2006a
	HvPAO1	Apoplast	-	-	Cervelli et al., 2001; Cona et al., 2006a
	HvPAO2	Vacuole	Spm, Spd	-	Cervelli et al., 2001; Cona et al., 2006a
	OsPAO7	Apoplast	Spm, Spd, N^1-acetyl-Spm	-	Liu et al., 2014b
Clade III	AtPAO5		Spm, Therm-Spm, Nor-Spm, N^1-acetyl-Spm	Polyamine homeostasis, plant growth, stress response	Ahou et al., 2014; Kim et al., 2014; Zarza et al., 2016
	SelPAO5		Therm-Spm, Spm, Nor-Spm, N^1-acetyl-Spm	-	Sagor et al., 2015
	OsPAO1	Cytosol	Spm, Therm-Spm, Nor-Spm, N^1-acetyl-Spm	Plant growth	Liu et al., 2014a,c
	BjPAO1		-	Shoot regeneration	Lim et al., 2006
	GhPAO4		-	Differentiation of embryogenic callus	Cheng et al., 2015
Clade IV	AtPAO2		Spm, Spd, Nor-Spm	Stress response	Moschou et al., 2008b; Takahashi et al., 2010; Fincato et al., 2011; Wimalasekera et al., 2015; Sagor et al., 2016
	AtPAO3	Peroxisomes	Spd, Spm, Nor-Spm	Pollen tip growth	Moschou et al., 2008b; Takahashi et al., 2010; Fincato et al., 2011; Wu et al., 2010
	AtPAO4		Spm	Senescence	Moschou et al., 2008b; Kamada-Nobusada et al., 2008; Takahashi et al., 2010; Fincato et al., 2011; Sequera-Mutiozabal et al., 2016
	OsPAO3		Spd, Nor-Spm	-	Ono et al., 2012
	OsPAO4		Spm, Therm-Spm, Nor-Spm	-	Ono et al., 2012
	OsPAO5		Spm, Therm-Spm, Nor-Spm	-	Ono et al., 2012

[a] *At, Arabidopsis thaliana; Bj, Brassica juncea; Ca, Cicer arietinum; Gh, Gossypium hirsutum; Hv, Hordeum vulgare; Md, Malus domestica; Nt, Nicotiana tabacum; Os, Oryza sativa; Sel, Selaginella lepidophylla; Sl, Solanum lycopersicum; Zm, Zea mays. ELAO, LSAO, and PSAO: CuAO from Euphorbia characias latex. Lens culinaris seedlings and Pisum sativum seedlings, respectively.*

Although, most of the so far characterized CuAOs, such as AtCuAOβ, AtCuAOγ1, AtCuAOα3, AtCuAOζ, PSAO, LSAO, and *Nicotiana tabacum* CuAO1 (NtDAO1), oxidize mainly Put, Cad, and Spd (Rossi et al., 1992; Tipping and McPherson, 1995; Møller and McPherson, 1998; Planas-Portell et al., 2013; Naconsie et al., 2014), MdAO1 of clade III shows preference for 1,3-diaminopropane (Dap), having no activity with Spd (Zarei et al., 2015a). Furthermore, AtCuAOζ and NtDAO1 oxidize also *N*-methyl-Put, though less efficiently than the non-methylated diamine. Thus, they differ from the*N. tabacum* *N*-methylputrescine oxidase (Nt-MPO1), which shows preference for *N*-methyl-Put and is involved in nicotine biosynthesis, although all three proteins are clustered together in clade III (Heim et al., 2007; Katoh et al., 2007; Dewey and Xie, 2013; Naconsie et al., 2014). This indicates that clade III consists of a heterogeneous group of CuAOs. Another remarkable finding is the higher catalytic activity of MdAO2 with monoamines, such as 2-phenylethylamine, tyramine and tryptamine, than with Put and Spd (Zarei et al., 2015a). Interestingly, tyramine is also a substrate of ELAO whereas tryptamine and other indoleamines are both substrates and inhibitors of LSAO (Medda et al., 1997; Pintus et al., 2013). It was speculated that 2-phenylacetaldehyde produced by MdAO2-mediated oxidation of 2-phenylethylamine may be converted in fruits to 2-phenylethanol, a volatile compound that is a major contributor to fruit flavor and flower fragrance. It is also possible that 4-hydroxyphenylacetaldehyde produced by tyramine oxidation is involved in benzylisoquinoline alkaloid biosynthesis in plants (Zarei et al., 2015a).

PAOs oxidize the secondary amino groups of a series of PAs and reaction products depend on the catalytic mechanism and substrate specificity. The apoplastic PAOs oxidize the carbon at the *endo*-side of the N^4 atom of Spd and Spm producing Dap, H_2O_2, and an aminoaldehyde (Tavladoraki et al., 1998; Cervelli et al., 2001; Liu et al., 2014b), whereas all the intracellular PAOs oxidize the carbon at the *exo*-side of the N^4 atom of Spd or Spm, to produce Put or Spd, respectively, together with H_2O_2 and 3-aminopropanal (Tavladoraki et al., 2006; Kamada-Nobusada et al., 2008;Moschou et al., 2008b; Fincato et al., 2011; Ahou et al., 2014; Kim et al., 2014; Liu et al., 2014a; Mo et al., 2015). Some of the intracellular PAOs are also able to oxidize Therm-Spm and norspermine (Nor-Spm) with the production of Spd and norspermidine (Nor-Spd), respectively (Tavladoraki et al., 2006; Fincato et al., 2011; Kim et al., 2014; Liu et al., 2014a). Recently, a *Selaginella lepidophylla* PAO (SelPAO5) was shown to produce Nor-Spd from Therm-Spm (Sagor et al., 2015). These differences in reaction products reflect differences in position and orientation of the substrate inside the catalytic site. Both the *exo*- and *endo*-mode of PA oxidation produce a biologically active diamine or triamine which can be converted to higher PAs. Indeed, even Dap, which has a role in the control of stomata movement through its acetylated form (Jammes et al., 2014), can be converted by aminopropyltransferases to Nor-Spd and subsequently to Nor-Spm, two PAs correlated to stress tolerance (Kuehn et al., 1990; Fuell et al., 2010; Sagor et al., 2015). In this way, all PAOs can be considered involved in PA back-conversion. This view changes the prevailing idea that the PAOs with an*endo*-mode of substrate cleavage are involved in PA terminal catabolism, thus attributing to CuAOs the role of PA terminal catabolism which permits nitrogen and carbon re-assimilation to various biochemical reactions (Moschou et al., 2012).

In Arabidopsis five PAOs are present (AtPAO1-AtPAO5), which are localized intracellularly and show an *exo*-mode of substrate oxidation. AtPAO1 and AtPAO5 present cytosolic localization and a preference for Spm, Therm-Spm, and Nor-Spm, as substrates (Tavladoraki et al., 2006; Ahou et al., 2014; Kim et al., 2014; Liu et al., 2014a). AtPAO5 accepts alsoN^1-acetyl-Spm as a substrate and appears to be a peculiar PAO, having a better activity as a dehydrogenase rather than as an oxidase (Ahou et al., 2014). AtPAO2, AtPAO3, and AtPAO4 are localized in the peroxisomes and oxidize both Spd and Spm (Kamada-Nobusada et al., 2008; Moschou et al., 2008b; Takahashi et al., 2010; Fincato et al., 2011; Sequera-Mutiozabal et al., 2016). Furthermore, AtPAO2, AtPAO3, and AtPAO4 present similar gene structures and tissue-specific expression patterns (root tips, guard cells and pollen grains; Takahashi et al., 2010; Fincato et al., 2012). Phylogenetic studies showed that PAOs are divided into four major clades (Liu et al., 2014a; Wang and Liu, 2015). Worth noticing is the fact that the PAOs of the same clade present some common characteristics (Table 1). In particular, clade I PAOs have cytosolic localization and oxidize specifically Spm but not Spd, while clade IV PAOs present peroxisomal localization and specificity for either Spm or Spd or both. The apoplastic PAOs of clade II show preference for both Spm and Spd and are characterized by a k_{cat} value at least 10-fold higher than that of the PAOs of the other clades. Clade III PAOs are cytosolic enzymes which recognize Spm, Therm-Spm, Nor-Spm, and N^1-acetyl Spm as substrates and are regulated by PAs (Ahou et al., 2014; Kim et al., 2014; Liu et al., 2014a; Wang and Liu, 2015). Furthermore, they present very low k_{cat} values as oxidases, which indicate that not only AtPAO5 but all AtPAO5 orthologs may have activity mainly as dehydrogenases (Ahou et al., 2014).

CELL-WALL AMINE OXIDASES: DRIVING ROS PRODUCTION IN THE APOPLASTIC "HUB"

Compelling evidence supports a key role for reactive oxygen species (ROS; superoxide anion, O·−2O2•-; H_2O_2; hydroxyl radical, ˙OH; singlet oxygen, $^1O_2$) and nitric oxide (NO) in orchestrating developmental processes, as well as in being involved in signaling of both local and systemic defense responses in plants. The apoplast is a major "hub" for these chemical species. Their accumulation in large amounts and the complexity of the regulatory mechanisms involved in their biosynthesis reflect the peculiar role of this compartment in physiological events that depend on temporarily regulated and spatially restricted ROS and NO signatures (Kärkönen and Kuchitsu, 2015). Indeed, ROS are key players in cell-wall loosening and stiffening, as well as in developmental cell death, and stress-related events, such as the two-phase oxidative burst, wound-healing and the hypersensitive response (De Pinto et al., 2012; Kärkönen and Kuchitsu, 2015). Furthermore, ROS generated in the apoplast may well influence Ca^{2+} transport across plasma membrane thus inducing a multiplicity of Ca^{2+}-mediated responses (Gilroy et al., 2014; Pottosin et al., 2014). Systems involved in ROS biosynthesis in the apoplast include plasma membrane NADPH oxidases and quinone reductases, cell-wall peroxidases (PODs), oxalate oxidases, amine oxidases (Kärkönen and Kuchitsu, 2015). Non enzymatic

·OH production can be also driven by Fenton-type reaction dependent on a reducing agent (for example ascorbate), transition metal ions and H_2O_2(Schopfer et al., 2002; Müller et al., 2009). Although, a very significative amount of data suggests NADPH oxidases and/or PODs to have a prevalent role in ROS production in response to microbial pathogens, wounding, as well as in development-related events, the contribution of other systems has been largely underestimated (Cona et al., 2006a,b;Monshausen et al., 2007; Angelini et al., 2008; Kärkönen and Kuchitsu, 2015; Roach et al., 2015). This consideration derives mainly from the analysis of literature data based on extensive use of diphenyleneiodonium (DPI), sodium azide or potassium cyanide and diethyldithiocarbamate (DDC) as diagnostic tools for the involvement of NADPH oxidase, POD or superoxide dismutase activity, respectively, in ROS production. However, these compounds are inhibitors of CuAOs (DDC, azide, cyanide) and PAOs (DPI) as well (Cona et al., 2006b; Medda et al., 2009). Noteworthy, a cross-talk has been recently suggested between AtPAO3 and NADPH oxidase activity, affecting ROS homeostasis and respiration rate in Arabidopsis (Andronis et al., 2014). On the other hand, several evidences support a role of PAs not only in ROS production, but also in that of NO (Tun et al., 2006; Pál et al., 2015). Indeed, loss-of-function *atcuaoγ1* and *atpao2*mutants are impaired in PA- and/or abscisic acid (ABA)-mediated NO production (Wimalasekera et al., 2011, 2015).

POLYAMINE OXIDATION IN THE APOPLAST: CELL-WALL MODIFICATIONS AND GROWTH REGULATION

Early studies in legumes, as well as in maize and tobacco plants evidenced a close correlation between tissue distribution of cell-wall associated CuAO, PAO, and POD activities (Federico and Angelini, 1991; Rea et al., 2002;Paschalidis and Roubelakis-Angelakis, 2005). These studies, also supported by the use of mechanism-based specific inhibitors, suggested a role of PA oxidation in providing H_2O_2 for POD activity during cell-wall cross-linking, lignification and/or suberization processes taking place in the course of organ development, de-etiolation, or wound-healing (Cona et al., 2003, 2006a, 2014; Angelini et al., 2008). A role for lignin synthesis in anther cell-walls has been also hypothesized for an *Oryza sativa* PAO (OsPAO7; Liu et al., 2014b). The major concern raised by these studies comes from the low levels or even absence of PAs in the apoplast under physiological growth conditions (Rea et al., 2004). However, this issue was overcome by the observation that PAs are secreted in the apoplastic milieu during specific developmental phases or under biotic and abiotic stress conditions (Yoda et al., 2003; Moschou et al., 2008a;Rodríguez et al., 2009). On the other hand, in *Zea mays*, PAO-mediated H_2O_2 production in the apoplast significantly contributes to leaf blade elongation, possibly through Fenton and/or Haber-Weiss type reactions driving ·OH synthesis (Rodríguez et al., 2009). A similar role has been suggested in *Glycine max* hypocotyls for the putative apoplastic GmCuAO1 (Delis et al., 2006). This contribution is particularly important under salt stress conditions which inhibit both NADPH oxidases and apoplastic POD activities and induce secretion of PAs in the apoplast (Rodríguez et al., 2009;Campestre et al., 2011; Shoresh et al., 2011).

APOPLASTIC CUAOS AND PAOS: DEVELOPMENTAL PCD AND ROOT XYLEM DIFFERENTIATION

It is well established that ROS have a key role in programmed cell death (PCD) in plants (De Pinto et al., 2012). Although, cell-wall maturation and lignification is intimately connected to the cell-death phase of xylem tissue differentiation, it is difficult to distinguish specific ROS contribution to each specific event (Bollhöner et al., 2012). Notably, transition between cell proliferation and tissue differentiation in the root is regulated independently by either hormonal balance (auxin and citokinin) or ROS (O·−2O2•- and H_2O_2) distribution (Tsukagoshi et al., 2010). Several lines of evidence links PA oxidation in the apoplast and concomitant ROS production with cell-wall maturation and developmental PCD in particular during xylem tissue differentiation. Møller and McPherson (1998) demonstrated that*AtCuAOβ* expression in root xylem tissues precedes and overlaps with lignin synthesis in Arabidopsis. Furthermore, although the roots of loss-of-function *atcuaoβ* mutants did not display evident changes as compared to wild-type plants during normal growth, methyl jasmonate induced early protoxylem differentiation in roots of wild-type plants, concomitantly with increased *AtCuAOβ* expression levels, decreased Put levels and H_2O_2 accumulation. As these events do not occur in *atcuaoβ* roots, a role of *AtCuAOβ* in protoxylem differentiation under stress condition can be suggested (Ghuge et al., 2015a,c). On the other hand, tobacco plants over-expressing a fungal endopolygalacturonase, releasing cell-wall derived oligogalacturonides and mediating both developmental events and defense responses, showed higher CuAO activity, lower Put level, H_2O_2 accumulation and an earlier root xylem differentiation. These events were reverted by 2-bromoethylamine, a mechanism-based CuAO inhibitor (Cona et al., 2014). Furthermore, it has been observed that active ZmPAO and H_2O_2 accumulate in root xylem and xylem parenchyma tissues early during their differentiation (Tisi et al., 2011). *N*-prenylagmatine, a specific PAO inhibitor, inhibited secondary cell-wall deposition, while exogenous Spd induced DNA fragmentation and nuclei condensation, thus suggesting a role for PAO in providing H_2O_2 during secondary wall deposition and developmental PCD in xylem tissue (Tisi et al., 2011). Moreover, over-expression of *ZmPAO1*, as well as down-regulation of the gene encoding S-adenosyl-L-methionine decarboxylase via RNAi in tobacco plants promoted vascular cell differentiation and induced PCD in root cap cells suggesting that the balance between intracellular PA anabolism and apoplastic catabolism is an integrated signaling system coordinating PCD or stress tolerance (Moschou et al., 2008a; Tisi et al., 2011).

PEROXISOMAL PAOS: INVOLVEMENT IN POLLEN TUBE AND ROOT GROWTH

PAs and ROS deriving from their oxidation regulate ion channels both in animals and plants during different physiological and stress-response processes directly influencing plasma membrane ion transport and/or acting as second messengers (Pegg, 2014; Pottosin et al., 2014). In particular, H_2O_2 produced through Spd oxidation by the peroxisomal AtPAO3, which is highly expressed in pollen grains also during pollen tube growth (Fincato et al., 2012), was shown to trigger the opening of hyperpolarization activated Ca^{2+}-permeable channels in pollen tubes, thus altering the tip-specific cytosolic Ca^{2+} gradient which plays a pivotal role in controlling pollen tube elongation (Wu et al., 2010). Indeed, two loss-of-function Arabidopsis *atpao3* mutants presented reduced pollen tube growth rate and seed production. Peroxisomal OsPAO3, OsPAO4, and OsPAO5 (Ono et al., 2012) are coherently expressed in anthers (Liu et al., 2014b). More studies are necessary to determine in detail the effect of PAs on pollen maturation, since the effect seems to be complex involving PA conjugation, ROS formation/scavenging, and cell-death events (Aloisi et al., 2015). Furthermore, based on the vigorous root growth of *AtPAO2* over-expressor plants and the hypersensitivity of *atpao2*loss-of-function mutant plants to ABA, it has been recently hypothesized a positive function of *AtPAO2* in Arabidopsis root growth (Wimalasekera et al., 2015).

CUAOS AND PAOS IN STOMATA MOVEMENT AND FRUIT RIPENING

ROS and calcium signatures in guard cells, as well as ion transport from the apoplast into the cytosol and from the cytosol for storage in the vacuole are important components of the regulatory network controlling stomata movements (Kollist et al., 2014; Murata et al., 2015). In *Vicia faba*, it has been shown that ABA-mediated stomata closure involves induction of an apoplastic CuAO activity as a source of H_2O_2, and that this activity is necessary to increase cytosolic Ca^{2+}levels in response to ABA (An et al., 2008). Consistently with these observations, the apoplastic AtCuAOβ was shown to be expressed in guard cells (Ghuge et al., 2015b). In addition, the peroxisomal AtCuAOζ which is expressed in guard cells is also involved in the ABA-mediated control of stomata opening (Qu et al., 2014). PAOs were also shown to contribute to the control of stomata movement in *Vitis vinifera* and *Arabidopsis* (Paschalidis et al., 2010; Hou et al., 2013).

PA catabolism has been also associated with grape and tomato fruit ripening (Agudelo-Romero et al., 2013; Tsaniklidis et al., 2016). Despite the increase of arginine decarboxylase expression levels during grape fruit ripening, the level of free and conjugated PAs was strongly decreased. This decrease was accompanied by up-regulation of two *CuAOs* and three *PAO* genes, one *CuAO* and two *PAOs* of them encoding for proteins with putative peroxisomal localization (Agudelo-Romero et al., 2013). Increase of CuAO and PAO activity, as well of H_2O_2 production during fruit maturation, was also observed. Furthermore, the relatively high expression levels of *Solanum lycopersicum* CuAO (SlCuAO1), which is clustered together with AtCuAOδ in clade IIb, was attributed to the implication of PA metabolism in physiological processes taking place during fruit ripening (Tsaniklidis et al., 2016). The up-regulation of CuAOs/PAOs during ripening may constitute a source of ROS for signaling events leading to the acceleration of the ripening process. It is also possible that PA catabolism interferes with plant hormonal pathways, such as ethylene and ABA (Agudelo-Romero et al., 2013). Alternatively, 4-aminobutanal produced through PA catabolism can be metabolized to γ-aminobutyric acid by an aminoaldehyde dehydrogenase (Zarei et al., 2015b) and enter into the Krebs cycle, thus constituting a link between nitrogen and carbon metabolism (Moschou et al., 2012).

CYTOSOLIC PAOS IN THE CONTROL OF THERM-SPM LEVELS

Loss-of-function Arabidopsis mutants for *AtPAO5* present increased levels of Therm-Spm, N^1-acetyl-Spm and Spm which are the substrates of the enzyme (Ahou et al., 2014; Kim et al., 2014; Sagor et al., 2016; Zarza et al., 2016). Interestingly, the expression levels of *AtPAO5*, as well as of its functional orthologue in rice (*OsPAO1*), are controlled by Therm-Spm and Spm (Ahou et al., 2014; Liu et al., 2014a,c). This indicates the necessity for a fine tune regulation of PA levels for proper growth and stress response, since levels above an upper limit or below a threshold may be detrimental for the plants. Indeed, the *atpao5* mutants present altered growth parameters at late developmental stages mainly due to the increased Therm-Spm levels (Kim et al., 2014), as have been also observed in mutants for *Therm-Spm synthase*and *S-adenosylmethionine decarboxylase* with reduced Therm-Spm levels (Clay and Nelson, 2005; Ge et al., 2006;Yoshimoto et al., 2016). Furthermore, the *atpao5* mutants were hypersensitive to low doses of exogenous Therm-Spm (Kim et al., 2014; Liu et al., 2014c).

INTRACELLULAR PAOS AND SOMATIC EMBRYOGENESIS

Some studies suggest that in *Gossypium hirsutum* an AtPAO1-like (GhPAO1) and an AtPAO5-like (GhPAO4) PAO may play a crucial role in the generation and differentiation of embryogenic callus during somatic embryogenesis (Cheng et al., 2015). Indeed, PAO activity levels significantly increased during conversion of embryogenic callus into somatic embryos, and inhibition of PAO activity by 1,8-diaminooctane resulted in brown and necrotic cultures, and a significant decrease in both fresh weight and somatic embryo number. Importantly, the negative effects of 1,8-diaminooctane were reversed by application of exogenous H_2O_2. Furthermore, in Arabidopsis, *AtPAO5* and its *B. juncea* ortholog (*BjPAO*) have a role in shoot regeneration from root cultures (Lim et al., 2006).

CONCLUDING REMARKS

Numerous recent studies have evidenced an extraordinary complexity in *CuAO* and *PAO* gene families regarding catalytic activity, subcellular localization, expression pattern and physiological roles of the encoded proteins. Indeed, important links to developmental and stress-related events are emerging for CuAOs and PAOs through ROS/NO production and regulation of specific PA levels.

ACKNOWLEDGMENTS

We apologize to authors whose papers could not be cited owing to space limitations. This work was supported by University "Roma Tre." We are grateful to Dr. Daniele Salvi (CIBIO-InBio, Portugal) for useful discussions regarding phylogenetic analysis. Moreover, we deeply thank Dr. Ilaria Fraudentali, Dr. Andrea Carucci, and Dr. Stefano Franchi (Department of Science, University "Roma Tre," Italy) for technical assistance in figure and table preparation.

REFERENCES

1. Agudelo-Romero, P., Bortolloti, C., Pais, M. S., Tiburcio, A. F., and Fortes, A. M. (2013). Study of polyamines during grape ripening indicate an important role of polyamine catabolism. *Plant Physiol. Biochem.* 67, 105–119. doi: 10.1016/j.plaphy.2013.02.024

2. Ahou, A., Martignago, D., Alabdallah, O., Tavazza, R., Stano, P., Macone, A., et al. (2014). A plant spermine oxidase/dehydrogenase regulated by the proteasome and polyamines. *J. Exp. Bot.* 65, 1585–1603. doi: 10.1093/jxb/eru016

3. Aloisi, I., Cai, G., Tumiatti, V., Minarini, A., and Del Duca, S. (2015). Natural polyamines and synthetic analogs modify the growth and the morphology of*Pyrus communis* pollen tubes affecting ROS levels and causing cell death. *Plant Sci.* 239, 92–105. doi: 10.1016/j.plantsci.2015.07.008

4. An, Z., Jing, W., Liu, Y., and Zhang, W. (2008). Hydrogen peroxide generated by copper amine oxidase is involved in abscisic acid-induced stomatal closure in *Vicia faba. J. Exp. Bot.* 59, 815–825. doi: 10.1093/jxb/erm370

5. Andronis, E. A., Moschou, P. N., Toumi, I., and Roubelakis-Angelakis, K. A. (2014). Peroxisomal polyamine oxidase and NADPH-oxidase cross-talk for ROS homeostasis which affects respiration rate in *Arabidopsis thaliana. Front. Plant Sci.* 5:132. doi: 10.3389/fpls.2014.00132

6. Angelini, R., Cona, A., Federico, R., Fincato, P., Tavladoraki, P., and Tisi, A. (2010). Plant amine oxidases "on the move": an update. *Plant Physiol. Biochem.* 48, 560–564. doi: 10.1016/j.plaphy.2010.02.001

7. Angelini, R., Tisi, A., Rea, G., Chen, M. M., Botta, M., Federico, R., et al. (2008). Involvement of polyamine oxidase in wound healing. *Plant Physiol.* 146, 162–177. doi: 10.1104/pp.107.108902

8. Bollhöner, B., Prestele, J., and Tuominen, H. (2012). Xylem cell death: emerging understanding of regulation and function. *J. Exp. Bot.* 63, 1081–1094. doi: 10.1093/jxb/err438

9. Boudart, G., Jamet, E., Rossignol, M., Lafitte, C., Borderies, G., Jauneau, A., et al. (2005). Cell wall proteins in apoplastic fluids of *Arabidopsis thaliana*rosettes: identification by mass spectrometry and bioinformatics. *Proteomics* 5, 212–221. doi: 10.1002/pmic.200400882

10. Campestre, M. P., Bordenave, C. D., Origone, A. C., Menéndez, A. B., Ruiz, O. A., Rodríguez, A. A., et al. (2011). Polyamine catabolism is involved in response to salt stress in soybean hypocotyls. *J. Plant Physiol.* 168, 1234–1240. doi: 10.1016/j.jplph.2011.01.007

11. Cervelli, M., Cona, A., Angelini, R., Polticelli, F., Federico, R., and Mariottini, P. (2001). A barley polyamine oxidase isoform with distinct structural features and subcellular localization. *Eur. J. Biochem.* 268, 3816–3830. doi: 10.1046/j.1432-1327.2001.02296.x

12. Cheng, W. H., Wang, F. L., Cheng, X. Q., Zhu, Q. H., Sun, Y. Q., Zhu, H. G., et al. (2015). Polyamine and its metabolite H_2O_2 play a key role in the conversion of embryogenic callus into somatic embryos in upland cotton (*Gossypium hirsutum* L). *Front. Plant Sci.* 6:1063. doi: 10.3389/fpls.2015.01063

13. Clay, N. K., and Nelson, T. (2005). Arabidopsis thickvein mutation affects vein thickness and organ vascularization, and resides in a provascular cell-specific spermine synthase involved in vein definition and in polar auxin transport. *Plant Physiol.* 138, 767–777. doi: 10.1104/pp.104.055756

14. Cona, A., Cenci, F., Cervelli, M., Federico, R., Mariottini, P., Moreno, S., et al. (2003). Polyamine oxidase, a hydrogen peroxide-producing enzyme, is up-regulated by light and down-regulated by auxin in the outer tissues of the maize mesocotyl. *Plant Physiol.* 131, 803–813. doi: 10.1104/pp.011379

15. Cona, A., Rea, G., Angelini, R., Federico, R., and Tavladoraki, P. (2006a). Functions of amine oxidases in plant development and defence. *Trends Plant Sci.* 11, 80–88. doi: 10.1016/j.tplants.2005.12.009

16. Cona, A., Rea, G., Botta, M., Corelli, F., Federico, R., and Angelini, R. (2006b). Flavin-containing polyamine oxidase is a hydrogen peroxide source in the oxidative response to the protein phosphatase inhibitor cantharidin in *Zea mays* L. *J. Exp. Bot.* 57, 2277–2289. doi: 10.1093/jxb/erj195

17. Cona, A., Tisi, A., Ghuge, S. A., Franchi, S., De Lorenzo, G., and Angelini, R. (2014). Wound healing response and xylem differentiation in tobacco plants over-expressing a fungal endopolygalacturonase is mediated by copper amine oxidase activity. *Plant Physiol. Biochem.* 82, 54–65. doi: 10.1016/j.plaphy.2014.05.004

18. Delis, C., Dimou, M., Flemetakis, E., Aivalakis, G., and Katinakis, P. (2006). A root- and hypocotyl-specific gene coding for copper-containing amine oxidase is related to cell expansion in soybean seedlings. *J. Exp. Bot.* 57, 101–111. doi: 10.1093/jxb/erj009

19. De Pinto, M. C., Locato, V., and De Gara, L. (2012). Redox regulation in plant programmed cell death. *Plant Cell Environ.* 35, 234–244. doi: 10.1111/j.1365-3040.2011.02387.x

20. Dewey, R. E., and Xie, J. (2013). Molecular genetics of alkaloid biosynthesis in *Nicotiana tabacum. Phytochemistry* 94, 10–27. doi: 10.1016/j.phytochem.2013.06.002

21. Federico, R., and Angelini, R. (1991). "Polyamine catabolism in plants," in *Biochemistry and Physiology of Polyamines in Plants*, eds R. D. Slocum and H. E. Flores (Boca Raton, FL: CRC Press), 41–56.

22. Fincato, P., Moschou, P. N., Ahou, A., Angelini, R., Roubelakis-Angelakis, K. A., and Federico, R. (2012). The members of *Arabidopsis thaliana* PAO gene family exhibit distinct tissue- and organ-specific expression pattern during seedling growth and flower development. *Amino Acids* 42, 831–841. doi: 10.1007/s00726-011-0999-7

23. Fincato, P., Moschou, P. N., Spedaletti, V., Tavazza, R., Angelini, R., and Federico, R. (2011). Functional diversity inside the Arabidopsis polyamine oxidase gene family. *J. Exp. Bot.* 62, 1155–1168. doi: 10.1093/jxb/erq341

24. Fuell, C., Elliot, K. A., Hanfrey, C. C., Franceschetti, M., and Michael, A. J. (2010). Polyamine biosynthetic diversity in plants and algae. *Plant Physiol. Biochem.* 48, 513–520. doi: 10.1016/j.plaphy.2010.02.008

25. Ge, C., Cui, X., Wang, Y., Hu, Y., Fu, Z., Zhang, D., et al. (2006). *BUD2*, encoding an S-adenosylmethionine decarboxylase, is required for Arabidopsis growth and development. *Cell Res.* 16, 446–456. doi: 10.1038/sj.cr.7310056

26. Ghuge, S. A., Carucci, A., Rodrigues-Pousada, R. A., Tisi, A., Franchi, S., Tavladoraki, P., et al. (2015a). The apoplastic copper AMINE OXIDASE1 mediates jasmonic acid-induced protoxylem differentiation in Arabidopsis roots. *Plant Physiol.* 168, 690–707. doi: 10.1104/pp.15.00121

27. Ghuge, S. A., Carucci, A., Rodrigues-Pousada, R. A., Tisi, A., Franchi, S., Tavladoraki, P., et al. (2015b). The MeJA-inducible copper amine oxidase *AtAO1*is expressed in xylem tissue and guard cells. *Plant Signal Behav.* 10:e1073872. doi: 10.1080/15592324.2015.1073872

28. Ghuge, S. A., Tisi, A., Carucci, A., Rodrigues-Pousada, R. A., Franchi, S., Tavladoraki, P., et al. (2015c). Cell wall amine oxidases: new players in root xylem differentiation under stress conditions. *Plants.* 4, 489–504. doi: 10.3390/plants4030489

29. Gilroy, S., Suzuki, N., Miller, G., Choi, W. G., Toyota, M., Devireddy, A. R., et al. (2014). A tidal wave of signals: calcium and ROS at the forefront of rapid systemic signaling. *Trends Plant Sci.* 19, 623–630. doi: 10.1016/j.tplants.2014.06.013

30. Heim, W. G., Sykes, K. A., Hildreth, S. B., Sun, J., Lu, R.-H., and Jelesko, J. G. (2007). Cloning and characterization of a *Nicotiana tabacum*methylputrescine oxidase transcript. *Phytochemistry* 68, 454–463. doi: 10.1016/j.phytochem.2006.11.003

31. Hou, Z.-H., Liu, G.-H., Wang, L.-X., and Liu, X. (2013). Regulatory function of polyamine oxidase-generated hydrogen peroxide in ethylene-induced stomatal closure in *Arabidopsis thaliana*. *J. Integr. Agricult.* 12, 251–262. doi: 10.1016/S2095-3119(13)60224-5

32. Jammes, F., Leonhardt, N., Tran, D., Bousserouel, H., Véry, A. A., Renou, J. P., et al. (2014). Acetylated 1,3-diaminopropane antagonizes abscisic acid-mediated stomatal closing in Arabidopsis. *Plant J.* 79, 322–333. doi: 10.1111/tpj.12564

33. Jiménez-Bremont, J. F., Marina, M., Guerrero-González, M., de, L., Rossi, F. R., Sánchez-Rangel, D., Rodríguez-Kessler, M., et al. (2014). Physiological and molecular implications of plant polyamine metabolism during biotic interactions. *Front Plant Sci.* 5:95. doi: 10.3389/fpls.2014.00095

34. Kamada-Nobusada, T., Hayashi, M., Fukazawa, M., Sakakibara, H., and Nishimura, M. (2008). A putative peroxisomal polyamine oxidase, AtPAO4, is involved in polyamine catabolism in *Arabidopsis thaliana*. *Plant Cell Physiol.* 49, 1272–1282. doi: 10.1093/pcp/pcn114

35. Kärkönen, A., and Kuchitsu, K. (2015). Reactive oxygen species in cell wall metabolism and development in plants. *Phytochemistry.* 112, 22–32. doi: 10.1016/j.phytochem.2014.09.016

36. Katoh, A., Shoji, T., and Hashimoto, T. (2007). Molecular cloning of N-methylputrescine oxidase from tobacco. *Plant Cell Physiol.* 48, 550–554. doi: 10.1093/pcp/pcm018

37. Kim, D. W., Watanabe, K., Murayama, C., Izawa, S., Niitsu, M., Michael, A. J., et al. (2014). Polyamine oxidase 5 regulates Arabidopsis growth through thermospermine oxidase activity. *Plant Physiol.* 165, 1575–1590. doi: 10.1104/pp.114.242610

38. Kollist, H., Nuhkat, M., and Roelfsema, M. R. (2014). Closing gaps: linking elements that control stomatal movement. *New Phytol.* 203, 44–62. doi: 10.1111/nph.12832

39. Kuehn, G. D., Rodriguez-Garay, B., Bagga, S., and Phillips, G. C. (1990). Novel occurrence of uncommon polyamines in higher plants. *Plant Physiol.* 94, 855–857. doi: 10.1104/pp.94.3.855

40. Lim, T. S., Chitra, T. R., Han, P., Pua, E. C., and Yu, H. (2006). Cloning and characterization of Arabidopsis and *Brassica juncea* flavin containing amine oxidases. *J. Exp. Bot.* 57, 4155–4169. doi: 10.1093/jxb/erl193

41. Liu, T., Kim, D. W., Niitsu, M., Berberich, T., and Kusano, T. (2014a). *Oryza sativa* polyamine oxidase 1 back-converts tetraamines, spermine and thermospermine, to spermidine. *Plant Cell Rep.* 33, 143–151. doi: 10.1007/s00299-013-1518-y

42. Liu, T., Kim, D. W., Niitsu, M., Berberich, T., and Kusano, T. (2014c). POLYAMINE OXIDASE 1 from rice (*Oryza sativa*) is a functional ortholog of Arabidopsis POLYAMINE OXIDASE 5. *Plant Signal Behav.* 9:e29773. doi: 10.4161/psb.29773

43. Liu, T., Kim, D. W., Niitsu, M., Maeda, S., Watanabe, S., Kamio, Y., et al. (2014b). Polyamine oxidase 7 is a terminal catabolism-type enzyme in *Oryza sativa* and is specifically expressed in anthers. *Plant Cell Physiol.* 55, 1110–1122. doi: 10.1093/pcp/pcu047

44. Marina, M., Sirera, F. V., Rambla, J. L., Gonzalez, M. E., Blázquez, M. A., Carbonell, J., et al. (2013). Thermospermine catabolism increases *Arabidopsis thaliana* resistance to *Pseudomonas viridiflava*. *J. Exp. Bot.* 64, 1393–13402. doi: 10.1093/jxb/ert012

45. McWilliam, H., Li, W., Uludag, M., Squizzato, S., Park, Y. M., Buso, N., et al. (2013). Analysis Tool Web Services from the EMBL-EBI. *Nucleic Acids Res.* 41, W597–W600. doi: 10.1093/nar/gkt376

46. Medda, R., Bellelli, A., Peč, P., Federico, R., Cona, A., and Floris, G. (2009). "Copper Amine oxidases from plants," in *Copper Amine Oxidases: Structure, Catalytic Mechanism and Role in Pathophysiology*, eds G. *Floris and B. Mondovì (Boca Raton, FL: Taylor and Francis Group, C.R.C. Press)*, 39–50.

47. Medda, R., Padiglia, A., Finazzi Agrò, A., Pedersen, J. Z., Lorrai, A., and Floris, G. (1997). Tryptamine as substrate and inhibitor of lentil seedling copper amine oxidase. *Eur. J. Biochem.* 250, 377–382.

48. Mo, H., Wang, X., Zhang, Y., Zhang, G., Zhang, J. F., and Ma, Z. Y. (2015). Cotton polyamine oxidase is required for spermine and camalexin signalling in the defence response to *Verticillium dahliae*. *Plant J.* 83, 962–975. doi: 10.1111/tpj.12941

49. Møller, S. G., and McPherson, M. J. (1998). Developmental expression and biochemical analysis of the *Arabidopsis atao1* gene encoding an H_2O_2-generating diamine oxidase. *Plant J.* 13, 781–791. doi: 10.1046/j.1365-313X.1998.00080.x

50. Monshausen, G. B., Bibikova, T. N., Messerli, M. A., Shi, C., and Gilroy, S. (2007). Oscillations in extracellular pH and reactive oxygen species modulate tip growth of Arabidopsis root hairs. *Proc. Natl. Acad. Sci. U.S.A.* 104, 20996–21001. doi: 10.1073/pnas.0708586104

51. Moschou, P. N., Paschalidis, K. A., Delis, I. D., Andriopoulou, A. H., Lagiotis, G. D., Yakoumakis, D. I., et al. (2008a). Spermidine exodus and oxidation in the apoplast induced by abiotic stress is responsible for H_2O_2 signatures that direct tolerance responses in tobacco. *Plant Cell* 20, 1708–1724. doi: 10.1105/tpc.108.059733

52. Moschou, P. N., Sanmartin, M., Andriopoulou, A. H., Rojo, E., Sanchez-Serrano, J. J., and Roubelakis-Angelakis, K. A. (2008b). Bridging the gap between plant and mammalian polyamine catabolism: a novel peroxisomal polyamine oxidase responsible for a full back-conversion pathway in Arabidopsis. *Plant Physiol.* 47, 1845–1857. doi: 10.1104/pp.108.123802

53. Moschou, P. N., Wu, J., Cona, A., Tavladoraki, P., Angelini, R., and Roubelakis-Angelakis, K. A. (2012). The polyamines and their catabolic products are significant players in the turnover of nitrogenous molecules in plants. *J. Exp. Bot.* 63, 5003–5015. doi: 10.1093/jxb/ers202

54. Müller, K., Linkies, A., Vreeburg, R. A., Fry, S. C., Krieger-Liszkay, A., and Leubner-Metzger, G. (2009). *In vivo* cell wall loosening by hydroxyl radicals during cress seed germination and elongation growth. *Plant Physiol.* 150, 1855–1865. doi: 10.1104/pp.109.139204

55. Murata, Y., Mori, I. C., and Munemasa, S. (2015). Diverse stomatal signaling and the signal integration mechanism. *Annu. Rev. Plant Biol.* 66, 369–392. doi: 10.1146/annurev-arplant-043014-114707

56. Naconsie, M., Kato, K., Shoji, T., and Hashimoto, T. (2014). Molecular evolution of N-methylputrescine oxidase in tobacco. *Plant Cell Physiol.* 55, 436–444. doi: 10.1093/pcp/pct179

57. Ono, Y., Kim, D. W., Watanabe, K., Sasaki, A., Niitsu, M., Berberich, T., et al. (2012). Constitutively and highly expressed *Oryza sativa* polyamine oxidases localize in peroxisomes and catalyze polyamine back conversion. *Amino Acids* 42, 867–876. doi: 10.1007/s00726-011-1002-3

58. Padiglia, A., Medda, R., Scanu, T., Longu, S., Rossi, A., and Floris, G. (2002). Structure and nucleotide sequence of *Euphorbia characias* copper/TPQ-containing amine oxidase gene. *J. Protein Chem.* 21, 435–441. doi: 10.1023/A:1021337302116

59. Pál, M., Szalai, G., and Janda, T. (2015). Speculation: polyamines are important in abiotic stress signaling. *Plant Sci.* 237, 16–23. doi: 10.1016/j.plantsci.2015.05.003

60. Paschalidis, K. A., and Roubelakis-Angelakis, K. A. (2005). Sites and regulation of polyamine catabolism in the tobacco plant. Correlations with cell division/ expansion, cell cycle progression, and vascular development. *Plant Physiol.* 138, 2174–2184. doi: 10.1104/pp.105.063941

61. Paschalidis, K. A., Toumi, I., Moschou, N. P., and Roubelakis-Angelakis, K. A. (2010). ABA-dependent amine oxidases-derived H_2O_2 affects stomata conductance. *Plant Signal Behav.* 5, 1153–1156. doi: 10.4161/psb.5.9.12679

62. Pegg, A. E. (2014). The function of spermine. *IUBMB Life* 66, 8–18. doi: 10.1002/iub.1237

63. Pintus, F., Spanò, D., Floris, G., and Medda, R. (2013). *Euphorbia characias* latex amine oxidase and peroxidase: interacting enzymes? *Protein J.* 32, 435–441. doi: 10.1007/s10930-013-9501-6

64. Planas-Portell, J., Gallart, M., Tiburcio, A. F., and Altabella, T. (2013). Copper containing amine oxidases contribute to terminal polyamine oxidation in peroxisomes and apoplast of *Arabidopsis thaliana*. *BMC Plant Biol.* 13:109. doi: 10.1186/1471-2229-13-109

65. Pottosin, I., Velarde-Buendía, A. M., Bose, J., Zepeda-Jazo, I., Shabala, S., and Dobrovinskaya, O. (2014). Cross-talk between reactive oxygen species and polyamines in regulation of ion transport across the plasma membrane: implications for plant adaptive responses. *J. Exp. Bot.* 65, 1271–1283. doi: 10.1093/jxb/ert423

66. Qu, Y., An, Z., Zhuang, B., Jing, W., Zhang, Q., and Zhang, W. (2014). Copper amine oxidase and phospholipase D act independently in abscisic acid (ABA)-induced stomatal closure in *Vicia faba* and Arabidopsis. *J. Plant Res.* 127, 533–544. doi: 10.1007/s10265-014-0633-3

67. Rea, G., de Pinto, M. C., Tavazza, R., Biondi, S., Gobbi, V., Ferrante, P., et al. (2004). Ectopic expression of maize polyamine oxidase and pea copper amine oxidase in the cell wall of tobacco plants. *Plant Physiol.* 134, 1414–1426. doi: 10.1104/pp.103.036764

68. Rea, G., Metoui, O., Infantino, A., Federico, R., and Angelini, R. (2002). Copper amine oxidase expression in defense responses to wounding and*Ascochyta rabiei* invasion. *Plant Physiol.* 128, 865–875. doi: 10.1104/pp.010646

69. Roach, T., Colville, L., Beckett, R. P., Minibayeva, F. V., Havaux, M., and Kranner, I. (2015). A proposed interplay between peroxidase, amine oxidase and lipoxygenase in the wounding-induced oxidative burst in *Pisum sativum* seedlings. *Phytochemistry* 112, 130–1388. doi: 10.1016/j.phytochem.2014.06.003

70. Rodríguez, A. A., Maiale, S. J., Menéndez, A. B., and Ruiz, O. A. (2009). Polyamine oxidase activity contributes to sustain maize leaf elongation under saline stress. *J. Exp. Bot.* 60, 4249–4262. doi: 10.1093/jxb/erp256

71. Rossi, A., Petruzzelli, R., and Agrò, A. F. (1992). cDNA-

derived amino-acid sequence of lentil seedlings' amine oxidase. *FEBS Lett.* 301, 253–257. doi: 10.1016/0014-5793(92)80251-B

72. Sagor, G. H., Inoue, M., Kim, D. W., Kojima, S., Niitsu, M., Berberich, T., et al. (2015). The polyamine oxidase from lycophyte *Selaginella lepidophylla*(SelPAO5), unlike that of angiosperms, back-converts thermospermine to norspermidine. *FEBS Lett.* 589, 3071–3078. doi: 10.1016/j.febslet.2015.08.045
73. Sagor, G. H. M., Zhang, S., Kojima, S., Simm, S., Berberich, T., and Kusano, T. (2016). Reducing cytoplasmic polyamine oxidase activity in Arabidopsis increases salt and drought tolerance by reducing reactive oxygen species production and increasing defense gene expression. *Front. Plant Sci.* 7:214. doi: 10.3389/fpls.2016.00214
74. Schopfer, P., Liszkay, A., Bechtold, M., Frahry, G., and Wagner, A. (2002). Evidence that hydroxyl radicals mediate auxin-induced extension growth.*Planta.* 214, 821–828. doi: 10.1007/s00425-001-0699-8
75. Sequera-Mutiozabal, M. I., Erban, A., Kopka J, Atanasov, K. E., Bastida, J., Fotopoulos, V., et al. (2016). Global metabolic profiling of Arabidopsis polyamine oxidase 4 (AtPAO4) loss-of-function mutants exhibiting delayed dark-induced senescence. *Front. Plant Sci.* 7:173. doi: 10.3389/fpls.2016.00173
76. Shoresh, M., Spivak, M., and Bernstein, N. (2011). Involvement of calcium-mediated effects on ROS metabolism in the regulation of growth improvement under salinity. *Free Radic Biol Med.* 51, 1221–1234. doi: 10.1016/j.freeradbiomed.2011.03.036
77. Strohm, A. K., Vaughn, L. M., and Masson, P. H. (2015). Natural variation in the expression of ORGANIC CATION TRANSPORTER 1 affects root length responses to cadaverine in Arabidopsis. *J. Exp. Bot.* 66, 853–862. doi: 10.1093/jxb/eru444
78. Takahashi, T., and Kakehi, J. (2010). Polyamines: ubiquitous polycations with unique roles in growth and stress responses. *Ann. Bot.* 105, 1–6. doi: 10.1093/aob/mcp259
79. Takahashi, Y., Cong, R., Sagor, G. H., Niitsu, M., Berberich, T., and Kusano, T. (2010). Characterization of five polyamine oxidase isoforms in *Arabidopsis thaliana*. *Plant Cell Rep.* 29, 955–965. doi: 10.1007/s00299-010-0881-1
80. Tamura, K., Peterson, D., Peterson, N., Stecher, G., Nei, M., and Kumar, S. (2011). MEGA5: molecular evolutionary genetics analysis using maximum likelihood, evolutionary distance, and maximum parsimony methods. *Mol. Biol. Evol.* 28, 2731–2739. doi: 10.1093/molbev/msr121
81. Tavladoraki, P., Cona, A., Federico, R., Tempera, G., Viceconte, N., Saccoccio, S., et al. (2012). Polyamine catabolism: target for antiproliferative therapies in animals and stress tolerance strategies in plants. *Amino Acids* 42, 411–426. doi: 10.1007/s00726-011-1012-1
82. Tavladoraki, P., Rossi, M. N., Saccuti, G., Perez-Amador, M. A., Polticelli, F., Angelini, R., et al. (2006). Heterologous expression and biochemical characterization of a polyamine oxidase from Arabidopsis involved in polyamine back conversion. *Plant Physiol.* 141, 1519–1532. doi: 10.1104/pp.106.080911
83. Tavladoraki, P., Schininà, M. E., Cecconi, F., Di Agostino, S., Manera, F., Rea, G., et al. (1998). Maize polyamine oxidase: primary structure from protein and cDNA sequencing. *FEBS Lett.* 426, 62–66. doi: 10.1016/S0014-5793(98)00311-1
84. Tiburcio, A. F., Altabella, T., Bitrián, M., and Alcázar, R. (2014). The roles of polyamines during the lifespan of plants: from development to stress. *Planta*240, 1–18. doi: 10.1007/s00425-014-2055-9
85. Tipping, A. J., and McPherson, M. J. (1995). Cloning and molecular analysis of the pea seedling copper amine oxidase. *J. Biol. Chem.* 270, 16939–16946.
86. Tisi, A., Federico, R., Moreno, S., Lucretti, S., Moschou, P. N., Roubelakis-Angelakis, K. A., et al. (2011). Perturbation of polyamine catabolism can strongly affect root development and xylem differentiation. *Plant Physiol.* 157, 200–215. doi: 10.1104/pp.111.173153
87. Tsaniklidis, G., Kotsiras, A., Tsafouros, A., Roussos, P. A., Aivalakis, G., Katinakis, P., et al. (2016). Spatial and temporal distribution of genes involved in polyamine metabolism during tomato fruit development. *Plant Physiol. Biochem.* 100, 27–36. doi: 10.1016/j.plaphy.2016.01.001
88. Tsukagoshi, H., Busch, W., and Benfey, P. N. (2010). Transcriptional regulation of ROS controls transition from proliferation to differentiation in the root.*Cell* 143, 606–616. doi: 10.1016/j.cell.2010.10.020
89. Tun, N. N., Santa-Catarina, C., Begum, T., Silveira, V., Handro, W., Floh, E. I., et al. (2006). Polyamines induce rapid biosynthesis of nitric oxide (NO) in*Arabidopsis thaliana* seedlings. *Plant Cell Physiol.* 47, 346–354. doi: 10.1093/pcp/pci252
90. Wang, W., and Liu, J. H. (2015). Genome-wide identification and expression analysis of the polyamine oxidase gene family in sweet orange (*Citrus sinensis*). *Gene* 555, 421–429. doi: 10.1016/j.gene.2014.11.042
91. Wimalasekera, R., Schaarschmidt, F., Angelini, R., Cona, A., Tavladoraki, P., and Scherer, G. F. (2015). POLYAMINE OXIDASE2 of Arabidopsis contributes to ABA mediated plant developmental processes. *Plant Physiol. Biochem.* 96, 231–240. doi: 10.1016/j.plaphy.2015.08.003
92. Wimalasekera, R., Villar, C., Begum, T., and Scherer, G. F. (2011). COPPER AMINE OXIDASE1 (CuAO1) of *Arabidopsis thaliana* contributes to abscisic acid- and polyamine-induced nitric oxide biosynthesis and abscisic acid signal transduction. *Mol. Plant.* 4, 663–678. doi: 10.1093/mp/ssr023
93. Wu, J., Shang, Z., Wu, J., Jiang, X., Moschou, P. N., Sun, W., et al. (2010). Spermidine oxidase-derived H_2O_2 regulates pollen plasma membrane hyperpolarization-activated Ca^{2+}-permeable channels and pollen tube growth. *Plant J.* 63, 1042–1053. doi: 10.1111/j.1365-313X.2010.04301.x
94. Yoda, H., Yamaguchi, Y., and Sano, H. (2003). Induction of hypersensitive cell death by hydrogen peroxide

produced through polyamine degradation in tobacco plants. *Plant Physiol.* 132, 1973–1981. doi: 10.1104/pp.103.024737

95. Yoshimoto, K., Takamura, H., Kadota, I., Motose, H., and Takahashi, T. (2016). Chemical control of xylem differentiation by thermospermine, xylemin, and auxin. *Sci. Rep.* 6, 21487. doi: 10.1038/srep21487

96. Zarei, A., Trobacher, C. P., Cooke, A. R., Meyers, A. J., Hall, J. C., and Shelp, B. J. (2015a). Apple fruit copper amine oxidase isoforms: peroxisomal MdAO1 prefers diamines as substrates, whereas extracellular MdAO2 exclusively utilizes monoamines. *Plant Cell Physiol.* 56, 137–147. doi: 10.1093/pcp/pcu155

97. Zarei, A., Trobacher, C. P., and Shelp, B. J. (2015b). NAD^+-aminoaldehyde dehydrogenase candidates for 4-aminobutyrate (GABA) and β-alanine production during terminal oxidation of polyamines in apple fruit. *FEBS Lett.* 589, 2695–2700. doi: 10.1016/j.febslet.2015.08.005

98. Zarza, X., Atanasov, K. E., Marco, F., Arbona, V., Carrasco, P., Kopka, J., et al. (2016). Polyamine oxidase 5 loss-of-function mutations in Arabidopsis thaliana trigger metabolic and transcriptional reprogramming and promote salt stress tolerance. *Plant Cell Environ.* doi: 10.1111/pce.12714. [Epub ahead of print].

CHAPTER 9

GLYCOSYLATION IS A MAJOR REGULATOR OF PHENYLPROPANOID AVAILABILITY AND BIOLOGICAL ACTIVITY IN PLANTS

Julien Le Roy, Brigitte Huss, Anne Creach, Simon Hawkins, and Godfrey Neutelings

Centre National de la Recherche Scientifique, UMR 8576-Unité de Glycobiologie Structurale et Fonctionnelle, Université de Lille, Lille, France

ABSTRACT

The phenylpropanoid pathway in plants is responsible for the biosynthesis of a huge amount of secondary metabolites derived from phenylalanine and tyrosine. Both flavonoids and lignins are synthesized at the end of this very diverse metabolic pathway, as well as many intermediate molecules whose precise biological functions remain largely unknown. The diversity of these molecules can be further increased under the action of UDP-glycosyltransferases (UGTs) leading to the production of glycosylated hydroxycinnamates and related aldehydes, alcohols and esters. Glycosylation can change phenylpropanoid solubility, stability and toxic potential, as well as influencing compartmentalization and biological activity. (De)-glycosylation therefore represents an extremely important regulation point in phenylpropanoid homeostasis. In this article we review recent knowledge on the enzymes involved in regulating phenylpropanoid glycosylation status and availability in different subcellular compartments. We also examine the potential link between monolignol glycosylation and lignification by exploring co-expression of lignin biosynthesis genes and phenolic (de)glycosylation genes. Of the different biological roles linked with their particular chemical properties, phenylpropanoids are often correlated with the plant's stress management strategies that are also regulated by glycosylation. UGTs can for instance influence the resistance of plants during infection by microorganisms and be involved in the mechanisms related to environmental changes. The impact of flavonoid glycosylation on the color of flowers, leaves, seeds and fruits will also be discussed. Altogether this paper underlies the fact that glycosylation and deglycosylation are powerful mechanisms allowing plants to regulate phenylpropanoid localisation, availability and biological activity.

INTRODUCTION

Terrestrialization was probably one of the most important evolutionary steps in the history of life. After the transition from water to land, a part of the green lineage was anchored in the soil and became sessile. Unable to move, plants are continuously subjected to UV radiations, large temperature variations and other extreme growing conditions currently accentuated by climate change such as drought, waterlogging and cold stresses as well as increased ozone levels and other industrial pollutants. Furthermore, plants are attacked by a wide range of pathogens including viruses, fungi, nematodes and bacteria, and are also targeted by herbivores that use plants for food. Finally, since the development of agriculture, plants must increasingly deal with different types of pesticides. Both biotic and abiotic stresses represent unfavorable growth conditions negatively affecting plant growth and survival and leading to dramatic crop losses. On the other hand, plants also depend on abiotic

Citation: Le Roy J, Huss B, Creach A, Hawkins S and Neutelings G (2016) Glycosylation Is a Major Regulator of Phenylpropanoid Availability and Biological Activity in Plants. Front. Plant Sci. 7:735. doi: 10.3389/fpls.2016.00735

and biotic agents for pollen and seed dispersal, on symbiotic microorganisms such as mycorrhizal fungi for water and mineral absorption, as well as rhizobia nitrogen-fixing bacteria in leguminous species. To manage both negative and positive interactions with the environment and other forms of life, individual plant species continually synthesize a complex and changing mixture of different secondary metabolites including phenylpropanoids, phenylpropanoid-derived compounds, terpenoids and alkaloids to respond to specific environmental situations (Timell, 1986; Cheynier et al., 2013; Mierziak et al., 2014; Pusztahelyi et al., 2015; Holbein et al., 2016; Ishihara et al., 2016). Although, a recent estimate put the number of secondary metabolites at more than 100,000 (Gachon et al., 2005) across plant species, it is likely that the real number is far higher. The increasing interest for metabolomics during the past decade has accelerated improvements in the sensitivity and specificity of spectrometric- and chromatographic-based technics as well as nuclear magnetic resonance (NMR) approaches.

Phenylpropanoids are a large group of secondary metabolites containing a phenyl group linked to a 3-C propane side chain. Both the position of the propenyl double bond and the substituents on the benzene ring result in different compounds with diverse biological activities (Koeduka et al., 2014). In many different species, the genes involved in the phenylpropanoid pathway belong to multigenic families (Ehlting et al., 1999) and are mainly regulated by MYB and NAC transcription factors (Sablowski et al., 1994; Zhong et al., 2006). The first enzymatic step is catalyzed by phenylalanine ammonia lyase (PAL), which converts phenylalanine issued from the shikimate pathway into cinnamic acid (Figure 1), which in turn is transformed to *p*-coumaric acid by cinnamate 4-hydrolase (C4H). Hydroxycinnamates (HCAs) derived from *p*-coumaric acid, CoA esters or hydroxycinnamaldehydes constitute the simplest phenylpropanoids (Strack, 2001). They can accumulate as ester- or amide-conjugates with monosaccharides, organic acids, lipids and amines and, when activated with CoenzymeA (CoA) or glucose, become acyl donors for modifications of secondary metabolites. HCAs are involved in the plant's response to pathogens and interactions with symbionts (Babst et al., 2014). Grass cell walls are characterized by large amounts of HCAs such as ferulic and *p*-coumaric acids. In these species, ferulic acid binds to lignin, polysaccharides and structural proteins and leads to their cross-linking but they are also present in smaller quantities in the cell walls of most dicots (for review see de Oliveira et al., 2015). In sugarcane, sinapic, *p*-coumaric and ferulic acids are deposited during the early stages of lignification, and levels of hydroxycinnamic acids were higher in parenchyma walls than in the vascular bundles (He and Terashima, 1990, 1991). They can also be integrated into lignins being involved in ligno-polysaccharide cross-links (Tobimatsu et al., 2012). The subsequent production of *p*-coumaroyl CoA by 4-coumaroyl CoA-ligase (4CL) constitutes an important branch point (Ehlting et al., 1999) leading to the biosynthesis of the different subfamilies of phenylpropanoid-derived molecules (Vogt, 2010), the most important being undoubtedly lignin, the second most abundant plant polymer on the Earth's surface.

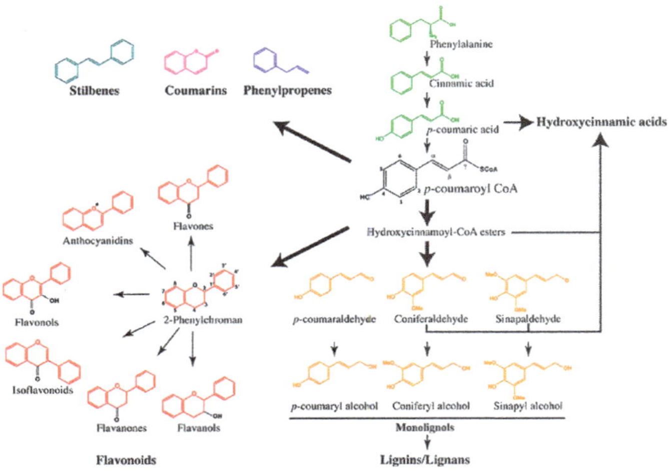

Figure 1. Schematic view of the phenylpropanoid biosynthesis. The general phenylpropanoid pathway begins by successive reactions resulting in the transformation of phenylalanine into *p*-coumaryl CoA, which is the common precursor of stilbenes, coumarins, phenylpropenes, 2-phenylchroman-containing flavonoids, and monolignols. The pathways leading to the production of hydroxycinnamic acids were obtained based on phenylpropanoid analysis in*Arabidopsis* mutants.

Lignin heteropolymers are derived mainly from three different monolignols: *p*-coumaryl alcohol, coniferyl alcohol, and sinapyl alcohol, synthesized in the cytoplasm through at least 10 different enzymatic reactions and are oxidized by laccases and/or peroxidases prior to polymerization into the lignin polymer (Freudenberg, 1968; Sterjiades et al., 1992;Boerjan et al., 2003). Instead of being exported to the apoplast, monolignols can also be stored as glycosylated conjugates in the vacuole (Dima et al., 2015). Lignins play a crucial role during the plant's life by providing cell integrity, mechanical support, water transport, assimilation through the vascular system and are also involved in defense mechanisms. (Neo)lignans are found in a wide range of species and represent a large group of diverse phenolics derived from the oxidative dimerization of two monolignols. Lignans are dimerized on their C3 side chains in a tail to tail structure whereas neolignans result from a head to tail dimerization (Cheynier et al., 2013). Flaxseed lignans have gained particular interest since some of them (e.g. SDG: secoisolariciresinol diglucoside) have potential antimutagenic, anti-inflammatory, anti-oxydant, antimicrobial and antiobesity effects (Imran et al., 2015).

Flavonoids are ubiquitously present across the plant kingdom. They are synthesized from the condensation of *p*-coumaroyl-CoA with three malonyl-CoA molecules by a chalcone synthase (CHS), which in turn produces a flavanone containing a 2-phenylchroman backbone. The same 2-phenylchroman backbone also gives rise to flavanols, isoflavonoids, flavonols, flavones and anthocyanidins. These compounds do not have a typical phenylpropanoid structure and are therefore considered as phenylpropanoid-derived compounds (Stafford, 1991). The different carbons of the flavonoid skeleton show numerous substitutions catalyzed by isomerases, reductases, hydroxylases, methylases, and glycosyltransferases leading to high structural diversity (Harborne and Williams, 2001; Ferrer et al., 2008).

Coumarins play roles in plant defense against fungi and possess both antioxidative and antimicrobial activities (Kai et al., 2006) as they can scavenge reactive oxygen intermediates produced during the hypersensitive response (Chong et al., 2002). The core structure derives from cinnamates via *ortho*-hydroxylation of the aromatic ring, *trans*/*cis*isomerization and lactonization. They are widely distributed in the plant kingdom and are classified into several categories depending on whether they harbor simple hydroxyl, alkoxy and/or alkyl groups in the benzene ring, or if additional benzene rings are present (Shimizu, 2014). Only 12 plant families have been listed to synthesize stilbenes from flavonoids precursors through stilbene synthases (Holl et al., 2013). There is a particular interest for *trans*-resveratrol synthesized in grapevine, cranberry, blueberry and peanut because of its pharmacological properties associated with protection toward cardiovascular, neurodegenerative diseases, cancer, and diabetes. Finally, phenylpropenes are volatiles emitted to attract both pollinators—through fragrances and seed dispersers through fruit scents. When they are produced in vegetative tissues, they also act as deterrents, toxicants and antifeedants to repel herbivores and bacterial pathogens.

The basic phenylpropanoid molecular backbone is synthesized in one of the numerous core metabolic pathways invented by plants during evolution, but the massive expansion of these different subfamilies of molecules is mainly due to the grafting of side chains on the basal structures. Among these modifications, glycosylation appears as a major reaction leading not only to an increase in structural diversity but also to an increase in the diversity of their properties. The glycosylation status of phenylpropanoids is regulated by the combination of regioselective glycosyltransferases (GTs) and glycoside hydrolases (GHs) with relatively broad substrate spectrum. Glycosyltransferases catalyze the transfer of sugar moieties from activated donor molecules to specific acceptor molecules such as sugars, lipids, proteins, or small molecules including phenylpropanoids. Information on GHs and GTs, as well as polysaccharide lyases and carbohydrate esterases, can be found in the Carbohydrate Active enZymes (CAZy) database (Davies et al., 2005) and a specific PlantCAZyme database is now also available (Ekstrom et al., 2014).

CAZy takes into account similarities in amino acid sequences, 3D-structures, sugar donors, transferred sugars, acceptors, catalytic mechanisms (inverting or retaining mechanisms) and more recently the combination of modules which can be catalytic or not (Lombard et al., 2014). Among the 98 GT families present in the CAZy database in may 2016, the GTs catalyzing secondary metabolite conjugation all belong to the GT1 family and are commonly named UDP-glycosyltransferases (UGTs) as UDP-sugars are used as the sugar donor. When they act on the conjugation of secondary metabolites, most UGTs perform O-glycosylation, but on some xenobiotics, UDP-glucose can act as sugar donor for S-, C- or N-glycosylations. Pollutants such as 2,4,5-trichlorophenyl and 3,4-dichloroaniline can potentially be metabolized by 44 different UGTs in *Arabidopsis* through O- and N-glucosylation, respectively, whereas one UGT (UGT72B1) was shown to act as a bifunctional enzyme carrying both activities (Brazier-Hicks et al., 2007). Conjugated xenobiotics may then be exported from the cell to the apoplast or imported into the vacuole. C-glucosylation of 2-hydroxyflavanone precursors of flavonoids was also reported for the rice OsCGT (Brazier-Hicks et al., 2009). S-glycosylation takes place during the biosynthesis of glucosinolates involved in plant defense against pathogen and herbivores (Grubb et al., 2004).

When an acceptor presents multiple possible binding sites for a sugar, UGTs exhibit regioselectivity by transferring the sugar to a specific position. This was demonstrated mainly for different coumarins (Lim et al., 2003) and flavonoids for which prediction of regioselectivity was assessed (Jackson et al., 2011). The glycosylation status is also regulated by glycoside hydrolases (GHs) involved in hydrolysis and/or rearrangement of glycosidic bonds. Among these, several β-glucosidase subfamilies can remove the non-reducing terminal ß-D-glucosyl from glucoconjugates. Most β-glucosidases are included in the CAZy GH1 family (Rouyi et al., 2014) but others can be found in GH3, GH5, GH9, GH30, and GH116 families. The different GH families contain enzymes showing different substrate specificities (Lombard et al., 2014) and most likely broad range specificities because of the high number of glycoconjugates within the plant compared to the number of ß-glucosidases (Ketudat Cairns et al., 2015).

In this present review, we will first summarize the current knowledge on UGTs and β-glucosidases related to the glycosylation status of phenylpropanoids and then we will evaluate the impact of sugar conjugation on their compartmentation. The function of these glucoconjugates will also be discussed in the context of lignification, development and the stress response.

CATALYTIC ACTIVITIES REGULATING PHENYLPROPANOID GLYCOSYLATION

UDP-Glycosyltransferases

UGTs are usually named with a specific nomenclature including family numbers 1–50 used for animals, 51–70 for yeasts, 71–100 for plants and 101–200 for bacteria followed by a letter for the sub-family defined by proteins sharing at least 60% homology, and finally an arabic number to describe the gene (Mackenzie et al., 1997; Ross et al., 2001). According to the nomenclature committee of the international union of biochemistry and molecular biology (IUBMB), they are also named as EC 2.4.1.x, EC 2.4.2.x, and 2.4.99.x. UGTs include enzymes able to transfer a sugar linked to uridine diphosphate (UDP) to a large range of acceptors. In plants, the nucleotidic sugar donor is mostly glucose but can also be galactose, xylose, rhamnose, arabinose, or glucuronic acid, the latter also being the main sugar derivative donor for UGTs in mammalians (Ross et al., 2001; Osmani et al., 2008; Yonekura-Sakakibara et al., 2008). The conjugation of acceptor molecules such as hormones, xenobiotics or secondary metabolites by UGTs allows the plant cell to modulate their biochemical proprieties and thus has a strong influence on their biological activity and compartment storage. Although, the primary sequences of the UGT proteins show only low similarities (Vogt and Jones, 2000), a highly conserved 44-amino acid sequence named plant secondary product glycosyltransferase (PSPG)-box believed to be involved in binding to the UDP moiety, is present in the C-terminal region (Mackenzie et al., 1997). This sequence is therefore commonly used as a query to identify the UGT genes in plant genomes by using the Blastp program in addition to position-specific weight matrix (PSWM) guided search and a hidden Markov model (HMMER) program

(Figure 2). The sequences can then be confirmed by using the MEME suite or the Pfam (http://pfam.sanger.ac.uk/search), SMART (http://smart.embl-heidelberg.de/) and Interpro (https://www.ebi.ac.uk/interpro/) databases. The very large spectrum of molecules potentially glycosylated by UGTs is in accordance with the high number of sequences identified at a genome-wide scale in plants as compared to other kingdoms including animals. However, there does not seem to be any relationship between the number of predicted UGT models and the taxonomic families such as liliopsida and eudicotyledons.

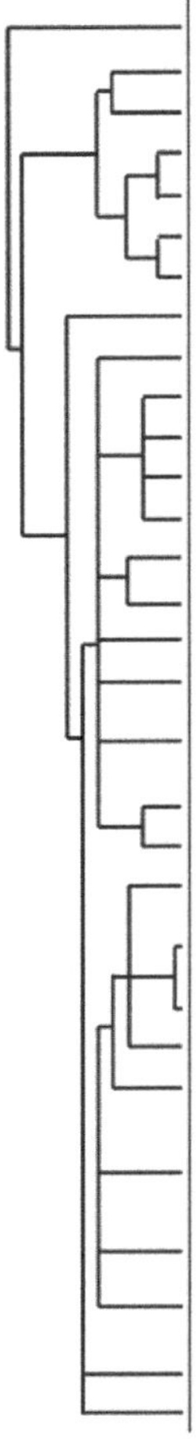

Plant species	Number of predicted UGTs	Identification method	References
Picea abies	219	HMMER	[4]
Oryza sativa	180/191/224/213	Blastp/HMMER	[2]/[4]/[3]/[10]
Brachypodium distachyon	143/163	HMMER	[10]/[4]
Zea mays	147/167/168	Blastp-Pfam-SMART/HMMER	[6]/[4]/[10]
Setaria italica	211	HMMER	[4]
Sorghum bicolor	180/188/201	Blastp/HMMER	[2]/[4]/[10]
Panicum virgatum	359	HMMER	[4]
Mimulus guttatus	100/147	Blastp/HMMER	[2]/[4]
Cucumis sativus	85/105/100	Blastp/HMMER	[2]/[10]/[4]
Cicer arietinum	96	Blastp- PSWM-HMMER	[8]
Glycine max	182/233/242	Blastp/HMMER	[2]/[4]/[10]
Medicago truncatula	182/187	HMMER	[4]/[10]
Phaseolus vulgaris	174	HMMER	[4]
Manihot esculenta	137/153	HMMER	[10]/[4]
Ricinus communis	109/117	HMMER	[4]/[10]
Linum usitatissimum	137/192	Blastp/HMMER	[1]/[4]
Populus trichocarpa	178/208/236	Blastp/HMMER	[2]/[4] [10]
Fragaria vesca	133	HMMER	[4]
Fragaria x ananassa	199	NS	[9]
Malus domestica	241/284	Blastp/HMMER	[2]/[4]
Prunus persica	162	HMMER	[4]
Brassica rapa	144	HMMER	[4]
Arabidopsis lyrata	116/118	HMMER	[4]/[10]
Arabidopsis thaliana	109	Blastp/PSI-BLAST	[7]
Capsella rubella	110	HMMER	[4]
Thellungiella halophila	87	HMMER	[4]
Carica papaya	56/68	HMMER	[10]/[4]
Gossypium arboreum	149	Blastp-PFAM-MEME	[5]
Gossypium hirsutum	196	Blastp-PFAM-MEME	[5]
Gossypium raimondii	142/152	Blastp-PFAM-MEME	[5]/[4]
Eucalyptus grandis	370	HMMER	[4]
Citrus clementina	152	HMMER	[4]
Citrus sinensis	126	HMMER	[4]
Vitis vinifera	74/102/181	Blastp/HMMER	[4]/[2]/[10]
Aquilegia coerulea	146	HMMER	[4]

Figure 2. Predicted UGT enzymes in different higher plants. Species are arranged in an order based on taxonomic families. [1], Barvkar et al. (2012); [2], Caputi et al. (2012); [3], Cao et al. (2008); [4], Ekstrom et al. (2014); [5], Huang et al. (2015); [6], Li et al. (2014); [7], Ross et al. (2001); [8], Sharma et al. (2014); [9], Song et al. (2015); [10], Yonekura-Sakakibara and Hanada (2011).

Among the numerous phenylpropanoids, flavonoids are probably among the best-characterized molecules in terms of glycosylation due to their numerous medical and commercial benefits. The basic structure of flavonoids consists of two phenyl groups joined by three carbons cyclized with oxygen (Figure 1). On this double ring, the carbons are annotated from 2 to 8 and on the third ring anchored on C2 they are annotated from 1′ to 6′. Successive glycosylation by UGTs is possible on ring positions bearing hydroxyl groups mostly on C3 and C7. In *Arabidopsis*, three homologous enzymes belonging to the UGT78D family catalyze the first 3-*O*-glycosylation steps on flavonols. These three enzymes share the same aglycone substrates kaempferol, quercetin and isorhamnetin but differ by their specificity toward the donor sugar, with UGT78D1 using UDP-rhamnose, UGT78D2 using UDP-glucose and UGT78D3 using UDP-arabinose (Yonekura-Sakakibara et al., 2008). The following step consists of a 7-*O*-glycosylation during which rhamnose is transferred by UGT89C1 or glucose by UGT73C6. Whether this transfer takes place or not, a secondary glycosylation step is also possible on the conjugated C3 by addition of a glucose on the flavonoid 3-O-glucoside by UGT79B6 (Yonekura-Sakakibara et al., 2014), or a rhamnose via a 1′-2″ link forming the triglycoside kaempferol 3-O-beta-[alpha-rhamnopyranosyl(1–>2)-glucopyranoside]-7-O-alpha-rhamnopyranoside. In the same metabolic pathway, recombinant protein or mutagenesis approaches showed that the production of anthocyanins also depended on the transfer of sugar moieties onto the carbon rings. Both UGT78D2 and UGT75C1 were identified in *Arabidopsis* as glucosyltransferases specific to the C3 and C5 positions, respectively (Tohge et al., 2005; Kubo et al., 2007) and UGT79B1 was shown to be involved in anthocyanin biosynthesis by converting cyanidin 3-*O*-glucoside to cyanidin 3-*O*-xylosyl(1→2)glucoside (Yonekura-Sakakibara et al., 2012). To a lesser extent the same enzyme is also capable of recognizing cyanidin 3-*O*-rhamnosyl (1→2) glucoside as substrate. Beyond the *Arabidopsis* model, numerous different UGT activities on flavonoids were identified in medicinal and aromatic plants: a flavonoid 7-*O*-glucuronosyltransferase in *Perilla frutescens* and *Scutellaria laeteviolacea* (Noguchi et al., 2009), a flavonoid glucoside: 6″-*O*-glycosyltransferase in *Catharanthus roseus* (Masada et al., 2009) or a flavonol 3-*O*-glucoside: 2″-*O*-glucosyltransferase in *Crocus sativus*(Trapero et al., 2012). The identification of the complete plant set of flavonoid-associated UGTs is still ongoing and major advances will undoubtedly be made with the growing number of full genome sequences available.

The identification and characterization of UGTs used to construct final complex molecular structures such as flavonoids or anthocyanins can be quite straightforward but is less obvious for those acting on more upstream intermediates such as hydroxycinnamic acids, aldehydes or alcohols. The role of these UGTs may indeed be difficult to reveal because of the proximity of related UGT genes in the *Arabidopsis* genome preventing generation of multiple mutants (Meißner et al., 2008). The use of techniques allowing fine gene targeting such as CRISPR-Cas9 may allow circumventing this constraint in the future. In addition the presence of overlapping substrates between several enzymes and the broad range of substrates recognized under *in vitro* conditions by each enzyme may also be a source of difficulty. One example is the TOGT1 recombinant enzyme derived from a tobacco gene and active toward ferulic, caffeic, cinnamic, *p*-coumaric, and*o*-coumaric acids as well as coniferyl alcohol (Fraissinet-Tachet et al., 1998) although it is important to be cautious with respect to the results obtained by *in vitro* activity approaches. In *Arabidopsis*, the UGT72E subfamily was extensively studied by the group of D. Bowles in the context of their possible implication in regulating lignin precursor availability (Lim et al., 2005). A series of substrates was tested on the recombinant proteins and results showed that UGT72E1 and UGT72E2 displayed catalytic activities on coniferyl and sinapyl aldehydes and that the monolignols coniferyl and sinapyl alcohols were glycosylated by UGT72E2 and UGT72E3 to the respective 4-O-ß -D-glucosides forms (coniferin and syringin; Lim et al., 2001, 2005). It should be noted that the enzymatic tests were performed on HCAs and on most of the intermediates of the lignification biosynthetic pathway but in general the CoA esters were omitted most likely because they have only recently become commercially available.

When used as a substrate by glucosyltransferases, caffeic acid can form caffeoyl-3-*O*-glucoside and caffeoyl-4-*O*-glucoside but also 1-*O*-caffeoylglucose, a high-energy glucose ester potentially used as a donor molecule during the formation of various hydroxycinnamic acid *O*-esters in plants (Mock and Strack, 1993; Lim et al., 2003). In the same way, 1-*O*-sinapoylglucose is formed by an UDP-glucose:sinapate glucosyltransferase (SGT) enzymatic activity and serves as a sinapoyl donor in various acyltransfer reactions leading to the accumulation of sinapate esters (Milkowski and Strack, 2010). During seed development in *Brassicaceae*, it is converted to sinapoylcholine (sinapine), the major seed phenylpropanoid that accumulates to levels of 1–2% of seed dry weight (Baumert et al., 2005), and is then further hydrolyzed in germinating seeds (Milkowski and Strack, 2010). In the seedling, sinapoylglucose is also channeled toward sinapoylmalate (2-*O*-sinapoyl-L-malate), a UV-B sunscreen protectant that accumulates in the vacuoles of the leaf epidermis and causes the adaxial surface to fluoresce in blue under UV light (Sinlapadech et al., 2007). In*Arabidopsis*, recombinant proteins of the four members of the UGT84A clade can convert HCA to 1-*O*-ß-glucose esters (Milkowski et al., 2000) and the overexpression of *UGT84A1, UGT84A2*, and *UGT84A3* leads to an increased level of sinapoylglucose in seeds and seedlings (Meißner et al., 2008). In *Brassica napus*, the Sinapoylglycosyltransferase (SGT) activity is due to UGT84A9, an enzyme with glucosylating capacities on sinapate, cinnamate, ferulate, 4-coumarate, and caffeate (Milkowski et al., 2000). The closely related UGT84A10 can also catalyze the formation of HCA glucose esters with preference for ferulate and sinapate when produced in *E. coli* (Mittasch et al., 2007). Recent work showed that the suppression of *UGT84A9* in RNAi lines induced strong metabolic modifications in late stages of seed development confirming the essential upstream role of the SGT activity on metabolites whose synthesis depends on 1-*O*-sinapoylglucose (Hettwer et al., 2016). A similar SGT activity in *Populus* has been assigned to GT1-316. Interestingly, the corresponding gene has a higher expression in plants grown under N-limiting conditions. Similarly, a stress response of the *Fragaria* × *ananassa FaGT2* gene was also observed after oxidative stress treatments in both strawberry fruit and cell cultures confirming that the final acceptors probably play an important role in the stress resistance mechanisms (Lunkenbein et al., 2006).

To reveal potential relations between genes involved in the phenylpropanoid backbone construction and those used by the plant for regulating the glycosylation status, we performed a mutual ranking classification using ATTED-II (Obayashi et al., 2009). This coexpression database (v8.0; March 2016) contains 58 experiments, 1388 GeneChips collected by AtGenExpress and about 20,000 publicly available files to provide gene-to-gene mutual ranks. We focused on a specific comparison between the known phenylpropanoid synthesis genes and the UGT genes found to be expressed in *Arabidopsis*. The complete results are presented as a heat map and a focus on five zones determined by hierarchical clustering containing a high number of coregulated genes is presented in Figure3. The UGT genes known to be involved in the glycosylation of the major phenylpropanoids are included in groups 1, 2, and 5. When the flavonoid biosynthesis pathway is considered (Figure 3, group 1), UGT78D1, UGT78D2, and UGT89C1 are all highly co-regulated with the four successive genes involved in the transformation of *p*-coumaroyl CoA to kaempferol/quercetin i.e., chalcone synthase (TT4), chalcone isomerase (TT5), flavanone 3-hydroxylase (F3H) and flavonol synthase (FLS). Only the three latest genes are coexpressed with *UGT78D3*. In the same group, the sinapoylglucose specific genes *UGT84A1* and *UGT84A2* are coexpressed with the flavonoid pathway genes, which is in accordance with the fact that this component is also a substrate for anthocyanin sinapoyltransferase (Fraser et al., 2007). *UGT75C1* and *UGT79B1* are present in group 2 but are not coregulated with the corresponding phenylpropanoid genes.

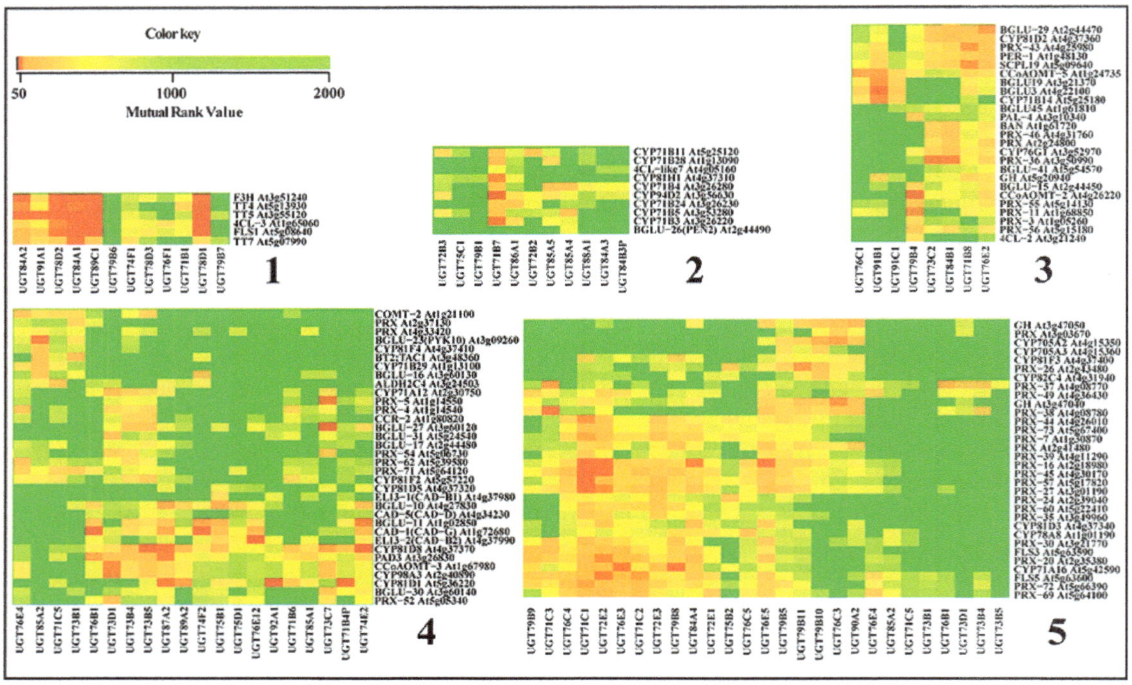

Figure 3. Co-expression of UGT and phenylpropanoid genes. Five relevant groups of coexpressed genes delimited by hierarchical clustering indicated on the left of the heatmap were extracted from the global heat map. Groups 1, 2, and 5 contain genes known to be involved in the glycosylation of the major phenylpropanoids and groups 3 and 4 contain genes with high coregulation values. When the mutual rank (MR) value <0, there is co-expression a50, co-expression between the genes is considered as strong; if 51 < *MR* < 1000, there is co-expression and if MR> 1001, there is no co-expression.

ß-Glucosidases

ß-glucosidases (EC 3.2.1.21) have been described in Archaea, Eubacteria, and Eukaryotes (Ketudat Cairns and Esen, 2010). In plants, they show many different quaternary structures and can form dimers (Dharmawardhana et al., 1995), tetramers (Hösel et al., 1978) or decamers (Fan and Conn, 1985). In *Cassava*, linamarases can also be present as homooligomers with different numbers of monomers forming very large aggregates with sizes up to 2000 kDa (Sermsuvityawong et al., 1995). They catalyze the hydrolysis of terminal, non-reducing β-D-glucosyl residues with release of β-D-glucose and the corresponding free aglycone from various substrates including glucosides, 1-*O*-glucosyl esters and oligosaccharides. ß-D-glucosidase activity is among the oldest enzymatic activity known as it was reported in the nineteenth century that almond emulsin was able to cleave the cyanogenic glucoside amygdalin (Wöhler and Liebig, 1837).

Plant glycosylated phenylpropanoids are not just substrates for endogenous (plant) enzymes but are also potential targets for ß-glucosidases from many different organisms. In yeast, hydrolytic activity toward 7- or 4′-*O*-glucosides of isoflavones, flavonols, flavones, and flavanones can be attributed to 3 genes and is probably related to metabolism of flavonoids from plant material within the culture broth (Schmidt et al., 2011). During human digestion, monoglycosylated and aglycone forms of flavonoids are more easily absorbed than the original multi-glycosylated compound (Vila-Real et al., 2011). The absorption of dietary flavonoid glycosides includes a critical deglycosylation step which is thought to occur in the oral cavity by microbiota showing β-glucosidase activity necessary for the delivery of the biologically active aglycones to epithelial cell surfaces (Walle et al., 2005). The plants themselves also show ß-glucosidase activity toward a large range of their own phenylpropanoids underlying the fact that the glycosylated forms of these molecules are unlikely to represent dead-end products. Several examples of the implication of deglycosylated phenylpropanoids in defense have been clearly illustrated (König et al., 2014). Nevertheless, this enzymatic activity has also been proposed to mediate the turnover of flavonol biglycosides and may therefore have important consequences on plant growth and development (Roepke and Bozzo, 2015).

On a genome-wide scale, the number of ß-glucosidase genes has only clearly been determined in a small number of species. In *Arabidopsis*, 47 GH1 ß-glucosidase genes annotated from *BGLU1* to *BGLU47* were shown to have a potential common evolutionary origin according to their position within the phylogenetic tree (Xu et al., 2004). The first global analysis of the rice GH1 family identified at least 31 genes (of 34 apparently functional genes) expressed in different organs and growth stages (Opassiri et al., 2006). In this species, duplication events occurring after the divergence with the *Arabidopsis thaliana* lineage gave rise to most of the identified genes. Although, the global number of predicted gene models is higher in maize compared to *Arabidopsis*, only 26 ß-glucosidase paralogs were identified in the inbred line B73. Gomez-Anduro et al. (2011) explained this difference by the fact that a glucosinolate-mediated biotic stress response is absent in maize, reducing the amount of glycosylated substrates. The number of GH1 genes compared to the global number of gene models was also low in *Sorghum bicolor* (Sekhwal et al., 2013) suggesting that this feature may be characteristic of liliopsida genomes, however this should be confirmed by an in depth analysis of a much larger number of plant genomes based on the CAZy annotation. In agreement with the large number and diversity of glycoconjugates present in a given plant and the limited number of ß-glucosidases, these enzymes present overlapping specificities. For example, the recombinant BGLU45 and BGLU46 from *Arabidopsis* hydrolyze the glycosylated monolignols syringin, coniferin, and *p*-coumaryl alcohol 4-O-ß-D-glucoside (Escamilla-Trevino et al., 2006) but BGLU45 preferentially hydrolyses syringin whereas BGLU46 presents a broader specificity to other phenolic glucosides. This promiscuity is exploited during *in vitro* assays since many ß-glucosidases are first tested with *p*-nitrophenyl-sugars before a more in depth characterization of the preferred cleaved sugar (Escamilla-Trevino et al., 2006; Baiya et al., 2014). On the contrary, other ß-glucosidases seem to show a more strict specificity for their natural glycosylated compounds (Ketudat Cairns et al., 2015). For instance, the *Arabidopsis* BGLU21, BGLU22, and BGLU23 exclusively cleave the coumarin scopolin into the aglycone scopoletin but do not hydrolyse *p*- or *o*-nitrophenyl-sugars (Ahn et al., 2010).

During lignin biosynthesis, plant cells produce the monolignols *p*-coumaryl alcohol, coniferyl alcohol and sinapyl alcohol. Their corresponding glucoconjugates *p*-coumaryl alcohol glucoside, coniferin, and syringin can be used as substrates by a specific group of coniferin ß-glucosidases. These coniferin β-glucosidases (EC 3.2.1.126) belong to the GH3 family and were first described in *Picea abies* seedlings (Marcinowski and Grisebach, 1978) and in *Cicer arietinum*suspension cells (Hösel et al., 1978). In both species, the enzymes deglycosylated coniferin as well as syringin. It was previously speculated that the role of these specific enzymes could be to hydrolyse monolignol glucosides prior to incorporation of the aglycones in the lignin macromolecule by oxidative radical coupling thereby regulating the amount of precursors for lignification. Lignins from gymnosperms are rich in G units (Weng and Chapple, 2010) and support for a possible role of β-glucosidases in lignification was provided by the observation that high amounts of coniferin are accumulated in the cambium together with the co-localization of coniferin ß-glucosidase and peroxidase activity in the differentiating xylem of *Pinus contorta* (Dharmawardhana et al., 1995). The presence of both glucosylated monolignols and ß-glucosidase activity was also previously reported in *Pinus banksiana* (Leinhos and Savidge, 1993). Despite such observations, the use of radioactive precursors in *Pinus contorta* indicated that coniferin was probably not the main source of the coniferyl alcohol incorporated in lignins (Kaneda et al., 2008). The presence of coniferin ß-glucosidase is not exclusive to gymnosperms and these enzymes have also been found in vegetative and reproductive tissues of *Oryza sativa* (Baiya et al., 2014), in the xylem of *Betula pendula* (Marjamaa et al., 2003) and in the interfascicular fibers and protoxylem of *Arabidopsis* (Chapelle et al., 2012). The relation between lignification and monolignol 4-O-ß -D-glucosides is further discussed below.

THE ROLE OF GLYCOSYLATION IN THE CELLULAR COMPARTMENTATION

One of the most predominant questions about phenylpropanoid biosynthesis concerns the spatial distribution of both precursors and final products within the cells. Many phenylpropanoid molecules necessary for plant development and/or defense are also toxic and an important challenge for plants is how to successfully manage the production/storage of these potentially dangerous molecules (Väisänen et al., 2015). While the association between PAL and C4H enzymes on microsomal membranes is believed to be responsible for the rapid transformation of phenylalanine to 4-coumaric acid in tobacco (Rasmussen and Dixon, 1999), much less evidence exists to support the existence of any large scale, tight metabolic channeling that could prevent accumulation of intermediates synthesized in the cytosol. The vacuole is one major compartment that has often been identified as an important site of storage for phenylpropanoid-derived products. Coniferin was shown to be accumulated in protoplasts of pine cambial cells, most likely in the vacuoles (Leinhos and Savidge, 1993) and recent phenolic profiling of vacuoles from *Arabidopsis* leaves identified coniferin, coniferin hexoside, and 34 other diverse glycosylated phenolics including derivatives of coniferyl alcohol, sinapyl alcohol, ferulic acid, and sinapic acid (Dima et al., 2015).

The cell wall is another major compartment where the oxidation machinery including peroxidases and laccases forms monolignol free-radicals that are subsequently cross-coupled in a non-stereospecific reaction to form lignin (Boerjan et al., 2003). In the apoplast, two monolignols can also undergo an enantioselective radical coupling reaction controlled by dirigent proteins to form lignans (Davin et al., 1997). Several lines of evidence support the biosynthesis of lignans in this compartment including the presence of the N-terminal signal peptide in dirigent proteins (Ralph et al., 2006), immunolocalization of these proteins in the cell wall (Burlat et al., 2001), and their presence in apoplastic protein extracts (Uzal et al., 2009). A previous report assuming that (neo)lignans were also located in vacuoles of *Linum* cell cultures (Kuhlmann et al., 2002) was confirmed recently by their detection in *Arabidopsis* leaf vacuoles (Dima et al., 2015). This paper also reported for the first time that in *Arabidopsis*, the combinatorial radical coupling reactions leading to (neo) lignans synthesis usually detected in the apoplast could also be detected in the cytoplasm. This observation greatly contributes to resolving the apparent paradox of how oligolignol hexosides arise in plant cells. Logically their formation requires an oxidative coupling reaction—usually thought to occur exclusively in the cell wall, followed by a glycosyl conjugation occurring in the cytoplasm. Nevertheless, this explanation does not totally exclude the possibility that a proportion of the cytoplasmic oligolignols may be formed in the apoplasm and subsequently transferred to the cytoplasm via specific (neo) lignin transporters as in mammalian systems (Miguel et al., 2014).

The glycosylation step is essential for the synthesis, transport, and sequestration of some phenylpropanoids. Because of their abundance and their role in flower coloration, the transport of glycosylated anthocyanins has been extensively studied over the past years (Yazaki, 2006). The involvement of Multidrug Resistance-associated Protein (MRP) also called ATP Binding Cassette subfamily C (ABCC) transporters in the vacuolar transport of such phenolic glucosides was first proposed in maize (Marrs et al., 1995) and later in carnation (Larsen et al., 2003) and *Arabidopsis* (Kitamura et al., 2004). However, these transporters are known to translocate a wide range of glutathione (GSH)-conjugated substrates through the membranes and anthocyanin-GSH conjugates have never been detected in plants (Ohkama-Ohtsu et al., 2011). Nevertheless, a potential co-transportation of GSH and glycosylated anthocyanidins was suggested by the heterologous expression of grapevine *ABCC1* in yeast (Francisco et al., 2013). A similar transportation system must still be validated in other species. In the lignification process, the glycosylation of hydroxycinnamic aldehydes and alcohols has been proposed to form part of an overall regulatory network of lignin precursor homoeostasis. In this model, ABC-like transporters located in the plasma and vacuolar membranes show differential affinities toward glycosylated and non-glycosylated molecules thereby controlling subcellular compartmentation (Miao and Liu, 2010). The 4-*O*-glycosylated forms are preferentially transported (in the presence of ATP) across the tonoplast into the vacuole whereas the specificity of the plasma membrane transporters is more relaxed with respect to ATP-dependence and favors the passage of aglycone forms. Given the existence of such preferential transport, it is tempting to speculate that unglycosylated lignin precursors are transported across the plasma membrane by specific ABC-like transporters and undergo free-radical polymerization while a surplus of monolignols could be glycosylated and transferred to the vacuole for storage by a transporter specific to glycoconjugates. *ABC* genes constitute large multigene families (approximately 130 members in *Arabidopsis*) and both structural and biochemical approaches are needed to fully characterize these transporters and evaluate the importance of glycosyl groups for substrate specificity. For example, the transporters characterized by Miao and Liu (2010) were located in *Arabidopsis* rosette leaves that contain only small amounts of lignins, and were shown to be mainly associated with the transport of *p*-coumaryl alcohol, a minor lignin component. More recently, Tsuyama et al. (2013) showed that a light membrane fraction prepared from differentiating xylem of hybrid and wild poplar, Japanese cypress and pine had clear ATP-dependent transport activity specific for coniferin suggesting the existence of a common transport mechanism dependent on a proton gradient created by a vacuolar-type H^+ ATPase (V-ATPase) in the lignifying tissues of both woody angiosperms and gymnosperms. Considering the fact that these V-ATPases were also located in the trans-Golgi network, it could be argued that coniferin may, in fact, be transported through the cell in small vesicles as suggested nearly 50 years ago by Pickett-Heaps (1968). It still remains necessary to determine whether the transport systems are universal or are dependent on molecules and/or species.

When considering the compartmentation of phenylpropanoids within the plant cell, the location of the enzymes implicated in their glycosylation status should also be taken into account. It seems clear that the absence of predicted signal peptides in plant UGT sequences supports the belief that they are located in the cytoplasm (Ross et al., 2001). The transfer of the glycosyl group on acceptor molecules can then be considered as exclusive to this compartment. On the other hand, the removal of sugar moieties by ß-glucosidase can potentially occur in a different compartment. Historically, the first ß-glucosidase capable of hydrolysing phenylpropanoid glycoconjugates was found in the cell wall fractions of roots and hypocotyls of *Picea abies* seedlings (Marcinowski and Grisebach, 1978) and cell suspensions of*Cicer arietinum* (Hösel et al., 1978). Since then, different ß-glucosidases have been identified in many plant species—generally by–omics approaches—but very few have been characterized at the biochemical level. It has been assumed for almost 40 years that the enzyme and the glycosylated substrate can be present in different locations and that contact between them occurs when the tissue or cells are damaged compromising cell integrity. This physical separation exists at the cellular level in sorghum seedlings where the cyanogenic glycoside dhurrin and a specific ß-glucosidase are located in different cells, the substrate in the epidermis and the enzyme in the mesophyll tissues (Kojima et al., 1979). However, it should be noted that due to the difficulty of physically separating these tissues, the results were obtained using protoplasts prepared from leaves and may therefore not represent the real situation *in planta*. In the case of phenylpropanoids, this separation generally occurs at the subcellular level. For example in*Melilota alba* leaves, coumarin (glycosylated form of 2-hydroxycinnamic acid) is located in the vacuole (also prepared from protoplasts) whereas the specific ß-glucosidase is found in the cell wall (Oba et al., 1981). The optimum pH determined for purified or recombinant ß-glucosidases exhibiting coniferin specificity in gymnosperms, monocots and dicots (Hösel et al., 1978; Marcinowski and Grisebach, 1978; Escamilla-Trevino et al., 2006; Baiya et al., 2014) generally ranges between 4.5 and 5.5 which is consistent with the acidic environment of the cell wall. The role of ß-glucosidases in cell wall lignification and plant defense responses is discussed further in the following section of this paper. The subcellular seperation is a critical need for molecules that are toxic under their aglycone form but may not be necessarily widespread. In rice, a ß-glucosidase showing *in vitro* activity against a *p*-coumaryl alcohol glucoside but not coniferin with an optimum pH of 6.5 was identified in the cytoplasm (Rouyi et al., 2014) suggesting that both glucosylation (by cytosolic UGTs) and deglucosylation (by cytosolic ß-glucosidases) may occur within the same cellular compartment (but not necessarily the same cell) at least in this species. Transglucosidases are GH1 enzymes that can transfer a sugar from a donor other than a nucleotide phosphate or phospholipid phosphate to an acceptor. Among these, the rice Os9BGlu31 enzyme is located in the vacuole of senescing flag leaves and developing seeds and is induced in seedlings in response to drought stress and different phytohormones (Luang et al., 2013). Subsequent functional characterization indicated that the enzyme showed the highest activity with an optimum pH of 4.5 toward 1-*O*-feruloyl-ß-D-glucose, as well as toward flavonoid glucosides but with lower efficiency. The presence of such an activity inside the vacuole would suggest that transported glycosylated phenylpropanoids can undergo further enzymatic modifications and so should not be considered as "dead-end" storage products. The detection of six GH1 glucosidases in *Arabidopsis*leaf vacuoles by proteomics also supports this idea (Carter et al., 2004).

In conclusion, a thorough understanding of the biological roles of (glycosylated) phenylpropanoids not only depends on detailed knowledge about their biosynthesis, but also on a better comprehension of their localisation and transport within, and between different tissues, cells and compartments.

AT THE CROSSROADS OF MONOLIGNOL GLYCOSYLATION AND LIGNIN BIOSYNTHESIS

There are still many uncertainties about whether pooled glucoconjugated monolignols—and by extension hydroxycinnamic acids and aldehydes—really constitute a stock for subsequent lignin construction. Terashima and Fukushima (1988) followed the incorporation of radiolabeled glycosylated monolignols in *Pinus thumbergii* by microautoradiography in softwood differentiating xylem. When incubated at concentrations of up to 1 mg glucoside in 100 μl, the labeled monolignols were incorporated in the cell wall leading the authors to conclude that glycosylated forms could contribute to the lignification process. Although, it is tempting to think that incorporation occurs after a simple hydrolysis of the sugar moieties, an alternative pathway involving the oxidization of coniferin to coniferaldehyde glucoside prior to the deglucosylation step, followed by the transformation to coniferyl alcohol was suggested by the use of radiolabeled coniferin in *Gingko biloba* shoots (Tsuji et al., 2005). Either way, these experiments show that exogenous coniferin can serve as a lignin precursor.

Lignin composition is modified when plants are submitted to a mechanical stress. Compression wood (CW) is formed on the lower part of stems and branches of inclined softwoods (Timell, 1986). When compared to normal wood (NW), CW tracheids contain less cellulose but more lignin with (in addition to G-units) a higher proportion of *p*-hydroxyphenyl units (H-units) that are almost absent in NW and opposite woods (Terashima and Fukushima, 1988; Donaldson, 2001). In addition to increased lignin content, Yoshinaga et al. (2015) also reported that coniferin was less abundant in compression wood than in normal wood and that the *p*-coumaryl alcohol glucoside was undetectable in the two Japanese softwoods *Chamaecyparis obtusa* and *Cryptomeria japonica*. In the light of these results the authors suggested that if glycosylated monolignols do indeed function as lignin precursors, then their incorporation as G- and H- subunits must occur immediately after their synthesis without being pooled in the cell. A possible relationship between the accumulation of glycoconjugated phenylpropanoids and lignin status has also been reported in flax. In this species, bast fibers located in the outer part of the stem possess very thick secondary cell walls containing low amounts (4% cell wall residue) of lignin compared to inner stem xylem cell walls containing high amounts of lignin (25% cell wall residue; Day et al., 2005). An ultra-high-performance liquid chromatography-Fourier transform ion cyclotron resonance-mass spectrometry (UHPLC-FT-ICR-MS) phenolic profiling approach showed that flax inner- and outer-stem tissues had different phenolic profiles (Huis et al., 2012). In total 81 different phenolic compounds including monolignols, oligolignols and (neo)lignans were identified and among these, 39 were glycosylated and present (except for one compound) in much higher amounts (up to 92-fold higher) in the outer stem tissues. Whether the observed differences in glycosylated phenolics content can be related to the differences in cell wall lignin content in the two stem tissues remains an interesting question for future research.

It is now well known that transgenic approaches targeting general phenylpropanoid or more specific monolignol genes can modify both lignin composition and quantity (Chabannes et al., 2001; Voelker et al., 2010). For example, inactivation of a caffeic acid *O*-methyltransferase (*COMT*) gene in *Panicum virgatum* decreased lignin content concomitantly with metabolites related to trans-sinapyl alcohol including the conjugation-product syringin (Fu et al., 2011; Tschaplinski et al., 2012). In *Arabidopsis*, a global view of the metabolic shifts in the lignin-related biosynthesis was obtained by a systems biology approach (Vanholme et al., 2012). These published data revealed that mutations in phenylpropanoid genes have different impacts on glycosylated monolignol production. In the *pal1, c4h, 4cl, ccoaomt1*, and *ccr1* mutants, the amounts of syringin and coniferin were significantly lower when compared to the wild type. All these mutants except for *pal1* also show reduced lignification. An increase in the syringin and coniferin levels was also reported in the *4cl2* and *f5h1* mutants respectively without any change in lignin contents. The overall effect depends on the mutated gene and the phenylpropanoid considered. A rather general tendency shows that the glycosylated metabolites beyond coniferaldehyde in the biosynthetic pathway tend to decrease in *pal1, c4h, 4cl1, ccoaomt1*, and *ccr1*and that the glycosylated metabolites produced before the mutated gene increase in *4cl1, ccoaomt1, ccr1*, and *comt*. The results obtained in this species clearly support the hypothesis that monolignols whose destiny is to become either glycosylated or polymerized as lignins are produced in a common metabolic pathway.

Comparison of lignin biosynthetic and UGT gene expression patterns reveals interesting co-expressions hinting at links between lignification and monolignol glycosylation. In Figure 3, the group 5 shows that two (*UGT72E1* and *UGT72E2*) of the three monolignol-associated UGT72E subfamily genes are coexpressed with the peroxidase *PRX49* and that all three members are coexpressed with *PRX72*. These two peroxidases were recently described as major actors of lignin polymerization in the cell wall (Herrero et al., 2013; Fernandez-Perez et al., 2015) and since both peroxidases and UGTs use the same substrates, the observed coexpression would strongly suggest the existence of a functional link between these enzymes. Further evidence for such an idea is provided by the effects of mutations in the triple mutant of 3 laccase genes (*LAC4, LAC11*, and *LAC17*) in *Arabidopsis* (Zhao et al., 2013). The growth of these plants is blocked in the early stages of development with (among other) phenotypic characters, a narrow root diameter and vascular development arrest. The authors attribute these severe growth perturbations to the lack of lignin and secondary cell walls in the mutants and to an overaccumulation of free forms of monolignols. Indeed levels of coniferin and syringin were both increased in the mutant along with those of sinapoylglucose and three kaempferol glycosides. Interestingly, the expression of most monolignol biosynthesis genes was not affected showing the lack of a negative feedback regulation on the transcription. On the other hand, the accumulation of monolignol glucosides in this triple mutant could be correlated to significant up-regulation of *UGT72E2* and *UGT72E3* potentially related to monolignol detoxification mechanisms in the absence of lignin polymerization.

The inversely proportional expression of *UGTs* and genes involved in lignin polymerization was also demonstrated in the *lbf1* flax mutant isolated from an ethyl methanesulfonate (EMS) population (Chantreau et al., 2014). In these plants, the bast fibers located in the external parts of the stems show ectopic lignification of their secondary cell walls. This additional production of lignin is accompanied by a higher *cinnamoyl-CoA reductase* (*CCR*), *caffeic acid O-methyltransferase* (*COMT*), *cinnamyl alcohol dehydrogenase* (*CAD*), and *PRX71* ortholog transcript accumulation indicating a possible increase in monolignol production and oxidation. In parallel, the reduction in the abundance of transcripts corresponding to a *UGT72E* ortholog observed in the outer tissues was explained by the reduced necessity for detoxification due to a higher incorporation of monolignols into the lignin polymer.

Altogether, these examples illustrate the fact that variations in lignin amounts can have a potential effect on monolignol glucoside production but the opposite is not necessarily true. The overexpression of *UGT72E2* and *UGT72E3* in*Arabidopsis* (Lanot et al., 2006, 2008) mostly led to increased accumulation of coniferin in light-grown roots but had little impact on syringin levels. Functional validation of both genes also came from the significant reduction in coniferin and syringin in down-regulated transgenic plants. Perhaps the most interesting result coming from these transgenic approaches is that regardless of the effects on monolignol glucoside levels, no impact on lignin was ever observed, even when the silencing of the three *UGT* genes by an RNAi approach led to a 90% reduction (Wang et al., 2013). These results strongly suggest that changes in monolignol glucosides do not affect lignin levels. Nevertheless, it is interesting to note that the heterologous overexpression of a poplar UGT gene in tobacco (Wang et al., 2012) led to an increase of lignin in stems. However, in this case the authors indicate that the corresponding recombinant enzyme was unable to glycosylate phenylpropanoid components suggesting that the observed effect was indirect.

Overexpression of *UGT72E2* and *UGT72E3* also led to the accumulation of coniferin and syringin in the leaves associated with a significant reduction in the amounts of sinapoyl malate. A link between monolignol glucosides and hydroxycinnamic acid esters was also suggested by the results obtained by Hemm et al. (2004) showing that the light-induced production of coniferin and syringin in roots is accompanied by up-regulation of the *4CL3* (At1g65060) gene involved in the production of sinapate esters (Nair et al., 2004). Interestingly, the analysis of the gene expression heatmap shown in Figure 3 (group 1) indicates that *4CL3* is coexpressed with the two glycosyltransferases *UGT84A1*and *UGT84A2* implicated in the production of sinapoyl glucose (Milkowski et al., 2000).

As stated above any link between lignin and phenylpropanoid glucoconjugates should not only take into account UGT activities and gene expression but also ß-glucosidase behavior. In the systems biology approach developed byVanholme et al. (2012), the decrease in the amounts of coniferin and syringin were nicely related to the lower *UGT72E2*and *UGT72E3* expression levels, but also to the increase in both *BGLU45* and *BGLU46* transcript accumulation. These two genes were shown to be specific for monolignol glycosides (Escamilla-Trevino et al., 2006) and the corresponding knockout mutants displayed a significant increase in stem coniferin content but, as for UGT overexpressors, no changes in lignin content were observed (Chapelle et al., 2012). A general overview of the effects of mutations of genes involved in monolignol production, polymerization or glycosylation is shown in Figure 4.

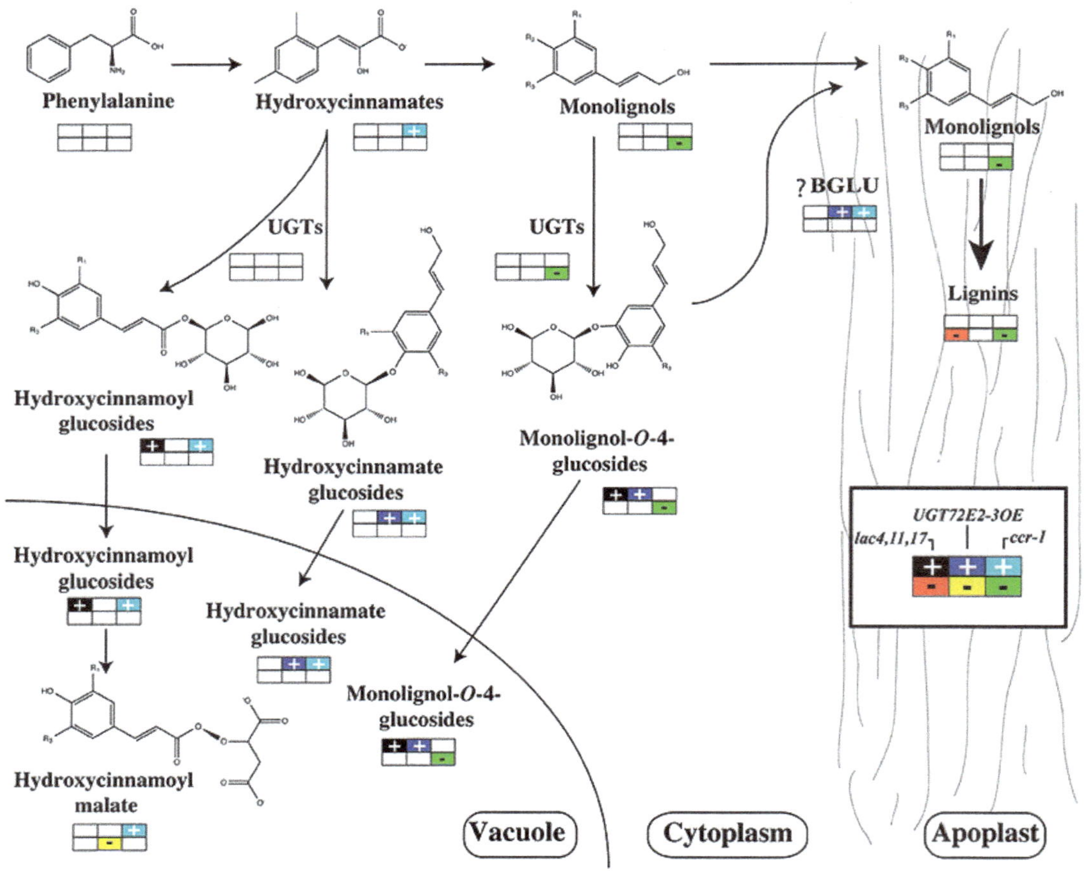

Figure 4. Effects of phenylpropanoid gene mutations or UGT gene overexpression. A schematic view of lignin, glucoconjugated monolignols, hydroxycinnamate glucosides, or hydroxycinnamoyl esters are presented in *ccr-1*and *lac4,11,17* mutants and *35S:UGT72E2-3* overexpressors. For metabolites and transcripts, a color code indicates an accumulation or a reduction compared to the wild-type. Data were extracted from Vanholme et al. (2012), Zhao et al. (2013), and Lanot et al. (2006, 2008) respectively.

In conclusion, although monolignol glucosides have long been considered as participants in the lignification process (Freudenberg and Harkin, 1963; Freudenberg and Torres-Serres, 1967) it seems unlikely that the plant actually uses these molecules stocked in the vacuole to synthesize lignin. In contrast, the perturbation of genes involved in lignification has a clear impact on the production of glycoconjugated monolignols. However, it would be important to check whether the results obtained in *Arabidopsis* are also valid for woody plants. Although, secondary cell wall formation and especially lignin formation have been extensively studied in poplar, few studies (Wang et al., 2012) have addressed the potential links between phenylpropanoid glycosylation and lignification in trees.

THE ROLE OF GLYCOSYLATION IN PHENYLPROPANOID FUNCTION

Glycosylation and Phenylpropanoid Stability

Many phenylpropanoids are toxic and unstable molecules and so rarely accumulate in their aglycone form in plant cells (Whetten and Sederoff, 1995; Alejandro et al., 2012; Väisänen et al., 2015). The mechanism of their toxicity is not clear but it has been suggested by Väisänen et al. (2015) that: (i) high concentrations of coniferyl alcohol could react with lipids or proteins in cellular environments in the presence of reactive oxygen species, (ii) since coniferyl alcohol could be easily degraded (e.g. to vanillin or ferulic acid) it is possible that its toxicity results from the accumulation of degradation products (iii) the availability of coniferyl-alcohol leads to the formation of a lignin-like compound and reduced growth of plant cells. More generally, aglycone forms of phenylpropanoids are nucleophilic molecules with high damage potentialities on diverse cellular components. The attachment of carbohydrate moieties will then reduce their reactivity and improve their stability (Jones and Vogt, 2001). In the case of monolignols, it is known that the conversion into their glycone forms by UGT activity reduces their toxicity (Bock, 2016). This is also the case of anthocyanidins for example, which only accumulate in cells in their glycosylated forms (Boss et al., 1996).

Glycosylation may not only prevent the toxicity of aglycone phenylpropanoids but can also contribute to the production of protectant molecules against reactive oxygen species (ROS). Flax seed coats contain the lignan secoisolariciresinol diglucoside (SDG) whereas their aglycone precursors pinoresinol, lariciresinol, and secoisolariciresinol are undetectable whatever the development stage considered (Hano et al., 2006). SDG was previously shown to prevent lipid peroxidation in a concentration-dependent manner (Prasad, 1997) and it is believed that it may play a similar role in the plant. This hypothesis is consistent with the concomitant presence of very high amounts of SDG and highly reactive polyunsaturated fats in the flax seeds (Lorenc-Kukula et al., 2005). Nevertheless, a peroxyl radical scavenging capacity comparable to that of glutathione has been described for the aglycone form of the isoflavone daidzein and the non-glycosylated lignan honokiol (Kim et al., 2010) suggesting that glycosylation is not essential for ROS protection capacity. Glycosylation can therefore contribute to reducing phenylpropanoid toxicity and increasing stability thereby explaining the widespread occurence of phenylpropanoids in plant development and resistance/tolerance to major biotic and abiotic stresses (Lim and Bowles, 2004; Bowles et al., 2005).

Phenylpropanoid Glycosylation and Plant Defense

The phenylpropanoid biosynthetic pathway is stimulated during the hypersensitive response (HR) induced by pathogen infection and is associated with the production of phenylpropanoid-derived phytoalexins such as coumarins and isoflavonoids (Hammond-Kosack and Jones, 1996; Dorey et al., 1997; Langlois-Meurinne et al., 2005). Chong et al. (2002) reported that the glycosylation of phenylpropanoids in tobacco played a significant role during Tobacco Mosaic Virus (TMV) infection and that the down-regulation of a tobacco glycosyltransferase gene (*TOGT1*) led to reduced TMV resistance. The authors suggested that scopolin (a glycoconjugated coumarin) is a storage form of the aglycone form scopoletin that functions as an antiviral agent potentially involved in scavenging ROS accumulated during the infection. Overexpression of the *TOGT* gene also enhanced resistance against the potato virus Y (PVY) in tobacco (Matros and Mock, 2004). Recent work also shows that sinapate esters, coniferin and the glycosylated pinoresinol and lariciresinol lignans accumulate when *Arabidopsis* leaves are infected by the soil-borne ascomycota *Verticillium longisporum* (König et al., 2014). When tested on agar plates, only the sinapoyl glucose and coniferyl alcohol, but not sinapyl alcohol, managed to inhibit the growth of the fungi, whereas coniferin specifically inhibited melanisation. The authors proposed that the growth-inhibiting differences observed between the two aglycone monolignols were solely due to the oxidization of coniferyl alcohol to ferulic acid, which was also shown to have an inhibitory effect on *V. longisporum*. Coniferin-accumulating transgenic plants over-expressing *UGT72E2* were less susceptible to the fungi. In this case, it was suggested that coniferin would operate as a storage form of coniferyl alcohol, which could further be transformed into ferulic acid. Plant species can also display glycosylation-related cultivar- and pathogen-dependent responses to biotic stresses. In wheat, genes encoding cytochrome P450 (CYP709C1) and a UGT were more strongly upregulated during the infection by *Fusarium graminearum* compared to *Magnaporthe grisea* and cultivar resistance to *F. graminearum* led to a stronger expression in incompatible (resistant) interactions with the chinese spring wheat cultivar Sumai 3 (Ha et al., 2016). Based on these observations the authors propose the existence of different resistance genes related to both fungi and confirm the necessity to combine genes in breeding programs to obtain multi-resistance plants as it was proposed by Schäfer et al. (1963). Deoxynivalenol produced by *Fusarium* is glycosylated by DOGT1 (UGT73C5) and the constitutive overexpression of the corresponding gene in *Arabidopsis* led to enhanced tolerance against this mycotoxin (Poppenberger et al., 2003). Other studies have underlined the importance of some UGT genes implicated in pathogen response and redox status during the pathogen infection. For example, it was reported that UGT73B3 and UGT73B5 are involved in the hypersensitive response to the bacteria *Pseudomonas syringae* pv tomato in *Arabidopsis* and participate in the regulation of the redox status of plant cells. However, the exact role of these UGT*in* *planta* remains unknown (Langlois-Meurinne et al., 2005; Simon et al., 2014).

There is now clear evidence that UGTs play an important role in disease resistance even though their precise contribution still remains unclear. However, it appears that the addition of a glycosyl group does not necessarily play a major role in determining the biological activity of phenylpropanoids and it may even block the biological activity of a particular molecule toward the pathogen. In this case, it is possible that glycosylation may play a role in maintaining the homeostasis of an ensemble of pathogen-specific molecules. Langenbach et al. (2013) suggested that disruption of the*UGT84A2/BRT1* gene can lead to an enhanced accumulation of certain sinapate derivatives or new metabolites absent from wild-type plants, which would confirm the important role of UGTs in maintaining a specific pool of phenylpropanoid-related metabolites.

Phenylpropanoid Glycosylation and Abiotic Stress

The *Arabidopsis* UGT84A2 and its *Brassica napus* ortholog UGT84A9 are both involved in the production of sinapoyl malate. This HCA glucose ester accumulates in vacuoles of sub-epidermal tissues and is believed to play a role in the protection against the harmful effects of UV-B radiation (Landry et al., 1995; Meißner et al., 2008; Dean et al., 2014). Li et al. (2010) isolated the *Arabidopsis* Pna-10 accession containing a 13-kb deletion that eliminates the *SNG1* gene responsible for the conversion of sinapoylglucose to sinapoylmalate, and the gene encoding sinapoylglucose:anthocyanin sinapoyltransferase (*SAT*). As a result, Pna-10 is unable to accumulate both sinapoylmalate and sinapoylated anthocyanins. Under field conditions, this mutation is not lethal suggesting that the presence of sinapoylmalate in the leaves may not be absolutely essential for plant protection. More recently, Hectors et al. (2014) highlighted the role of the rhamnosylated kaempferol and quercetin glycosides during UV acclimation in*Arabidopsis*. The concentration of these compounds increased as a result of UV-stress (and related oxidative stress) and could be correlated with the upregulation of the flavonol-7-O-rhamnosyltransferase *UGT89C1*. The biological role of these flavonoid glycosides is not clear as they are less effective anti-oxidants that the corresponding aglycone forms (Vogt and Jones, 2000; Gachon et al., 2005). The authors suggest that the accumulation of flavonol glycosides constitutes a reserve of flavonols which could be easily mobilized at any given time and especially under UV stress conditions.

A relationship between aglycone/glycone flavonoids and oxidative stress in *Arabidopsis* was also shown by Kim et al. (2010). In this case, the loss of function of three genes *UGT73B1, UGT73B2*, and *UGT73B3* involved in the glycosylation of flavonoids, led to a greater tolerance to oxidative stress whereas overexpression of *UGT73B2* increased the sensitivity to ROS. The significant role of HCA glycosylation was also highlighted in senescent and water stressed plants (Fini et al., 2012; Torras-Claveria et al., 2012). An accumulation of the coumarin esculetin and quercetin 3-O-glucosides was observed in water stressed *Fraxinus ornus* leaves simultaneously with a decrease in antioxidant enzyme activities (Fini et al., 2012). The authors suggested that these glycosides could act as H_2O_2 scavengers during water stress. These results are also consistent with those of Kylli et al. (2008) who showed that sinapic and ferulic acid glycoside esters were efficient antioxidants. More recently, Babst et al. (2014) showed an up-regulation of *UGT84A17* expression under environmental stress (in particular under N limitation) in transgenic *Populus*, leading to the accumulation of hydroxycinnamate glucose esters (especially caffeoyl-4-coumaroyl- and cinnamoyl-glucose esters). This accumulation was to the detriment of flavonoid glucoside synthesis, which also requires hydroxycinnamates in their free form as intermediates. These authors highlighted a metabolic trade-off associated with stress dependent phenylpropanoid glycosylation (Dauwe et al., 2007). Finally, a novel *Arabidopsis* glycosyltransferase gene *UGT85A5* (Sun et al., 2013) and more recently *UGT85U1* (Ahrazem et al., 2015) are both significantly induced by salt stress. Nevertheless, their connection with the phenylpropanoid pathway and their substrate affinity require more detailed characterization.

Although, the antioxidant proprieties of phenolic molecules have been known for a long time, the precise role(s) of glycosylation and, by extension the role of UGTs and ß-glucosidases in abiotic stresses is difficult to understand because of the existence of numerous signal pathways that can positively or negatively interact with each other. Such cross-talk is sometimes illustrated by the production of common glycosylated phenylpropanoids under different stresses. For example roots of hydroponically-grown *Arabidopsis* plants produce the glycosylated coumarin scopolin and monolignol coniferin when they are submitted to oxidative stress, root wounding, or nitrate deprivation (Ward et al., 2011). A major challenge is therefore to deepen our knowledge on the response mechanisms of plants to environmental changes in order to understand how these glycoconjugates can contribute to plant protection.

Phenylpropanoid Glycosylation: Aroma, Taste, and Color of Plant Products

Flavonoid glycosylation plays a significant role in determining flower, leaf, seed, and fruit color necessary for attracting animals for flower pollination and/or seed dispersion (Tanaka et al., 2008). For example, strawberry fruits contain anthocyanins (glycosides of anthocyanidins) that give them the attractive red color in the ripe fruit (Griesser et al., 2008). Six anthocyanidins are generally present and their glycosylation is usually driven by UDP-glucose:flavonoid-O-glycosyl-transferase (UFGT) (Jaakola, 2013). However, the mechanism is quite complex and takes place in a species-specific manner. These anthocyanidins are most frequently O-glycosylated at the C3- followed by the C5-and/or C7-position (Tanaka et al., 2008). The glycosyl moieties of anthocyanins are themselves usually later modified by aromatic and/or aliphatic acyl moities. After these modifications, anthocyanins become stable compounds and accumulate in vacuoles. The range of these different mechanisms causes a variety of flower and fruit colors (Vogt and Jones, 2000;Grotewold, 2006; He et al., 2015; Li et al., 2016). In the case of strawberry fruits, Griesser et al. (2008) demonstrated that FaGT1, an anthocyanidin-3-O-glucosyltransferase, is a key enzyme in the phenylpropanoid biosynthesis because it channels the flavonoid pathway into anthocyanins and not into the glycosylated flavonols. Anthocyanidin 3-O-glucosyltransferases have also been characterized in many ornamental commercial plants (Tanaka et al., 1996; Yamazaki et al., 2002; Bowles et al., 2005; Sasaki and Nakayama, 2015). Recently, Matsuba et al. (2010) have shown that a novel glucosylation reaction on anthocyanins occurs with 1-*O*-β-D-vanilly-glucose as the sugar donor molecule in the petals of carnation and *Delphinium*.

The glycosylation of flavonoids may also be responsible for fruit flavor and aroma (Frydman et al., 2013; Tikunov et al., 2013; Hjelmeland and Ebeler, 2015). In Citrus, bitter species (e.g. pummelo, grapefruit) accumulate bitter flavonone 7-O-neohesperidosides (neohesperidin and naringin) which are synthesized through 1,2 rhamnosyltransferase activity and non-bitter species (e.g. oranges, mandarins) accumulate flavanone-7-O-rutinosides (hesperidin and narirutin) synthesized by 1,6 rhamnosyltransferases (Frydman et al., 2013). The roles of these different glycosylated flavonoids have not yet been determined but Del Rio et al. (2004) suggested that they are involved in plant defense against pathogens. Phenylpropanoid volatiles or volatile organic compounds (VOCs) are responsible for fruit aroma (Hjelmeland and Ebeler, 2015). In the case of green tomatoes, the aroma is termed "smoky" (Tikunov et al., 2013) and results from the presence of volatile phenylpropanoid diglycosides (with a hexose pentose sugar moiety), which can be easily cleaved. In mature fruits, glycosylation is more complex and triglycosides are synthetized (with a dihexose-pentose moiety) that are less easily cleaved. Tikunov et al. (2013) have characterized a glycosyltransferase named NON-SMOKY GLYCOSYLTRANSFERASE1 (NSGT1) in ripe fruits unable to produce and release the "smoky" aroma. NSGT1 is unable to glycosylate the aglycone form of phenylpropanoids and is only involved in further elongation of the glycosidic moiety of glycosides. The physiological and ecological roles of these aromas have not yet been elucidated. Some authors propose that they may repel (in the case of green fruits) or attract (in ripe fruits) animals (Borges et al., 2008). Glycosyltransferase activities would therefore indirectly affect seed dispersion.

CONCLUSIONS

In response to a wide range of developmental and environmental cues, plants continuously produce (and modify) an extremely large pool of different phenylpropanoid-based secondary metabolites thereby contributing to the success of these autotrophic and, in many cases, sessile organisms. Many plant phenylpropanoids are also of economic interest for humans and a better understanding of the different factors controlling their biosynthesis and compartmentation should facilitate production and plant improvement. It is now becoming clear that the synthesis, localisation and biological activity of plant phenylpropanoids is not only regulated via the transcriptional and post-transcriptional regulation of different biosynthetic genes/enzymes, but also involves the reversible glycosylation of these molecules by regioselective glycosyltransferases (GTs) and glycosyl hydrolases (GHs). A major future research challenge will be to obtain a better understanding of how plants regulate the individual members of these multigenic GT/GH families during development and in response to abiotic and biotic stress. Given the wide diversity of phenylpropanoid structures across the plant kingdom it will be important to obtain information on a number of different species. In addition it will also be necessary to explore the links between GT/GH regulation and the regulation of other genes/enzymes involved in determining phenylpropanoid availability and activity (e.g. phenylpropanoid biosynthesis, peroxidases/laccases, transporters, UDP-sugar production). Such knowledge will be particularly valuable for maintaining/improving cultivated plant production in a context of increased atmospheric CO_2 levels, global warming and modified pathogen biodiversity.

REFERENCES

1. Ahn, Y. O., Shimizu, B., Sakata, K., Gantulga, D., Zhou, C., Bevan, D. R., et al. (2010). Scopolin-hydrolyzing beta-glucosidases in roots of *Arabidopsis.Plant Cell Physiol.* 51, 132–143. doi: 10.1093/pcp/pcp174
2. Ahrazem, O., Rubio-Moraga, A., Trapero-Mozos, A., Climent, M. F., Gomez-Cadenas, A., and Gomez-Gomez, L. (2015). Ectopic expression of a stress-inducible glycosyltransferase from saffron enhances salt and oxidative stress tolerance in *Arabidopsis* while alters anchor root formation. *Plant Sci* 234, 60–73. doi: 10.1016/j.plantsci.2015.02.004
3. Alejandro, S., Lee, Y., Tohge, T., Sudre, D., Osorio, S., Park, J., et al. (2012). AtABCG29 is a monolignol transporter involved in lignin biosynthesis. *Curr. Biol.* 22, 1207–1212. doi: 10.1016/j.cub.2012.04.064
4. Babst, B. A., Chen, H. Y., Wang, H. Q., Payyavula, R. S., Thomas, T. P., Harding, S. A., et al. (2014). Stress-responsive hydroxycinnamate glycosyltransferase modulates phenylpropanoid metabolism in Populus. *J. Exp. Bot.* 65, 4191–4200. doi: 10.1093/jxb/eru192
5. Baiya, S., Hua, Y., Ekkhara, W., and Ketudat Cairns, J. R. (2014). Expression and enzymatic properties of rice (*Oryza sativa* L.) monolignol beta-glucosidases. *Plant Sci.* 227, 101–109. doi: 10.1016/j.plantsci.2014.07.009
6. Barvkar, V. T., Pardeshi, V. C., Kale, S. M., Kadoo, N. Y., and Gupta, V. S. (2012). Phylogenomic analysis of UDP glycosyltransferase 1 multigene family in*Linum usitatissimum* identified genes with varied expression patterns. *BMC Genomics* 13:175. doi: 10.1186/1471-2164-13-175
7. Baumert, A., Milkowski, C., Schmidt, J., Nimtz, M., Wray, V., and Strack, D. (2005). Formation of a complex pattern of sinapate esters in *Brassica napus*seeds, catalyzed by enzymes of a serine carboxypeptidase-like acyltransferase family? *Phytochemistry* 66, 1334–1345. doi: 10.1016/j.phytochem.2005.02.031
8. Bock, K. W. (2016). The UDP-glycosyltransferase (UGT) superfamily expressed in humans, insects and plants: animal-plant arms-race and co-evolution.*Biochem. Pharmacol.* 99, 11–17. doi: 10.1016/j.bcp.2015.10.001
9. Boerjan, W., Ralph, J., and Baucher, M. (2003). Lignin biosynthesis. *Annu. Rev. Plant Biol.* 54, 519–546. doi: 10.1146/annurev.arplant.54.031902.134938
10. Borges, R. M., Bessiere, J. M., and Hossaert-McKey, M. (2008). The chemical ecology of seed dispersal in monoecious and dioecious figs. *Funct. Ecol.* 22, 484–493. doi: 10.1111/j.1365-2435.2008.01383.x
11. Boss, P. K., Davies, C., and Robinson, S. P. (1996). Expression of anthocyanin biosynthesis pathway genes in red and white grapes. *Plant Mol. Biol.* 32, 565–569.

12. Bowles, D., Isayenkova, J., Lim, E.-K., and Poppenberger, B. (2005). Glycosyltransferases: managers of small molecules. *Curr. Opin. Plant Biol.* 8, 254–263. doi: 10.1016/j.pbi.2005.03.007
13. Brazier-Hicks, M., Evans, K. M., Gershater, M. C., Puschmann, H., Steel, P. G., and Edwards, R. (2009). The C-glycosylation of flavonoids in cereals. *J. Biol. Chem.* 284, 17926–17934. doi: 10.1074/jbc.M109.009258
14. Brazier-Hicks, M., Offen, W. A., Gershater, M. C., Revett, T. J., Lim, E. K., Bowles, D. J., et al. (2007). Characterization and engineering of the bifunctional N- and O-glucosyltransferase involved in xenobiotic metabolism in plants. *Proc. Natl. Acad. Sci. U.S.A.* 104, 20238–20243. doi: 10.1073/pnas.0706421104
15. Burlat, V., Kwon, M., Davin, L. B., and Lewis, N. G. (2001). Dirigent proteins and dirigent sites in lignifying tissues. *Phytochemistry* 57, 883–897. doi: 10.1016/S0031-9422(01)00117-0
16. Cao, P. J., Bartley, L. E., Jung, K. H., and Ronald, P. C. (2008). Construction of a rice glycosyltransferase phylogenomic database and identification of rice-diverged glycosyltransferases. *Mol. Plant* 1, 858–877. doi: 10.1093/mp/ssn052
17. Caputi, L., Malnoy, M., Goremykin, V., Nikiforova, S., and Martens, S. (2012). A genome-wide phylogenetic reconstruction of family 1 UDP-glycosyltransferases revealed the expansion of the family during the adaptation of plants to life on land. *Plant J.* 69, 1030–1042. doi: 10.1111/j.1365-313X.2011.04853.x
18. Carter, C., Pan, S., Zouhar, J., Avila, E. L., Girke, T., and Raikhel, N. V. (2004). The vegetative vacuole proteome of *Arabidopsis thaliana* reveals predicted and unexpected proteins. *Plant Cell* 16, 3285–3303. doi: 10.1105/tpc.104.027078
19. Chabannes, M., Barakate, A., Lapierre, C., Marita, J. M., Ralph, J., Pean, M., et al. (2001). Strong decrease in lignin content without significant alteration of plant development is induced by simultaneous down-regulation of cinnamoyl CoA reductase (CCR) and cinnamyl alcohol dehydrogenase (CAD) in tobacco plants. *Plant J.* 28, 257–270. doi: 10.1046/j.1365-313X.2001.01140.x
20. Chantreau, M., Portelette, A., Dauwe, R., Kiyoto, S., Cronier, D., Morreel, K., et al. (2014). Ectopic lignification in the flax lignified bast fiber1 mutant stem is associated with tissue-specific modifications in gene expression and cell wall composition. *Plant Cell* 26, 4462–4482. doi: 10.1105/tpc.114.130443
21. Chapelle, A., Morreel, K., Vanholme, R., Le-Bris, P., Morin, H., Lapierre, C., et al. (2012). Impact of the absence of stem-specific beta-glucosidases on lignin and monolignols. *Plant Physiol.* 160, 1204–1217. doi: 10.1104/pp.112.203364
22. Cheynier, V., Comte, G., Davies, K. M., Lattanzio, V., and Martens, S. (2013). Plant phenolics: recent advances on their biosynthesis, genetics, and ecophysiology. *Plant Physiol. Biochem.* 72, 1–20. doi: 10.1016/j.plaphy.2013.05.009
23. Chong, J., Baltz, R., Schmitt, C., Beffa, R., Fritig, B., and Saindrenan, P. (2002). Downregulation of a pathogen-responsive tobacco UDP-Glc:phenylpropanoid glucosyltransferase reduces scopoletin glucoside accumulation, enhances oxidative stress, and weakens virus resistance. *Plant Cell*14, 1093–1107. doi: 10.1105/tpc.010436
24. Dauwe, R., Morreel, K., Goeminne, G., Gielen, B., Rohde, A., Van Beeumen, J., et al. (2007). Molecular phenotyping of lignin-modified tobacco reveals associated changes in cell-wall metabolism, primary metabolism, stress metabolism and photorespiration. *Plant J.* 52, 263–285. doi: 10.1111/j.1365-313X.2007.03233.x
25. Davies, G. J., Gloster, T. M., and Henrissat, B. (2005). Recent structural insights into the expanding world of carbohydrate-active enzymes. *Curr. Opin. Struct. Biol.* 15, 637–645. doi: 10.1016/j.sbi.2005.10.008
26. Davin, L. B., Wang, H.-B., Crowell, A. L., Bedgar, D. L., Martin, D. M., Sarkanen, S., et al. (1997). Stereoselective bimolecular phenoxy radical coupling by an auxiliary (dirigent) protein without an active center. *Science* 275, 362–367.
27. Day, A., Ruel, K., Neutelings, G., Cronier, D., David, H., Hawkins, S., et al. (2005). Lignification in the flax stem: evidence for an unusual lignin in bast fibers. *Planta* 222, 234–245. doi: 10.1007/s00425-005-1537-1
28. Dean, J. C., Kusaka, R., Walsh, P. S., Allais, F., and Zwier, T. S. (2014). Plant sunscreens in the UV-B: ultraviolet spectroscopy of jet-cooled sinapoyl malate, sinapic acid, and sinapate ester derivatives. *J. Am. Chem. Soc.* 136, 14780–14795. doi: 10.1021/ja5059026
29. Del Rio, D., Stewart, A. J., Mullen, W., Burns, J., Lean, M. E., Brighenti, F., et al. (2004). HPLC-MSn analysis of phenolic compounds and purine alkaloids in green and black tea. *J. Agric. Food Chem.* 52, 2807–2815. doi: 10.1021/jf0354848
30. de Oliveira, D. M., Finger-Teixeira, A., Rodrigues Mota, T., Salvador, V. H., Moreira-Vilar, F. C., Correa Molinari, H. B., et al. (2015). Ferulic acid: a key component in grass lignocellulose recalcitrance to hydrolysis. *Plant Biotechnol. J.* 13, 1224–1232. doi: 10.1111/pbi.12292
31. Dharmawardhana, D. P., Ellis, B. E., and Carlson, J. E. (1995). A beta-glucosidase from lodgepole pine xylem specific for the lignin precursor coniferin.*Plant Physiol.* 107, 331–339.
32. Dima, O., Morreel, K., Vanholme, B., Kim, H., Ralph, J., and Boerjan, W. (2015). Small glycosylated lignin oligomers are stored in *Arabidopsis* leaf vacuoles. *Plant Cell* 27, 695–710. doi: 10.1105/tpc.114.134643
33. Donaldson, L. A. (2001). Lignification and lignin topochemistry - an ultrastructural view. *Phytochemistry* 57, 859–873. doi: 10.1016/S0031-9422(01)00049-8
34. Dorey, S., Baillieul, F., Pierrel, M.-A., Saindrenan, P., Fritig, B., and Kauffmann, S. (1997). Spatial and temporal induction of cell death, defense genes, and accumulation of Salicylic Acid in tobacco leaves reacting hypersensitively to a fungal glycoprotein elicitor. *Mol. Plant Microbe Inter.* 10, 646–655. doi: 10.1094/MPMI.1997.10.5.646

35. Ehlting, J., Buttner, D., Wang, Q., Douglas, C. J., Somssich, I. E., and Kombrink, E. (1999). Three 4-coumarate:coenzyme A ligases in *Arabidopsis thaliana* represent two evolutionarily divergent classes in angiosperms. *Plant J.* 19, 9–20.

36. Ekstrom, A., Taujale, R., McGinn, N., and Yin, Y. (2014). PlantCAZyme: a database for plant carbohydrate-active enzymes. *Database (Oxford)*2014:bau079. doi: 10.1093/database/bau079

37. Escamilla-Trevino, L. L., Chen, W., Card, M. L., Shih, M. C., Cheng, C. L., and Poulton, J. E. (2006). *Arabidopsis thaliana* beta-Glucosidases BGLU45 and BGLU46 hydrolyse monolignol glucosides. *Phytochemistry* 67, 1651–1660. doi: 10.1016/j.phytochem.2006.05.022

38. Fan, T. W., and Conn, E. E. (1985). Isolation and characterization of two cyanogenic beta-glucosidases from flax seeds. *Arch. Biochem. Biophys.* 243, 361–373.

39. Fernandez-Perez, F., Pomar, F., Pedreno, M. A., and Novo-Uzal, E. (2015). Suppression of *Arabidopsis* peroxidase 72 alters cell wall and phenylpropanoid metabolism. *Plant Science* 239, 192–199. doi: 10.1016/j.plantsci.2015.08.001

40. Ferrer, J. L., Austin, M. B., Stewart, C. Jr., and Noel, J. P. (2008). Structure and function of enzymes involved in the biosynthesis of phenylpropanoids. *Plant Physiol. Biochem.* 46, 356–370. doi: 10.1016/j.plaphy.2007.12.009

41. Fini, A., Guidi, L., Ferrini, F., Brunetti, C., Di Ferdinando, M., Biricolti, S., et al. (2012). Drought stress has contrasting effects on antioxidant enzymes activity and phenylpropanoid biosynthesis in Fraxinus ornus leaves: an excess light stress affair? *J. Plant Physiol.* 169, 929–939. doi: 10.1016/j.jplph.2012.02.014

42. Fraissinet-Tachet, L., Baltz, R., Chong, J., Kauffmann, S., Fritig, B., and Saindrenan, P. (1998). Two tobacco genes induced by infection, elicitor and salicylic acid encode glucosyltransferases acting on phenylpropanoids and benzoic acid derivatives, including salicylic acid. *FEBS Lett.* 437, 319–323. doi: 10.1016/s0014-5793(98)01257-5

43. Francisco, R. M., Regalado, A., Ageorges, A. S., Burla, B. J., Bassin, B., Eisenach, C., et al. (2013). ABCC1, an ATP binding cassette protein from grape berry, transports Anthocyanidin 3-O-Glucosides. *Plant Cell* 25, 1840–1854. doi: 10.1105/tpc.112.102152

44. Fraser, C. M., Thompson, M. G., Shirley, A. M., Ralph, J., Schoenherr, J. A., Sinlapadech, T., et al. (2007). Related *Arabidopsis* serine carboxypeptidase-like sinapoylglucose acyltransferases display distinct but overlapping substrate specificities. *Plant Physiol.* 144, 1986–1999. doi: 10.1104/pp.107.098970

45. Freudenberg, K. (1968). *Lignin Biosynthesis*. Berlin; Heidelberg; New York, NY: Springer.

46. Freudenberg, K., and Harkin, J. M. (1963). The glucosides of cambial sap of spruce. *Phytochemistry* 2, 189–193. doi: 10.1016/S0031-9422(00)82980-5

47. Freudenberg, K., and Torres-Serres, J. (1967). Conversion of the phenylalanines in lignin-component glucosides. *Justus Liebigs Ann. Chem.* 703, 225–230.

48. Frydman, A., Liberman, R., Huhman, D. V., Carmeli-Weissberg, M., Sapir-Mir, M., Ophir, R., et al. (2013). The molecular and enzymatic basis of bitter/non-bitter flavor of citrus fruit: evolution of branch-forming rhamnosyltransferases under domestication. *Plant J.* 73, 166–178. doi: 10.1111/tpj.12030

49. Fu, C., Mielenz, J. R., Xiao, X., Ge, Y., Hamilton, C. Y., Rodriguez, M. Jr., et al. (2011). Genetic manipulation of lignin reduces recalcitrance and improves ethanol production from switchgrass. *Proc. Natl. Acad. Sci. U.S.A.* 108, 3803–3808. doi: 10.1073/pnas.1100310108

50. Gachon, C. M., Langlois-Meurinne, M., and Saindrenan, P. (2005). Plant secondary metabolism glycosyltransferases: the emerging functional analysis. *Trends Plant Sci.* 10, 542–549. doi: 10.1016/j.tplants.2005.09.007

51. Gomez-Anduro, G., Ceniceros-Ojeda, E. A., Casados-Vazquez, L. E., Bencivenni, C., Sierra-Beltran, A., Murillo-Amador, B., et al. (2011). Genome-wide analysis of the beta-glucosidase gene family in maize (*Zea mays* L. var B73). *Plant Mol. Biol.* 77, 159–183. doi: 10.1007/s11103-011-9800-2

52. Griesser, M., Hoffmann, T., Bellido, M. L., Rosati, C., Fink, B., Kurtzer, R., et al. (2008). Redirection of flavonoid biosynthesis through the down-regulation of an anthocyanidin glucosyltransferase in ripening strawberry fruit. *Plant Physiol.* 146, 1528–1539. doi: 10.1104/pp.107.114280

53. Grotewold, E. (2006). The genetics and biochemistry of floral pigments. *Annu. Rev. Plant Biol.* 57, 761–780. doi: 10.1146/annurev.arplant.57.032905.105248

54. Grubb, C. D., Zipp, B. J., Ludwig-Muller, J., Masuno, M. N., Molinski, T. F., and Abel, S. (2004). *Arabidopsis* glucosyltransferase UGT74B1 functions in glucosinolate biosynthesis and auxin homeostasis. *Plant J.* 40, 893–908. doi: 10.1111/j.1365-313X.2004.02261.x

55. Ha, X., Koopmann, B., and von Tiedemann, A. (2016). Wheat blast and fusarium head blight display contrasting interaction patterns on ears of wheat genotypes differing in resistance. *Phytopathology* 106, 270–281. doi: 10.1094/PHYTO-09-15-0202-R

56. Hammond-Kosack, K. E., and Jones, J. D. (1996). Resistance gene-dependent plant defense responses. *Plant Cell* 8, 1773–1791.

57. Hano, C., Martin, I., Fliniaux, O., Legrand, B., Gutierrez, L., Arroo, R. R., et al. (2006). Pinoresinol-lariciresinol reductase gene expression and secoisolariciresinol diglucoside accumulation in developing flax (*Linum usitatissimum*) seeds. *Planta* 224, 1291–1301. doi: 10.1007/s00425-006-0308-y

58. Harborne, J. B., and Williams, C. A. (2001). Anthocyanins and other flavonoids. *Nat. Prod. Rep.* 18, 310–333. doi: 10.1039/b006257j

59. He, F., Chen, W. K., Yu, K. J., Ji, X. N., Duan, C. Q., Reeves, M. J., et al. (2015). Molecular and biochemical characterization of the UDP-glucose: Anthocyanin 5-O-glucosyltransferase from *Vitis amurensis*. *Phytochemistry* 117, 363–372. doi: 10.1016/j.phytochem.2015.06.023

60. He, L., and Terashima, N. (1990). Formation and structure of lignin in Monocotyledons. III. heterogeneity of sugarcane (*Saccharum officinarum L.*) lignin with respect to the composition of structural units in different morphological regions. *J. Wood Chem. Technol.* 10, 435–459. doi: 10.1080/02773819008050251

61. He, L., and Terashima, N. (1991). Formation and structure of lignin in Monocotyledons IV. Deposition process and structural diversity of the lignin in the cellwall of sugarcane and rice plant studied by Ultraviolet Microscopic Spectroscopy. *Holzforschung* 45, 191–198. doi: 10.1515/hfsg.1991.45.3.191

62. Hectors, K., Van Oevelen, S., Geuns, J., Guisez, Y., Jansen, M. A., and Prinsen, E. (2014). Dynamic changes in plant secondary metabolites during UV acclimation in *Arabidopsis thaliana*. *Physiol. Plant.* 152, 219–230. doi: 10.1111/ppl.12168

63. Hemm, M. R., Rider, S. D., Ogas, J., Murry, D. J., and Chapple, C. (2004). Light induces phenylpropanoid metabolism in *Arabidopsis* roots. *Plant J.* 38, 765–778. doi: 10.1111/j.1365-313X.2004.02089.x

64. Herrero, J., Esteban-Carrasco, A., and Zapata, J. M. (2013). Looking for *Arabidopsis thaliana* peroxidases involved in lignin biosynthesis. *Plant Physiol. Biochem.* 67, 77–86. doi: 10.1016/j.plaphy.2013.02.019

65. Hettwer, K., Böttcher, C., Frolov, A., Mittasch, J., Albert, A., von Roepenack-Lahaye, E., et al. (2016). Dynamic metabolic changes in seeds and seedlings of *Brassica napus* (oilseed rape) suppressing UGT84A9 reveal plasticity and molecular regulation of the phenylpropanoid pathway. *Phytochemistry* 124, 46–57. doi: 10.1016/j.phytochem.2016.01.014

66. Hjelmeland, A., and Ebeler, S. (2015). Glycosidically bound volatile aroma compounds in grapes and wine: a review. *Am. J. Enol. Vitic.* 66, 1–11. doi: 10.5344/ajev.2014.14104

67. Holbein, J., Grundler, F. M., and Siddique, S. (2016). Plant basal resistance to nematodes: an update. *J. Exp. Bot.* 67, 2049–2061. doi: 10.1093/jxb/erw005

68. Holl, J., Vannozzi, A., Czemmel, S., D'Onofrio, C., Walker, A. R., Rausch, T., et al. (2013). The R2R3-MYB transcription factors MYB14 and MYB15 regulate stilbene biosynthesis in *Vitis vinifera*. *Plant Cell* 25, 4135–4149. doi: 10.1105/tpc.113.117127

69. Hösel, W., Surholt, E., and Borgmann, E. (1978). Characterization of beta-glucosidase isoenzymes possibly involved in lignification from chick pea (*Cicer arietinum* L.) cell suspension cultures. *Eur. J. Biochem.* 84, 487–492. doi: 10.1111/j.1432-1033.1978.tb12190.x

70. Huang, J., Pang, C., Fan, S., Song, M., Yu, J., Wei, H., et al. (2015). Genome-wide analysis of the family 1 glycosyltransferases in cotton. *Mol. Genet. Genomics* 290, 1805–1818. doi: 10.1007/s00438-015-1040-8

71. Huis, R., Morreel, K., Fliniaux, O., Lucau-Danila, A., Fenart, S., Grec, S., et al. (2012). Natural hypolignification is associated with extensive oligolignol accumulation in flax stems. *Plant Physiol.* 158, 1893–1915. doi: 10.1104/pp.111.192328

72. Imran, M., Ahmad, N., Anjum, F. M., Khan, M. K., Mushtaq, Z., Nadeem, M., et al. (2015). Potential protective properties of flax lignan secoisolariciresinol diglucoside. *Nutr. J.* 14, 1–7. doi: 10.1186/s12937-015-0059-3

73. Ishihara, H., Tohge, T., Viehover, P., Fernie, A. R., Weisshaar, B., and Stracke, R. (2016). Natural variation in flavonol accumulation in *Arabidopsis* is determined by the flavonol glucosyltransferase BGLU6. *J. Exp. Bot.* 67, 1505–1517. doi: 10.1093/jxb/erv546

74. Jaakola, L. (2013). New insights into the regulation of anthocyanin biosynthesis in fruits. *Trends Plant Sci.* 18, 477–483. doi: 10.1016/j.tplants.2013.06.003

75. Jackson, R., Knisley, D., McIntosh, C., and Pfeiffer, P. (2011). Predicting Flavonoid UGT Regioselectivity. *Adv. Bioinformat.* 2011:506583. doi: 10.1155/2011/506583

76. Jones, P., and Vogt, T. (2001). Glycosyltransferases in secondary plant metabolism: tranquilizers and stimulant controllers. *Planta* 213, 164–174. doi: 10.1007/s004250000492

77. Kai, K., Shimizu, B., Mizutani, M., Watanabe, K., and Sakata, K. (2006). Accumulation of coumarins in *Arabidopsis thaliana*. *Phytochemistry* 67, 379–386. doi: 10.1016/j.phytochem.2005.11.006

78. Kaneda, M., Rensing, K. H., Wong, J. C., Banno, B., Mansfield, S. D., and Samuels, A. L. (2008). Tracking monolignols during wood development in lodgepole pine. *Plant Physiol.* 147, 1750–1760. doi: 10.1104/pp.108.121533

79. Ketudat Cairns, J. R., and Esen, A. (2010). beta-Glucosidases. *Cell. Mol. Life Sci.* 67, 3389–3405. doi: 10.1007/s00018-010-0399-2

80. Ketudat Cairns, J. R., Mahong, B., Baiya, S., and Jeon, J. S. (2015). beta-Glucosidases: multitasking, moonlighting or simply misunderstood? *Plant Sci.*241, 246–259. doi: 10.1016/j.plantsci.2015.10.014

81. Kim, S. J., Kwon do, Y., Kim, Y. S., and Kim, Y. C. (2010). Peroxyl radical scavenging capacity of extracts and isolated components from selected medicinal plants. *Arch. Pharm. Res.* 33, 867–873. doi: 10.1007/s12272-010-0609-3

82. Kitamura, S., Shikazono, N., and Tanaka, A. (2004). TRANSPARENT TESTA 19 is involved in the accumulation of both anthocyanins and proanthocyanidins in *Arabidopsis*. *Plant J.* 37, 104–114. doi: 10.1046/j.1365-313X.2003.01943.x

83. Koeduka, T., Sugimoto, K., Watanabe, B., Someya, N., Kawanishi, D., Gotoh, T., et al. (2014). Bioactivity of natural O-prenylated phenylpropenes from Illicium anisatum leaves and their derivatives against spider mites and fungal pathogens. *Plant Biol.* 16, 451–456. doi: 10.1111/plb.12054

84. Kojima, M., Poulton, J. E., Thayer, S. S., and Conn, E. E. (1979). Tissue distributions of dhurrin and of enzymes involved in its metabolism in leaves of*Sorghum bicolor*. *Plant Physiol.* 63, 1022–1028.

85. König, S., Feussner, K., Kaever, A., Landesfeind, M., Thurow, C., Karlovsky, P., et al. (2014). Soluble

phenylpropanoids are involved in the defense response of *Arabidopsis* against *Verticillium longisporum*. *New Phytol.* 202, 823–837. doi: 10.1111/nph.12709

86. Kubo, H., Nawa, N., and Lupsea, S. A. (2007). Anthocyaninless1 gene of *Arabidopsis thaliana* encodes a UDP-glucose:flavonoid-3-O-glucosyltransferase.*J. Plant Res.* 120, 445–449. doi: 10.1007/s10265-006-0067-7
87. Kuhlmann, S., Kranz, K., Lücking, B., Alfermann, A. W., and Petersen, M. (2002). Aspects of cytotoxic lignan biosynthesis in suspension cultures of *Linum nodiflorum*. *Phytochem. Rev.* 1, 37–43. doi: 10.1023/a:1015876001812
88. Kylli, P., Nousiainen, P., Biely, P., Sipila, J., Tenkanen, M., and Heinonen, M. (2008). Antioxidant potential of hydroxycinnamic acid glycoside esters. *J. Agric. Food Chem.* 56, 4797–4805. doi: 10.1021/jf800317v
89. Landry, L. G., Chapple, C. C., and Last, R. L. (1995). *Arabidopsis* mutants lacking phenolic sunscreens exhibit enhanced ultraviolet-B injury and oxidative damage. *Plant Physiol.* 109, 1159–1166.
90. Langenbach, C., Campe, R., Schaffrath, U., Goellner, K., and Conrath, U. (2013). UDP-glucosyltransferase UGT84A2/BRT1 is required for *Arabidopsis*nonhost resistance to the Asian soybean rust pathogen *Phakopsora pachyrhizi*. *New Phytol.* 198, 536–545. doi: 10.1111/nph.12155
91. Langlois-Meurinne, M., Gachon, C. M., and Saindrenan, P. (2005). Pathogen-responsive expression of glycosyltransferase genes UGT73B3 and UGT73B5 is necessary for resistance to *Pseudomonas syringae* pv tomato in *Arabidopsis*. *Plant Physiol.* 139, 1890–1901. doi: 10.1104/pp.105.067223
92. Lanot, A., Hodge, D., Jackson, R. G., George, G. L., Elias, L., Lim, E. K., et al. (2006). The glucosyltransferase UGT72E2 is responsible for monolignol 4-O-glucoside production in *Arabidopsis thaliana*. *Plant J.* 48, 286–295. doi: 10.1111/j.1365-313X.2006.02872.x
93. Lanot, A., Hodge, D., Lim, E. K., Vaistij, F. E., and Bowles, D. J. (2008). Redirection of flux through the phenylpropanoid pathway by increased glucosylation of soluble intermediates. *Planta* 228, 609–616. doi: 10.1007/s00425-008-0763-8
94. Larsen, E. S., Alfenito, M. R., Briggs, W. R., and Walbot, V. (2003). A carnation anthocyanin mutant is complemented by the glutathione S-transferases encoded by maize Bz2 and petunia An9. *Plant Cell Rep.* 21, 900–904. doi: 10.1007/s00299-002-0545-x
95. Leinhos, V., and Savidge, R. A. (1993). Isolation of protoplasts from developing xylem of Pinus banksiana and Pinus strobus. *Can. J. Forest Res.* 23, 343–348. doi: 10.1139/x93-050
96. Li, X., Bergelson, J., and Chapple, C. (2010). The *Arabidopsis* Accession Pna-10 is a naturally occurring sng1 deletion mutant. *Mol. Plant* 3, 91–100. doi: 10.1093/mp/ssp090
97. Li, X. J., Zhang, J. Q., Wu, Z. C., Lai, B., Huang, X. M., Qin, Y. H., et al. (2016). Functional characterization of a glucosyltransferase gene, *LcUFGT1*, involved in the formation of cyanidin glucoside in the pericarp of *Litchi chinensis*. *Physiol. Plant.* 156, 139–149. doi: 10.1111/ppl.12391
98. Li, Y., Li, P., Wang, Y., Dong, R., Yu, H., and Hou, B. (2014). Genome-wide identification and phylogenetic analysis of Family-1 UDP glycosyltransferases in maize (*Zea mays*). *Planta* 239, 1265–1279. doi: 10.1007/s00425-014-2050-1
99. Lim, E. K., and Bowles, D. J. (2004). A class of plant glycosyltransferases involved in cellular homeostasis. *EMBO J.* 23, 2915–2922. doi: 10.1038/sj.emboj.7600295
100. Lim, E. K., Higgins, G. S., Li, Y., and Bowles, D. J. (2003). Regioselectivity of glucosylation of caffeic acid by a UDP-glucose:glucosyltransferase is maintained in planta. *Biochem. J.* 373, 987–992. doi: 10.1042/bj20021453
101. Lim, E. K., Jackson, R. G., and Bowles, D. J. (2005). Identification and characterisation of *Arabidopsis* glycosyltransferases capable of glucosylating coniferyl aldehyde and sinapyl aldehyde. *FEBS Lett.* 579, 2802–2806. doi: 10.1016/j.febslet.2005.04.016
102. Lim, E. K., Li, Y., Parr, A., Jackson, R., Ashford, D. A., and Bowles, D. J. (2001). Identification of glucosyltransferase genes involved in sinapate metabolism and lignin synthesis in *Arabidopsis*. *J. Biol. Chem.* 276, 4344–4349. doi: 10.1074/jbc.M007263200
103. Lombard, V., Golaconda Ramulu, H., Drula, E., Coutinho, P. M., and Henrissat, B. (2014). The carbohydrate-active enzymes database (CAZy) in 2013.*Nucleic Acids Res.* 42, D490–D495. doi: 10.1093/nar/gkt1178
104. Lorenc-Kukula, K., Amarowicz, R., Oszmianski, J., Doermann, P., Starzycki, M., Skala, J., et al. (2005). Pleiotropic effect of phenolic compounds content increases in transgenic flax plant. *J. Agric. Food Chem.* 53, 3685–3692. doi: 10.1021/jf047987z
105. Luang, S., Cho, J. I., Mahong, B., Opassiri, R., Akiyama, T., Phasai, K., et al. (2013). Rice Os9BGlu31 is a transglucosidase with the capacity to equilibrate phenylpropanoid, flavonoid, and phytohormone glycoconjugates. *J. Biol. Chem.* 288, 10111–10123. doi: 10.1074/jbc.M112.423533
106. Lunkenbein, S., Bellido, M., Aharoni, A., Salentijn, E. M., Kaldenhoff, R., Coiner, H. A., et al. (2006). Cinnamate metabolism in ripening fruit. Characterization of a UDP-glucose:cinnamate glucosyltransferase from strawberry. *Plant Physiol.* 140, 1047–1058. doi: 10.1104/pp.105.074955
107. Mackenzie, P. I., Owens, I. S., Burchell, B., Bock, K. W., Bairoch, A., Belanger, A., et al. (1997). The UDP glycosyltransferase gene superfamily: recommended nomenclature update based on evolutionary divergence. *Pharmacogenetics* 7, 255–269.
108. Marcinowski, S., and Grisebach, H. (1978). Enzymology of lignification: cell wall beta-glucosidase for coniferin from spruce (*Picea abies*) seedlings. *Eur. J. Biochem.* 87, 37–44.
109. Marjamaa, K., Lehtonen, M., Lundell, T., Toikka, M., Saranpaa, P., and Fagerstedt, K. V. (2003). Developmental lignification and seasonal variation in beta-glucosidase

and peroxidase activities in xylem of Scots pine, Norway spruce and silver birch. *Tree Physiol.* 23, 977–986. doi: 10.1093/treephys/23.14.977

110. Marrs, K. A., Alfenito, M. R., Lloyd, A. M., and Walbot, V. (1995). A glutathione S-transferase involved in vacuolar transfer encoded by the maize gene Bronze-2. *Nature* 375, 397–400.

111. Masada, S., Terasaka, K., Oguchi, Y., Okazaki, S., Mizushima, T., and Mizukami, H. (2009). Functional and structural characterization of a flavonoid glucoside 1,6-glucosyltransferase from Catharanthus roseus. *Plant Cell Physiol.* 50, 1401–1415. doi: 10.1093/pcp/pcp088

112. Matros, A., and Mock, H. P. (2004). Ectopic expression of a UDP-glucose:phenylpropanoid glucosyltransferase leads to increased resistance of transgenic tobacco plants against infection with Potato Virus Y. *Plant Cell Physiol.* 45, 1185–1193. doi: 10.1093/pcp/pch140

113. Matsuba, Y., Sasaki, N., Tera, M., Okamura, M., Abe, Y., Okamoto, E., et al. (2010). A novel glucosylation reaction on anthocyanins catalyzed by acyl-glucose-dependent glucosyltransferase in the petals of carnation and delphinium. *Plant Cell* 22, 3374–3389. doi: 10.1105/tpc.110.077487

114. Meißner, D., Albert, A., Böttcher, C., Strack, D., and Milkowski, C. (2008). The role of UDP-glucose:hydroxycinnamate glucosyltransferases in phenylpropanoid metabolism and the response to UV-B radiation in *Arabidopsis thaliana*. *Planta* 228, 663–674. doi: 10.1007/s00425-008-0768-3

115. Miao, Y.-C., and Liu, C.-J. (2010). ATP-binding cassette-like transporters are involved in the transport of lignin precursors across plasma and vacuolar membranes. *Proc. Natl. Acad. Sci. U.S.A.* 107, 22728–22733. doi: 10.1073/pnas.1007747108

116. Mierziak, J., Kostyn, K., and Kulma, A. (2014). Flavonoids as important molecules of plant interactions with the environment. *Molecules* 19, 16240–16265. doi: 10.3390/molecules191016240

117. Miguel, V. N., Otero, J. A., García-Villalba, R. O., Tomás-Barberán, F., Espín, J. C., Merino, G., et al. (2014). Role of ABCG2 in Transport of the Mammalian Lignan Enterolactone and its Secretion into Milk in Abcg2 Knockout Mice. *Drug Metab. Dispos.* 42, 943–946. doi: 10.1124/dmd.113.055970

118. Milkowski, C., Baumert, A., and Strack, D. (2000). Identification of four *Arabidopsis* genes encoding hydroxycinnamate glucosyltransferases. *FEBS Lett.*486, 183–184. doi: 10.1016/S0014-5793(00)02270-5

119. Milkowski, C., and Strack, D. (2010). Sinapate esters in *brassicaceous* plants: biochemistry, molecular biology, evolution and metabolic engineering.*Planta* 232, 19–35. doi: 10.1007/s00425-010-1168-z

120. Mittasch, J., Strack, D., and Milkowski, C. (2007). Secondary product glycosyltransferases in seeds of *Brassica napus*. *Planta* 225, 515–522. doi: 10.1007/s00425-006-0360-7

121. Mock, H.-P., and Strack, D. (1993). Energetics of the uridine 5′-diphosphoglucose: Hydroxycinnamic acid acyl-glucosyltransferase reaction.*Phytochemistry* 32, 575–579. doi: 10.1016/S0031-9422(00)95139-2

122. Nair, R. B., Bastress, K. L., Ruegger, M. O., Denault, J. W., and Chapple, C. (2004). The *Arabidopsis thaliana* REDUCED EPIDERMAL FLUORESCENCE1 gene encodes an aldehyde dehydrogenase involved in ferulic acid and sinapic acid biosynthesis. *Plant Cell* 16, 544–554. doi: 10.1105/tpc.017509

123. Noguchi, A., Horikawa, M., Fukui, Y., Fukuchi-Mizutani, M., Iuchi-Okada, A., Ishiguro, M., et al. (2009). Local differentiation of sugar donor specificity of flavonoid glycosyltransferase in Lamiales. *Plant Cell* 21, 1556–1572. doi: 10.1105/tpc.108.063826

124. Oba, K., Conn, E. E., Canut, H., and Boudet, A. M. (1981). Subcellular localization of 2-(beta-d-Glucosyloxy)-Cinnamic Acids and the related beta-glucosidase in Leaves of Melilotus alba Desr. *Plant Physiol.* 68, 1359–1363.

125. Obayashi, T., Hayashi, S., Saeki, M., Ohta, H., and Kinoshita, K. (2009). ATTED-II provides coexpressed gene networks for *Arabidopsis*. *Nucleic Acids Res.* 37, D987–D991. doi: 10.1093/nar/gkn807

126. Ohkama-Ohtsu, N., Sasaki-Sekimoto, Y., Oikawa, A., Jikumaru, Y., Shinoda, S., Inoue, E., et al. (2011). 12-Oxo-Phytodienoic Acid–glutathione conjugate is transported into the vacuole in *Arabidopsis*. *Plant Cell Physiol.* 52, 205–209. doi: 10.1093/pcp/pcq181

127. Opassiri, R., Pomthong, B., Onkoksoong, T., Akiyama, T., Esen, A., and Ketudat Cairns, J. R. (2006). Analysis of rice glycosyl hydrolase family 1 and expression of Os4bglu12 beta-glucosidase. *BMC Plant Biol.* 6:33. doi: 10.1186/1471-2229-6-33

128. Osmani, S. A., Bak, S., Imberty, A., Olsen, C. E., and Moller, B. L. (2008). Catalytic key amino acids and UDP-sugar donor specificity of a plant glucuronosyltransferase, UGT94B1: molecular modeling substantiated by site-specific mutagenesis and biochemical analyses. *Plant Physiol.* 148, 1295–1308. doi: 10.1104/pp.108.128256

129. Pickett-Heaps, J. D. (1968). Xylem wall deposition. radioautographic investigations using lignin precursors. *Protoplasma* 65, 181–205.

130. Poppenberger, B., Berthiller, F., Lucyshyn, D., Sieberer, T., Schuhmacher, R., Krska, R., et al. (2003). Detoxification of the Fusarium mycotoxin deoxynivalenol by a UDP-glucosyltransferase from *Arabidopsis thaliana*. *J. Biol. Chem.* 278, 47905–47914. doi: 10.1074/jbc.M307552200

131. Prasad, K. (1997). Hydroxyl radical-scavenging property of secoisolariciresinol diglucoside (SDG) isolated from flax-seed. *Mol. Cell Biochem.* 168, 117–123.

132. Pusztahelyi, T., Holb, I. J., and Pocsi, I. (2015). Secondary metabolites in fungus-plant interactions. *Front. Plant Sci.* 6:573. doi: 10.3389/fpls.2015.00573

133. Ralph, S., Park, J.-Y., Bohlmann, J., and Mansfield, S. D. (2006). Dirigent proteins in conifer defense: gene discovery, phylogeny, and differential wound- and insect-induced expression of a family of DIR and DIR-like genes in spruce (*Picea* spp.). *Plant Mol. Biol.* 60, 21–40. doi: 10.1007/s11103-005-2226-y

134. Rasmussen, S., and Dixon, R. A. (1999). Transgene-mediated and elicitor-induced perturbation of metabolic channeling at the entry point into the phenylpropanoid pathway. *Plant Cell* 11, 1537–1552.

135. Roepke, J., and Bozzo, G. G. (2015). *Arabidopsis thaliana* beta-glucosidase BGLU15 attacks flavonol 3-O-beta-glucoside-7-O-alpha-rhamnosides. *Phytochemistry* 109, 14–24. doi: 10.1016/j.phytochem.2014.10.028

136. Ross, J., Li, Y., Lim, E., and Bowles, D. J. (2001). Higher plant glycosyltransferases. *Genome Biol.* 2, REVIEWS3004. doi: 10.1186/gb-2001-2-2-reviews3004

137. Rouyi, C., Baiya, S., Lee, S.-K., Mahong, B., Jeon, J.-S., Ketudat-Cairns, J. R., et al. (2014). Recombinant expression and characterization of the cytoplasmic rice β-Glucosidase Os1BGlu4. *PLoS ONE* 9:e96712. doi: 10.1371/journal.pone.0096712

138. Sablowski, R. W., Moyano, E., Culianez-Macia, F. A., Schuch, W., Martin, C., and Bevan, M. (1994). A flower-specific Myb protein activates transcription of phenylpropanoid biosynthetic genes. *EMBO J.* 13, 128–137.

139. Sasaki, N., and Nakayama, T. (2015). Achievements and perspectives in biochemistry concerning anthocyanin modification for blue flower coloration. *Plant Cell Physiol.* 56, 28–40. doi: 10.1093/pcp/pcu097

140. Schäfer, J., Caldwell, R., Patterson, F., and Compton, L. (1963). Wheat leaf rust combinations. *Phytopathology* 53, 569–573.

141. Schmidt, S., Rainieri, S., Witte, S., Matern, U., and Martens, S. (2011). Identification of a *Saccharomyces cerevisiae* glucosidase that hydrolyzes flavonoid glucosides. *Appl. Environ. Microbiol.* 77, 1751–1757. doi: 10.1128/AEM.01125-10

142. Sekhwal, M. K., Sharma, V., and Sarin, R. (2013). Annotation of glycoside hydrolases in *Sorghum bicolor* using proteins interaction approach. *J. Proteome Sci. Comput. Biol.* 2:2. doi: 10.7243/2050-2273-2-2

143. Sermsuvityawong, K., Svasti, M. R. J., Sawang-Areetrakul, P., Kisamanonta, P., and Chulavatnatol, M. (1995). Aggregation of *Cassava linamarase*. *J. Sci. Soc. Thailand* 21, 283–292.

144. Sharma, R., Rawat, V., and Suresh, C. G. (2014). Genome-wide identification and tissue-specific expression analysis of UDP-glycosyltransferases genes confirm their abundance in *Cicer arietinum* (Chickpea) genome. *PLoS ONE* 9:e109715. doi: 10.1371/journal.pone.0109715

145. Shimizu, B. (2014). 2-Oxoglutarate-dependent dioxygenases in the biosynthesis of simple coumarins. *Front. Plant Sci.* 5:549. doi: 10.3389/fpls.2014.00549

146. Simon, C., Langlois-Meurinne, M., Didierlaurent, L., Chaouch, S., Bellvert, F., Massoud, K., et al. (2014). The secondary metabolism glycosyltransferases UGT73B3 and UGT73B5 are components of redox status in resistance of *Arabidopsis* to *Pseudomonas syringae* pv. tomato. *Plant Cell Environ.* 37, 1114–1129. doi: 10.1111/pce.12221

147. Sinlapadech, T., Stout, J., Ruegger, M. O., Deak, M., and Chapple, C. (2007). The hyper-fluorescent trichome phenotype of the brt1 mutant of *Arabidopsis* is the result of a defect in a sinapic acid: UDPG glucosyltransferase. *Plant J.* 49, 655–668. doi: 10.1111/j.1365-313X.2006.02984.x

148. Song, C., Gu, L., Liu, J., Zhao, S., Hong, X., Schulenburg, K., et al. (2015). Functional characterization and substrate promiscuity of UGT71 Glycosyltransferases from Strawberry (Fragaria x ananassa). *Plant Cell Physiol.* 56, 2478–2493. doi: 10.1093/pcp/pcv151

149. Stafford, H. A. (1991). Flavonoid evolution: an enzymic approach. *Plant Physiol.* 96, 680–685.

150. Sterjiades, R., Dean, J. F., and Eriksson, K. E. (1992). Laccase from sycamore maple (*Acer pseudoplatanus*) polymerizes monolignols. *Plant Physiol.* 99, 1162–1168.

151. Strack, D. (2001). Enzymes involved in hydroxycinnamate metabolism. *Methods Enzymol.* 335, 70–81. doi: 10.1016/S0076-6879(01)35232-1

152. Sun, Y.-G., Wang, B., Jin, S.-H., Qu, X.-X., Li, Y.-J., and Hou, B.-K. (2013). Ectopic expression of *Arabidopsis* Glycosyltransferase *UGT85A5* enhances salt stress tolerance in tobacco. *PLoS ONE* 8:e59924. doi: 10.1371/journal.pone.0059924

153. Tanaka, Y., Sasaki, N., and Ohmiya, A. (2008). Biosynthesis of plant pigments: anthocyanins, betalains and carotenoids. *Plant J.* 54, 733–749. doi: 10.1111/j.1365-313X.2008.03447.x

154. Tanaka, Y., Yonekura, K., Fukuchi-Mizutani, M., Fukui, Y., Fujiwara, H., Ashikari, T., et al. (1996). Molecular and biochemical characterization of three anthocyanin synthetic enzymes from Gentiana triflora. *Plant Cell Physiol.* 37, 711–716.

155. Terashima, N., and Fukushima, K. (1988). Heterogeneity in formation of lignin—XI: an autoradiographic study of the heterogeneous formation and structure of pine lignin. *Wood Sci. Technol.* 22, 259–270. doi: 10.1007/bf00386021

156. Tikunov, Y. M., Molthoff, J., de Vos, R. C., Beekwilder, J., van Houwelingen, A., van der Hooft, J. J., et al. (2013). Non-smoky glycosyltransferase1 prevents the release of smoky aroma from tomato fruit. *Plant Cell* 25, 3067–3078. doi: 10.1105/tpc.113.114231

157. Timell, T. (1986). *Compression Wood in Gymnosperms, Vol 1.* Berlin; Heidelberg; New York, NY; Tokyo: Springer.

158. Tobimatsu, Y., Elumalai, S., Grabber, J. H., Davidson, C. L., Pan, X., and Ralph, J. (2012). Hydroxycinnamate conjugates as potential monolignol replacements: *in vitro* lignification and cell wall studies with rosmarinic acid. *ChemSusChem* 5, 676–686. doi: 10.1002/cssc.201100573

159. Tohge, T., Nishiyama, Y., Hirai, M. Y., Yano, M., Nakajima, J., Awazuhara, M., et al. (2005). Functional genomics by integrated analysis of metabolome and transcriptome of *Arabidopsis* plants over-expressing an MYB transcription factor. *Plant J.* 42, 218–235. doi: 10.1111/j.1365-313X.2005.02371.x

160. Torras-Claveria, L., Jauregui, O., Codina, C., Tiburcio, A. F., Bastida, J., and Viladomat, F. (2012). Analysis of phenolic compounds by high-performance liquid chromatography coupled to electrospray ionization tandem mass spectrometry in senescent and water-stressed tobacco. *Plant Sci.* 182, 71–78. doi: 10.1016/j.plantsci.2011.02.009

161. Trapero, A., Ahrazem, O., Rubio-Moraga, A., Jimeno, M. L., Gomez, M. D., and Gomez-Gomez, L. (2012). Characterization of a glucosyltransferase enzyme involved in the formation of kaempferol and quercetin sophorosides in *Crocus sativus*. *Plant Physiol.* 159, 1335–1354. doi: 10.1104/pp.112.198069

162. Tschaplinski, T. J., Standaert, R. F., Engle, N. L., Martin, M. Z., Sangha, A. K., Parks, J. M., et al. (2012). Down-regulation of the caffeic acid O-methyltransferase gene in switchgrass reveals a novel monolignol analog. *Biotechnol. Biofuels* 5:71. doi: 10.1186/1754-6834-5-71

163. Tsuji, Y., Chen, F., Yasuda, S., and Fukushima, K. (2005). Unexpected behavior of coniferin in lignin biosynthesis of Ginkgo biloba L. *Planta* 222, 58–69. doi: 10.1007/s00425-005-1517-5

164. Tsuyama, T., Kawai, R., Shitan, N., Matoh, T., Sugiyama, J., Yoshinaga, A., et al. (2013). Proton-dependent coniferin transport, a common major transport event in differentiating xylem tissue of woody plants. *Plant Physiol.* 162, 918–926. doi: 10.1104/pp.113.214957

165. Uzal, E. N., Gomez-Ros, L. V., Hernandez, J. A., Pedreno, M. A., Cuello, J., and Ros Barcelo, A. (2009). Analysis of the soluble cell wall proteome of gymnosperms. *J. Plant Physiol.* 166, 831–843. doi: 10.1016/j.jplph.2008.11.009

166. Väisänen, E. E., Smeds, A. I., Fagerstedt, K. V., Teeri, T. H., Willför, S. M., and Kärkönen, A. (2015). Coniferyl alcohol hinders the growth of tobacco BY-2 cells and *Nicotiana benthamiana* seedlings. *Planta* 242, 747–760. doi: 10.1007/s00425-015-2348-7

167. Vanholme, R., Storme, V., Vanholme, B., Sundin, L., Christensen, J. H., Goeminne, G., et al. (2012). A systems biology view of responses to lignin biosynthesis perturbations in *Arabidopsis*. *Plant Cell* 24, 3506–3529. doi: 10.1105/tpc.112.102574

168. Vila-Real, H., Alfaia, A. J., Bronze, M. R., Calado, A. R., and Ribeiro, M. H. (2011). Enzymatic synthesis of the flavone glucosides, prunin and isoquercetin, and the aglycones, naringenin and quercetin, with selective alpha-L-Rhamnosidase and beta-D-Glucosidase Activities of Naringinase. *Enzyme Res.*2011:692618. doi: 10.4061/2011/692618

169. Voelker, S. L., Lachenbruch, B., Meinzer, F. C., Jourdes, M., Ki, C., Patten, A. M., et al. (2010). Antisense down-regulation of 4CL expression alters lignification, tree growth, and saccharification potential of field-grown poplar. *Plant Physiol.* 154, 874–886. doi: 10.1104/pp.110.159269

170. Vogt, T. (2010). Phenylpropanoid biosynthesis. *Mol. Plant* 3, 2–20. doi: 10.1093/mp/ssp106

171. Vogt, T., and Jones, P. (2000). Glycosyltransferases in plant natural product synthesis: characterization of a supergene family. *Trends Plant Sci.* 5, 380–386. doi: 10.1016/s1360-1385(00)01720-9

172. Walle, T., Browning, A. M., Steed, L. L., Reed, S. G., and Walle, U. K. (2005). Flavonoid glucosides are hydrolyzed and thus activated in the oral cavity in humans. *J. Nutr.* 135, 48–52.

173. Wang, Y., Chantreau, M., Sibout, R., and Hawkins, S. (2013). Plant cell wall lignification and monolignol metabolism. *Front. Plant Sci.* 4:220. doi: 10.3389/fpls.2013.00220

174. Wang, Y. W., Wang, W. C., Jin, S. H., Wang, J., Wang, B., and Hou, B. K. (2012). Over-expression of a putative poplar glycosyltransferase gene, PtGT1, in tobacco increases lignin content and causes early flowering. *J. Exp. Bot.* 63, 2799–2808. doi: 10.1093/jxb/ers001

175. Ward, J. L., Baker, J. M., Llewellyn, A. M., Hawkins, N. D., and Beale, M. H. (2011). Metabolomic analysis of *Arabidopsis* reveals hemiterpenoid glycosides as products of a nitrate ion-regulated, carbon flux overflow. *Proc. Natl. Acad. Sci. U.S.A.* 108, 10762–10767. doi: 10.1073/pnas.1018875108

176. Weng, J. K., and Chapple, C. (2010). The origin and evolution of lignin biosynthesis. *New Phytol.* 187, 273–285. doi: 10.1111/j.1469-8137.2010.03327.x

177. Whetten, R., and Sederoff, R. (1995). Lignin biosynthesis. *Plant Cell* 7, 1001–1013. doi: 10.1105/tpc.7.7.1001

178. Wöhler, F., and Liebig, J. (1837). Ueber die Bildung des Bittermandelöls. *Annalen Pharmacie* 22, 1–24.

179. Xu, Z., Escamilla-Trevino, L., Zeng, L., Lalgondar, M., Bevan, D., Winkel, B., et al. (2004). Functional genomic analysis of *Arabidopsis thaliana* glycoside hydrolase family 1. *Plant Mol. Biol.* 55, 343–367. doi: 10.1007/s11103-004-0790-1

180. Yamazaki, M., Yamagishi, E., Gong, Z., Fukuchi-Mizutani, M., Fukui, Y., Tanaka, Y., et al. (2002). Two flavonoid glucosyltransferases from Petunia hybrida: molecular cloning, biochemical properties and developmentally regulated expression. *Plant Mol. Biol.* 48, 401–411. doi: 10.1023/A:1014043214943

181. Yazaki, K. (2006). ABC transporters involved in the transport of plant secondary metabolites. *FEBS Lett.* 580, 1183–1191. doi: 10.1016/j.febslet.2005.12.009

182. Yonekura-Sakakibara, K., Fukushima, A., Nakabayashi, R., Hanada, K., Matsuda, F., Sugawara, S., et al. (2012). Two glycosyltransferases involved in anthocyanin modification delineated by transcriptome independent component analysis in *Arabidopsis thaliana*. *Plant J.* 69, 154–167. doi: 10.1111/j.1365-313X.2011.04779.x

183. Yonekura-Sakakibara, K., and Hanada, K. (2011). An evolutionary view of functional diversity in family 1 glycosyltransferases. *Plant J.* 66, 182–193. doi: 10.1111/j.1365-313X.2011.04493.x

184. Yonekura-Sakakibara, K., Nakabayashi, R., Sugawara, S., Tohge, T., Ito, T., Koyanagi, M., et al. (2014). A flavonoid 3-O-glucoside:2″-O-glucosyltransferase responsible for terminal modification of pollen-specific flavonols in *Arabidopsis thaliana*. *Plant J.* 79, 769–782. doi: 10.1111/tpj.12580

185. Yonekura-Sakakibara, K., Tohge, T., Matsuda, F., Nakabayashi, R., Takayama, H., Niida, R., et al. (2008). Comprehensive flavonol profiling and transcriptome coexpression analysis leading to decoding gene-metabolite correlations in *Arabidopsis*. *Plant Cell* 20, 2160–2176. doi: 10.1105/tpc.108.058040

186. Yoshinaga, A., Kamitakahara, H., and Takabe, K. (2015). Distribution of coniferin in differentiating normal and compression woods using MALDI mass spectrometric imaging coupled with osmium tetroxide vapor treatment. *Tree Physiol*. doi: 10.1093/treephys/tpv116. [Epub ahead of print].

187. Zhao, Q., Nakashima, J., Chen, F., Yin, Y., Fu, C., Yun, J., et al. (2013). Laccase is necessary and nonredundant with peroxidase for lignin polymerization during vascular development in *Arabidopsis*. *Plant Cell* 25, 3976–3987. doi: 10.1105/tpc.113.117770

188. Zhong, R., Demura, T., and Ye, Z. H. (2006). SND1, a NAC domain transcription factor, is a key regulator of secondary wall synthesis in fibers of*Arabidopsis*. *Plant Cell* 18, 3158–3170. doi: 10.1105/tpc.106.047399

CHAPTER 10

THE DEFENSE METABOLITE, ALLYL GLUCOSINOLATE, MODULATES ARABIDOPSIS THALIANA BIOMASS DEPENDENT UPON THE ENDOGENOUS GLUCOSINOLATE PATHWAY

Marta Francisco[1,2], Bindu Joseph[1], Hart Caligagan[1], Baohua Li[1], Jason A. Corwin[1], Catherine Lin[1], Rachel Kerwin[1], Meike Burow[3], and Daniel J. Kliebenstein[1,3]

[1]Department of Plant Sciences, University of California, Davis, CA, USA

[2]Group of Genetics, Breeding and Biochemistry of Brassicas, Department of Plant Genetics, Misión Biológica de Galicia, Spanish Council for Scientific Research, Pontevedra, Spain

[3]DynaMo Center of Excellence, Copenhagen Plant Science Centre, University of Copenhagen, Frederiksberg, Denmark

ABSTRACT

Glucosinolates (GSLs) play an important role in plants as direct mediators of biotic and abiotic stress responses. Recent work is beginning to show that the GSLs can also inducing complex defense and growth networks. However, the physiological significance of these GSL-induced responses and the molecular mechanisms by which GSLs are sensed and/or modulate these responses are not understood. To identify these potential mechanisms within the plant and how they may relate to the endogenous GSLs, we tested the regulatory effect of exogenous allyl GSL application on growth and defense metabolism across sample of Arabidopsis thaliana accessions. We found that application of exogenous allyl GSL had the ability to initiate changes in plant biomass and accumulation of defense metabolites that genetically varied across accessions. This growth effect was related to the allyl GSL side-chain structure. Utilizing this natural variation and mutants in genes within the GSL pathway we could show that the link between allyl GSL and altered growth responses are dependent upon the function of known genes controlling the aliphatic GSL pathway.

INTRODUCTION

To ward off biotic attackers, plants utilize a wide array of physical defenses, such as trichomes and wax layers that deter access to the tissue. Another layer of defense is provided by a diversity of chemicals that affect the biotic attacker's physiology and reproductive potential (Levin, 1973; Feeny, 1976). These defenses are hypothesized to be costly to the plant, diverting energy and resources away from primary metabolic functions and toward generating these structures and compounds (Karban and Baldwin, 1997; Baldwin, 1998; Mauricio, 1998; Cipollini et al., 2003; Paul-Victor et al., 2010; Züst et al., 2011). As well as the potential metabolic cost, there are strong ecological costs that may be as strong or stronger than the metabolic cost, e.g., an effective defense against one organism may actually generate susceptibility against a different organism (Karban and Baldwin, 1997; Heil, 2002; Kliebenstein and Rowe, 2008). An example of this are the glucosinolate (GSL) defense metabolites in the order Brassicales that provide resistance to generalist herbivores and concurrently provide signals that attract and benefit infestation by specialist herbivores (Renwick and Chew, 1994; Griffiths et al., 2001; Lambrix et al., 2001; Kliebenstein et al., 2002; Ratzka et al., 2002; Renwick et al., 2006; Pfalz et al., 2007, 2009; Beekwilder et al., 2008; Hansen et al., 2008). Similarly ecological costs are being noted where resistance genes against one pathogen are beginning to be found as susceptibility genes against counter-adapted pathogens (Lorang et al., 2004, 2007).

Citation: Francisco M, Joseph B, Caligagan H, Li B, Corwin JA, Lin C, Kerwin R, Burow M and Kliebenstein DJ (2016) The Defense Metabolite, Allyl Glucosinolate, Modulates Arabidopsis thaliana Biomass Dependent upon the Endogenous Glucosinolate Pathway. Front. Plant Sci. 7:774. doi: 10.3389/fpls.2016.00774

These countervailing costs and benefits of defense, requires that the plant properly calibrates its defense within any given environment to optimize the ratio of all benefits to all costs.

One method to dynamically balance the plant's defense portfolio is an integrated regulatory system that senses as many aspects of the biotic environment as possible. This system would then integrate these inputs from the biotic environment into the signaling network to properly modulate the defense outputs. This regulatory architecture involves a myriad of interconnections between the different pathways, allowing integration of different perception signals and generation of a coordinated output (Kunkel and Brooks, 2002; Kovac et al., 2009; Horstman et al., 2013;Zhang et al., 2013). The current model of plant defense signaling as a hierarchical regulatory system is, however, unlikely to properly explain the full complexity of plant defense networks. Current systems engineering theory suggests that purely hierarchical systems, like most plant network models, are unstable due to a lack of negative feedback and output integration (Kerwin et al., 2011; Pourcel et al., 2013; Bonawitz et al., 2014). Therefore, a truly stable system must have some form of feedback that connects system outputs to inputs to optimize effectiveness. Evidence that defense metabolites can provide this feedback regulation in plants is beginning to accumulate (Kerwin et al., 2011;Pourcel et al., 2013; Bonawitz et al., 2014). Recent work has suggested that plant specialized metabolites, such as GSLs and phenylpropanoids, have downstream regulatory influences on the plant in which they are synthesized (Kerwin et al., 2011; Pourcel et al., 2013; Bonawitz et al., 2014). In *Arabidopsis thaliana*, an indolic GSL compound alters defense signaling, suggesting that defense outputs can feedback modulate upstream regulatory processes. Similarly, in *Raphanus sativus*, a GSL hydrolysis product can directly modulate plant physiology by affecting the TIR1 auxin receptor (Hasegawa et al., 2000; Yamada et al., 2003; Clay et al., 2009). Genetic evidence also suggests that the aliphatic GSL pathway influences a myriad of other pathways. For example, a number of GSL mutants have significant growth defects, suggesting the potential for cross-talk between the GSL pathway and hormone metabolism (Delarue et al., 1998; Barlier et al., 2000; Hansen et al., 2001; Chen et al., 2003; Mikkelsen et al., 2004). More directly, the introduction of a functional AOP2, a biosynthetic enzyme in the aliphatic GSL pathway, into a naturally occurring AOP2 knockout genotype altered flowering, jasmonyl-isoleucine (JA-ILE) mediated defense signaling and oscillatory behavior of the circadian clock (Wentzell et al., 2007; Kerwin et al., 2011; Burow et al., 2015). The results indicate that at least the AOP2 RNA interconnects with cellular signaling and metabolic pathways via unknown mechanisms (Burow et al., 2015). However, it is unclear if GSL metabolites alone can modulate plant physiology. Thus, there is the potential that specific GSL metabolites, in contrast to the RNAs, may provide direct feedback regulation within the plant, but there is a need to test whether the metabolite controls these processes and to identify the genes and mechanisms that may facilitate this.

While the above efforts have started to show that GSLs can modulate regulatory pathways, our knowledge of the underpinning molecular mechanisms by which the GSL pathway is linked to other pathways in the plant, and how the connections translate into physiological output of plant growth, development, and defense is limited. To begin testing if a GSL metabolite produced by AOP2 can function as a feedback signal that alters plant biomass and defense metabolism, we fed exogenous allyl GSL (also known as 2-propenyl GSL or sinigrin) to *A. thaliana* and uncovered a wide range of heritable effects upon growth and endogenous GSL accumulation. Fifty micrometer allyl GSL was introduced exclusively to the roots and the compound was transported up to the leaf where it accumulated as less than 5% of the endogenous pool in any of the tested accessions. Using a population of 96 natural *A. thaliana* accessions and mutant genotypes in the GSL pathway we showed that exogenous allyl GSL has the capacity to differentially affect plant biomass and metabolite content of Arabidopsis dependent upon the environment and endogenous GSL genetic variation. Future work is required to test if allyl GSL or a related derivative is the active component and where the site(s) of activity are located.

MATERIALS AND METHODS

Plant Material and Exogenous Allyl GSL Feeding Experiment

Arabidopsis thaliana seeds were surface-sterilized (1-min, 70% ethanol soaking followed by a 20-min, 50% sodium hypochlorite), rinsed (five times) in sterile, distilled water. They were then placed on petri dishes containing half-strength Murashige and Skoog (MS) salt medium (CAISSON, MSP01-1LT) adjusted to pH 5.8, containing 0.8% agar and 1% sucrose (control). To study the effect of exogenous allyl GSL on plant biomass and metabolite content, 0.22 μm filter sterilized allyl GSL 100 mM stock solution (Sigma S1647-1G) was added to the autoclaved MS (at 55°C) to a final concentration of 50 μM (treatment). To identify suitable genetic screening conditions, we initially tested different sucrose concentrations (0-1-2%) by planting 20 seeds in a 55 mm Petri dish from each of seven *A. thaliana* accessions (Bay-0, Col-0, Ler-0, Tsu-1, Cvi-1, Kas, and Sha) with three independent biological replications for each treatment. The entire above experiment was replicated in triplicate providing an N of 180 seedlings per accession per treatment. All plantings were independently randomized.

The study was increased to a survey of a 96 *A. thaliana* natural accessions (Nordborg et al., 2002, 2005; Borevitz et al., 2007; Atwell et al., 2010; Chan et al., 2010a,b, 2011; Table S1) and GSL genes mutant genotypes (*AOP2, gsm1, myb28/29, myb28, myb29, gsox1/3*, and *gsox3*; Table 1; Haughn et al., 1991; Kliebenstein et al., 2001a; Hansen et al., 2007; Sønderby et al., 2007, 2010; Li et al., 2008). For natural accessions, seeds were placed in 36 grid square 100 × 15 mm plates. Five plants per accession were grown in a randomized partial block design (one seed per grid square). Seeds were planted on control MS and allyl-containing MS to provide five measurements per accession per treatment. For GSL gene mutant genotypes, six plants per line were grown in a randomized partial block design (two seed per grid square and 10 blocks) providing 60 measurements per genotype per treatment. After planting on media, plates were stratified for 3 days in the dark at 4°C to break dormancy. Plates were then transferred to a growth chamber under long-day conditions (16 h light at 100–120 μEi, 20°C). Any seedlings with leaf contact to the agar were removed from the analysis to ensure that root-to-shoot transport had occurred. At 15 days post germination, the rosette of each seedling was harvested from the plates, weighed to record the plant fresh weight (fw), then placed into a 96-deep well tube containing 90% methanol for GSL extraction and analyzed for GSL content as described below.

Table 1. Description of the single and double mutants on GSL genes.

GSL mutant gene	Mutant name	Gene ATG #	Phenotype
MYB28	*myb28*	At5g61420	Reduced levels of aliphatic GSL
MYB29	*myb29*	AT5g07690	Reduced levels of aliphatic GSL
MYB28/MYB29	*myb28/29*	At5g61420/ AT5g07690	Absence of aliphatic GSL
MAM1	*gsm1*	At5g23010	Accumulation of C3 GSL
GSOX1/GSOX3	*gsox1/3*	At1g65860/ At1g62560	Accumulation of methylthioalkyl GSL
GSOX3	*gsox3*	At1g62560	Accumulation of methylthioalkyl GSL
AOP2	*AOP2*	At4g03060	Accumulation of alkenyl GSL

Analysis of GSL Content

GSLs of excised shoots were measured using a previously described high-throughput analytical system (Kliebenstein et al., 2001a,b,c). Briefly, rosettes of all seedlings were individually removed from plates with forceps, weighed and placed in a single well of 96-well microliter plate containing 400 μL of 90% methanol and one 3.8 mm stainless steel ball-bearing. Tissues were homogenized for 3 min in a paint shaker, centrifuged, and the supernatants transferred to a 96-well filter plate with 50 μL of DEAE sephadex and washed once with water. The sephadex-bound GSL were eluted by overnight, room temperature incubation with 110 μL of sulfatase. Individual desulfo-GSLs within each sample was separated and detected by HPLC-DAD, identified, and quantified by comparison to purified standards. The GSL traits are reported as μmol g of fw of each plant. All seedlings were measured individually and GSL abundance was normalized to the fresh weight. In addition to the content of individual GSLs, we developed a series of summation and ratio traits based on prior knowledge of the GSL pathways (Table S2; Kliebenstein, 2007; Wentzell et al., 2007).

Quantitative Real-Time PCR

RNA from Col-0 genotype was extracted from three pools of three seedlings for each treatment (MS and MS + allyl) with Sigma Spectrum Plant Total RNA kit, treated with Sigma DNAse1 and reverse transcribed with iScript (Bio-rad). Expression was assayed by quantitative real-time PCR using SYBR Green and the data was normalized to *UBC* expression of each pool. The following primers were used: *UBC* (At5g25760), 5′-CTGAGCCGGACAGTCCTCTTAACTG-3′ and 5′-CGGCGAGGC GTGTATACATTTGTG-3′; *MYB28* (At5g 61420), 5′-TC CCCAAAAAGCTGGGTTGAAA-3′ and 5′-TTTA AGGTAGTTGGTCCATCGCA-3′; *MYB29* (At5g07690), 5′-GA ACACGCATCTCAAAAAGCTCCTG-3′ and 5′-ACTTTGGAG AGATGGAACCCGATTG-3′; *MAM3* (At5g23020), 5′-CG CTGATCTGAAGGCATTAGTGGTG-3′ and 5′-GCGGAAATCTGAGGGCTTGACATA-3′; *CYP83A1* (At4g13770), 5′-TCTCGCCGCGGTTCTCCTTT-3′ and 5′-GCCCATCCAGCGAAGAAGCGT-3′; *GS-OX1* (At1g65860), 5′-GCCGGTTAACGGGAAATGGAGTGT-3′ and 5′-ATTTCATGGGCGGCGAAACCAA-3′. Gene expression levels are presented as mean fold difference between treated and untreated as obtained across two independent experiments with three biological replicates per experiment.

Statistical Analyses

To test how the plant biomass and GSL responses to allyl treatment interact with the growth media, we conducted a three-way ANOVA using the factors accession, allyl treatment (MS and MS + allyl), MS media (0, 1, and 2% Sucrose) as well as the interactions between these factors. Plate was tested for significance as a random effect in a mixed model but not found to significantly alter the results and hence dropped from the model. The least-square means of each plant biomass and GSL phenotype per each accession within each condition were obtained using this model. ANOVA was also utilized to test for the effect of exogenous allyl GSL on plant biomass and GSL content of different GSL gene mutant lines. Each mutant was tested in an individual ANOVA against the wild-type (WT) Col-0 genotype. Multiple comparisons were made *post-hoc* using Tukey's *t*-test with $P \leq 0.05$ within the model.

To directly test if the GSL profile of the different accessions was influencing the response to exogenous allyl GSL, we utilized the mean phenotypes for each accession within each treatment. Using their GSL chemotype, we assigned each accession their appropriate genotype at the *GS-AOP* (methylsulfinyl, AOP1; alkenyl, AOP2 and hydroxyalkyl, AOP3) and*GS-Elong* (3C vs. 4C) loci. None of the studied accessions displayed the AOP3 C4 chemotype. We then conducted an ANOVA using the *GS-AOP* genotype (AOP1, AOP2, and AOP3), *GS-Elong* genotype (3C and 4C) and allyl treatment as factors. We also explicitly tested the various interaction of these factors in the model; *GS-AOP* × *GS-Elong, GS-AOP* × allyl treatment, *GS-Elong* × allyl treatment and *GS-AOP* × *GS-ELong* × allyl treatment.

To test for correlations in the concentrations of individual GSL compounds or classes to the changes in plant biomass we used Spearman›s correlation test within a stepwise linear regression model. For these analyses, we tested for a correlation of the relative plant biomass response with the individual relative GSL responses. All the relative responses for each trait were calculated as: (MS + Allyl) − (MS)12[(MS + Allyl) + (MS)](MS + Allyl) - (MS)12[(MS + Allyl) + (MS)] (MS stands for MS media with (MS+allyl) or without (MS) exogenous allyl). For the stepwise model, only the variables with a $P = 0.5$ for entry and $P = 0.1$ for removal were kept in the model. All statistical analyses were conducted using SAS.

RESULTS

Allyl GSL Fed Plant Biomass under Different Sucrose Concentration

Previous work has suggested that Arabidopsis and Raphanus GSLs can influence plant physiology (Hasegawa et al., 2000; Yamada et al., 2003; Kerwin et al., 2011). Thus, we proceeded to test if the allyl GSL that is produced in numerous Brassica species, including many but not all natural Arabidopsis accessions, could influence plant development.

Because sucrose is well known to condition many growth phenotypes by influencing various regulatory processes, we grew Arabidopsis seedlings in MS media containing 50 μM allyl GSL with different sucrose concentrations (0, 1, and 2%) to identify optimal conditions for testing any observed responses (Figure 1, Table S3). This concentration of allyl GSL is approximately the endogenous concentration found in allyl GSL producing Arabidopsis accessions (Kliebenstein et al., 2001a,b,c; Wentzell et al., 2007; Chan et al., 2010b). To simultaneously test if any observed effects differed across accessions, we used seven Arabidopsis accessions that differ in their endogenous GSL profile. We measured the biomass of at least 180 individual seedlings at 15 days post-germination for each accession × sucrose × treatment combination spread across three independent experiments.

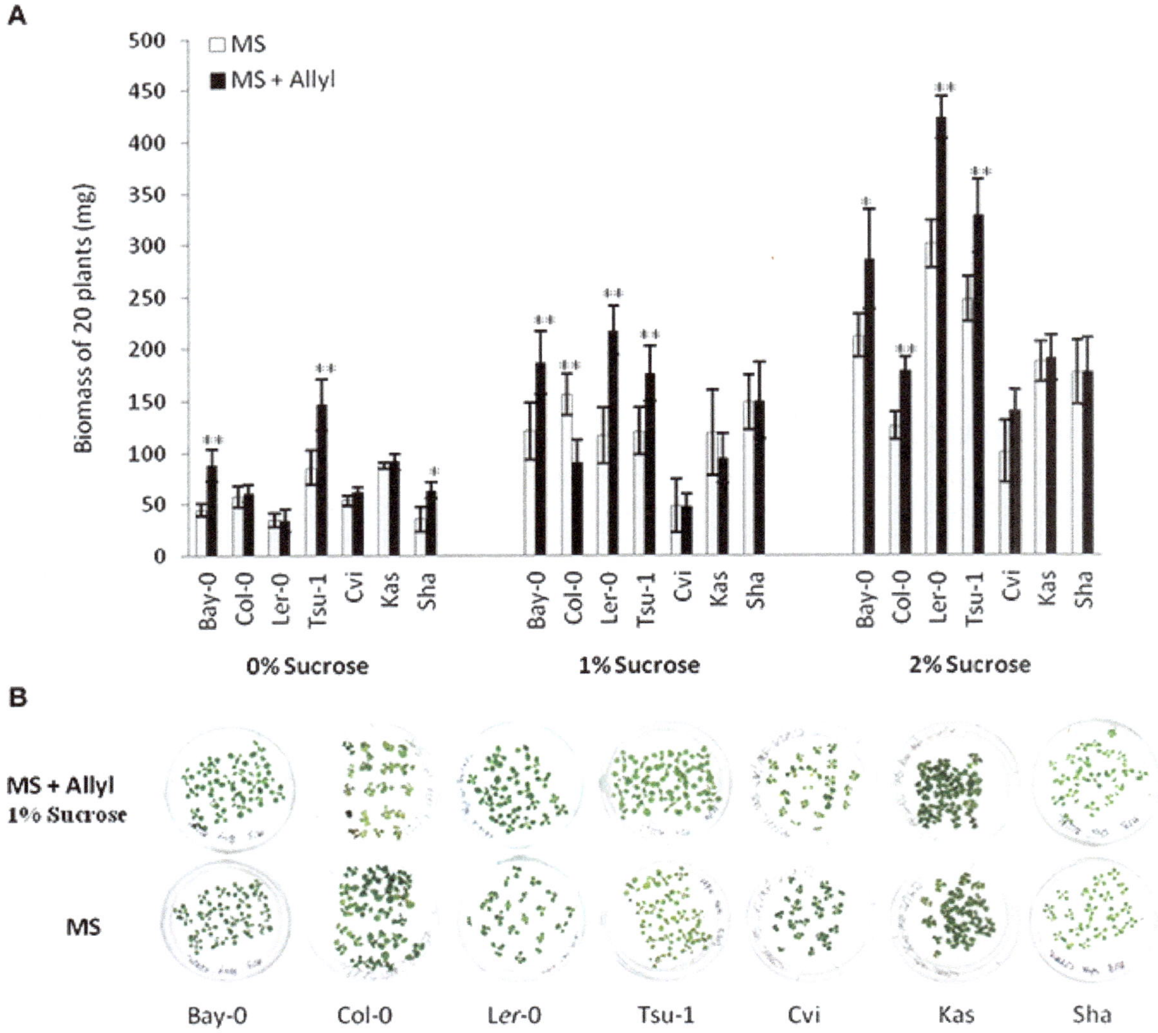

Figure 1. Differential plant biomass responses of Arabidopsis accessions to exogenous allyl GSL is sugar dependent. (A) Quantification of biomass of 15-day-old seedlings from seven *A. thaliana natural* accessions fed with 50 μM of allyl GSL using MS media differing in sucrose concentrations. The accessions Bay-0, Col-0, L*er*-0, and Tsu-1 do not synthesize endogenous allyl GSL, while the accessions Cvi, Kas, and Sha produce endogenous allyl GSL. Asterisks indicates significant effect of the exogenous allyl GSL treatment on that accession under that condition, $*P \leq 0.05$ and $**P \leq 0.01$ from the ANOVA analysis (Table S3). The error bars represent standard deviation from three independent experiments with each bar showing the average biomass of at least 60 individual plants across these experiments. (B) Representative photographs of seven *Arabidoposis* accessions feed with or without 50 μM of allyl GSL at 1% of sucrose concentration.

The addition of allyl GSL within the media led to increased biomass of most accessions, which was conditional upon the sucrose concentration with increasing effects as sucrose increased (Figure 1). In general, the accessions that do not synthesize endogenous allyl GSL had the largest response to exogenous allyl GSL application (Bay-0, Tsu-1, L*er*-0, and Col-0). Addition of allyl GSL to the media increased the biomass of Bay-0 and Tsu-1 at all sucrose concentrations and L*er*-0 at 1 and 2% sucrose. These three accessions produce predominantly 3-hydroxypropyl (3OHP) GSL. Another non-allyl GSL producing accession, Col-0, which produces primarily 4-methylsulfinylbutyl (4MSB), responded differently to addition of allyl GSL and sucrose, with growth being promoted at 2% sucrose but inhibited at 1%. In contrast, accessions that can synthesize allyl GSL, such as Cvi and Kas, showed no significant plant biomass response to exogenous allyl GSL feeding. The exception to this was Sha, which responded with increased biomass upon exogenous allyl application, but only when sucrose was not present in the media. Together this shows that allyl GSL has the capacity to affect plant biomass of Arabidopsis accessions in a manner that is dependent upon the growing conditions and endogenous GSL synthesis capacity.

Allyl GSL Effects are related to the Side-Chain Structure

To assess if plant biomass responses are solely dependent on the core sulfate/thioglucose structure of generic GSLs or is related to the side-chain of allyl GSL, we tested the growth effect of 4MSB upon the *Ler*-0 accession. For this test, we choose *Ler*-0 since this accession showed the largest change in plant biomass when it was feed with allyl GSL. Moreover, *Ler*-0 does not have the enzymes necessary to synthesize allyl or 4MSB GSL. Thus, *Ler*-0 is naïve to both GSLs and any plant responses to these GSL could be directly interpreted in comparison to the negative control. Using 1% sucrose that maximized the *Ler*-0 response to allyl GSL confirmed that allyl GSL stimulated biomass accumulation in these conditions (Figure 2). In contrast, 4MSB GSL lead to a significant decrease in biomass accumulation in Ler-0 under these conditions. As both GSLs have glucose and sulfate as a core component of their structure, this argues against these two GSLs have different affects and as such, these effects are not solely determined by the common sulfate/thioglucose structure that shared by the two GSLs. The fact that allyl GSL and 4MSB GSL produce opposing effects on *Ler*-0 biomass suggests that they are perceived via different mechanisms that are related to the differences in the two GSLs side-chains.

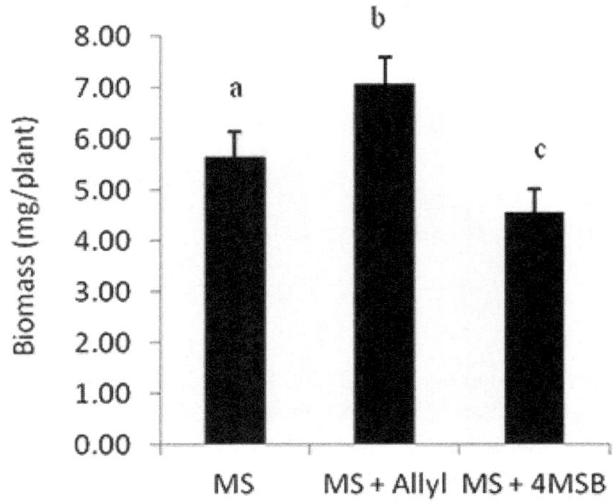

Figure 2. Differential plant biomass responses of *Ler*-0 accession fed with 50 **μM** of allyl and 4MSB GSLs using MS media at 1% of sucrose concentration. The bar chart represents the mean fw (mg tissue/plant) and the error bars represent the standard deviation within each treatment (MS + allyl GSL; MS + 4MSB GSL). Letters indicate significant differences ($P \leq 0.05$) between treatments using ANOVA and *post-hoc t*-test.

Exogenous Allyl GSL Is Taken Up by the Plant

Previous studies have shown that exogenously applied GSL can be transported and accumulate within the plant (Brudnell et al., 1999; Iqbal and Mollers, 2003; Hansen et al., 2008; Andersen et al., 2013). To test if the exogenous allyl GSL can be taken up by the plant roots and accumulate in the leaves, we measured the GSL levels within the above Arabidopsis accessions grown in the presence and absence of exogenous allyl GSL on MS with 0, 1, and 2% sucrose. Individual plants for these assays were chosen to ensure that there was no leaf contact with the media and any detected allyl GSL had to be taken up by the roots and transported to the leaves. The seedlings were measured individually for GSL accumulation and not pooled.

The four Arabidopsis accessions within our experiment that do not produce endogenous allyl GSL are missing the necessary enzyme to make this compound. They either contain a five basepair frameshift deletion abolishing the protein or have a local inversion that removes the AOP2 promoter abolishing expression (Kliebenstein et al., 2001a;Chan et al., 2011). Therefore, in these accessions, all measured allyl GSL in the leaves of these seedlings must have been obtained from the exogenous feeding and provide a test for allyl GSL uptake and foliar accumulation. These accessions were able to take up allyl GSL from the media and this overall accumulation was dependent upon the sucrose concentration (Figure 3A). In general, allyl GSL accumulation increases with increasing sucrose content in all accessions, except for Col-0, which reached the highest allyl GSL foliar accumulation at 1% sucrose (Figure 3A). Interestingly, the highest variation in foliar accumulation of the exogenous allyl GSL across the accessions occurred with sucrose at 1%. At these conditions, the mean foliar accumulation of allyl GSL within accessions that cannot make allyl GSL was 0.28 μmol/g of fw. This concentration of exogenous allyl GSL was less than 5% of the total aliphatic GSL within these accessions (Figure 3B), showing that these treatment conditions have not artificially flooded the system. Further, this suggests that the lack of a plant biomass response in the allyl GSL producing accessions may be because they have sufficient internal allyl GSL such that the exogenous amount does not stimulate an additional response (Figure 1).

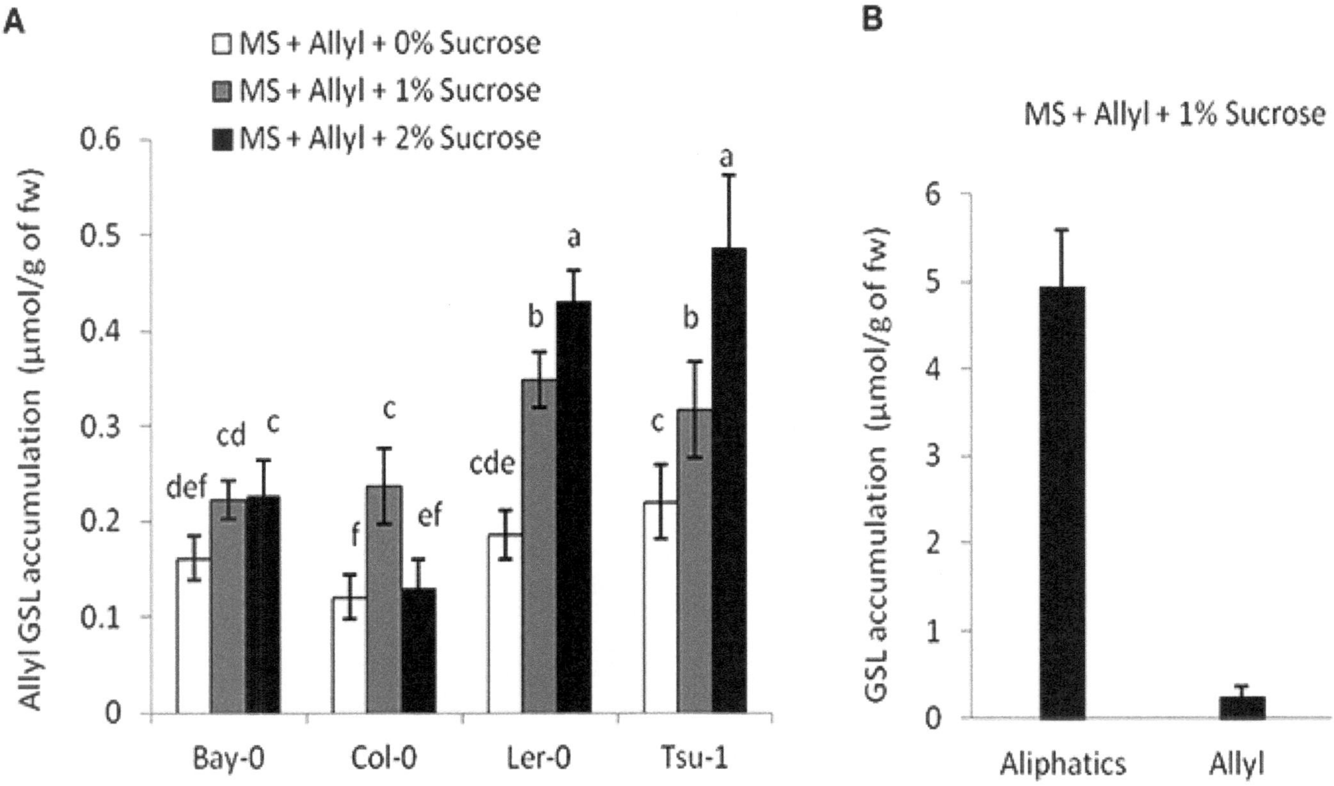

Figure 3. Allyl GSL accumulation. (A) The average of exogenous allyl GSL accumulation within leaves of 15-day-old seedlings was directly measured using HPLC. Shown are the accessions that are genetically incapable of synthesizing endogenous allyl GSL. Thus, detected allyl must have been taken up from the surrounding media, transported via the vasculature to the leaf and provides a direct measurement of the accumulation. Averages with the same letter are not significantly different at $P \leq 0.05$ using a Tukey's *post hoc t*-test within the ANOVA. The error bars represent standard deviation from three independent experiments where 12 plants were separately measured per experiment. (B) Average of the total endogenous aliphatic GSL and accumulated allyl content within the accessions that do not synthesize endogenous allyl GSL grown in MS with allyl at 1% of sucrose on the growing media. The error bars represent standard deviation across accessions and experiments as described in part (A).

Exogenous Allyl GSL Effects Endogenous GSL Accumulation

To test endogenous GSL levels may also respond to exogenous allyl GSL application, we measured GSL content from the seven Arabidopsis accessions seedlings (Bay-0, Tsu-1, L*er*-0, Col-0, Kas, Sha, and Cvi) fed with allyl GSL as well as from the control samples. These were from the same individual seedlings measured for biomass and all values are adjusted to the seedlings' biomass. These analyses detected 14 aliphatic GSL compounds and three indolic GSL traits. Since the accessions have different GSL profiles we focused on three GSL phenotypes that are measurable in all evaluated accessions (short-chain GSL accumulation, long-chain GSL accumulation, and indolic GSL accumulation). The accumulation of short-chain, long-chain and indolic GSLs showed statistically different responses to exogenous allyl GSL treatment across the accessions and sucrose concentrations (Table S4). The largest effect of exogenous allyl GSL on endogenous GSL accumulation was identified at 1% of sucrose (Figure 4). At these conditions, changes in endogenous GSL accumulation were larger than could be accounted for by the additive effect of exogenous GSL application and also affected a range of GSLs that cannot be synthesized from the allyl GSL, such as but-3-enyl GSL, 3OHP, 4MSB, and 8-methylsulfinyloctyl GSL (8 MSO) (Figure 4). Interestingly, while all the short-chain GSL positively responded to exogenous allyl GSL treatment in all accessions, the Cvi and Kas accessions had no long-chain GSL response to allyl GSL treatment. More dramatic was the indolic GSL accumulation where allyl GSL treatment induced the Bay and Col-0 accessions while repressing accumulation in L*er*-0 and Sha. Thus, exogenous allyl GSL causes diverse responses in the accumulation of endogenous GSL accumulation even in compounds that have little to no biosynthetic relationship (Figure 4).

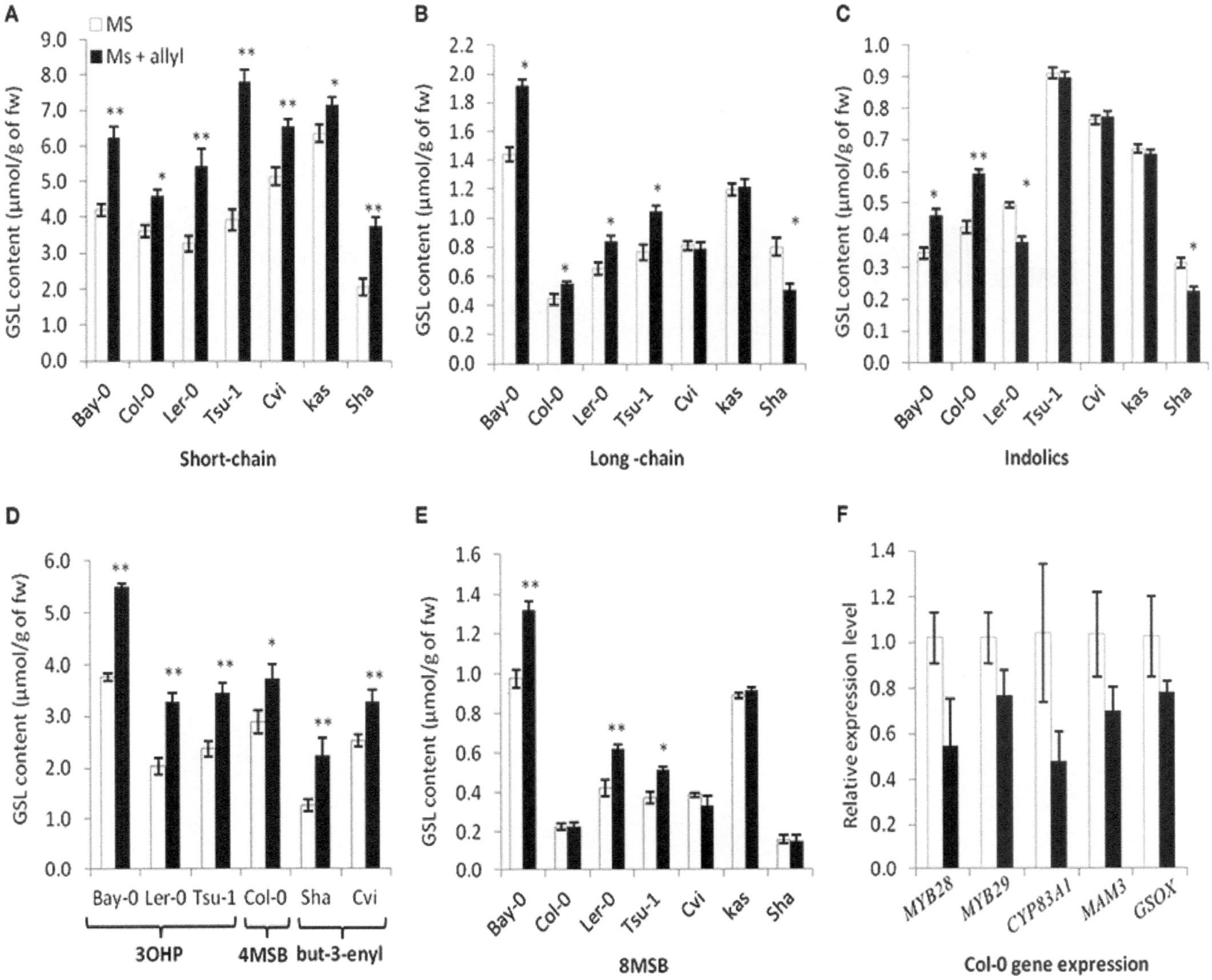

Figure 4. Differential GSL responses of Arabidopsis accessions to exogenous allyl GSL. Quantification of GSLs in 15-day-old seedlings from seven *A. thaliana natural* accessions fed with 50 μM of allyl GSL using MS media with 1% sucrose. Asterisks indicate a significant effect of the exogenous allyl GSL treatment on the accumulation of the shown GSL within the specific accession using a *post-hoc* Tukey test within the ANOVA analysis (*$P \leq 0.05$ and **$P \leq 0.01$). The error bars represent standard deviation from three independent experiments with each bar showing the average biomass of at least 12 individual plants across these experiments. The GSL are as follows (A) Short-chain GSL content. (B) Long-chain GSL content. (C) Indolic GSL content. (D) Accumulation of the dominant short-chain GSL within each accession as shown by 3-hydroxypropyl (3OHP), 4-methylsulfinylbutyl (4MSB) and but-3-enyl GSL. (E) 8-methylsulfinyloctyl GSL (8 MSB) content. (F) qRT-PCR expression levels of five selected GSL biosynthetic genes in seedlings of Col-0 grown in the presence and absence of exogenous 50 μM of allyl GSL at 1% sucrose. The error bars represent standard deviation from two independent experiments with three biological replicates per experiment. None of the differences were statistically significant (ANOVA, $P \leq 0.05$).

Understanding Relationships between GSL Content and Plant Biomass Responses to Exogenous Allyl GSL

To begin identifying the potential mechanism(s) by which allyl GSL can stimulate changes in Arabidopsis biomass and defense, we expanded the study to a population of 96 natural Arabidopsis accessions. To maximize the potential phenotypic variance, we used the 1% sucrose concentration where we could identify both positive and negative effects on plant biomass and indolic GSL accumulation. The distribution of fresh weight across the accessions showed that individual accessions displayed both positive and negative changes in biomass in response to exogenous allyl GSL (Figure 5; Table S5). Thus, there is genetic variation for the plant biomass response to exogenous allyl GSL application in *A. thaliana*. GSL analysis detected 14 aliphatic GSL compounds and three indolic GSL compounds from which we focused on the traits described above (Wentzell et al., 2007; Chan et al., 2010b). As previously observed, exogenous allyl GSL treatment also altered endogenous GSL accumulation with varying effects across the Arabidopsis accessions (Figure 5, Table S5).

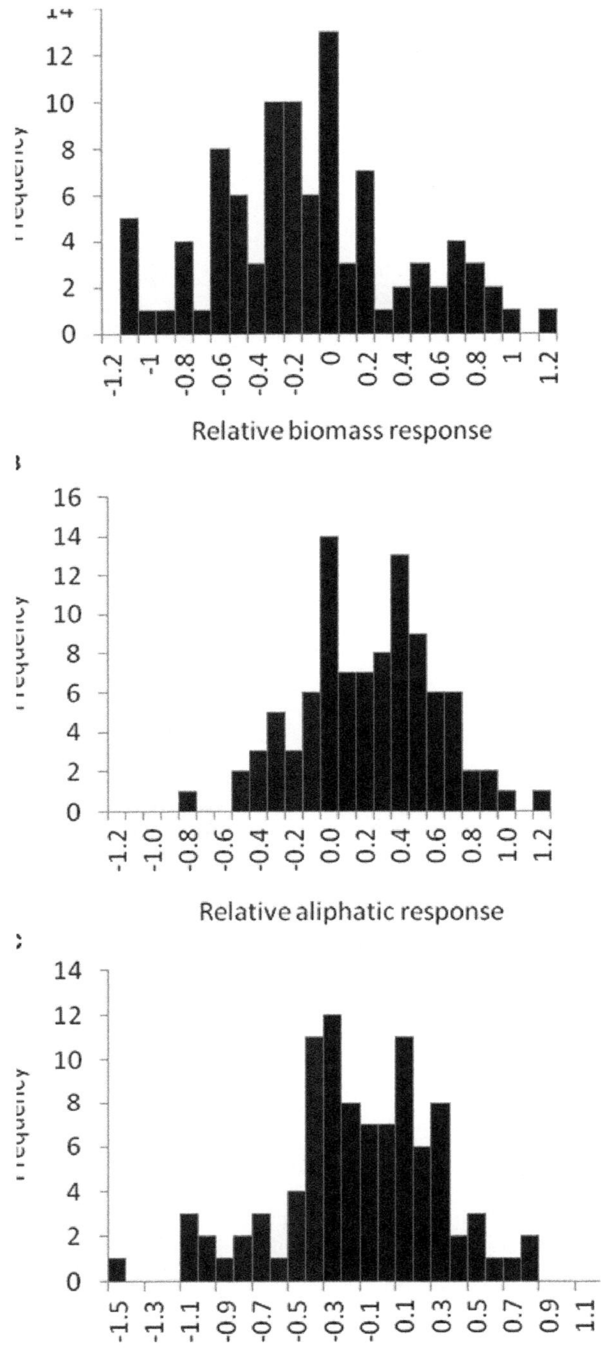

Figure 5. Natural variation in Arabidopsis biomass and GSL acumulation in response to exogenous allyl GSL. Histograms showing the frequency distribution of the relative responses to exogenous allyl GSL (50 μM of allyl GSL using MS media with 1% sucrose) across the 96 Arabidopsis natural accessions for the following phenotypes (A)plant biomass (B) total aliphatic GSL (C) total indolic GSL. The relative response of each phenotype to the allyl treatment within each accession was determined as (Phenotype in plants treated with allyl—phenotype in plants from the control treatement)/((0.5 × [Phenotype in plants treated with allyl + phenotype in plants from the control treatement]).

The 96 Arabidopsis accessions have known and validated polymorphisms within key GSL loci allowing us to test if genetic variation in GSL loci altered the response to exogenous allyl GSL. For this analysis, we conducted an ANOVA using the GSL functional haplotype at the two major chemotype loci across the accessions as factors in the linear model. All accessions were assigned their functional haplotype at the *GS-Elong* locus, controlling the production of GSL with three carbon (C3) or four carbon (C4) side chains and *GS-AOP*, controlling the side-chain structure via three alleles AOP1, the null methylsulfinylalkyl allele; AOP2, the alkenyl allele and AOP3 the hydroxypropyl allele (Kliebenstein et al., 2001a,c). The ANOVA using the accessions showed that plant biomass varies significantly among treatments and GSL genotypes (Table 2). Further, there was a significant interaction of *GS-Elong* with allyl GSL treatment. In general, allyl GSL treatment had a stronger effect on plant biomass accumulation in genotypes which predominantly accumulate C3 GSL than those with C4 GSL (Figure 6). Thus, it appears that there is an association of the ability of the exogenous allyl GSL to affect plant biomass with the endogenous GSL genetic variation. However, the normal distribution of plant biomass responses across this Arabidopsis collection of accessions, with positive and negative responses (Figure 5), suggests that any underlying mechanism(s) are highly pleiotropic. Further, the variation of response within each chemotype shows that there are genes varying independently of the GSL loci that determine the final phenotypic response in each accession.

Table 2. Effect of GSL haplotype upon the interaction of exogenous allyl treatment and plant biomass.

Source	Degrees of freedom	Sums of squares	*F*-Value	*P*
GS-AOP	2	2.4E-03	19.5	< 0.0001
GS-Elong	1	2.0E-05	3.3	0.0788
Allyl Treatment	1	4.4E-04	7.3	0.0107
GS-AOP × *GS-Elong*	1	1.1E-03	18.9	0.0001
GS-AOP × Allyl Treatment	2	1.2E-05	1.0	0.3767
GS-Elong × Allyl Treatment	1	2.5E-04	4.1	0.0494
GS-AOP × *GS-Elong* × Allyl Treatment	1	5.3E-06	0.1	0.7683
Block	4	1.3E-04	0.5	0.7050
Error	913	7.6E-02		

ANOVA was used to test the effect of the GSL haplotype within the 96 Arabidopsis natural accessions upon the biomass response to 50 μM of exogenously applied allyl GSL. The accessions were given their GSL genotypes at the two major structural loci, GS-AOP (alkenyl, hydroxyalkyl, and methylsulfinyl) and GS-Elong locus (3C vs. 4C). These were then treated as factors within the ANOVA. Type III Sums-of-squares are presented.

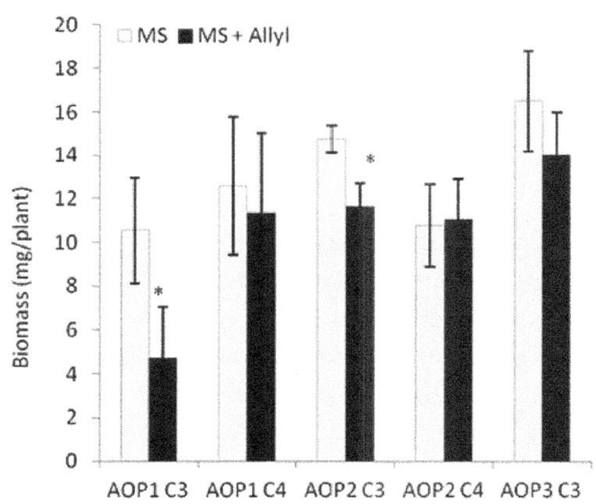

Figure 6. GSL chemotype interacts with exogenous Allyl treatment to influence plant biomass of*Arabidopsis* accessions. Quantifica-

tion of fw (mg tissue/plant) from 15-day-old seedlings from 96 *Arabidopsis*natural accessions fed with 50 μM of allyl GSL using MS media with 1% sucrose. The population was grouped by their GSL chemotype profile generated by the combination of variation at *GS-AOP* (alkenyl, hydroxyalkyl, and methylsulfinyl)*GS-Elong* (3C vs. 4C). The sub-populations generated are labeled by their predominant glucosinolate; AOP1 C3 accumulates 3-methylsulfinylpropyl, AOP1 C4 accumulates 4-methylsulfinylbutyl, AOP2 C3 accumulates allyl. AOP2 C4 accumulates But-3-enyl and AOP3 C3 accumulates 3-hydroxypropyl (Table S1). None of the studied accessions displayed the AOP3 C4 chemotype accumulating 4-hydroxybutyl. The bar chart represents the mean fw and the error bars represent the standard deviation among the accessions within each chemotype. *shows GSL haplotypes that had a significant difference in plant biomass response to exogenous allyl GSL using *post hoc* Tukey's *t*-test with $P \leq 0.05$ from the ANOVA analysis.

To extend this analysis to include all of the measure GSL, we tested for correlations between relative plant biomass response and the individual relative GSL traits responses to exogenous allyl GSL treatment. We found significant correlations between plant biomass response and 15 GSL traits. Among them, the strongest correlations were between plant biomass response and total aliphatic GSL accumulation (R^2 = 0.49, $P \leq 0.001$) and the ratio of methylthioalkyl/methylsulfinylalkyl GSL (R^2 = −0.50, $P \leq 0.001$). Both the methylthioalkyl and methylsulfinyalkyl GSL are upstream of allyl GSL in the aliphatic GSL pathway indicating a key role for early components of the GSL pathway in the response to the allyl GSL treatment (Table S6).

The above analysis suggests that there may be multiple links between endogenous GSL variation and the response to exogenous allyl GSL with regards to plant biomass accumulation. To create a model linking how the different GSL compounds may combine to influence plant biomass responses to ally GSL, we used a stepwise regression analysis testing all the different GSL traits. This generated a model where variation in the response of eight GSL traits, seven aliphatics and one indolic, explained 43% of the variability in plant biomass response to exogenous allyl GSL across the accessions (Table 3). 8MSO was the compound with the strongest influence on plant biomass (R^2 = 0.19) followed by 8-methylthiooctyl (8MTO) (R^2 = 0.06). The other GSL traits, 4-hydroxybutyl (4OHB), 7-methylsulfinylheptyl (7MSH), allyl, 7-methylthioheptyl (7MTH), 4MSB, and N-methoxy-indol-3-ylmethyl (NMI3M) explained small but significant proportions of overall variance (Table 3). Interestingly, the correlation of plant biomass response with allyl GSL response was due to differences in endogenous allyl GSL content across the accessions that carried a functional AOP2 enzyme and is not a response to differential accumulation of exogenous allyl GSL. In concordance with our correlation analysis, the relationships between the two long chain methylthioalkyl GSL (8MTO and 7MTH) and plant biomass responses were negative, while the other traits were positively correlated (Figure 7). This suggests that variation in GSLs across the accessions is playing a role in the plant biomass response to exogenous allyl GSL treatment.

Table 3. Stepwise regression model linking differential GSL and plant biomass responses to exogenous allyl GSL treatment.

GSL	Parameter estimate	Standard error	Partial R^2	Mode R^2	*F*	Pr > F
8MSO	0.31	0.10	0.19	0.19	9.0	0.0036
8MTO	−0.30	0.13	0.06	0.25	5.3	0.0232
4OHP	−0.18	0.07	0.04	0.29	7.2	0.0090
7MSH	0.14	0.06	0.04	0.34	4.6	0.0358
Allyl	0.11	0.05	0.03	0.37	5.2	0.0248
7MTH	−0.19	0.08	0.02	0.39	5.3	0.0241
4MSB	0.10	0.05	0.02	0.41	3.7	0.0575
MI3M	0.15	0.09	0.02	0.43	3.3	0.0733
Intercept	−0.29	0.07			16.1	0.0001

Shown are the GSL selected as significantly correlated to differential plant biomass accumulation in response to exogenous allyl GSL across the accessions. The GSL variables are listed in order of their predictive strength with their individual contribution and total model R^2 as well as significance within the combined model. See Table S2 for abbreviations.

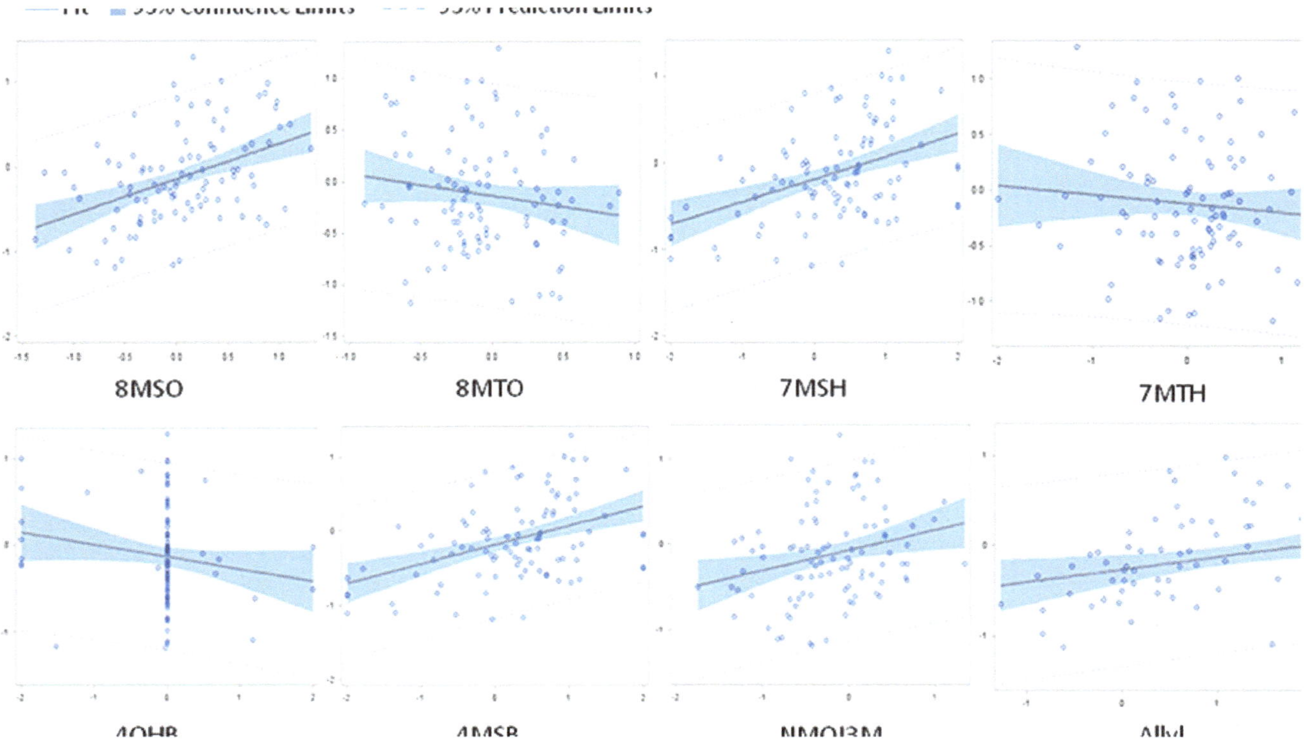

Figure 7. Model results comparing relative plant biomass response to relative GSL content response to exogenous allyl GSL treatment. Scatter plots with 95% confidence intervals on the partial correlations between relative plant biomass response (y axis) and the relative GSL content differences (x axis) for traits showing significance in the stepwise linear regression analysis (Table 3). See Table S2 for abbreviations.

The above effects of exogenous allyl on endogenous GSL accumulation suggest that there may be a transcriptional effect on the genes responsible for GSL biosynthesis. To test this hypothesis, we used qRT-PCR to measure the expression of five aliphatic GSL related genes (*MYB28, MYB29, CYP83A1, MAM3*, and *GS-OX1*). Using two independent experiments and three biological replicates per experiment, we harvested Col-0 leaves grown in the presence and absence of exogenous allyl GSL and measured transcript accumulation for the above genes by qRT-PCR. Analysis of this data identified no statistically significant differences between the treated and untreated samples for any genes (ANOVA, $P > 0.05$; Figure 4F). As such, there is no statistical support to argue that allyl GSL leads to direct alterations in transcript accumulation for aliphatic GSL genes. There are other reported instances where transcript abundance and metabolite abundance do not correlate within the aliphatic GSLs. For example, single knockouts in *MYB28* and *MYB29* generate the same reduction in metabolite accumulation but have disparate effects on transcript abundance (Sønderby et al., 2007, 2010). This suggests that whole genome transcriptomics will be required to test for any transcriptional effects of allyl GSL application.

Evaluation of Known GSL Mutant Genotypes Treated with Exogenous Allyl GSL

To directly test whether alterations in the major genes of the GSL biosynthetic pathway controlling GSL structure and accumulation may play a role in the plant response to exogenous allyl treatment, we compared the response of seven GSL mutants to exogenous allyl GSL application (Tables S7, S8). This analysis showed that the genes controlling endogenous GSL accumulation and structure also influence how the plants biomass responds to the application of allyl GSL. The *gsm1* mutant in MAM1 at the *GS-Elong* locus abolished the biomass response of Col-0 to exogenous allyl GSL (Figure 8A). The introduction of a functional AOP2 enzyme into Col-0, a natural knockout, significantly decreased control plant biomass as if the plant was now responding to the endogenous allyl GSL. Interestingly, this line had a small but significant positive response to exogenous allyl GSL suggesting that the introduction of AOP2 and/or endogenous allyl GSL synthesis can shift Col-0 from having a negative response to exogenous allyl GSL to having a positive effect on biomass accumulation (Figure 8A). The *myb28/myb29* double knockout was similar to *gsm1* in that it abolished the Col-0 biomass response to exogenous allyl GSL treatment (Figure 8A).

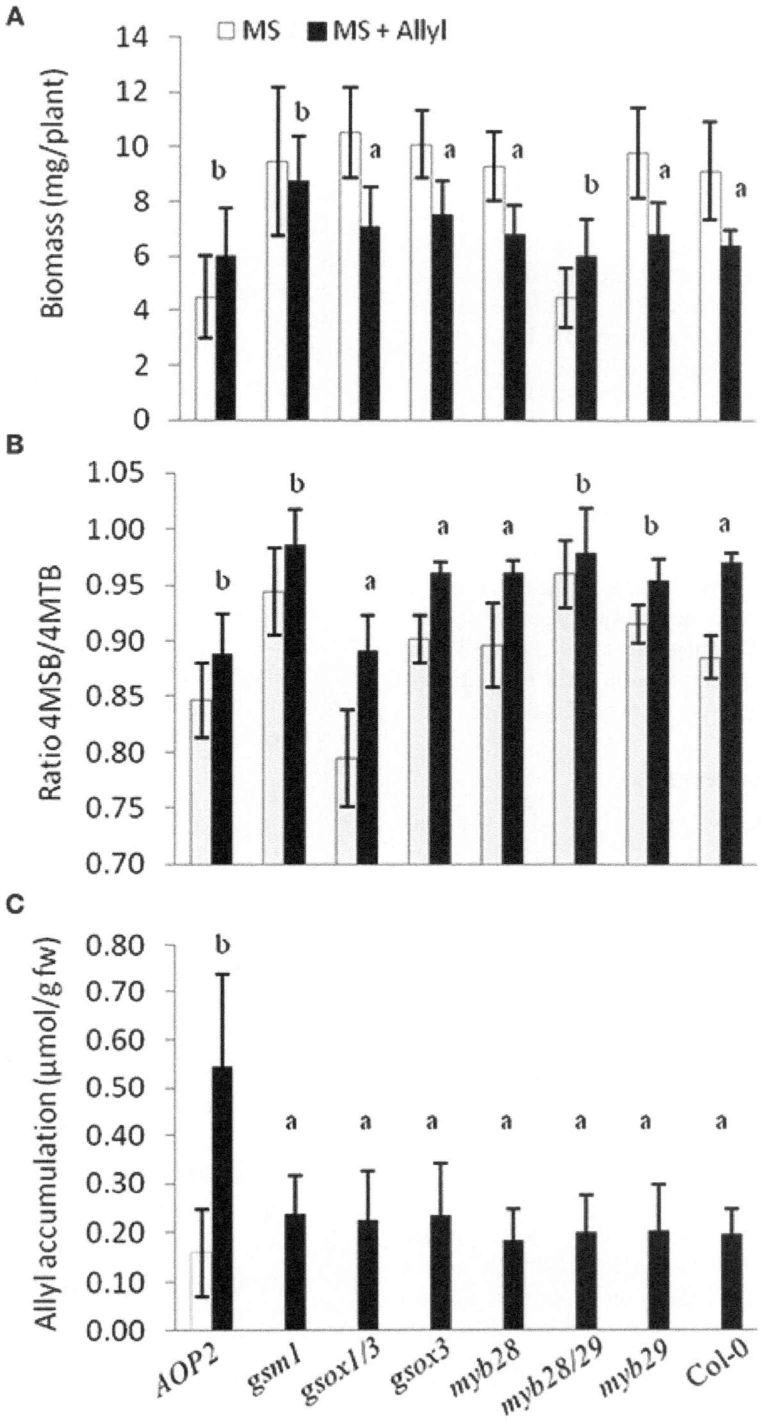

Figure 8. Plant biomass responses and GSL content variation among GSL mutant genotypes treated with exogenously applied allyl GSL. (A) Quantification of 15-day-old fw (mg/plant) seedlings from seven GSL mutant genotypes and wild-type (Col-0) fed with 50 μM of allyl glucosinolate using MS media with 1% sucrose. (B) Ratio of 4-methylsulfinylbutyl (4MSB)/4-methylthiobutyl (4MTB) calculated as 4MSB/(4MSB + 4MTB). **(C)** Allyl GSL accumulation average of the evaluated genotypes. The bar chart represents the mean and standard deviation. Each genotype within each treatment has a minimum of 60 independent plants measured. These plants were grown in a randomized block design with six individuals per genotype per treatment and 10 separate independent experiments. Means with the same letter show if the genotype's response to the treatment was statistically similar to Col-0 (a) or different from Col-0 (b) at $P \leq 0.05$ from the two-way ANOVA analysis (Table S8).

In addition to altering biomass, the GSL mutants also identified differential GSL responses to exogenous allyl GSL in comparison to the Col-0 parent. Col-0 responds to exogenous allyl GSL by altering the oxidation status of the 4C GSL away from the 4MTB and toward the 4MSB (Figure 8B). Genotypes that abolished the plant biomass response to exogenous allyl GSL also tended to change internal GSL accumulation in response to the treatment. The altered response of these genotypes was not due to differences in allyl accumulation, suggesting that these effects are not simply via altered GSL metabolism (Figure 8C). Thus, known genes within the aliphatic GSL pathway can either abolish or invert both the plant biomass and GSL response of Col-0 to exogenous allyl GSL application.

DISCUSSION

Stabilizing dynamic response systems such as plant responses to biotic attackers typically requires some level of feedback wherein the response output modulates the upstream system. There is a growing body of knowledge regarding the vast suite of receptors present in the plant that measure diverse aspects of the biotic environment, including bacterial, fungal, insect, plant, and almost any form of antagonistic organism (Dangl and Jones, 2001; Jones and Dangl, 2006; Gouhier-Darimont et al., 2013). The perceived signals then feed into a downstream regulatory system that is often considered to involve key defense signaling hormones, including salicylic acid and JA-ILE, among others (Glazebrook et al., 2003; Glazebrook, 2005; Jung et al., 2009). However, plant defense outputs like secondary metabolites are not routinely tested for the ability to modulate these systems. In our study, we show that exogenous allyl GSL or some other component derived from allyl GSL can modulate both plant biomass and defense metabolite accumulation in *A. thaliana*. This effect is naturally variable and dependent upon the environmental conditions in which the plant is grown. Thus, it appears that, at the very least, allyl GSL and potentially other defense metabolites have the capacities to modulate plant biomass and defense outputs in the plant.

Endogenous Role of Exogenous GSL

A number of lines of evidence support the idea that our application of exogenous allyl GSL is identifying an endogenous response. First, previous studies confirmed on the AOP2 gene have linked the level of endogenous allyl GSL to alterations in growth and development that were confirmed in with the application of exogenous allyl GSL (Kerwin et al., 2011; Züst et al., 2011). In our studies, applying 50 μM allyl GSL, fed to the roots led to altered plant biomass while accumulating to less than 5% of the endogenous pool, suggesting that we have not flooded the system (Figure 3). This suggests the plant biomass and GSL respond to a level of GSL that are within the endogenous range.

The application of 4MSB and allyl GSL produced different plant biomass responses in the same genotype under the same environment (Figure 2). This suggests that the effect is not caused by the common sulfate/thioglucose structure of these compounds but is instead related to the side-chain of the compounds. This could be caused by the intact GSL metabolite or potentially by compounds derived from the intact GSL. One potential mechanism is the activation/degradation of the GSLs by the myrosinases that in the indolic GSLs is known to affect their potential signaling roles (Bednarek et al., 2009; Clay et al., 2009). The myrosinases and associated nitrile specifier proteins can cause the GSLs to be converted into isothiocyanates, oxazolidine-2-thiones, nitriles, epithionitriles, and thiocyanates (Lambrix et al., 2001; Burow et al., 2006). For the two tested GSLs (allyl and 4MSB) myrosinases to treat them identically with release of the sugar and sulfate (Halkier and Gershenzon, 2006). In the leaves of Arabidopsis, the allyl GSL can be converted to the epithionitrile while the 4MSB can be converted to the simple nitrile depending upon the genotype (Lambrix et al., 2001; Wentzell and Kliebenstein, 2008). However, in the roots, there is no epithionitrile formation and allyl and 4MSB are equally converted to the same structural derivatives (nitrile and isothiocyanate) in all tested accessions (Wentzell and Kliebenstein, 2008). This suggests that if the difference between the two compounds is related to epithionitrile vs. nitrile formation that this would have to occur within the leaf after transport from the root. In some accessions, both the allyl and 4MSB GSLs are converted into potentially toxic isothiocyanates upon tissue disruption. However, some accessions that create predominantly the isothiocyanate, like Col-0, show a positive biomass response to allyl GSL which would suggest that this is not a toxin mediated effect. Arguing further against the idea that myrosinase mediate activation is the key player is the fact that endogenous myrosinase activation of aliphatic GSLs requires tissue disruption and our assay is being conducted on intact and undisturbed plants wherein there should be no co-occurrence of myrosinase and exogenous GSL. Together, this suggests that 4MSB and allyl GSL cause different responses but the specific active components mediating this response and the active site(s) in which this response is stimulated need to be elucidated by further experiments.

Interestingly, foliar analysis of the treated seedlings showed that the exogenous allyl GSL is absorbed by the roots and transported into the leaves. This indicates that allyl GSL must move into the root vasculature, at which point it would intermingle with the endogenously transported GSL (Nour-Eldin et al., 2012; Andersen et al., 2013). The major GSL transporters in the vasculature do not distinguish between different GSL structures (Nour-Eldin et al., 2012; Andersen et al., 2013). Thus, the exogenously applied allyl GSL will follow the same transport and accumulation routes as the endogenous GSL (Nour-Eldin et al., 2012; Andersen et al., 2013). In combination, these points support the idea that the Arabidopsis responses to exogenous allyl GSL treatment we observe go through the GSL metabolite, and not only AOP2 RNA, being the result of an endogenous process. Further work to identify the mechanisms behind exogenous allyl GSL induced responses is required to prove this hypothesis.

Sucrose Dependency of Allyl GSL Effects

The ability of exogenous allyl GSL to accumulate within Arabidopsis and to also alter plant biomass was dependent upon sucrose concentration. One possibility is that sucrose is merely altering the energy status of the plant and the sucrose impact on allyl GSL effects is simply indirect. An alternative idea arises from recent work that is showing that sucrose can function as a signaling molecule that influences a large array of developmental processes through interacting with diverse pathways including cell cycle, cell wall signaling, hormonal regulation, vacuolar transporters, programmed cell death pathways, ribosomal biogenesis and the regulation of translation (Ruan, 2012; Tognetti et al., 2013). This raises the possibility that allyl GSL effects somehow intersect with the sucrose signaling pathways. Supporting this possibility is the observation that some accessions show opposing exogenous GSL responses depending upon sucrose which indicates that there is some cross-talk between intracellular pathways elicited by sucrose and those endogenous signals related to exogenous GSL application. Interestingly, our results suggest that sucrose responses are naturally variable within Arabidopsis and that it may be possible to find additional players within this response network using quantitative genetics. Future work will have to identify the mechanistic basis of how allyl GSL alters biomass and defense compound accumulation to assess how cross-talk may be occurring.

Response of Plant Biomass and GSL Traits to Exogenous Allyl GSL Treatment Are Linked

Feeding exogenous allyl GSL to 96 Arabidopsis accessions identified a wide range of genetic diversity in the plant biomass and GSL responses (Figure 5). Utilizing this natural variation, we were able to develop a model in which we could link the variation in GSL responses to the plant biomass responses (Figure 7). This model indicated a negative correlation between plant biomass response and the ratio of methylthioalkyl/methylsulfinylalkyl GSL and positive correlation between plant biomass response and total aliphatic GSL accumulation (Table S6). These results suggested that accessions that increase their biomass in response to exogenous allyl GSL application also increase their GSL content per unit biomass. Additionally, there is a structural specificity to this relationship in that increase in methylthioalkyl GSL is negatively associated with plant biomass.

In addition, the model indicated a correlation between multiple aliphatic GSLs and the plant biomass accumulation response. To test if endogenous GSL accumulation alters biomass accumulation, we utilized a T-DNA insertion line that abolishes the enzyme responsible for various chain elongated aliphatic GSLs accumulation, MAM1 (*gsm1*). This mutant abolished the response to exogenous allyl GSL, suggesting that MAM1 plays a role in this response in *A. thaliana*(Figure 8; Haughn et al., 1991). Another mutant line, *myb28/myb29* double knockout that dramatically reduces the accumulation of most aliphatic GSLs also did not respond to exogenous allyl GSL (Sønderby et al., 2007, 2010). Both the natural variation in endogenous allyl GSL levels as well as the feeding of exogenous allyl GSL suggested that the total level of allyl GSL impacts plant biomass (Figures 1, 5). In agreement with this, the introduction of a functional *AOP2*gene into Col-0 accession, a natural *AOP2* knockout, led to a significant reduction in biomass that was also associated with a loss of responsiveness to exogenous allyl GSL (Figure 8). Thus, the combination of data strongly supports the hypothesis that the response to exogenous allyl GSL mimics the biomass response to endogenous allyl GSL. Thus, GSL biosynthetic and regulatory genes can either abolish or invert the biomass response in Col-0 to exogenous allyl GSL suggesting that endogenous perturbations in GSL structure and accumulation affect this response. This interplay of exogenous allyl GSL with endogenous GSL further suggests that this response is an in planta component of how the plant connects defense to physiology.

In conclusion, Arabidopsis can respond to allyl GSL by modifying its biomass accumulation and its endogenous GSL pool dependent upon the environment and endogenous GSL genetic variation. This process utilizes the endogenous GSL biosynthetic pathway, key developmental regulatory genes and probably a set of uncharacterized genes. Further genome-wide association studies will help to elucidate the regulatory network and candidate genes controlling plant biomass response variation to exogenous allyl GSL.

REFERENCES

1. Andersen, T. G., Nour-Eldin, H. H., Fuller, V. L., Olsen, C. E., Burow, M., and Halkier, B. A. (2013). Integration of biosynthesis and long-distance transport establish organ-specific glucosinolate profiles in vegetative Arabidopsis. *Plant Cell* 25, 3133–3145. doi: 10.1105/tpc.113.110890
2. Atwell, S., Huang, Y., Vilhjalmsson, B. J., Willems, G., Horton, M., Li, Y., et al. (2010). Genome-wide association study of 107 phenotypes in a common set of *Arabidopsis thaliana* in-bred lines. *Nature* 465, 627–631. doi: 10.1038/nature08800
3. Baldwin, I. T. (1998). Jasmonate-induced responses are costly but benefit plants under attack in native populations. *Proc. Natl. Acad. Sci. U.S.A.* 95, 8113–8118.
4. Barlier, I., Kowalczyk, M., Marchant, A., Ljung, K., Bhalerao, R., Bennett, M., et al. (2000). The SUR2 gene of *Arabidopsis thaliana* encodes the cytochrome P450CYP83B1, a modulator of auxin homeostasis. *Proc. Natl. Acad. Sci. U.S.A.* 97, 14819–14824. doi: 10.1073/pnas.260502697
5. Bednarek, P., Piślewska-Bednarek, M., Svatoš, A., Schneider, B., Doubský, J., Mansurova, M., et al. (2009). A glucosinolate metabolism pathway in living plant cells mediates Broad-Spectrum Antifungal defense. *Science* 323, 101–106. doi: 10.1126/science.1163732
6. Beekwilder, J., van Leeuwen, W., van Dam, N. M., Bertossi, M., Grandi, V., Mizzi, L., et al. (2008). The impact of the absence of aliphatic glucosinolates on insect herbivory in Arabidopsis. *PLoS ONE* 3:e2068. doi: 10.1371/journal.pone.0002068
7. Bonawitz, N. D., Kim, J. I., Tobimatsu, Y., Ciesielski, P. N., Anderson, N. A., Ximenes, E., et al. (2014). Disruption of mediator rescues the stunted growth of a lignin-deficient Arabidopsis mutant. *Nature* 509, 376–380. doi: 10.1038/nature13084
8. Borevitz, J. O., Hazen, S. P., Michael, T. P., Morris, G. P., Baxter, I. R., Hu, T. T., et al. (2007). Genome-wide patterns of single-feature polymorphism in*Arabidopsis thaliana*. *Proc. Natl. Acad. Sci. U.S.A.* 104, 12057–12062. doi: 10.1073/pnas.0705323104
9. Brudnell, A. J. P., Griffiths, H., Rossiter, J. T., and Baker, D. A. (1999). The phloem mobility of glucosinolates. *J. Exp. Bot.* 50, 745–756.
10. Burow, M., Atwell, S., Francisco, M., Kerwin, R. E., Halkier, B. A., and Kliebenstein, D. J. (2015). The glucosinolate biosynthetic gene AOP2 Mediates feed-back regulation of Jasmonic Acid signaling in Arabidopsis. *Mol. Plant* 8, 1201–1212. doi: 10.1016/j.molp.2015.03.001
11. Burow, M., Markert, J., Gershenzon, J., and Wittstock, U. (2006). Comparative biochemical characterization of nitrile-forming proteins from plants and insects that alter myrosinase-catalysed hydrolysis of glucosinolates. *FEBS J.* 273, 2432–2446. doi: 10.1111/j.1742-4658.2006.05252.x

12. Chan, E. K., Rowe, H. C., Hansen, B. G., and Kliebenstein, D. J. (2010a). The complex genetic architecture of the metabolome. *PLoS Genet.* 6:e1001198. doi: 10.1371/journal.pgen.1001198

13. Chan, E. K. F., Rowe, H. C., and Kliebenstein, D. J. (2010b). Understanding the evolution of defense metabolites in *Arabidopsis thaliana* using genome-wide association mapping. *Genetics* 185, 991–1007. doi: 10.1534/genetics.109.108522

14. Chan, E. K., Rowe, H. C., Corwin, J. A., Joseph, B., and Kliebenstein, D. J. (2011). Combining genome-wide association mapping and transcriptional networks to identify novel genes controlling glucosinolates in *Arabidopsis thaliana*. *PLoS Biol.* 9:e1001125. doi: 10.1371/journal.pbio.1001125

15. Chen, S. X., Glawischnig, E., Jorgensen, K., Naur, P., Jorgensen, B., Olsen, C. E., et al. (2003). CYP79F1 and CYP79F2 have distinct functions in the biosynthesis of aliphatic glucosinolates in Arabidopsis. *Plant J.* 33, 923–937. doi: 10.1046/j.1365-313X.2003.01679.x

16. Cipollini, D., Purrington, C. B., and Bergelson, J. (2003). Costs of induced responses in plants. *Basic Appl. Ecol.* 4, 79–89. doi: 10.1078/1439-1791-00134

17. Clay, N. K., Adio, A. M., Denoux, C., Jander, G., and Ausubel, F. M. (2009). Glucosinolate Metabolites required for an Arabidopsis innate immune response. *Science* 323, 95–101. doi: 10.1126/science.1164627

18. Dangl, J. L., and Jones, J. D. G. (2001). Plant pathogens and integrated defence responses to infection. *Nature* 411, 826–833. doi: 10.1038/35081161

19. Delarue, M., Prinsen, E., Van Onckelen, H., Caboche, M., and Bellini, C. (1998). *Sur2* mutations of *Arabidopsis thaliana* define a new locus involved in the control of auxin homeostasis. *Plant J.* 14, 603–611.

20. Feeny, P. (1976). "Plant apparency and chemical defense," in *Biochemical Interaction between Plants and Insects: Proceedings of the Fifteenth Annual Meeting of the Phytochemical Society of North America*, eds J. W. Wallace and R. L. Mansell (New York, NY: Plenum Press), 1–40.

21. Glazebrook, J. (2005). Contrasting mechanisms of defense against biotrophic and necrotrophic pathogens. *Annu. Rev. Phytopathol.* 43, 205–227. doi: 10.1146/annurev.phyto.43.040204.135923

22. Glazebrook, J., Chen, W. J., Estes, B., Chang, H. S., Nawrath, C., Metraux, J. P., et al. (2003). Topology of the network integrating salicylate and jasmonate signal transduction derived from global expression phenotyping. *Plant J.* 34, 217–228. doi: 10.1046/j.1365-313X.2003.01717.x

23. Gouhier-Darimont, C., Schmiesing, A., Bonnet, C., Lassueur, S., and Reymond, P. (2013). Signalling of *Arabidopsis thaliana* response to Pieris brassicae eggs shares similarities with PAMP-triggered immunity. *J. Exp. Bot.* 64, 665–674. doi: 10.1093/jxb/ers362

24. Griffiths, D. W., Deighton, N., Birch, A. N., Patrian, B., Baur, R., and Stadler, E. (2001). Identification of glucosinolates on the leaf surface of plants from the Cruciferae and other closely related species. *Phytochemistry* 57, 693–700. doi: 10.1016/S0031-9422(01)00138-8

25. Halkier, B. A., and Gershenzon, J. (2006). Biology and biochemistry of glucosinolates. *Annu. Rev. Plant Biol.* 57, 303–333. doi: 10.1146/annurev.arplant.57.032905.105228

26. Hansen, B. G., Kerwin, R. E., Ober, J. A., Lambrix, V. M., Mitchell-Olds, T., Gershenzon, J., et al. (2008). A novel 2-oxoacid-dependent dioxygenase involved in the formation of the goiterogenic 2-hydroxybut-3-enyl glucosinolate and generalist insect resistance in Arabidopsis. *Plant Physiol.* 148, 2096–2108. doi: 10.1104/pp.108.129981

27. Hansen, B. G., Kliebenstein, D. J., and Halkier, B. A. (2007). Identification of a flavin-monooxygenase as the S-oxygenating enzyme in aliphatic glucosinolate biosynthesis in Arabidopsis. *Plant J.* 50, 902–910. doi: 10.1111/j.1365-313X.2007.03101.x

28. Hansen, C. H., Wittstock, U., Olsen, C. E., Hick, A. J., Pickett, J. A., and Halkier, B. A. (2001). Cytochrome P450CYP79F1 from Arabidopsis catalyzes the conversion of dihomomethionine and trihomomethionine to the corresponding aldoximes in the biosynthesis of aliphatic glucosinolates. *J. Biol. Chem.*276, 11078–11085. doi: 10.1074/jbc.M010123200

29. Hasegawa, T., Yamada, K., Kosemura, S., Yamamura, S., and Hasegawa, K. (2000). Phototropic stimulation induces the conversion of glucosinolate to phototropism-regulating substances of radish hypocotyls. *Phytochemistry* 54, 275–279. doi: 10.1016/S0031-9422(00)00080-7

30. Haughn, G. W., Davin, L., Giblin, M., and Underhill, E. W. (1991). Biochemical genetics of plant secondary metabolites in *Arabidopsis thaliana*. The glucosinolates. *Plant Physiol.* 97, 217–226.

31. Heil, M. (2002). Ecological costs of induced resistance. *Curr. Opin. Plant Biol.* 5, 345–350. doi: 10.1016/S1369-5266(02)00267-4

32. Horstman, A., Willemsen, V., Boutilier, K., and Heidstra, R. (2013). AINTEGUMENTA-LIKE proteins: hubs in a plethora of networks. *Trends Plant Sci.*19, 146–157. doi: 10.1016/j.tplants.2013.10.010

33. Iqbal, M. C. M., and Mollers, C. (2003). Uptake and distribution of sinigrin in microspore derived embryos of *Brassica napus* L. *J. Plant Physiol.* 160, 961–966. doi: 10.1078/0176-1617-00527

34. Jones, J. D. G., and Dangl, J. L. (2006). The plant immune system. *Nature* 444, 323–329. doi: 10.1038/nature05286

35. Jung, H. W., Tschaplinski, T. J., Wang, L., Glazebrook, J., and Greenberg, J. T. (2009). Priming in Systemic Plant Immunity. *Science* 324, 89–91. doi: 10.1126/science.1170025

36. Karban, R., and Baldwin, I. T. (1997). *Induced Responses to Herbivory*. Chicago, IL: University of Chicago Press.

37. Kerwin, R. E., Jiménez-Gómez, J. M., Fulop, D., Harmer, S. L., Maloof, J. N., and Kliebenstein, D. J. (2011). Network quantitative trait loci mapping of circadian clock outputs identifies metabolic pathway-to-

clock linkages in Arabidopsis. *Plant Cell* 23, 471–485. doi: 10.1105/tpc.110.082065

38. Kliebenstein, D. J. (2007). "Metabolomics and plant quantitative trait locus analysis - the optimum genetical genomics platform?" in *Concepts in Plant Metabolomics*, eds B. J. Nikolau and E. S. Wurtele (Dordrect: Springer), 29–45.

39. Kliebenstein, D., Lambrix, V., Reichelt, M., Gershenzon, J., and Mitchell-Olds, T. (2001a). Gene duplication and the diversification of secondary metabolism: side chain modification of glucosinolates in *Arabidopsis thaliana*. *Plant Cell* 13, 681–693. doi: 10.1105/tpc.13.3.681

40. Kliebenstein, D. J., Gershenzon, J., and Mitchell-Olds, T. (2001b). Comparative quantitative trait loci mapping of aliphatic, indolic and benzylic glucosinolate production in *Arabidopsis thaliana* leaves and seeds. *Genetics* 159, 359–370.

41. Kliebenstein, D. J., Kroymann, J., Brown, P., Figuth, A., Pedersen, D., Gershenzon, J., et al. (2001c). Genetic control of natural variation in *Arabidopsis thaliana* glucosinolate accumulation. *Plant Physiol.* 126, 811–825. doi: 10.1104/pp.126.2.811

42. Kliebenstein, D. J., and Rowe, H. C. (2008). Ecological costs of biotrophic versus necrotrophic pathogen resistance, the hypersensitive response and signal transduction. *Plant Sci.* 174, 551–556. doi: 10.1016/j.plantsci.2008.03.005

43. Kliebenstein, D., Pedersen, D., Barker, B., and Mitchell-Olds, T. (2002). Comparative analysis of quantitative trait loci controlling glucosinolates, myrosinase and insect resistance in *Arabidopsis thaliana*. *Genetics* 161, 325–332.

44. Kovac, J., Husse, J., and Oster, H. (2009). A time to fast, a time to feast: The crosstalk between metabolism and the circadian clock. *Mol. Cells* 28, 75–80. doi: 10.1007/s10059-009-0113-0.

45. Kunkel, B. N., and Brooks, D. M. (2002). Cross talk between signaling pathways in pathogen defense. *Curr. Opin. Plant Biol.* 5, 325–331. doi: 10.1016/S1369-5266(02)00275-3

46. Lambrix, V., Reichelt, M., Mitchell-Olds, T., Kliebenstein, D. J., and Gershenzon, J. (2001). The Arabidopsis epithiospecifier protein promotes the hydrolysis of glucosinolates to nitriles and influences trichoplusia ni herbivory. *Plant Cell* 13, 2793–2807. doi: 10.1105/tpc.010261

47. Levin, D. A. (1973). Role of trichomes in plant defense. *Q. Rev. Biol.* 48, 3–15. doi: 10.1086/407484

48. Li, J., Hansen, B. G., Ober, J. A., Kliebenstein, D. J., and Halkier, B. A. (2008). Subclade of flavin-monooxygenases involved in Aliphatic Glucosinolate Biosynthesis. *Plant Physiol.* 148, 1721–1733. doi: 10.1104/pp.108.125757

49. Lorang, J. M., Carkaci-Salli, N., and Wolpert, T. J. (2004). Identification and characterization of victorin sensitivity in *Arabidopsis thaliana*. *Mol. Plant Microbe Interac.* 17, 577–582. doi: 10.1094/MPMI.2004.17.6.577

50. Lorang, J. M., Sweat, T. A., and Wolpert, T. J. (2007). Plant disease susceptibility conferred by a "resistance" gene. *Proc. Natl. Acad. Sci. U.S.A.* 104, 14861–14866. doi: 10.1073/pnas.0702572104

51. Mauricio, R. (1998). Costs of resistance to natural enemies in field populations of the annual plant *Arabidopsis thaliana*. *Am. Natural.* 151, 20–28.

52. Mikkelsen, M. D., Naur, P., and Halkier, B. A. (2004). Arabidopsis mutants in the C-S lyase of glucosinolate biosynthesis establish a critical role for indole-3-acetaldoxime in auxin homeostasis. *Plant J.* 37, 770–777. doi: 10.1111/j.1365-313X.2004.02002.x

53. Nordborg, M., Borevitz, J. O., Bergelson, J., Berry, C. C., Chory, J., Hagenblad, J., et al. (2002). The extent of linkage disequilibrium in *Arabidopsis thaliana*. *Nat. Genet.* 30, 190–193. doi: 10.1038/ng813

54. Nordborg, M., Hu, T. T., Ishino, Y., Jhaveri, J., Toomajian, C., Zheng, H., et al. (2005). The Pattern of Polymorphism in *Arabidopsis thaliana*. *PLoS Biol.*3:e196. doi: 10.1371/journal.pbio.0030196

55. Nour-Eldin, H. H., Andersen, T. G., Burow, M., Madsen, S. R., Jorgensen, M. E., Olsen, C. E., et al. (2012). NRT/PTR transporters are essential for translocation of glucosinolate defence compounds to seeds. *Nature* 488, 531–534. doi: 10.1038/nature11285

56. Paul-Victor, C., Züst, T., Rees, M., Kliebenstein, D. J., and Turnbull, L. A. (2010). A new method for measuring relative growth rate can uncover the costs of defensive compounds in *Arabidopsis thaliana*. *New Phytol.* 187, 1102–1111. doi: 10.1111/j.1469-8137.2010.03325.x

57. Pfalz, M., Vogel, H., and Kroymann, J. (2009). The gene controlling the Indole Glucosinolate Modifier1 quantitative trait locus alters indole glucosinolate structures and aphid resistance in Arabidopsis. *Plant Cell* 21, 985–999. doi: 10.1105/tpc.108.063115

58. Pfalz, M., Vogel, H., Mitchell-Olds, T., and Kroymann, J. (2007). Mapping of QTL for resistance against the crucifer specialist herbivore *Pieris brassicae* in a new Arabidopsis inbred line population, Da(1)-12 × Ei-2. *PLoS ONE* 2:e578. doi: 10.1371/journal.pone.0000578

59. Pourcel, L., Irani, N. G., Koo, A. J. K., Bohorquez-Restrepo, A., Howe, G. A., and Grotewold, E. (2013). A chemical complementation approach reveals genes and interactions of flavonoids with other pathways. *Plant J.* 74, 383–397. doi: 10.1111/tpj.12129

60. Ratzka, A., Vogel, H., Kliebenstein, D. J., Mitchell-Olds, T., and Kroymann, J. (2002). Disarming the mustard oil bomb. *Proc. Natl. Acad. Sci. U.S.A.* 99, 11223–11228. doi: 10.1073/pnas.172112899

61. Renwick, J. A. A., Haribal, M., Gouinguene, S., and Stadler, E. (2006). Isothiocyanates stimulating oviposition by the diamondback moth, Plutella xylostella. *J. Chem. Ecol.* 32, 755–766. doi: 10.1007/s10886-006-9036-9

62. Renwick, J., and Chew, F. (1994). Oviposition behavior in Lepidoptera. *Annu. Rev. Entomol.* 39, 377–400.

63. Ruan, Y.-L. (2012). Signaling role of sucrose metabolism in development. *Mol. Plant* 5, 763–765. doi: 10.1093/mp/sss046

64. Sønderby, I. E., Burow, M., Rowe, H. C., Kliebenstein, D. J., and Halkier, B. A. (2010). A complex interplay of three R2R3 MYB transcription factors determines the profile of aliphatic glucosinolates in Arabidopsis. *Plant Physiol.* 153, 348–363. doi: 10.1104/pp.109.149286

65. Sønderby, I. E., Hansen, B. G., Bjarnholt, N., Ticconi, C., Halkier, B. A., and Kliebenstein, D. J. (2007). A systems biology approach identifies a R2R3 MYB gene subfamily with distinct and overlapping functions in regulation of aliphatic glucosinolates. *PLoS ONE* 2:e1322. doi: 10.1371/journal.pone.0001322

66. Tognetti, J. A., Pontis, H. G., and Martínez-Noël, G. M. A. (2013). Sucrose signaling in plants: a world yet to be explored. *Plant Signal. Behav.* 8:e23316. doi: 10.4161/psb.23316

67. Wentzell, A. M., and Kliebenstein, D. J. (2008). Genotype, age, tissue, and environment regulate the structural outcome of glucosinolate activation. *Plant Physiol.* 147, 415–428. doi: 10.1104/pp.107.115279

68. Wentzell, A. M., Rowe, H. C., Hansen, B. G., Ticconi, C., Halkier, B. A., and Kliebenstein, D. J. (2007). Linking metabolic QTLs with network and cis-eQTLs controlling biosynthetic pathways. *PLoS Genet.* 3, 1687–1701. doi: 10.1371/journal.pgen.0030162

69. Yamada, K., Hasegawa, T., Minami, E., Shibuya, N., Kosemura, S., Yamamura, S., et al. (2003). Induction of myrosinase gene expression and myrosinase activity in radish hypocotyls by phototropic stimulation. *J. Plant Physiol.* 160, 255–259. doi: 10.1078/0176-1617-00950

70. Zhang, C., Xie, Q., Anderson, R. G., Ng, G., Seitz, N. C., Peterson, T., et al. (2013). Crosstalk between the Circadian clock and innate immunity in Arabidopsis. *PLoS Pathogens* 9:e1003370. doi: 10.1371/journal.ppat.1003370.

71. Züst, T., Joseph, B., Shimizu, K. K., Kliebenstein, D. J., and Turnbull, L. A. (2011). Using knockout mutants to reveal the growth costs of defensive traits.*Proc. R. Soc. B Biol. Sci.* 278, 2598–2603. doi: 10.1098/rspb.2010.2475

CHAPTER 11

TRANSGENIC STUDIES REVEAL THE POSITIVE ROLE OF LEEIL-1 IN REGULATING SHIKONIN BIOSYNTHESIS IN LITHOSPERMUM ERYTHRORHIZON HAIRY ROOTS

Rongjun Fang[1,2,3], Ailan Zou[1,3], Hua Zhao[1,3], Fengyao Wu[1,3], Yu Zhu[1,3], Hu Zhao[1,3], Yonghui Liao[1,3], Ren-Jie Tang[4], Yanjun Pang[1,3], Rongwu Yang[1,3], Xiaoming Wang[1,3], Jinliang Qi[1,3,4], Guihua Lu[1,3], and Yonghua Yang[1,3]

[1]State Key Laboratory of Pharmaceutical Biotechnology, NJU-NJFU Joint Institute of Plant Molecular Biology, School of Life Sciences, Nanjing University, Nanjing 210046, People's Republic of China.

[2]Jiangsu University of Science and Technology, Zhenjiang 212003, People's Republic of China.

[3]Co-Innovation Center for Sustainable Forestry in Southern China, Nanjing Forestry University, Nanjing 210037, People's Republic of China.

[4]Department of Plant and Microbial Biology, University of California, Berkeley, CA 94720, USA.

ABSTRACT

Background

The phytohormone ethylene (ET) is a key signaling molecule for inducing the biosynthesis of shikonin and its derivatives, which are secondary metabolites in Lithospermum erythrorhizon. Although ETHYLENE INSENSITIVE3 (EIN3)/EIN3-like proteins (EILs) are crucial transcription factors in ET signal transduction pathway, the possible function of EIN3/EIL1 in shikonin biosynthesis remains unknown. In this study, by targeting LeEIL-1 (L. erythrorhizon EIN3-like protein gene 1) at the expression level, we revealed the positive regulatory effect of LeEIL-1 on shikonin formation.

RESULTS

The mRNA level of LeEIL-1 was significantly up-regulated and down-regulated in the LeEIL-1-overexpressing hairy root lines and LeEIL-1-RNAi hairy root lines, respectively. Specifically, LeEIL-1 overexpression resulted in increased transcript levels of the downstream gene of ET signal transduction pathway (LeERF-1) and a subset of genes for shikonin formation, excretion and/or transportation (LePAL, LeC4H-2, Le4CL-1, HMGR, LePGT-1,LeDI-2, and LePS-2), which was consistent with the enhanced shikonin contents in the LeEIL-1-overexpressing hairy root lines. Conversely, LeEIL-1-RNAi dramatically repressed the expression of the above genes and significantly reduced shikonin production.

Citation: Fang R, Zou A, Zhao H, Wu F, Zhu Y, Zhao H, Liao Y, Tang RJ, Pang Y, Yang R, Wang X, Qi J, Lu G, and Yang Y (2016). Transgenic studies reveal the positive role of LeEIL-1 in regulating shikonin biosynthesis in Lithospermum erythrorhizon hairy roots. BMC Plant Biology, 16:121. DOI: 10.1186/s12870-016-0812-6

CONCLUSIONS

The results revealed that LeEIL-1 is a positive regulator of the biosynthesis of shikonin and its derivatives in L. erythrorhizon hairy roots. Our findings gave new insights into the molecular regulatory mechanism of ET in shikonin biosynthesis. LeEIL-1 could be a crucial target gene for the genetic engineering of shikonin biosynthesis.

BACKGROUND

The roots of medicinal plant *Lithospermum erythrorhizon* can specifically accumulate shikonin and its derivatives which have significant anti-inflammatory, antitumor, and antimicrobial activities [1, 2, 3, 4]. These red naphthoquinone pigments are also excellent natural dyes widely used in the industries of fabric, food and cosmetics [5, 6].

In the last decade, the cell or hairy root culture systems of *L. erythrorhizon* have been successfully used to produce these valuable compounds through a two-stage culture system, i.e., (1) *L. erythrorhizon* cells or hairy roots are cultured in B5 multiplication medium under light for fast amplification, and (2) transferred into M9 production medium to produce shikonin pigments in darkness [7, 8, 9]. This excellent system has become a promising tool to better understand the metabolism of shikonin pigments. Genes encoding pivotal enzymes or regulators for shikonin biosynthesis, excretion and/or transportation have also been cloned and characterized, such as the *L. erythrorhizon* phenylalanine ammonia-lyase gene (*LePAL*) [10], the *L. erythrorhizon* cinnamic acid 4-hydroxylase gene (*LeC4H*) [11], the 3-hydroxy-3-methylglutaryl-coenzyme A reductase gene (*HMGR*) [12], the *L. erythrorhizon* p-hydroxybenzoate:geranyltransferase gene (*LePGT*) [13, 14,15], the *L. erythrorhizon* pigment callus-specific gene (*LePS-2*) [16], and the *L. erythrorhizon* dark-inducible gene (*LeDI-2*) [17, 18]. Moreover, several factors, such as light [8, 19, 20], mineral elements [21, 22], fungal elicitor [23], culture medium [9], nitric oxide [24], methyl jasmonate [25], and ET [26], have been described to be crucial regulators of shikonin biosynthesis.

ETHYLENE INSENSITIVE3 (EIN3)/EIN3-like proteins (EILs) are plant-specific transcription factors which regulate various ET responses [27, 28, 29]. EIN3/EIL1 is a key integration node between ET and other signals in the complex molecular signaling network [30]. By binding to specific promoter elements, EIN3/EIL1 activates or represses the expression of target genes responsible for ET signaling, and thus modulates multiple ET-related responses of plants, such as those for development, phenotype, and adaptation to environmental stresses [31, 32, 33]. However, the knowledge of the relationship between EIN3/EIL and plant secondary metabolism remains limited.

In the medicinal plant *L. erythrorhizon*, ET is an important regulator of shikonin biosynthesis [26]. Furthermore, an optimal concentration of endogenous ET was presumed to be pivotal for shikonin formation [34]. *LeEIL-1*, a homolog of *Arabidopsis EIN3*, has been isolated from *L. erythrorhizon* cells. It was speculated to be important for ET-regulated shikonin biosynthesis [35]. However, the function of *LeEIL-1* in shikonin biosynthesis at the molecular level remains unknown. Functional study on *LeEIL-1* could be useful for elucidating the relationship among *LeEIL-1*, ET, and shikonin production.

In this study, two transgenic strategies, overexpression and RNA interference (RNAi), were applied to induce*LeEIL-1*-overexpression and *LeEIL-1*-RNAi transgenic hairy roots. The relationship between the expression pattern of *LeEIL-1* and shikonin production was investigated to offer new insights into the understanding of the possible function of *LeEIL-1* in shikonin biosynthesis.

RESULTS

Hairy Root Induction, Cultivation, and Identification

For more detailed understanding of the function of *LeEIL-1* in shikonin formation, both RNAi and overexpression transgenic strategies were applied in this study. The plasmids of pBI121-*LeEIL-1*-Overexpression (EO) and pBI121-*LeEIL-1*-RNAi (Ei) were constructed and verified based on the pBI121-enhanced green fluorescent protein gene empty vector (pBI121-*eGFP*) (EV) (Fig. 1a; Additional file 1: Figure S1). The *Agrobacterium rhizogenes* strain 15834 (WT) and the *A. rhizogenes* 15834 containing EV, Ei or EO plasmid were employed to infect the nodes of aseptic seedlings [36]. Over 15 transgenic lines of hairy root had been successfully induced for each construct (i.e., EV, Ei, and EO) and WT (Fig. 1b).

For stock culture, all hairy roots (WT, EV, Ei, and EO) were transferred into B5 solid medium (hormone-free and antibiotics-free) at 26–28°C under subdued light. The remarkable growth polymorphisms of four types of hairy roots were observed at the infection sites (Fig. 1b) or in the stock culture medium (Additional file 2: Figure S2). EO hairy root lines were displayed obviously red either at the infection sites of seedling nodes or in sub-cultured stock medium under subdued light. No obvious red hairy root lines were observed in WT, Ei, and EV. Many black segments were observed in most Ei. The growth rates of EO and Ei were relative slow in comparison with those of WT and EV in the stock culture medium or multiplication culture medium, with Ei exhibiting the slowest growth rate (Additional file 3: Figure S3). This finding indicated that the growth of *L. erythrorhizon* hairy roots might be affected by excessively low or high expression level of *LeEIL-1*. We also found that callusing and regeneration phenomena easily occurred in the Ei cultured in B5 medium (Fig. 1c andd) compared with other hairy root lines. Hence, we speculated that the repressed mRNA level of *LeEIL-1*changed the development and phenotype of hairy roots to some extent.

To confirm the transformation of hairy roots, the DNA samples from all transgenic hairy roots (WT, EV, Ei, and EO) were used as template for PCR amplification of the tagging *rolC* gene of *A. rhizogenes* 15834 [37, 38], and the DNA of untransformed *L. erythrorhizon* seedling was used as negative control. Results showed that the*rolC* gene was only amplified from WT, EV, Ei, and EO hairy roots, (Fig. 1e), which confirms the success of hairy root transformation of *L. erythrorhizon*.

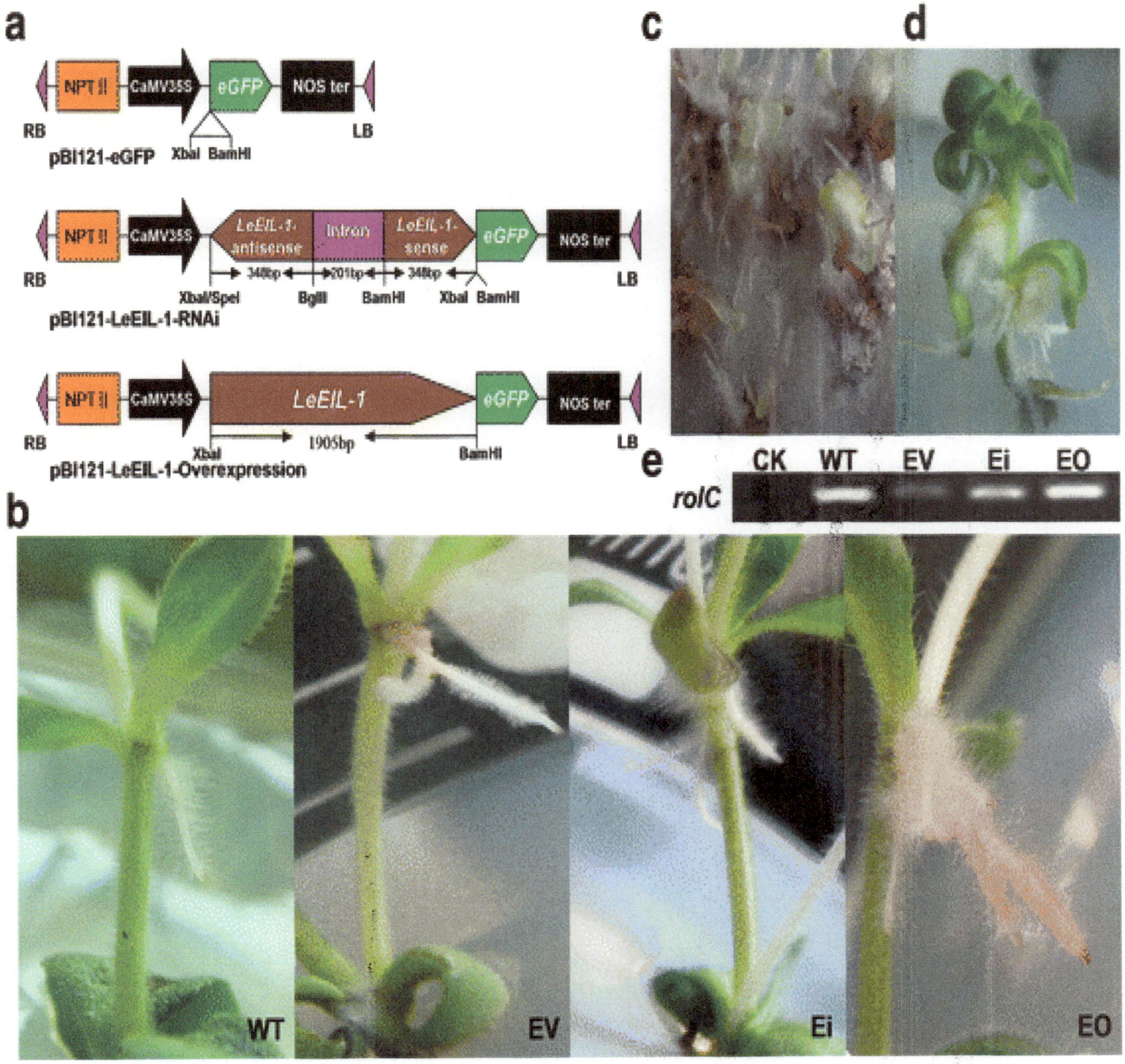

Figure 1. Induction, verification, and growth polymorphism observation of four different types of hairy root lines. (**a**) Structure and restriction maps of the transformation vectors pBI121-*eGFP* (EV), pBI121-*LeEIL-1*-RNAi (Ei), and pBI121-*LeEIL-1*-Overexpression (EO); (**b**) Growth polymorphism of the induced *L. erythrorhizon* hairy roots by using the method of pricking nodes with a needle. Differentiated callus, leaf (**c**), and regenerated shoots (**d**) from the Ei hairy root lines on the stock culture medium; (**e**) PCR analysis of the *rolC* gene in hairy roots. The untransformed *L. erythrorhizon*seedling was used as the negative control (CK).

Confocal scanning laser microscopy analysis was performed on eGFP-tagged cells to identify all the hairy roots and clarify the subcellular localization of LeEIL-1. Different patterns of eGFP localization were visualized in hairy roots. No fluorescence could be observed in WT hairy roots (Fig. 2a), whereas uniform and intense signals of fluorescence appeared in the nucleus, cell wall and cytoplasm of EV hairy roots (Fig. 2b). In Ei hairy root lines, only weak fluorescence signal was discerned (Fig. 2c). In EO hairy root lines, strong signals emitted by eGFP were detected predominantly in the nucleus (Fig. 2d), where EIN3/EILs localized [39]. Based on these findings, we speculated that co-expression or co-localization of the fusion protein eGFP:LeEIL-1 may occur in the nucleus of EO but may be effectively suppressed in Ei hairy roots.

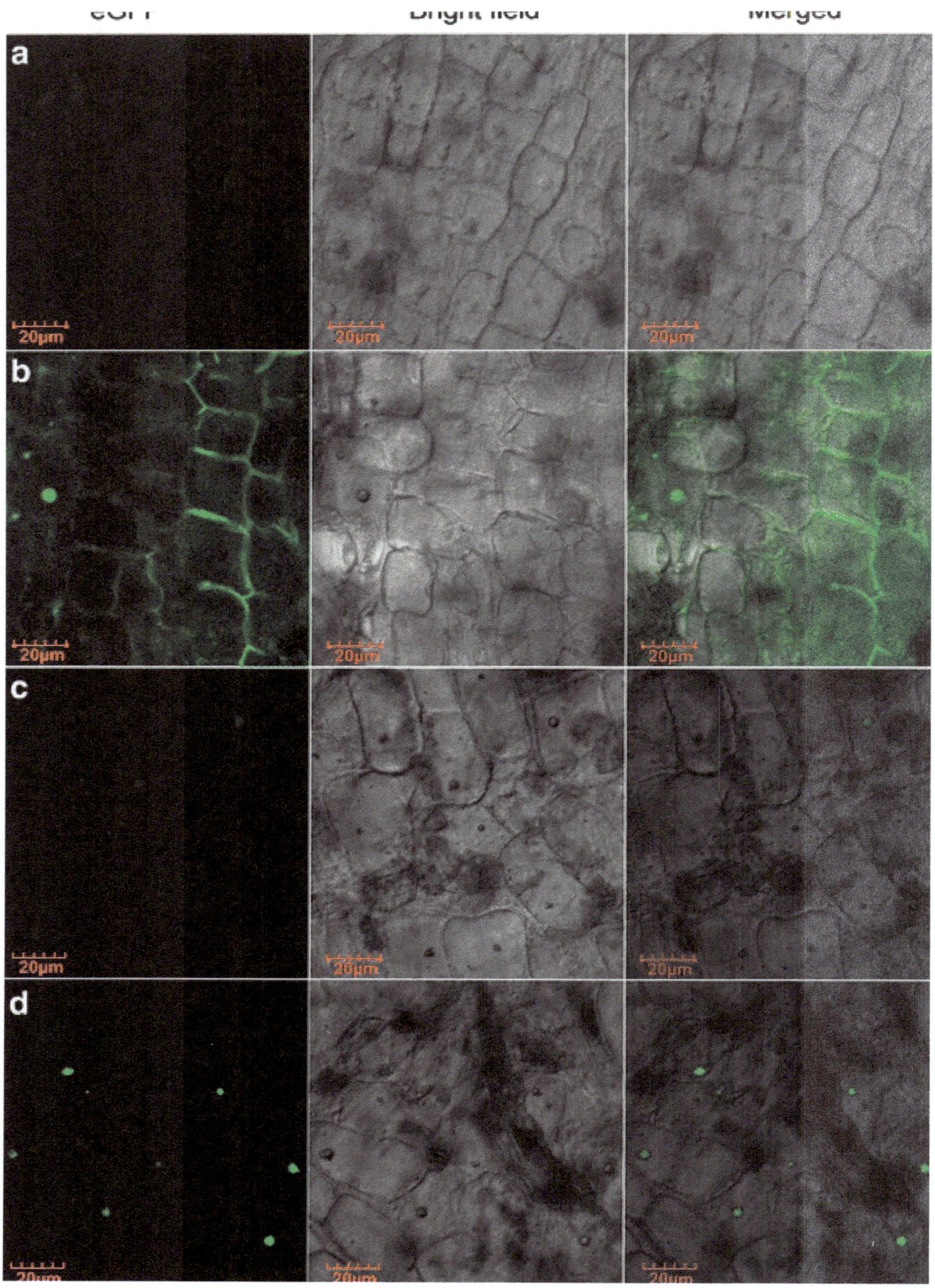

Figure 2. Confocal analyses of eGFP in WT (**a**), EV (**b**), Ei (**c**), and EO (**d**) hairy roots. *Scale bar* =20 μm; excitation wavelength=488 nm; emission wavelength of eGFP=510 nm.

Expression Patterns of *LeEIL-1* in the Two-stage Culture System

In order to assess the transgenic effects, expression patterns of *LeEIL-1* in four types of hairy root were characterized by using real-time PCR method. Each five lines of Ei and EO cultured in B5 medium were randomly selected; each three lines of WT and EV were used as controls. Similar transcript levels were observed in three WT hairy root lines (WT-1, WT-2, and WT-3). Compared with that in WT hairy root lines, *LeEIL-1* did not significantly change in EV hairy root lines ($P>0.05$). However, in Ei hairy root lines, *LeEIL-1* was significantly down-regulated compared with that in WT or EV ($P<0.01$). In EO hairy root lines, *LeEIL-1*significantly increased ($P<0.01$) (Fig. 3a). Therefore, the transcript level of *LeEIL-1* actually significantly increased in hairy roots of *LeEIL-1*-overexpressing and decreased in *LeEIL-1*-RNAi hairy root lines.

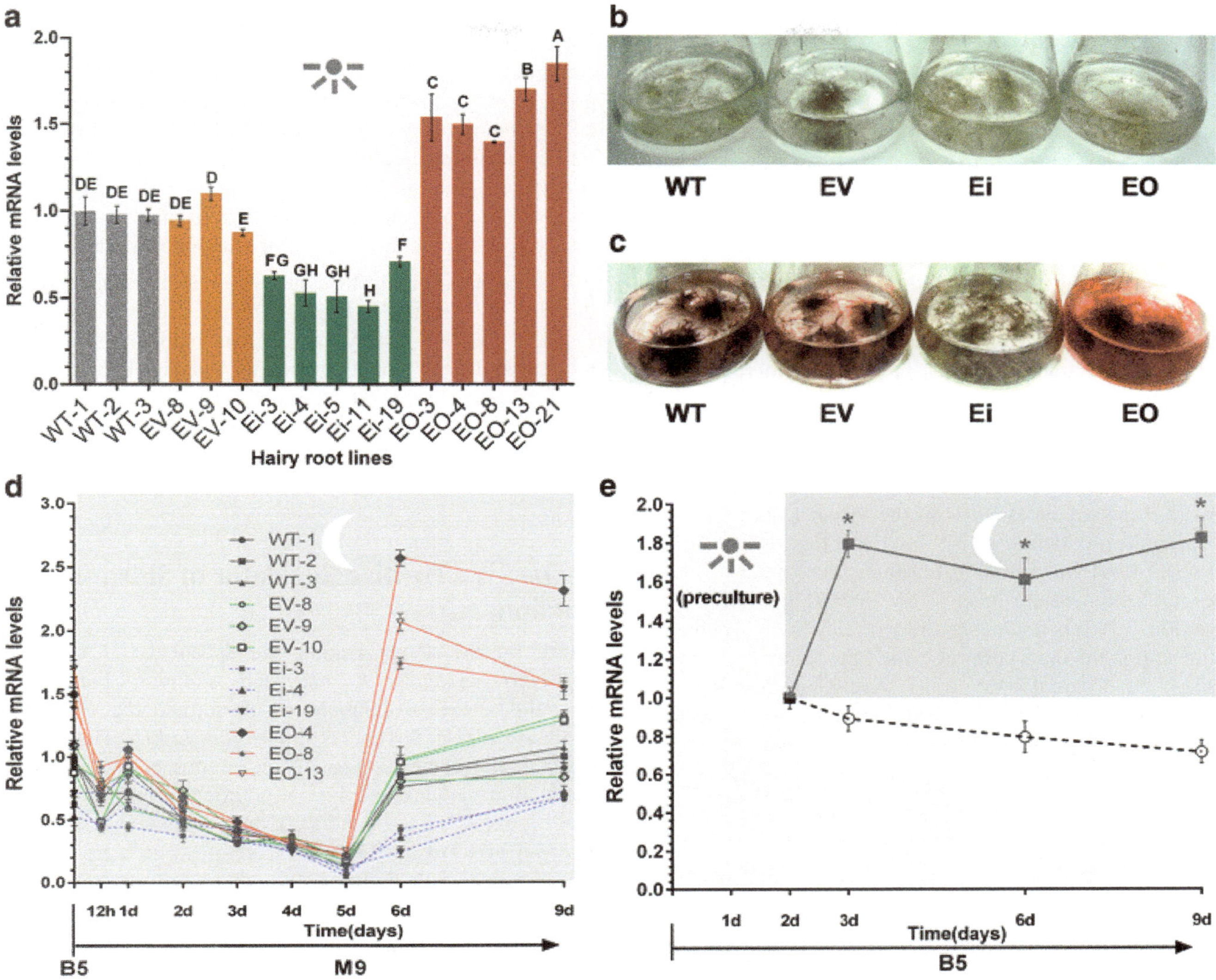

Figure 3. *LeEIL-1* Expression patterns and visual inspection of WT, EV, Ei, and EO hairy root lines. (a) Transcript levels of *LeEIL-1* in the randomly selected hairy root lines of WT, EV, Ei, and EO cultured in B5 medium under light at 26–28°C for 15 days with constant shaking at 80 rpm. A representative example from two biological experiments was shown; data represent means ± SD (n=3) and the *bars* with different *capital letters* indicate significant differences at $P<0.01$ (Least Significant Difference); (b) Phenotypes of WT, EV, Ei, and EO hairy roots in conical flasks containing hormone-free B5 liquid medium for multiplication culture; (c) Visual inspection for the color changes of WT, EV, Ei, and EO hairy root lines in M9 for 6 days (20 ml medium/50 ml flask); (d) Dynamic expression pattern analysis of*LeEIL-1* when hairy root lines were transferred from B5 into M9 medium. The values of Ei and EO lines are significantly different from those of WT or EV lines at the time points of 6 or 9 days, respectively (Least Significant Difference, $P<0.05$); (e) Expression patterns of *LeEIL-1* of the WT-1 during the dark/light transition, in which hairy roots were pre-cultured under light for 2 days and then transferred into darkness. The *asterisk* indicates that the mean value in the dark was significantly different from that under light at 3, 6 or 9 days time point, respectively (Student's t-test, $P<0.05$). The *sunshine logo* and *light gray* represent light period for B5 culture condition, and the*moon logo* and *deep gray* represent night period for M9 culture condition.

We speculated that the transformation of *LeEIL-1* genetically influences the biosynthesis of shikonin. Shikonin and its derivatives are red specific pigments, and color changes of the pigments in hairy roots as well as in M9 medium excreted from hairy roots were detected upon visual inspection [23]. Time-course analysis was also performed to assess shikonin pigment accumulation in hairy roots of WT, EV, Ei, and EO through visual inspection. No pigment was observed in B5 multiplication medium under light (Fig. 3b). The red color significantly changed when transferred hairy roots from B5 multiplication medium into M9 production medium for 1–6 days, reached the highest levels at 6–9 days, and exhibited minimal changes thereafter, with the highest yield found in EO and the lowest level in Ei from 1 to 9 days (Fig. 3c). The time course accumulation analyses of shikonin in different types of hairy roots were performed using the randomly selected four typical lines, i.e., WT-1, EV-9, Ei-19 and EO-13. From 1 to 6 days, the concentration of shikonin in hairy roots above increased rapidly, and then reached to a relative high level at the time point of 6 days. Thereafter, the content of shikonin in different hairy roots began to accumulate in different patterns. Significantly enhanced content of shikonin was observed in EO-13, while dramatically reduced in Ei-19 in comparison with that in EV-9 or WT-1 at each time point of 3, 6, 9 and 12 days ($P<0.05$) (Additional file 4: Figure S4).

Based on the time-course analysis of shikonin formation, each three lines of WT, EV, Ei, and EO were selected and transferred from B5 multiplication medium into M9 production medium for the expression patterns analysis of *LeEIL-1*. Dynamic expression patterns of *LeEIL-1* in the M9 production medium (12 h and 1–9 days) were detected, similar to the expression pattern of *LeEIL-1* in *L. erythrorhizon* callus cells in our previous report [35]. Remarkable changes occurred when transferred hairy roots of WT, EV, Ei, and EO from B5 multiplication medium into M9 production medium for 9 days. The mRNA level of *LeEIL-1* in each three lines of WT, EV, Ei, and EO reduced dramatically at 12 h, increased at 1 day in most hairy root lines, constantly declined, and then reached the lowest level at 5 days.

The transcript levels of *LeEIL-1* significantly exhibited varied increase rates in each type of hairy roots cultured in M9 from 5 to 9 days (Fig. 3d). In hairy roots of WT, EV, and Ei, the *LeEIL-1* transcripts presented a rapid increase from 5 to 6 days, and then continually increased for 6–9 days. The *LeEIL-1* transcript level in EO hairy root lines, however, slightly reduced from 6 to 9 days and maintained a relatively high level compared with those in WT, EV, and Ei hairy root lines at 9 days. Consistent with the visual inspection results on shikonin production in M9 medium, the highest *LeEIL-1* transcript level was found in EO lines, the lowest level in Ei lines, and the medium level in EV or WT lines from 5 to 9 days (Fig. 3d). It is obviously, the sixth day in the M9 medium is the important time point at which the expression level of *LeEIL-1* in different hairy roots should be compared. This result indicated that *LeEIL-1* expression might be concurrent with the accumulation of shikonin and its derivatives.

Interestingly, we noticed that *LeEIL-1* was down-regulated at the early stage within 5 days no matter it was located downstream of CaMV-35s promoter or not. In the two-stage culture system, the antibiotic- and hormone-free B5 medium was used as the growth medium for multiplying hairy roots under constant light condition, whereas the hormone-containing M9 medium was employed for hairy roots to produce shikonin and its derivatives under continuous dark culture condition. We speculated that the dramatic differences (medium components and culture condition) between these two stages might affect the expression of *LeEIL-1*at the early stage when transferred *L. erythrorhizon* hairy roots from B5 multiplication medium into M9 production medium.

Since shikonin is biosynthesized in the dark, we also detected the effect of light signal on the expression pattern of *LeEIL-1*. The WT-1 hairy root line was randomly selected and pre-cultured in B5 for 2 days under constant light condition. Similar to the expression pattern of *LeERF-1*, which is another ET response factor reported in our previous study [40], *LeEIL-1* was induced by dark (Fig. 3e). This result was also consistent with the dark-induced mRNA levels of *EIN3* in *Arabidopsis* [41].

LeEIL-1 is a Positive Regulator of Shikonin Production

Based on the above results, we speculated that *LeEIL-1* positively mediates ET-regulated shikonin biosynthesis. To validate the speculation, we performed the relationship analysis between the expression level of *LeEIL-1*(Fig. 4a) and the shikonin contents (Fig. 4b) in various hairy roots in M9 production medium for 6 days. In addition, as for different hairy root types (WT, EV, Ei and EO), only the expression levels of *LeEIL-1* affected by RNAi or overexpresison were different; all other factors, including M9 medium, culture condition (in darkness at 26°C), were same for them in the research system.

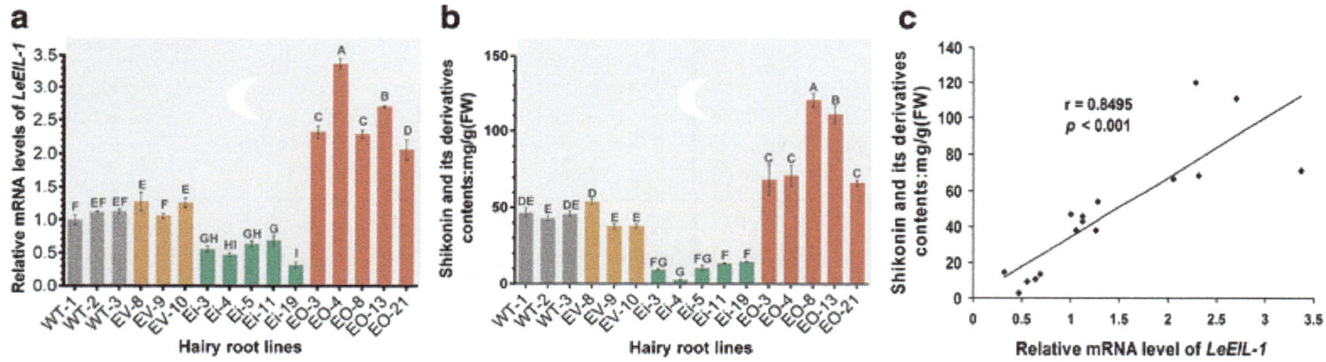

Figure 4. *LeEIL-1*expression levels (a) and shikonin production (b) of the randomly selected hairy root lines. WT, EV, Ei, and EO hairy root lines cultured in M9 production medium for 6 days. The values are means ± SD ($n=3$), and the *bars* with different *capital letters* indicate significant differences at $P<0.01$ (Least Significant Difference). The *deep gray* and*moon logo* indicate the hairy roots were cultured in the dark. Scatter diagram showing the significantly positive linear relationship between *LeEIL-1* expression level and shikonin production ($r=0.8495$; $P<0.001$) (c).

The results showed that both the *LeEIL-1* expression levels and shikonin production were significantly increased in EO hairy root lines, but dramatically declined in Ei in comparison with those in WT or EV control lines ($P<0.01$) (Fig. 4a and b). Thus, a significantly positive linear correlation between shikonin content and*LeEIL-1* expression level was deduced based on the results of all lines detected above ($r=0.8495$; $P<0.001$) (Fig. 4c). Furthermore, a number of high-yield shikonin hairy roots had been obtained from EO (Additional file5: Figure S5), suggesting the strategy of *LeEIL-1*-overexpression has great potential to obtain hairy roots with high-yield shikonin production.

LeEIL-1 Confers Shikonin Production by Regulating the Expression of Key Genes for Shikonin Biosynthesis

As EIN3/EILs act as key positive regulators at the most downstream of ET signaling transduction pathway [39,42], we deduced that the expression of *LeEIL-1* is possibly associated with the downstream target genes which regulate shikonin formation. Further research on genes involved in ET signaling transduction pathway and shikonin biosynthesis pathway under the mechanism of overexpression and RNAi of *LeEIL-1* will bridge the gap between shikonin production and accumulation of *LeEIL-1* transcripts.

Since there was no difference for shikonin production and expression level of *LeEIL-1* between EV and WT, and to avoid the transgenic effects of plasmid, three pBI121 vector containing lines (i.e., EV-9, Ei-19, and EO-13) cultured in M9 for 6 days were selected for further analyses of the expression of several key genes implicated in shikonin biosynthesis. These genes include the speculated downstream gene of *LeEIL-1* (*LeERF-1*), the key genes involved in biosynthetic pathway of shikonin (*LePAL, LeC4H-2, Le4CL-1, HMGR, LePGT-1*) (Fig. 5a), and the genes possibly responsible for the transportation, stabilization and/or excretion of shikonin pigments (*LeDI-2* and *LePS-2*).

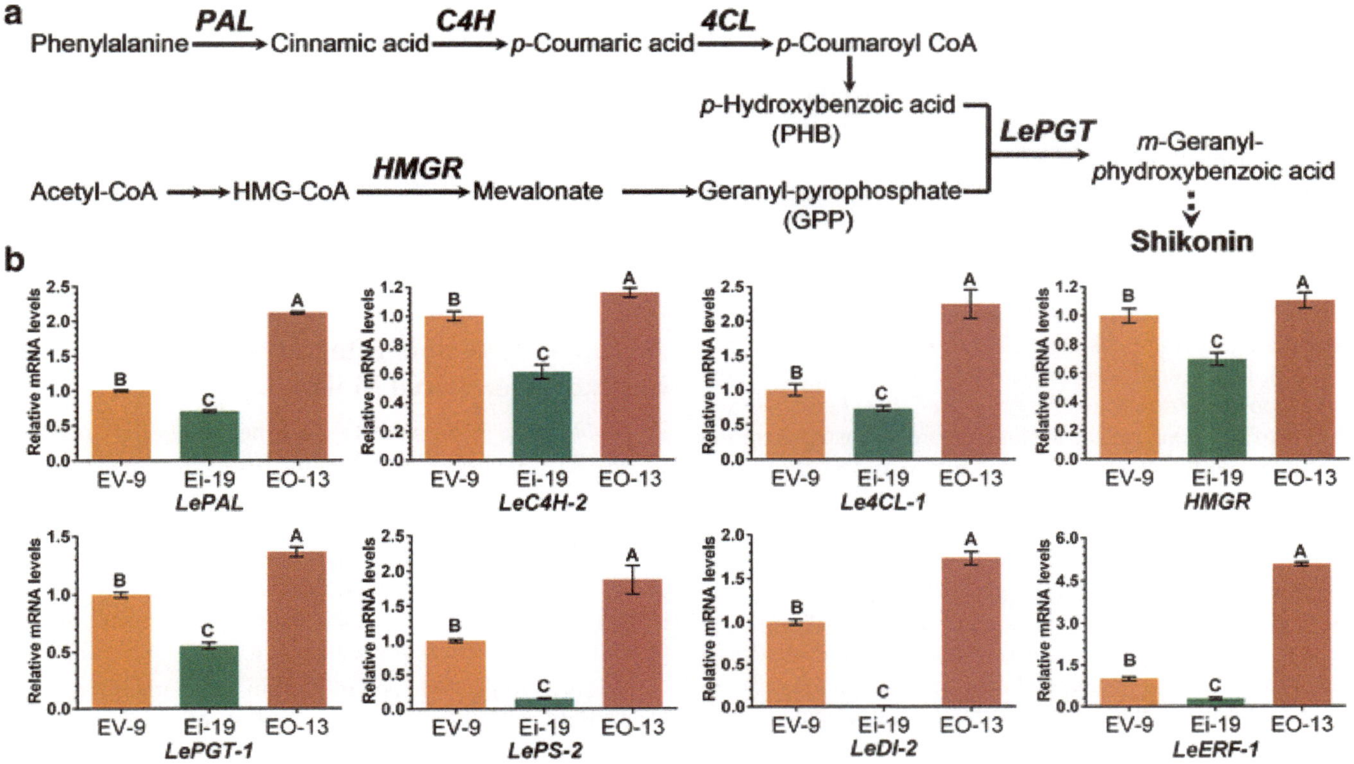

Figure 5. Expression patterns of shikonin biosynthesis-related genes regulated by *LeEIL-1*. (**a**) The metabolic pathway outline for the biosynthesis of shikonin and its derivatives. The genes encoding crucial enzymes of this pathway are marked in *bold* and *italic* fonts; (**b**) Transcript levels of the key shikonin biosynthesis-related gene in EV, Ei, and EO hairy root lines cultured in M9 medium for 6 days. A representative example from two biological experiments is shown; the values are means ± SD ($n=3$) and the *bars* with different *capital letters* indicate significant differences at $P<0.01$ (Least Significant Difference).

The expression level of *LeERF-1* was significantly up-regulated in EO-13 but down-regulated in Ei-19 compared with that in the EV-9 control ($P<0.01$). Similarly, *LeEIL-1* overexpression significantly up-regulated the mRNA accumulation of *LePAL, LePGT-1, LeC4H-2, LePS-2, Le4CL-1, LeDI-2* and *HMGR* in EO hairy root line ($P<0.01$). By contrast, RNAi-mediated knockdown of *LeEIL-1* in Ei hairy roots significantly suppressed the expression of all genes, especially *LeDI-2* ($P<0.01$) (Fig. 5b). We speculated that the up- or down-regulated transcripts of*LeEIL-1* in turn altered the transcription of most genes in the biosynthetic pathway of shikonin. In the hairy roots cultured in M9, expression patterns of these shikonin biosynthesis-related genes (*LePAL, LePGT-1,LeC4H-2, LePS-2, Le4CL-1, LeDI-2* and *HMGR*) matched to the expression pattern of *LeEIL-1*, as well as shikonin production at the time point of 6 days (Fig. 4a and b). This observation indicated that these shikonin biosynthesis-related genes were induced by *LeEIL-1* and might be involved in the *LeEIL-1*-regulated shikonin production.

LeDI-2 gene does not participate in the biosynthesis of shikonin, however, it possibly functions in the stabilization and/or transport of shikonin [8, 17]. *LePS-2* is possibly a key gene functioning in the trapping and/or intra-cell wall excretion of shikonin pigments [16]. These studies collectively indicated that the regulation of *LeEIL-1* in shikonin accumulation possibly also includes the event of excretion, stabilization, and/or transportation of shikonin and its derivatives after biosynthesis. Hence, a certain cooperative relationship is necessary between *LeEIL-1* and genes participating in shikonin biosynthesis in *L. erythrorhizon*.

Taken together, *LeEIL-1* confers shikonin production through the regulation of key genes participating in shikonin biosynthesis.

DISCUSSION

LeEIL-1 is a Pivotal Target Gene Contributing to Efficient Shikonin Formation

In the signaling network of plant, EIN3/EIL1 acts as a key integration node between ET and other signals [30]. EIN3/

EIL1 activates the expression of downstream genes, in which ERFs are the typical ones [39, 43]. ERFs then bind to the GCC box *cis*-elements of many genes regulated by ET, and function positively by activating ET responses [44].

In the medicinal plant *L. erythrorhizon*, shikonin and its derivatives are specifically accumulated in the dark but are suppressed under white or blue light condition. Consistent with the mRNA level of dark-inducible genes*LeERF-1* [40] and *LeDI-2* [17], *LeEIL-1* was also up-regulated in the dark. This finding indicated that light signal acts as a pivotal regulator of *LeEIL-1* during the process of shikonin biosynthesis.

As the first precursor for shikonin biosynthesis, *p*-hydroxybenzoic acid (PHB) is synthesized in the phenylpropanoid pathway. PHB is regulated by sequential enzymes, namely, PAL, C4H, and 4CL [10, 45]. The other precursor is geranyl pyrophosphate (GPP), which is synthesized in the isoprenoid pathway catalyzed by HMGR [12]. The formation of *m*-geranyl-phydroxybenzoic acid (GBA) is the pivotal step of shikonin formation. In this process, the substrates PHB and GPP were catalyzed by LePGT to form GBA, thereby synthesizing shikonin pigmens in *L. erythrorhizon* [15, 46] (Fig. 5a). In this study, *LeEIL-1* overexpression enhanced the expression of the downstream gene of *LeEIL-1* (*LeERF-1*) and a subset of key genes related to shikonin biosynthesis, excretion and/or transportation and stabilization (i.e., *LePAL*, *LePGT-1*, *LeC4H-2*, *LePS-2*, *Le4CL-1*,*LeDI-2* and *HMGR*) (Fig. 5b), and thus increased shikonin accumulation. Conversely, RNAi of *LeEIL-1* effectively decreased shikonin accumulation and down-regulated the above genes. Herein, we speculate that both the isoprenoid pathway and the phenylpropanoid pathway for shikonin biosynthesis were regulated by *LeEIL-1*, but further direct evidences should be provided in future studies. Moreover, the regulatory process also includes the transportation and/or excretion and stabilization of shikonin and its derivatives after formation. In a word, *LeEIL-1* is one of the most important contributors possibly in the ET-regulated shikonin biosynthesis in *L. erythrorhizon*.

Although the up-regulated transcripts of some EIN3/EIL members contribute to constitutive ET responses, exogenous ET can not affect the EIN3/EIL mRNA levels [47, 48]; nevertheless, the abundance of the EIN3/EIL protein considerably increased upon ET exposure because of the strong stabilization of this protein [42]. The accumulation of EIN3/EIL protein is primarily regulated at post-transcriptional level in vivo, and EIN2 is essential in regulating ET-induced EIN3/EIL1 [49, 50]. Based on the results of previous and present studies, we propose a model for describing the *LeEIL-1*-regulated shikonin production possibly in the ET signaling cascade (Fig. 6); in this model, which is similar to that described by Li et al. (2013), dark-induced *LeEIL-1* acts as a positive regulator in ET-regulated shikonin formation. Therefore, hairy roots with high-yield of shikonin could be obtained by using the *LeEIL-1*-overexpression strategy (Additional file 5: Figure. S5).

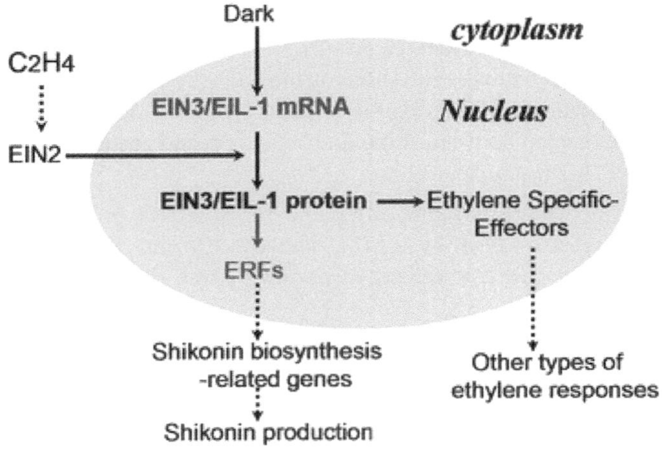

Figure 6. A proposed model illustrates the function of EIN3/EIL-1 in regulating shikonin production. The *dotted lines*represent regulatory steps, in which a direct physical link between upstream and downstream components has yet to be demonstrated.

LeEIL-1 is a Possible Integration Node between ET and Other Signal Molecules

In *Arabidopsis*, EIN3 and its close homolog EIL1 regulate myriad ET responses [39, 47]. In the life cycle of plant, EIN3/EIL1 acts as a regulator of ET signaling as well as myriad processes, such as the development and stress responses [29, 32], plant immunity defenses [28], and crosstalk between ET and jasmonate [30]. Moreover, EIN3/EIL1 is also involved in plant morphological phenotype changes [43], the de-etiolation of seedlings, and leaf senescence [41, 51].

In this study, the polymorphism and growth rate of hairy roots changed under the condition of either *LeEIL-1*-overexpression or *LeEIL-1*-RNAi (Fig. 1b; Additional file 2: Figure. S2). Meanwhile, callusing, differentiation and regeneration phenomena easily occurred in Ei hairy root lines cultured in B5 solid or liquid medium (Fig. 1c and d), compared with that in other hairy root lines cultured under the same condition. Thus, we speculate that *LeEIL-1* possibly regulates other inputs or ET specific effectors in multiple layers in the ET signaling cascade, thereby inducing the cascade of ET responses for cell development, differentiation, and regeneration. It will thus be interesting to investigate whether *LeEIL-1* acts as one of the convergence points between ET and other signal molecules in the complex network of ET transduction pathway in *L. erythrorhizon*.

CONCLUSIONS

Although the biosynthetic pathway of shikonin and its derivatives has been basically well-characterized, the molecular mechanism and key target regulators of shikonin formation are largely unknown. In the present study, by using an excellent two-stage hairy root culture system as the basal condition for shikonin production, we provided evidences for the positive role of *LeEIL-1* in regulating shikonin formation. In hairy root lines of EO, *LeEIL-1*-overexpressing significantly enhanced shikonin content in comparison with that in Ei, as well as the control lines of WT or EV under the same culture conditions. Our findings paved the way for future work aiming at extending our knowledge of the molecular regulatory mechanism of ET in shikonin biosynthesis. Furthermore, we offered a key target gene which has great potential for efficient shikonin biosynthesis in genetic engineering of *L. erythrorhizon*. Future studies focusing on the direct function of *LeEIL-1* or other *LeEILs* in the ET signaling transduction pathway in *L. erythrorhizon* will give more evidences for the regulation of ethylene on shikonin biosynthesis.

METHODS

Plant Materials and Treatment

The seeds of *L. erythrorhizon* Sieb. *et* Zucc were collected in the field location of Ulanhot city, Inner Mongolia Autonomous Region of China (N46°04′13.55″, E122°07′42.15″) in accordance with local legislation, and no specific permission was required for this study. Seed stratification and sterilization, as well as the seedling culture condition were essentially as we previously reported [52]. The robust seedlings were used for hairy root induction.

Plasmid Construction and Transformation

The plasmid of *LeEIL-1*-overexpression or *LeEIL-1*-RNAi was constructed based on the plant expression vector pBI121-*eGFP*. XbaI/BamHI were selected for the digestion of pBI121-*eGFP* vector, the coding region of *LeEIL-1*(1905 bp) was cloned and introduced into the cleavage site, with the cDNA expression cassette between the CaMV35s promoter and the *eGFP*, thereby generated the plasmid of pBI121-*LeEIL-1*-Overexpression [52].

To construct the pBI121-*LeEIL-1*-RNAi plasmid, a strategy similar to the previous reports was adopted [52, 53,54]. The intron-containing intermediate vector pUCCRNAi was used [55]. A sequence of *LeEIL-1* (348 bases) was selected based on the bioinformatics analyses of *LeEIL-1*. The sense sequence was cloned and inserted into pUCCRNAi in positive orientation, thereby generated a vector of pUCCRNAi-intron-F. The anti-sense sequence was also amplified and inserted into pUCCRNAi-intron-F in reverse orientation. This process generated plasmid of pUCCRNAi-R-intron-F. To obtain the target inverted repeat sequence, the SpeI/XbaI were selected for the digestion of pUCCRNAi-R-intron-F plasmid. Finally, the sequences obtained were inserted in the pBI121-*eGFP* vector, thereby generated the plasmid of pBI121-*LeEIL-1*-RNAi [52, 56].

All plasmids (pBI121-*LeEIL-1*-RNAi, pBI121-*LeEIL-1*-Overexpression and pBI121-*eGFP*) were transformed into the *Escherichia coli* strain TOP10 for amplification and PCR verification. Finally, these recombined expression plasmids were introduced into *A. rhizogenes* strain ATCC15834 as previously reported [52, 57, 58]. All primers for the construction of plasmids and verification of ATCC15834 carrying vectors are listed in Additional file 6: Table S1.

Induction and Cultivation of Hairy Roots

The hairy root system was used for shikonin biosynthesis and expression analysis of *LeEIL-1* via the two-stage culture system [8, 9]. The infection strains used for hairy root induction were as follows: *A. rhizogenes* strain 15834 (WT), and *A. rhizogenes* 15834 strain carrying the plasmid pBI121-*eGFP* (EV), pBI121-*LeEIL-1*-Overexpression (EO) or pBI121-*LeEIL-1*-RNAi (Ei). All stains were inoculated in YEB liquid medium in a rotary shaker (120 rpm, 26–28°C) for about 36 h. Kanamycin (50 mg l^{-1}) was added to the YEB medium for antibiotic resistance screening of EV, Ei and EO. When the OD_{600} of infection strains reached 0.5, acetosyringone (AS) at 100 μM was supplemented.

Hairy roots were then induced with the method as we previously reported [36, 52]. In brief, seedling nodes were pricked with a needle soaked in the infection medium. Then the vaccinated seedlings were transplanted to MS solid medium under subdued light or dark condition till the hairy roots appeared after 2 weeks, and over 15 lines for each type of hairy root were generated. At the infection sites, individual hairy root of about 1 cm was excised. After the elimination of residual *Agrobacterium* with cefotaxime, hairy roots were cultured in B5 solid medium (antibiotic- and hormone-free). The conditions of stock cultivation were set as follow: under subdued light at 26–28°C.

The two-stage culture system was adopted for hairy root cultivation as previously described [7, 8, 9, 21]. Hairy roots cultured on B5 solid medium were cut, and then transferred into B5 multiplication liquid medium (antibiotic- and hormone-free) for fast amplification under continuous white light at 26–28°C with a constant shaking rate of 80 rpm. Hairy roots were subcultured every 15 days (Fig. 3b).

For shikonin biosynthesis, hairy roots were transferred from B5 multiplication medium into M9 production medium and cultured in the dark at 26–28°C with a constant shaking rate of 80 rpm [21, 36, 52, 59] (Fig. 3c).

Subcellular Localization Analysis of LeEIL-1

The confocal laser scanning fluorescence microscope (LSFM, FV10-ASW, Olympus, Japan) was used for the detection of eGFP fluorescence. Excitation and emission wavelength were 488 and 510 nm, respectively [39,60].

Real-time PCR Analysis

Hairy root lines cultured in B5 multiplication medium or in M9 production medium were selected for mRNA level analysis. TRIzol reagent (TaKaRa Biotech, Shiga, Japan) was used to extract total RNA from hairy roots. First-strand cDNA was synthesized by using 1 μg of the total RNA and M-MLV reverse transcriptase (Promega, Madison, WI, USA). The real-time PCR was performed by using CFX96TM (Bio-Rad, USA) and the SYBR Green real-time PCR Master Mix (Toybo Co., Ltd., Osaka, Japan). Primers for real-time PCR analysis were designed based on the sequences in the GenBank database [24] (Additional file 7: Table S2). The glyceraldehyde-3-phosphate dehydrogenase gene (*GAPDH*), an internal reference gene, was selected as in our previous reports [24, 36, 52]. The thermal program was performed using the following parameters: denatured at 95°C for 1 min, followed by 40 amplification cycles (95°C for 15 s, 55°C for 15 s, 72°C for 45 s). Melting curves were performed after 40 cycles to confirm the specificity of the reactions. Relative expression level was calculated using the ΔΔCt method [24, 35, 36, 40, 52]. At least two to three biological replicates and three technical replicates were performed for the data analysis.

Measurement of Shikonin Contents

Measurement of shikonin contents was conducted as we previously reported [19, 20, 24, 36, 52]. In brief, shikonin and its derivatives were extracted with petrol ether and measured by using the WFZ UV-2800H spectrophotometer (Unico, Shanghai, China). Absorption of these pigments was detected at the characteristic wavelength of 520 nm. Shikonin contents were determined using a standard curve as we previously reported (shikonin content $=41.66 \times OD_{520} \times$ dilution fold) [19, 20, 24, 52]. The total shikonin content including the pigments extracted from both the hairy roots and the M9 production medium was expressed as mg/g fresh weight (FW) of hairy root.

ACKNOWLEDGEMENTS

This work was supported by the grant from the National Natural Science Foundation of China (NSFC) (31170275, 31071082, 31171161, and 31470384), the Program for Changjiang Scholars and Innovative Research Team in University (IRT_14R27), the Project of New Century Excellent Talents in University (NCET-11-0234), and the fund for University Ph.D. Program from the Ministry of Education of China (20120091110037).

ADDITIONAL FILES

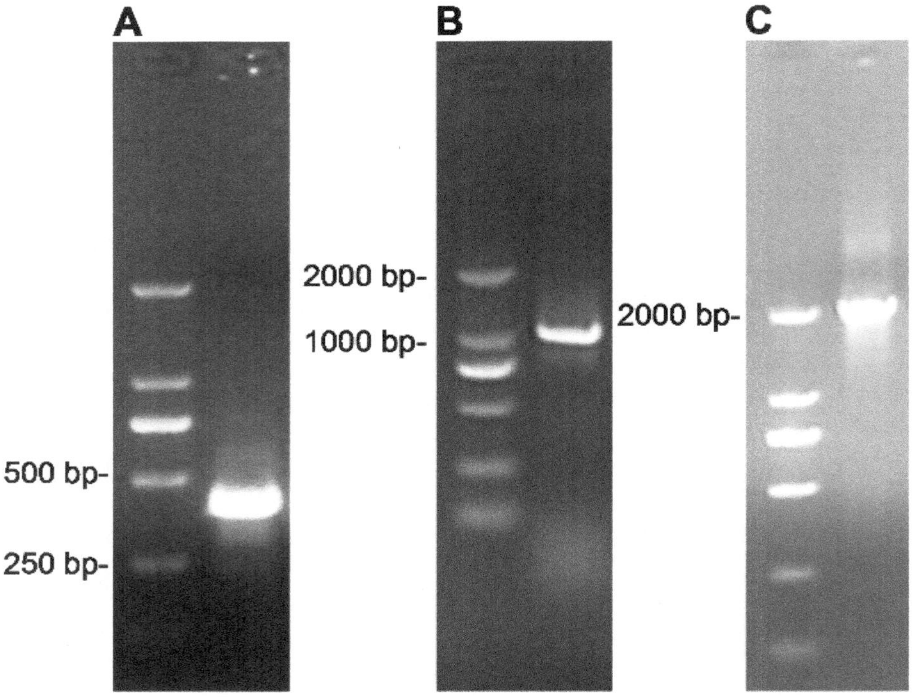

Figure S1. Verification of the constructed recombinant vectors. (A) Identification of the *eGFP* gene in the pBI121-eGFP vector by using the primer pair eGFP-F/R; (B) Amplification of the inserted target sequence in the pBI121-*LeEIL-1*-RNAi vector by using the primer pair 35S-F/GFP-R. (C) Amplification of the inserted target sequence of *LeEIL-1* in the pBI121-*LeEIL-1*-Overexpression vector by using the primer pair 35S-F/GFP-R. Primer sets are listed in Additional file 6: Table S1. (TIF 456 kb)

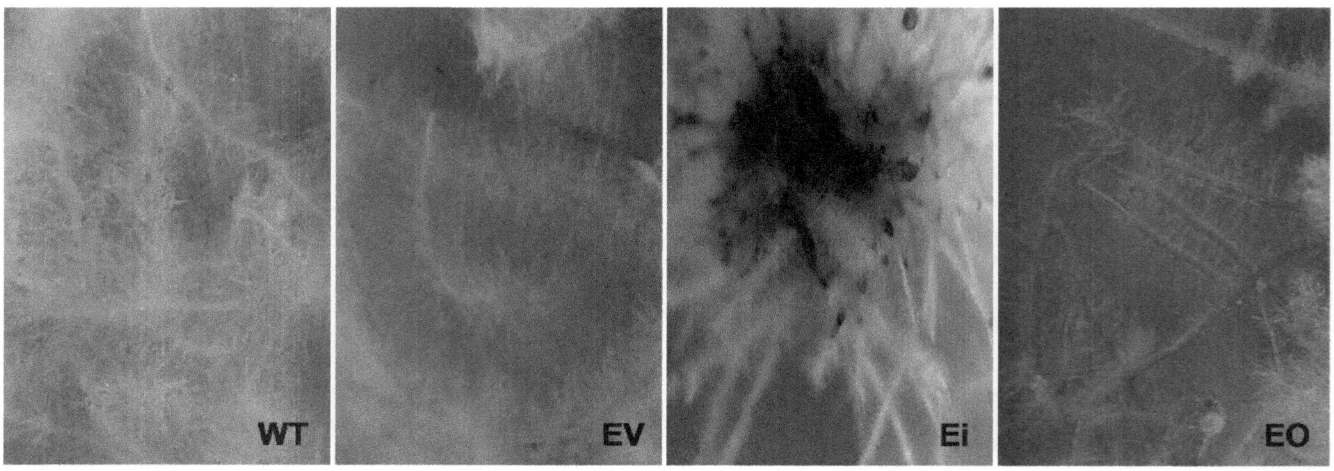

Figure S2. Stock culture of WT (A), EV (B), Ei (C) and EO (D) hairy roots of *L. erythrorhizon* on B5 solid medium (antibiotics-free and hormone-free). (TIF 3277 kb)

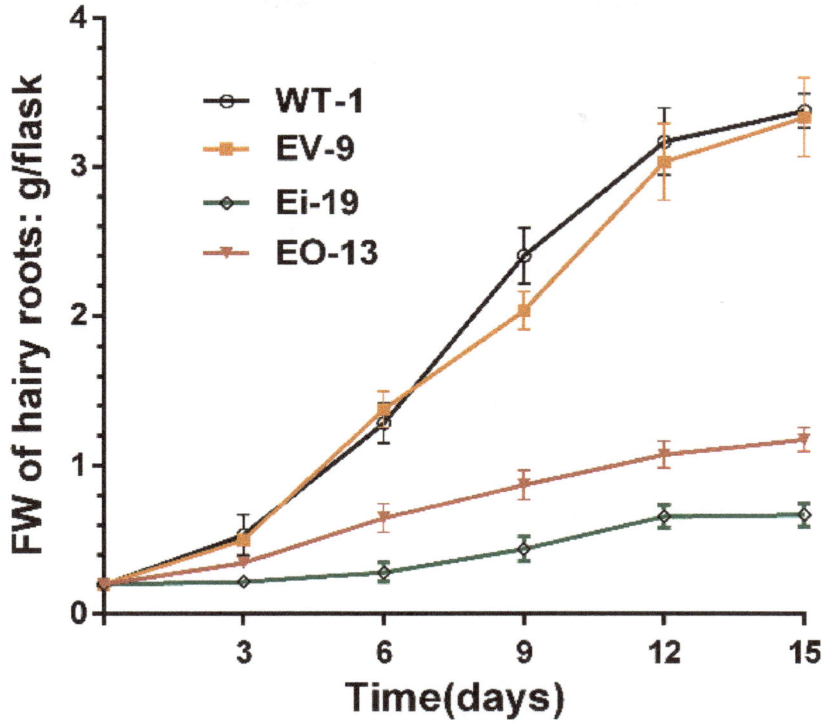

Figure S3. The time course of fresh weight (FW) of WT-1, EV-9, Ei-19 and EO-13 cultured in B5 multiplication medium (continuous light at 26–28°C with a constant shaking rate of 80 rpm). The inoculum size of 0.2 g hairy roots was subcultured in 20 ml medium/50 ml flask. Three biological replicates were performed; the value of Ei-19 or EO-13 is significantly different from that control line of WT-1 or EV-9 at each time point from 3 to 15 days, respectively (Student's *t*-test,$P<0.05$). (TIF 83 kb)

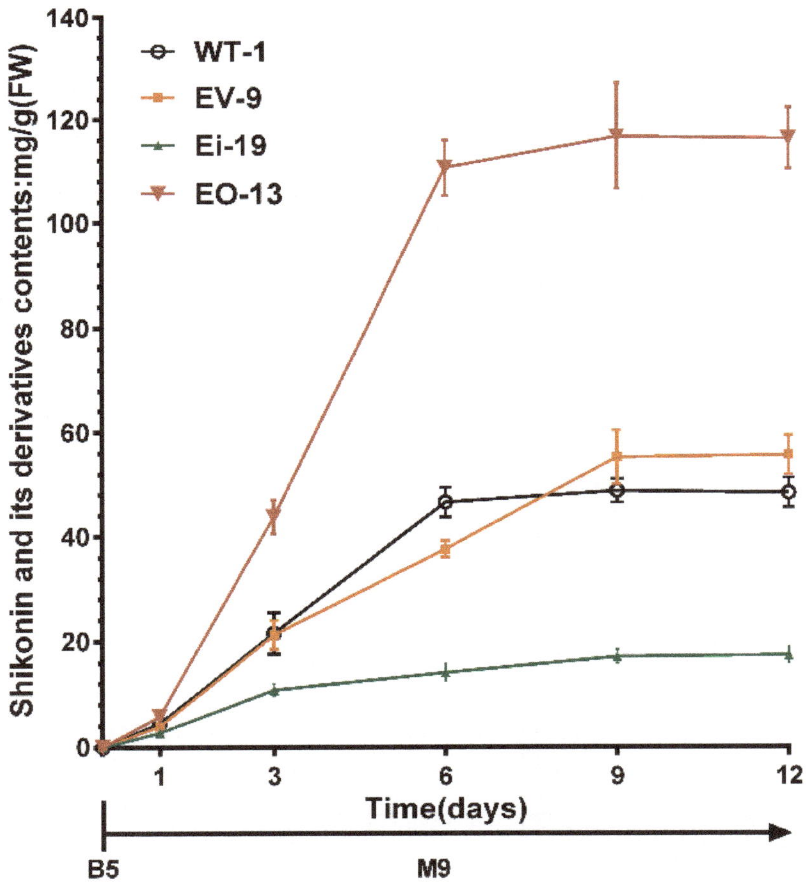

Figure S4. Time-course accumulation of shikonin in four typical hairy root lines. Value of Ei-19 or EO-13 is significantly different from that of the control line WT-1 or EV-9 at each time point from 3 to 12 days, respectively (Student's *t*-test, $P<0.05$). (TIF 125 kb)

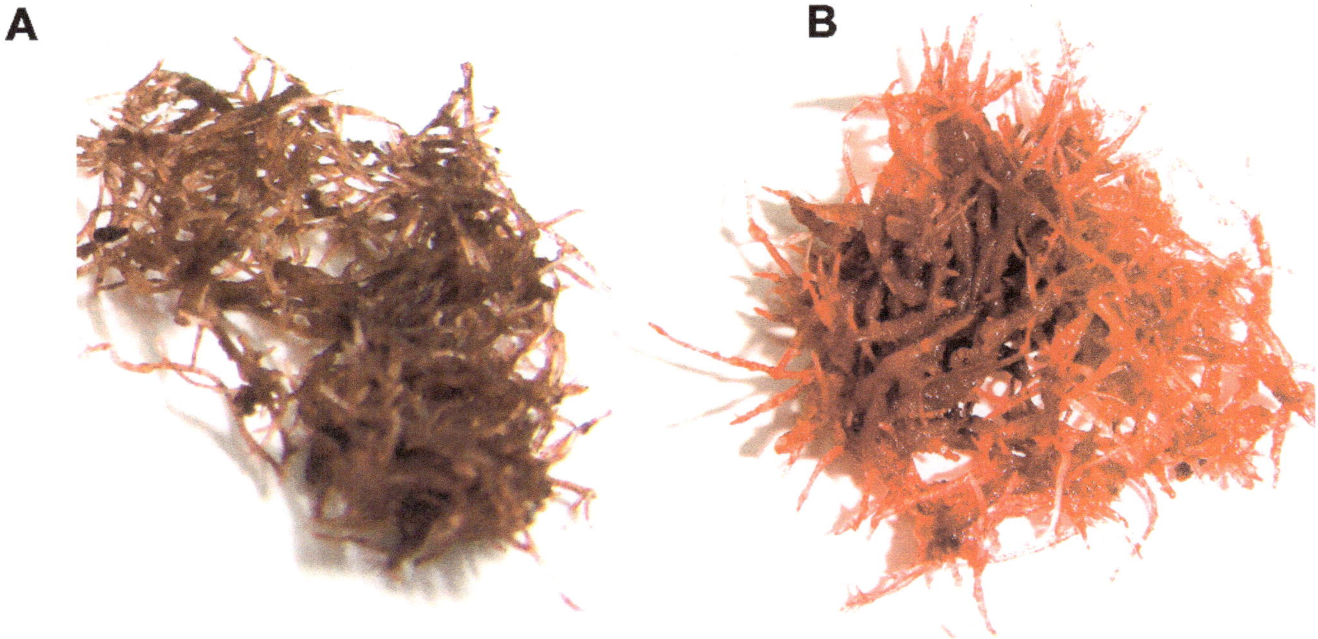

Figure S5. Comparison of growth characteristics between the typical low-yield shikonin line (A) of Ei and high-yield shikonin line (B) of EO. Hairy roots were cultured in M9 production medium in the dark for 6 days (20 ml medium/50 ml flask). (TIF 3133 kb)

Table S1. Primers for *LeEIL-1* cDNA cloning and vector construction and verification. (DOC 39 kb)

Primer type	Size of target sequence (bp)	**Primer name**	**Sequence (5'→3')**
Cloning of coding region of *LeEIL-1* and construction of *LeEIL-1*-overexpression vector	1905	LeEIL-1-OE-F	CTAGTCTAGAATGATGATGTTCGAGGAGATGGGGT
		LeEIL-1-OE-R	CTAGTCTAGAAAACCAGATGGATTCGTCCTGCTTT
Intron cloning for the construction of *LeEIL-1*-RNAi vector	213	Intron-F	GGTACGGACCGTACTACTCTATTCGTT
		Intron-R	CCTATATAATTTAAGTGGAAAAAAAGGTTAAC
Sense sequence cloning for the construction of *LeEIL-1*-RNAi vector	348	LeEIL-1-RNAi-SF	CATGGATCC CCTCATACCTTGCAAGAGCTTCAAGACA
		LeEIL-1-RNAi-SR	CATTCTAGA TGCCAGCCATGTCGCACTCTCCTTCGCT
Anti-sense sequence cloning for the construction of *LeEIL-1*-RNAi vector	348	LeEIL-1-RNAi-AF	CATACTAGTTGCCAGCCATGTCGCACTCTCCTTCGCT
		LeEIL-1-RNAi-AR	CATAGATCT CCTCATACCTTGCAAGAGCTTCAAGACA
Verification of the target sequence in vectors of pBI121-*LeEIL-eGFP*/ pBI121-*LeEIL*-1-RNAi	2072/1076	35S-F	CACTATCCTTCGCAAGACCCT
		GFP-R	GCTGAACTTGTGGCCGTTT
Verification of *Agrobacterium rhizogenes* ATCC15834 and hairy roots	626	rolC-F	CTCCTGACATCAAACTCGTC
		rolC-R	TGCTTCGAGTTATGGGTACA
Verification of *E. coli* TOP10 or *Agrobacterium rhizogenes* ATCC15834 harboring the constructed vectors	319	eGFP-F	ACCCTCGTGACCACCCTGAC
		eGFP-R	AGTTCACCTTGATGCCGTTC

Table S2. Real-time PCR primers for gene expression analysis. (DOC 34 kb)

Real-time PCR primer type	Gene name	GenBank Accession number	Sense primer (5'→3')	Anti-sense primer (5'→3')
For ethylene signaling pathway related genes	*LeEIL-1*	FJ890314	GCTCCTCAGGATGAACCAG	TCCTTACCCTTGCTGCTCATCTC
	LeERF-1	GQ246186.1	GCCAATTCATTCCATCGC	GTCCCAAATGTTCCAAGC
For shikonin biosynthesis pathway-related genes	*LePAL*	D83075.1	GCAGCCATGCTTGTTAGGA	GGTGTTGAGGAACTTGGTGATA
	LeC4H-2	AB055508.1	AGGGTGAAATCAACGAGG	ATACGGAGGCGAAGGGTC
	Le4CL-1	D49366.1	CCTATTCGTCAGGGACTACAG	GCACATCACAACATCTTCGT
	HMGR	X74783.1	TTGGAGTTCCTACAGAATGAGT	GCGGCTGGTTTCTTGTCA
	LePGT-1	AB055078.1	CTCTTAGGCTCCTCTGCT	CGTCGTCCACCTTATCTT
	LeDI-2	D45901.1	TAATCTCGTTACGGCTCAA	GCAAAGGCAAGCACCAAC
	LePS-2	AB047916.1	ATCAACGGAACGGACCAAGT	TTGAAGGCATAACGGACACC

REFERENCES

1. Chen X, Yang L, Zhang N, Turpin JA, Buckheit RW, Osterling C, et al. Shikonin, a component of Chinese herbal medicine, inhibits chemokine receptor function and suppresses human immunodeficiency virus type 1. Antimicrob Agents Chemother. 2003; 47: 2810–6.
2. Zhang Y, Qian R-Q, Li P-P. Shikonin, an ingredient of *Lithospermum erythrorhizon*, down-regulates the expression of steroid sulfatase genes in breast cancer cells. Cancer Lett. 2009; 284: 47–54.
3. Chen J, Xie J, Jiang Z, Wang B, Wang Y, Hu X. Shikonin and its analogs inhibit cancer cell glycolysis by targeting tumor pyruvate kinase-M2. Oncogene. 2011; 30: 4297–306.
4. Wang Z, Liu T, Gan L, Wang T, Yuan X, Zhang B, et al. Shikonin protects mouse brain against cerebral ischemia/reperfusion injury through its antioxidant activity. Europ J Pharmacol. 2010; 643: 211–7.
5. Papageorgiou V. Naturally occurring isohexenylnaphthazarin pigments: a new class of drugs. Planta Med. 1980; 38: 193–203.
6. Papageorgiou VP, Assimopoulou AN, Couladouros EA, Hepworth D, Nicolaou K. The chemistry and biology of alkannin, shikonin, and related naphthazarin natural products. Angew Chem Int Ed. 1999; 38: 270–301.
7. Shimomura K, Sudo H, Saga H, Kamada H. Shikonin production and secretion by hairy root cultures of *Lithospermum erythrorhizon*. Plant Cell Rep. 1991; 10: 282–5.
8. Yazaki K, Matsuoka H, Ujihara T, SATO F. Shikonin biosynthesis in *Lithospermum erythrorhizon*: Light-induced negative regulation of secondary metabolism. Plant Biotechnol. 1999; 16: 335–42.
9. Touno K, Harada K, Yoshimatsu K, Yazaki K, Shimomura K. Shikonin derivative formation on the stem of cultured shoots in *Lithospermum erythrorhizon*. Plant Cell Rep. 2000; 19: 1121–6.
10. Yazaki K, Kataoka M, Honda G, Severin K, Heide L. cDNA cloning and gene expression of phenylalanine ammonia-lyase in*Lithospermum erythrorhizon*. Biosci Biotechnol Biochem. 1997; 61: 1995–2003.
11. Yamamura Y, Ogihara Y, Mizukami H. Cinnamic acid 4-hydroxylase from *Lithospermum erythrorhizon*: cDNA cloning and gene expression. Plant Cell Rep. 2001; 20: 655–62.
12. Lange BM, Severin K, Bechthold A, Heide L. Regulatory role of microsomal 3-hydroxy-3-methylglutaryl-coenzyme A reductase for shikonin biosynthesis in *Lithospermum erythrorhizon* cell suspension cultures. Planta. 1998; 204: 234–41.
13. Yazaki K, Kunihisa M, Fujisaki T, Sato F. Geranyl Diphosphate: 4-Hydroxybenzoate Geranyltransferase from*Lithospermum erythrorhizon* CLONING AND CHARACTERIZATION OF A KEY ENZYME IN SHIKONIN BIOSYNTHESIS. J Biol Chem. 2002; 277: 6240–6.
14. Ohara K, Muroya A, Fukushima N, Yazaki K. Functional characterization of LePGT1, a membrane-bound prenyltransferase involved in the geranylation of *p*-hydroxybenzoic acid. Biochem J. 2009; 421: 231–41.
15. Ohara K, Mito K, Yazaki K. Homogeneous purification and characterization of LePGT1- a membrane-bound aromatic substrate prenyltransferase involved in secondary metabolism of *Lithospermum erythrorhizon*. FEBS J. 2013; 280: 2572–80.
16. Yamamura Y, Sahin FP, Nagatsu A, Mizukami H. Molecular cloning and characterization of a cDNA encoding a novel apoplastic protein preferentially expressed in a shikonin-producing callus strain of *Lithospermum erythrorhizon*. Plant Cell Physiol. 2003; 44: 437–46.
17. Yazaki K, Matsuoka H, Sato F. cDNA cloning and functional analysis of *LEDI-2*, a gene preferentially expressed in the dark in *Lithospermum* cell suspension cultures. Plant Physiol. 1997; 114: 1348.
18. Touno K, Harada K, Yoshimatsu K, Yazaki K, Shimomura K. Histological observation of red pigment formed on shoot stem of *Lithospermum erythrorhizon*. Plant Biotech. 2000; 17: 127–30.

19. Liu Z, Qi J-L, Chen L, Zhang M-S, Wang X-Q, Pang Y-J, et al. Effect of light on gene expression and shikonin formation in cultured *Onosma paniculatum* cells. Plant Cell Tiss Org Cult. 2006; 84: 38–48.

20. Zhang W-J, Su J, Tan M-Y, Liu G-L, Pang Y-J, Shen H-G, et al. Expression analysis of shikonin-biosynthetic genes in response to M9 medium and light in *Lithospermum erythrorhizon* cell cultures. Plant Cell Tiss Organ Cult. 2010; 101: 135–42.

21. Fujita Y, Hara Y, Suga C, Morimoto T. Production of shikonin derivatives by cell suspension cultures of *Lithospermum erythrorhizon*. Plant Cell Rep. 1981; 1: 61–3.

22. Liu Z, Li Y, Yang T, Su J, Zhang M, Tian R, et al. Shikonin accumulation is related to calcium homeostasis in *Onosma paniculata* cell cultures. Phyton-Ann Rei Bot. 2011; 51: 103–13.

23. Brigham LA, Michaels PJ, Flores HE. Cell-specific production and antimicrobial activity of naphthoquinones in roots of*Lithospermum erythrorhizon*. Plant Physiol. 1999; 119: 417–28.

24. Wu S-J, Qi J-L, Zhang W-J, Liu S-H, Xiao F-H, Zhang M-S, et al. Nitric oxide regulates shikonin formation in suspension-cultured *Onosma paniculatum* cells. Plant Cell Physiol. 2009; 50: 118–28.

25. Yazaki K, Takeda K, Tabata M. Effects of methyl jasmonate on shikonin and dihydroechinofuran production in Lithospermum cell cultures. Plant Cell Physiol. 1997; 38: 776–82.

26. Touno K, Tamaoka J, Ohashi Y, Shimomura K. Ethylene induced shikonin biosynthesis in shoot culture of *Lithospermum erythrorhizon*. Plant Physiol Biochem. 2005; 43: 101–5.

27. Potuschak T, Lechner E, Parmentier Y, Yanagisawa S, Grava S, Koncz C, et al. EIN3-dependent regulation of plant ethylene hormone signaling by two *Arabidopsis* F box proteins: EBF1 and EBF2. Cell. 2003; 115: 679–89.

28. Chen H, Xue L, Chintamanani S, Germain H, Lin H, Cui H, et al. ETHYLENE INSENSITIVE3 and ETHYLENE INSENSITIVE3-LIKE1 repress *SALICYLIC ACID INDUCTION DEFICIENT2* expression to negatively regulate plant innate immunity in*Arabidopsis*. Plant Cell. 2009; 21: 2527–40.

29. Peng J, Li Z, Wen X, Li W, Shi H, Yang L, et al. Salt-induced stabilization of EIN3/EIL1 confers salinity tolerance by deterring ROS accumulation in *Arabidopsis*. PLoS Genet. 2014; 10, e1004664.

30. Zhu Z, An F, Feng Y, Li P, Xue L, Mu A, et al. Derepression of ethylene-stabilized transcription factors (EIN3/EIL1) mediates jasmonate and ethylene signaling synergy in *Arabidopsis*. Proc Natl Acad Sci U S A. 2011; 108: 12539–44.

31. Boutrot F, Segonzac C, Chang KN, Qiao H, Ecker JR, Zipfel C, et al. Direct transcriptional control of the *Arabidopsis*immune receptor FLS2 by the ethylene-dependent transcription factors EIN3 and EIL1. Proc Natl Acad Sci U S A. 2010; 107: 14502–7.

32. Shi Y, Tian S, Hou L, Huang X, Zhang X, Guo H, et al. Ethylene signaling negatively regulates freezing tolerance by repressing expression of *CBF* and type-A *ARR* genes in *Arabidopsis*. Plant Cell. 2012; 24: 2578–95.

33. Chang J, Clay JM, Chang C. Association of cytochrome b_5 with ETR1 ethylene receptor signaling through RTE1 in*Arabidopsis*. Plant J. 2014; 77: 558–67.

34. Qi JL, Zhang WJ, Liu SH, Wang H, Sun DY, Xu GH, et al. Expression analysis of light-regulated genes isolated from a full-length-enriched cDNA library of *Onosma paniculatum* cell cultures. J Plant Physiol. 2008; 165: 1474–82.

35. Zou A, Zhang W, Pan Q, Zhu S, Yin J, Tian R, et al. Cloning, characterization, and expression of *LeEIL-1*, an *Arabidopsis*EIN3 homolog, in *Lithospermum erythrorhizon*. Plant Cell Tiss Org Cult. 2011; 106: 71–9.

1. Zhao H, Chang Q, Zhang D, Fang R, Wu F, Wang X, et al. Overexpression of *LeMYB1* enhances shikonin formation by up-regulating key shikonin biosynthesis-related genes in *Lithospermum erythrorhizon*. Biol Plant. 2015; 59: 429–35.

2. Batra J, Dutta A, Singh D, Kumar S, Sen J. Growth and terpenoid indole alkaloid production in *Catharanthus roseus* hairy root clones in relation to left- and right-termini-linked Ri T-DNA gene integration. Plant Cell Rep. 2004; 23: 148–54.

3. Grąbkowska R, Królicka A, Mielicki W, Wielanek M, Wysokińska H. Genetic transformation of *Harpagophytum procumbens* by *Agrobacterium rhizogenes*: iridoid and phenylethanoid glycoside accumulation in hairy root cultures. Acta Physiol Plant. 2010; 32: 665–73.

4. An F, Zhao Q, Ji Y, Li W, Jiang Z, Yu X, et al. Ethylene-induced stabilization of ETHYLENE INSENSITIVE3 and EIN3-LIKE1 is mediated by proteasomal degradation of EIN3 binding F-box 1 and 2 that requires EIN2 in *Arabidopsis*. Plant Cell. 2010; 22: 2384–401.

5. Zhang W, Zou A, Miao J, Yin Y, Tian R, Pang Y, et al. *LeERF-1*, a novel AP2/ERF family gene within the B3 subcluster, is down-regulated by light signals in *Lithospermum erythrorhizon*. Plant Biol. 2011; 13: 343–8.

6. Li Z, Peng J, Wen X, Guo H. *Ethylene-insensitive3* is a senescence-associated gene that accelerates age-dependent leaf senescence by directly repressing *miR164* transcription in *Arabidopsis*. Plant Cell. 2013; 25: 3311–28.

7. Guo H, Ecker JR. Plant responses to ethylene gas are mediated by $SCF^{EBF1/EBF2}$-dependent proteolysis of EIN3 transcription factor. Cell. 2003; 115: 667–77.

8. Solano R, Stepanova A, Chao Q, Ecker JR. Nuclear events in ethylene signaling: a transcriptional cascade mediated by ETHYLENE-INSENSITIVE3 and ETHYLENE-RESPONSE-FACTOR1. Genes Dev. 1998; 12: 3703–14.

9. Fujimoto SY, Ohta M, Usui A, Shinshi H, Ohme-Takagi M. Arabidopsis ethylene-responsive element binding factors act as transcriptional activators or

repressors of GCC box–mediated gene expression. Plant Cell. 2000; 12: 393–404.

10. Yazaki K, Ogawa A, Tabata M. Isolation and characterization of two cDNAs encoding 4-coumarate: CoA ligase in*Lithospermum* cell cultures. Plant Cell Physiol. 1995; 36: 1319–29.

11. Fujisaki T, Ohara K, Yazaki K. Characterization of an aromatic substrate prenyltransferase, LePGT-1, involved in secondary metabolism. Wood Res. 2003; 90: 1–2.

12. Chao Q, Rothenberg M, Solano R, Roman G, Terzaghi W, Ecker JR. Activation of the ethylene gas response pathway in*Arabidopsis* by the nuclear protein ETHYLENE-INSENSITIVE3 and related proteins. Cell. 1997; 89: 1133–44.

13. Yanagisawa S, Yoo S-D, Sheen J. Differential regulation of EIN3 stability by glucose and ethylene signalling in plants. Nature. 2003; 425: 521–5.

14. Ji Y, Guo H. From endoplasmic reticulum (ER) to nucleus: EIN2 bridges the gap in ethylene signaling. Mol Plant. 2013; 6: 11–4.

15. Miyata K, Kawaguchi M, Nakagawa T. Two distinct EIN2 genes cooperatively regulate ethylene signaling in *Lotus japonicus*. Plant Cell Physiol. 2013; 54: 1469–77.

16. Zhong S, Zhao M, Shi T, Shi H, An F, Zhao Q, et al. EIN3/EIL1 cooperate with PIF1 to prevent photo-oxidation and to promote greening of *Arabidopsis* seedlings. Proc Natl Acad Sci U S A. 2009; 106: 21431–6.

17. Fang R, Wu F, Zou A, Zhu Y, Zhao H, Zhao H, et al. Transgenic analysis reveals *LeACS-1* as a positive regulator of ethylene-induced shikonin biosynthesis in *Lithospermum erythrorhizon* hairy roots. Plant Mol Biol. 2016; 90: 345–58.

18. Manamohan M, Chandra GS, Asokan R, Deepa H, Prakash M, Kumar NK. One-step DNA fragment assembly for expressing intron-containing hairpin RNA in plants for gene silencing. Anal Biochem. 2013; 433: 189–91.

19. Rastogi S, Kumar R, Chanotiya CS, Shanker K, Gupta MM, Nagegowda DA, et al. 4-Coumarate: CoA ligase partitions metabolites for eugenol biosynthesis. Plant Cell Physiol. 2013; 54: 1238–52.

20. Gan D, Zhang J, Jiang H, Jiang T, Zhu S, Cheng B. Bacterially expressed dsRNA protects maize against SCMV infection. Plant Cell Rep. 2010; 29: 1261–8.

21. Zhu M, Chen G, Zhou S, Tu Y, Wang Y, Dong T, et al. A new tomato NAC (NAM/ATAF1/2/CUC2) transcription factor, SlNAC4, functions as a positive regulator of fruit ripening and carotenoid accumulation. Plant Cell Physiol. 2014; 55: 119–35.

22. Höfgen R, Willmitzer L. Storage of competent cells for *Agrobacterium* transformation. Nucleic Acids Res. 1988; 16: 9877.

23. Bowra S, Vincze E. Transformation of rhizobia with broad-host-range plasmids by using a freeze-thaw method. Appl Environ Microb. 2006; 72: 2290–3.

24. Sommer S, Köhle A, Yazaki K, Shimomura K, Bechthold A, Heide L. Genetic engineering of shikonin biosynthesis hairy root cultures of *Lithospermum erythrorhizon* transformed with the bacterial *ubiC* gene. Plant Mol Biol. 1999; 39: 683–93.

25. Yu P, Yuan J, Deng X, Ma M, Zhang H. Subcellular targeting of bacterial *CusF* enhances Cu accumulation and alters root to shoot Cu translocation in Arabidopsis. Plant Cell Physiol. 2014; 55: 1568–81.

CHAPTER 12

GENOME-WIDE IDENTIFICATION OF MICRORNAS IN POMEGRANATE (PUNICA GRANATUM L.) BY HIGH-THROUGHPUT SEQUENCING

Thangasamy Saminathan[1], Abiodun Bodunrin[1], Nripendra V. Singh[2], Ramajayam Devarajan[3], Padma Nimmakayala[1], Moersfelder Jeff[4], Mallikarjuna Aradhya[4], and Umesh K. Reddy[1]

[1]Department of Biology, Gus R. Douglass Institute, West Virginia State University, Institute, WV 25112-1000, USA.

[2]ICAR-National Research Center on Pomegranate, Kegaon, Solapur, Maharashtra 413255, India.

[3]ICAR-Indian Institute of Oil Palm Research, Pedavegi, West Godavari, Andhra Pradesh 534450, India.

[4]National Clonal Germplasm Repository, USDA-ARS, University of California, Davis, CA 95616, USA.

ABSTRACT

Background

MicroRNAs (miRNAs), a class of small non-coding endogenous RNAs that regulate gene expression post-transcriptionally, play multiple key roles in plant growth and development and in biotic and abiotic stress response. Knowledge and roles of miRNAs in pomegranate fruit development have not been explored.

Results

Pomegranate, which accumulates a large amount of anthocyanins in skin and arils, is valuable to human health, mainly because of its antioxidant properties. In this study, we developed a small RNA library from pooled RNA samples from young seedlings to mature fruits and identified both conserved and pomegranate-specific miRNA from 29,948,480 high-quality reads. For the pool of 15- to 30-nt small RNAs, ~50 % were 24 nt. The miR157 family was the most abundant, followed by miR156, miR166, and miR168, with variants within each family. The base bias at the first position from the 5' end had a strong preference for U for most 18- to 26-nt sRNAs but a preference for A for 18-nt sRNAs. In addition, for all 24-nt sRNAs, the nucleotide U was preferred (97 %) in the first position. Stem-loop RT-qPCR was used to validate the expression of the predominant miRNAs and novel miRNAs in leaves, male and female flowers, and multiple fruit developmental stages; miR156, miR156a, miR159a, miR159b, and miR319b were upregulated during the later stages of fruit development. Higher expression of miR156 in later fruit developmental may positively regulate anthocyanin biosynthesis by reducing SPL transcription factor. Novel miRNAs showed variation in expression among different tissues.

Citation: Saminathan T, Bodunrin A, Singh NV, Devarajan R, Nimmakayala P, Jeff M, Aradhya M, and Reddy UK (2016). Genome-wide identification of microRNAs in pomegranate (Punicagranatum L.) by high-throughput sequencing. BMC Plant Biology, 16:122. DOI: 10.1186/s12870-016-0807-3

These novel miRNAs targeted different transcription factors and hormone related regulators. Gene ontology and KEGG pathway analyses revealed predominant metabolic processes and catalytic activities, important for fruit development. In addition, KEGG pathway analyses revealed the involvement of miRNAs in ascorbate and linolenic acid, starch and sucrose metabolism; RNA transport; plant hormone signaling pathways; and circadian clock.

Conclusion

Our first and preliminary report of miRNAs will provide information on the synthesis of biochemical compounds of pomegranate for future research. The functions of the targets of the novel miRNAs need further investigation.

BACKGROUND

Pomegranate (*Punica granatum* L.), one of the two species within the genus *Punica*, producing a non-climacteric fruit with a low respiration rate [1], is a tropical and subtropical attractive deciduous shrub. Pomegranate was previously placed within its own family Punicaceae, but recent phylogenetic studies have shown that it belongs to Lythraceae. It is one of the oldest edible fruits and has grown naturally from Iran to the Himalayas in northern India since ancient times, although it is native to Iran [2, 3, 4]. Although pomegranate is widely cultivated, the five major producers are India, Iran, China, the United States and Turkey [5].

The plant is tolerant of various soil conditions and grows well under sunlight and mild winters [6]. The fruit is a round or spherical in shape, with a fleshy, tubular calyx and leathery skin often deep pink or rich red in color [7]. The inside of the fruit is separated by membranous walls into compartments packed with sac-like structures filled with fleshy juicy, red, pink or whitish pulp called arils, and each aril sac contains one white or red, angular, soft or hard seed [6, 7].

In recent years, pomegranate has become popular for its medicinal properties and its nutritional benefit in the human diet. Pomegranate is a nutrient-dense food source rich in phytochemical compounds. It contains high levels of flavonoids and polyphenols, potent antioxidants offering protection against heart disease and cancer. Because of the health-promoting traits from both the edible and nonedible parts of the fruit in treating a wide range of human diseases such as cancer, diabetes, obesity, Alzheimer disease, and hypertension, pomegranate is considered an important commercial and valuable fruit crop across the world [8, 9]. Metabolome analysis revealed that parts of pomegranate including the fruit peel, juice, root and bark, flowers, leaves and seed contain almost 40 biochemical compounds that are beneficial in different therapies [10]. The compounds include gallotannins, ellagic acid, flavonoids, antioxidants, terpenoids and alkaloids [11, 12, 13].

The color of the pomegranate fruit including arils develops from the presence of anthocyanins, water-soluble flavonoid pigments, mostly orange to red and purple/blue [14]. In addition to playing significant roles in plant defense mechanisms [15], anthocyanins are considered valuable to human health because of high antioxidant activity [16], and fruit arils, the edible part of pomegranate, contain the highest quantity of anthocyanins [17]. The biochemical pathway of anthocyanin production has been well documented in numerous plant species, with the involvement of chalcone synthase, chalcone isomerase, and leucoanthocyanidin [18].

In Arabidopsis, the anthocyanin pathway is regulated at the transcription level by transcriptional regulators such as the R2R3-MYB domain, WD40 repeat, and a basic helix-loop-helix (bHLH) [19, 20, 21]. The WD40-repeat gene is a functional homologue of Arabidopsis *TTG1* and is involved in regulating anthocyanin biosynthesis during pomegranate fruit development [22]. Recently, anthocyanin biosynthetic genes in red and white pomegranate were cloned and characterized [23] and the expression of key regulatory genes of anthocyanin biosynthesis in pomegranate was analyzed [24].

Plants have two major classes of small regulatory non-coding RNAs. They are small interfering RNAs (siRNAs) and microRNAs (miRNAs), both generated from double-stranded RNA precursors into 20- to 24-nt molecules with the help of Dicers or Dicer-like (DCL) [25]. Many basic aspects of plant development and stresses are controlled by miRNA families [26]. Most of the miRNAs are coded by genes spanning 100–400 nt and further processed by the RNA-induced silencing complex containing Argonaute (AGO) proteins. At the end of processing, depending on the presence of the type of AGO effector protein, the targets can be degraded at the mRNA level or inhibited at the translation level [27]. Bioinformatics analyses revealed at least 21 conserved miRNA families, including miR156, miR159, and miR160, in angiosperms. Plants contain more non-conserved than conserved miRNAs [28], and high-throughput sequencing led to the discovery of non-conserved miRNAs from divergent plant species such as cucurbits, grape, barley and apple [29, 30, 31, 32, 33, 34]. miRNAs play key roles in different crops for development and stress response, regulation of anthocyanin accumulation in tomato [35], mediation of nitrogen starvation adaptation in *Arabidopsis thaliana* [36], and elongation of fiber in cotton [37].

Although pomegranate is an important fruit crop with many medicinal properties, the information on miRNAs in pomegranate is lacking. In this study, we report the profiling of miRNAs from seedling to fruit with use of Illumina HiSeq 2000 RNA sequencing and expression analysis of specific miRNAs in leaves, flowers and during fruit development. miR157 was the most abundant miRNA, followed by miR156, miR166, and others. Among different small RNAs (sRNAs), those of 24 nt were most abundant. We found 28 novel miRNAs along with predicted precursor structures and participating pathways. The results from this study could provide valuable information to further reveal the regulatory roles in pomegranate.

METHODS

Plant Materials

Young leaves, male and female flowers and arils of developing fruits (developmental stages I to VI described in Fig. 1) were collected in 2014 from the cultivar 'Al-sirin-nar' grown in the USDA pomegranate germplasm collection at the Wolfskill experimental orchard in Winters, CA, USA (38°50'34.48" N; 121°97'83.02" W), were immediately frozen in liquid nitrogen, and were finally stored at −70 °C. For each tissue type, we have collected leaves, flowers, and fruits of different stages from three independent trees. And these three independent trees were considered as biological replications for stem-loop RT-qPCR experiments.

Figure 1. Morphological features of pomegranate fruit development stages. Harvested fruit at different developmental stages from days after pollination divided into six stages. Scale bar: 2 cm.

Collection of Arils from Mature Fruits to Grow Seedlings

Arils of physiologically mature 'Al-sirin-nar' fruits were removed by gently opening the fruits and extracting the arils with the help of air and water. The extracted pomegranate arils were immersed in a bath of cold water, and all other elements of the fruit were washed away. All extracted arils were separated from all other fruit parts, leaving them pristine, whole, and untouched, and then were washed and air-dried. The arils were sown in peat moss pads to grow young seedlings.

RNA Extraction

Total RNA from 10-day-old seedlings was extracted as described [38] by using TRIzol reagent (Invitrogen, Carlsbad, CA) and the RNA MiniPrep kit (Zymo Research, Irvine, CA). Total RNA from leaves, flowers and fruits of different developmental stages was extracted using a modified CTAB-LiCl method [39]. For fruit samples, we used only separated arils for all developmental stages. About 200 mg of finely ground sample in liquid nitrogen for each tissue was used for extraction. Extraction buffer I, II and other solutions were prepared as suggested [39]. The chloroform: isoamyl alcohol (24:1) and LiCl steps were repeated three times. Finally, the RNA pellet was dissolved in 40 μL RNase-free water. All RNA samples were purified with use of the RNA Clean & Concentrator kit with on-column digestion of genomic DNA by using DNase I (Zymo Research, Irvine, CA). RNA integrity number > 8.0 was confirmed by use of the 2100 Bioanalyzer (Agilent Technologies, Santa Clara, CA). For global miRNA transcriptome profiling, an equimolar concentration of total RNA extracted from three biological replications of all samples was pooled and sent for RNA sequencing. Total RNA from all three biological replications was independently used in stem-loop RT-qPCR.

Small RNA Sequencing

sRNA samples were sequenced by the Beijing Genomics Institute (BGI, Hong Kong) with the Illumina HiSeq 2000 platform. The construction of the sRNA library and sequencing consisted of the following steps [40]. After extracting the total RNA from the samples, sRNAs of 18~30 nt were gel-purified, 5' RNA adapter-ligated and gel-purified, 3' RNA adapter-ligated and gel-purified, then underwent RT-PCR and gel purification. Finally, the library products were ready for sequencing by using Illumina HiSeq 2000.

sRNAs from deep sequencing covered almost every kind of RNA, including miRNAs, siRNAs, piwi-interacting RNAs (piRNAs), ribosomal RNAs (rRNAs), transfer RNAs (tRNAs), small nuclear RNAs (snRNAs), small nucleolar RNAs (snoRNAs), repeat-associated sRNAs and degraded tags of exons or introns. The sRNA digitization analysis based on high-throughput sequencing involved use of sequencing by synthesis (SBS), which can decrease the loss of nucleotides caused by the secondary structure. This HiSeq method is robust and also strong because of its requirement for small sample quantity, high throughput, and high accuracy with a simply operated automatic platform. Such analysis resulted in millions of sRNA sequence tags from the pomegranate RNA sample.

RNA-seq Bioinformatics Analysis and miRNA Prediction

After sequencing, raw sequence reads (FASTQ files) were processed into clean reads, then filtered to discard low-quality adapter contaminative tags, and the remaining reads with lengths < 18 nt were discarded. Usually, the sRNA is 18 to 30 nt (miRNA, 21 or 22 nt; siRNA, 24 nt; and piRNA, 30 nt). All unique clean reads, specifically non-redundant ones, were considered for further analysis, including non-coding RNA identification and proper annotation. First, clean reads of sRNAs such as rRNAs, small cytoplasmic RNAs (scRNAs), snoRNAs, snRNAs, and tRNAs were identified by a BLASTall search against the Rfam (v10.1) and GenBank databases. miRNAs were identified by mapping sRNA reads against poplar genome sequences by using SOAP2 [41]. The SOAP2 output was filtered with use of in-house filter tool to identify the candidate sequences as miRNA precursors by analyzing a mapping pattern of one or more blocks of aligned small RNAs with perfect matches [42]. The secondary structures of candidate sequences were checked by applying stringent criteria as suggested [43]. To determine conserved miRNAs, clean reads were compared with known plant miRNAs deposited at miRBase [44]. Those with non-perfect matches were considered variants of known miRNAs. Other sequences that did not map to known miRNAs and other kinds of sRNAs were considered un-annotated sequences for novel miRNA prediction. To obtain the miRNA predicted precursor structure, the sequences were analyzed by using TurboFold [45] http://rna.urmc.rochester.edu/RNAstructure.html) and guide and star sequences were obtained.

Target Prediction, Functional Annotation and Pathway Analysis

The target prediction method involved loading miRNA reads in a FASTA format file containing sRNA sequences to search for targets from a known poplar (*Populus trichocarpa*) transcript database by using the suggested rules [46, 47]. Specifically, criteria for choosing miRNA/target duplexes were 1) less than four mismatches between sRNA and the target, 2) less than two adjacent mismatches in the miRNA/target duplex, 3) no adjacent mismatches in positions 2–12 of the miRNA/target duplex (5' of miRNA), 4) no mismatches in positions 10–11 of the miRNA/target duplex, 5) less than 2.5 mismatches in positions 1–12 of the miRNA/target duplex (5' of miRNA), and 6) minimum free energy (MFE) of the miRNA/target duplex ≥74 % of the MFE of the miRNA bound to its perfect complement. To investigate the putative functions of potential target genes, the target sequences from poplar were annotated by using the databases Gene Ontology (GO) and Kyoto Encyclopedia of Genes and Genomes (KEGG) Orthology (KO) [48, 49]. The GO results were classified into three independent groups: cellular component, molecular function, and biological process. KO pathways were grouped into different metabolism functions and signal transduction.

Validation of miRNA Variants and Novel miRNAs by Stem-loop RT-qPCR

Stem-loop RT-qPCR was used to confirm the differential expression of miRNAs and their variants across leaves, flowers, and fruit developmental stages. About 1 µg DNA-free total RNA was hybridized with miRNA-specific stem-loop RT primers for six miRNA families and six novel miRNAs, and the hybridized molecules were reverse-transcribed into cDNAs with use of the Superscript III kit (Thermo Fisher Scientific, Waltham, MA USA). The forward miRNA-specific primer for the mature miRNA sequences and the universal reverse primer for the stem-loop sequences were designed (Table S6). For each reaction, 1 µL cDNA, 10 µL 2X FastStart SYBR Green (Roche), and primers were mixed. PCR runs were 95 °C for 10s, 60 °C for 30s with the StepOnePlus Real-Time PCR System (Applied Biosystems, Foster City, CA, USA). The expression of miRNAs was normalized to that in leaves in all three biological replications. 5.8S ribosomal RNA was used as reference to calculate relative gene expression by the $2^{-\Delta\Delta Ct}$ method [50].

RESULTS AND DISCUSSION

Pomegranate fruit contains a variety of natural compounds such as phenolics, alkaloids, terpenoids, and fatty acids and has a role in numerous health-promoting activities [51]. Both fruit peels and arils are used to extract natural compounds such as punicalagin (derivative of gallic acid and glucose) and anthocyanins (class of water-soluble phenolic compounds responsible for the pink to red fruit) [52]. Many reports describe the benefits of pomegranate natural products for humans, but lack of genomic information is a major bottleneck in genomic research of pomegranate. In this study, we profiled the conserved and novel miRNAs in pomegranate and discuss their different biochemical pathways.

Fruit Development and Collection of Tissues

Pomegranate fruit development is divided into different stages. The fruit growth pattern depends on the cultivar as well as location and season [53, 54]. We divided the developmental stages of Al-sirin-nar as follows (Fig. 1): stage 1, approximately 8–10 days from initial flowering (petal drop stage); stage 2, approximately 10 days from stage 1 (fruit has begun to expand, but no color change); stage 3, approximately 12–15 days later (fruit has swelled more and is just starting to change from red to green); stage 4, approximately 15–18 days later (fruit has expanded from pear shape to more rounded shape, more green from red); stage 5, approximately 15 days later (continued expansion of fruit, color continues to change from red to green); and stage 6, approximately 15 days later (continued expansion of fruit, color continues to change from red to green), the calyx remains red, referred to as the "lipstick" stage. The process takes 75 to 85 days from initial flowering to stage 6. After stage 6, the fruit becomes glossy red and contains rosy-pink arils with a sweet tart taste. To profile the overall miRNA expression, we collected leaves, male and female flowers and fruit tissues from different stages. Throughout the fruit developmental stages, the color development of the peel (fading of dark red) and arils inside the fruit (accumulation of dark red) is the reverse. So, the anthocyanin is increasingly accumulating in arils during the later stages of fruit development.

During fruit development, pomegranate accumulates a variety of phytochemical compounds [55] that function as a defense mechanism. The edible part is 50 % of the fruit: 40 % and 10 % are arils and seeds, respectively. Arils contain mostly water (85 %), 10 % sugar (glucose and fructose), organic acids (citric acid, ascorbic acid, malic acid), and the bioactive compounds anthocyanins (phenolics and flavonoids) [56]. In addition, the seed cover contains six types of glucosides, with delphinidin-3,5-diglucoside the main anthocyanin in juice [57]. Pigmentation of fruit peel and arils is an important quality indicator of fruit. Al-sirin-nar fruit peel is rosy-red as compared with dark red for 'Wonderful', and the color of peel and arils is not related [2].

High-throughput Sequencing and Annotation of Small RNAs

Total RNA was extracted from young seedlings and other tissues and pooled for building a small RNA library for further sequencing. About 30 million reads were generated by using Illumina HiSeq 2000 (Table 1). From 29.95 million high-quality reads after removing 5'- and 3' adapters, insert nulls, sRNAs < 18 nt, and poly A reads, 99 % clean reads was obtained. A total of 8,603,217 (28.97 %) reads in all categories were unique to pomegranate. Because the genome sequence of pomegranate is not available and poplar is a deciduous flowering tree with full genome information, we used the poplar genome as a reference for mapping the clean reads with use of SOAP2 [41].

Table 1. Overview of miRNA sequencing reads

Read type	Count	Percent
Total reads	30000000	-
High-quality reads	29948480	100.00 %
3'adapter-null reads	43776	0.15 %
Insert-null reads	5228	0.02 %
5'adapter contaminants	92262	0.31 %
<18-nt reads	115751	0.39 %
PolyA reads	1429	0.00 %
Clean reads	29690034	99.14 %

Approximately 8.3 % (2,480,745) of the reads were mapped to the known non-coding RNAs, including scRNAs, tRNAs, snoRNAs, and snRNAs (Fig. 2a). Among all sRNAs, 23.9 % belonged to miRNAs, 0.7 % unique to pomegranate. However, the number of reads in each category differed when matched to the Rfam and GenBank databases. Particularly, the number of rRNA specific reads was high (2,258,108) in GenBank but low (463,788) in Rfam. The number of other sRNAs, including miRNAs, was more or less similar in both Rfam and Genbank databases (Fig. 2b, c). Most of the known and novel small RNA reads identified in pomegranate were 24 nt (~50 %), followed by 21 nt (21.8 %), 23 nt (8.95 %), 20 nt (6.93 %), and 22 nt (6.31 %). Other sRNAs between 15 and 29 nt were not significantly abundant (Fig. 3; Table S1). The 24-nt small RNAs also exist in many plant species such as maize, Arabidopsis, tomato, barrel clover, and trifoliate oranges [40,58, 59, 60, 61]. Thus, the 24-nt small RNAs may also be involved in critical functions in pomegranate as in other plants.

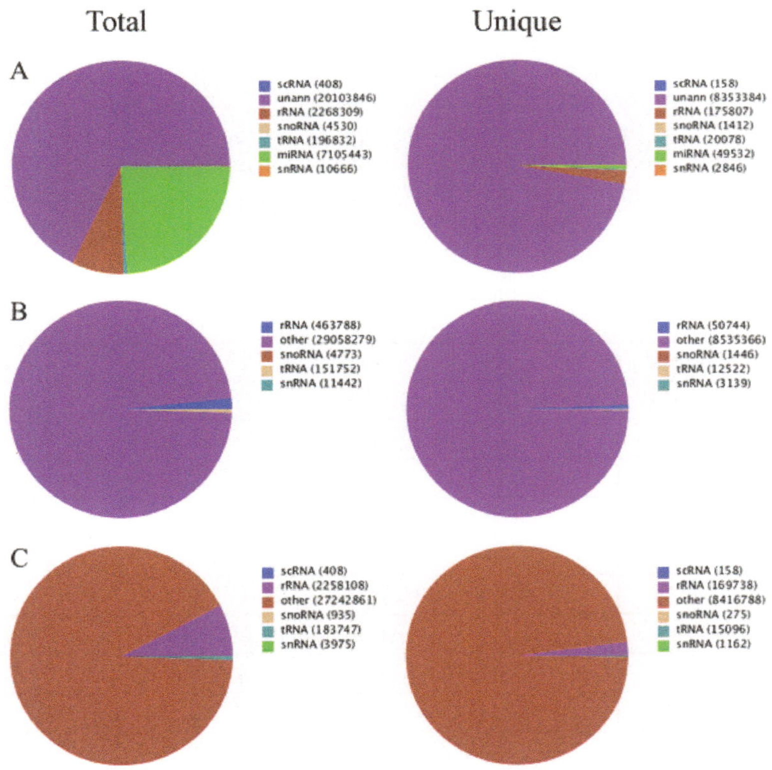

Figure 2. Distribution of small RNAs by annotation categories. Pie chart shows pomegranate small RNAs matching data in the non-coding RNA database. **a** Cumulative distribution of different non-coding small RNA categories pooled from Rfam and NCBI. **b** Small RNAs matching Rfam non-coding RNAs. **c** Small RNAs matching GenBank non-coding RNAs. Each small RNA database shows differences in subcategories depending on the availability of existing data.

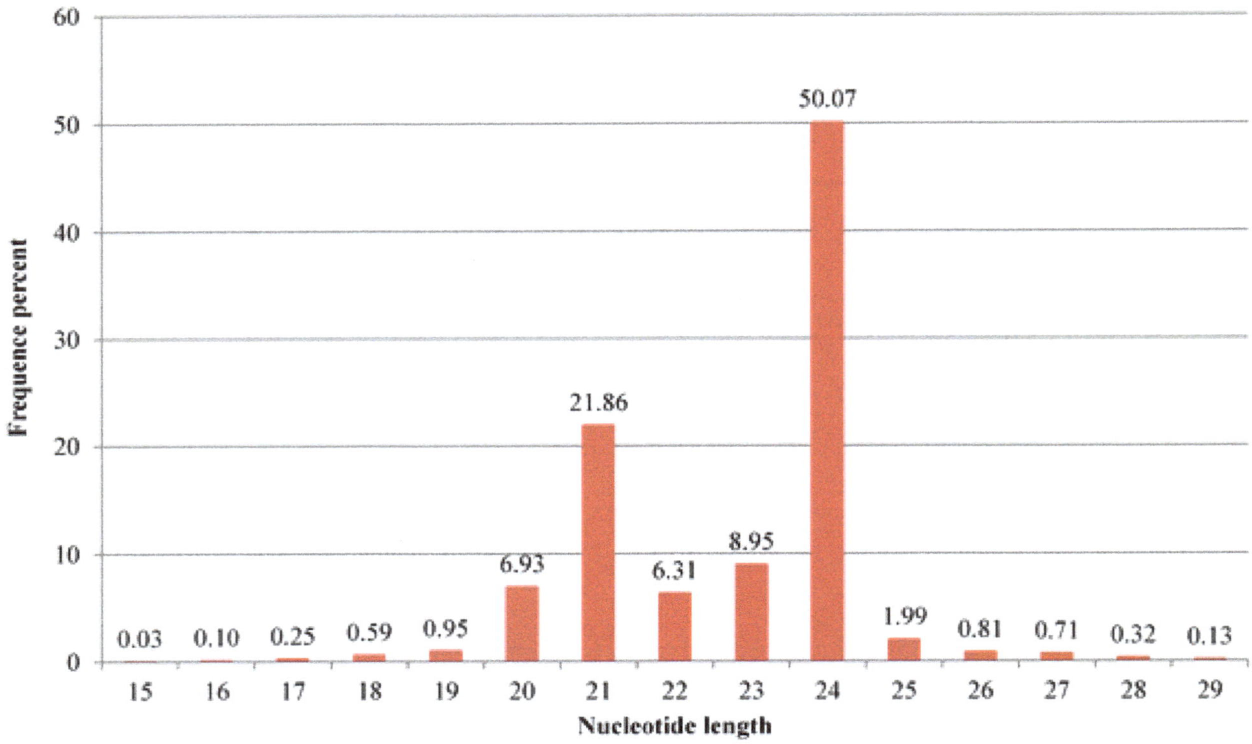

Figure 3. Length distribution of all small RNA tags.

Identification of Conserved miRNAs in Pomegranate

miRNAs in plant systems can be identified by examining the potential fold-back precursor structure containing a ~21-nt sequence within one arm of the hairpin structure. To identify the known miRNAs and obtain miRNA counts, we used the base bias on the first position of identified miRNAs and on each position separately in the pomegranate library; clean reads of sRNA tags were aligned to the miRNA precursor/ mature miRNA of plant and animals deposited in miRBase 20.0 (http://www.mirbase.org) [62]. The results gave information on alignment, including the structure of known miRNA precursors, lengths and counts. A total of 30 known miRNA families from our library matched miRBase, containing 28,645 entries (Table 2). Analysis of read counts for known miRNA families indicated that the expression frequency varied significantly from 4,015,427 to 511 among different miRNA families. Known miRNA families with less than 500 reads were ignored. Each miRNA family featured various counts with its own variants. MiR157 was the most abundant family (4,015,427) followed by miR156 (1,632,172), miR166, miR168, miR167, miR535, miR169, and miR390. The number of variants of well-known miRNAs in pomegranate was high for miR156 followed by miR157, miR159 and miR160. These miRNAs showed variation for a few families in pomegranate despite high counts (Table S2).

Table 2. Details of conserved miRNA families in pomegranate

miRNA family	Counts	Sequence	miRBase database
miR157	4015427	TTGACAGAAGATAGAGAGCAC	ath-miR157a
miR156	1632172	TGACAGAAGAGAGTGAGCAC	csi-miR156
miR166	454425	TCGGACCAGGCTTCATTCCCC	pvu-miR166a
miR168	299953	TCGCTTGGTGCAGGTCGGGAA	ath-miR168a-5p
miR167	159932	TGAAGCTGCCAGCATGATCTGA	ccl-miR167a
miR535	104111	TGACAACGAGAGAGAGCACGC	ppt-miR535a
miR169	84700	TGAGCCAAGAATGACTTGCCGG	cme-miR169t
miR390	33324	AAGCTCAGGAGGGATAGCGCC	ath-miR390a-5p
miR479	19760	CGTGATGTTGGTTCGGCTCATC	ghr-miR479
miR171	10381	CGAGCCGAATCAATATCACTC	csi-miR171b
miR2916	8924	GGGGCTCGAAGACGATCAGATA	peu-miR2916
miR482	4825	TTCCCAAGGCCGCCCATTCCGA	mdm-miR482a-3p
miR160	4710	GCGTATGAGGAGCCATGCATA	ptc-miR160b-3p

miR4414	4472	TGTGAATGATGCGGGAGATAC	mtr-miR4414b
miR159	4088	TTTGGATTGAAGGGAGCTCTA	ptc-miR159a
miR164	2755	TGGAGAAGCAGGGCACGTGCA	ptc-miR164a
miR6300	2507	GTCGTTGTAGTATAGTGGT	gma-miR6300
miR319	2113	TAGCTGCCGACTCATTCATCCA	ppe-miR319b
miR894	1830	GTTCGTTTCACGTCGGGTTCACCA	ppt-miR894
miR408	1620	CTGGGAACAGGCAGGGCATGG	ptc-miR408-5p
miR172	1608	AGAATCTTGATGATGCTGCAT	ptc-miR172a
miR162	1581	TCGATAAACCTCTGCATCCAG	ptc-miR162a
miR396	1566	GCTCAAGAAAGCTGTGGGAAA	ath-miR396b-3p
miR3639	1389	AAATGACTTCTGAACGGCAAAAC	vvi-miR3639-5p
miR6248	792	TAATTGTGGATGGAGGTAT	osa-miR6248
miR1171	721	TGGGAATGGAGTGGAGTGGAGTAG	cre-miR1171
miR858	658	TTCGTTGTCTGTTCGACCTTG	ath-miR858b
miR5653	533	TGAGAGTTGAGTTGAGTTGAGTTT	ath-miR5653
miR530	513	TGCATTTGCACCTGCACCTTA	ptc-miR530a
miR4415	511	AAGGTTGTGATTGGAATTAATGGC	gma-miR4415b-5p

Because of their high sequence similarity and conserved targets, miR156 and miR157 were grouped into a single family. Cleavage of the Squamosa promoter binding protein-like (SPL) by miR156/157 has been confirmed in different crops including Arabidopsis [63] and rice [64, 65]. In our studies, miR157 was the largest miRNA family among all families. This finding contrasts with recent reports of pear fruit development [66] and peanut [67], showing miR156 as the most abundant. MiR157 may have unique targets and common targets between miR156/miR157. In addition to families, variants of each family showed differential expression. The number of miRNAs was counted and normalized to total reads of sRNAs. The total counts for each family variant varied greatly. The expression of miRNA families of miR157a, miR156, miR157b, miR156a, miR156g, miR159a and miR160b was high in our pooled pomegranate sample. In contrast, a few other families and variants showed less expression (Table S2). The abundance of each family also varied. When the miRNAs were predicted from miRBase, different family members exactly matched known miRNAs from different plants such as Arabidopsis, rice, grapes, poplar tree, maize, and soybean.

Novel miRNAs and Their Identification in Pomegranate

To reveal the novel miRNA candidates from the pomegranate small RNA library, we used MIREAP and explored the characteristic hairpin structure of miRNA precursors. Only secondary structures with the lowest free energy and a high degree of pairing were included as miRNA precursors. Precursors forming hairpin structures for the 10 novel miRNAs (Table 3) were predicted with an average minimum folding free energy of −55.82 kcal/mol, from −73.1 to −31.94 kcal/mol (Table S3). The counts of novel miRNAs ranged from 115 (PgmiR25) to 4807 (PgmiR08). The length of precursors of the novel miRNAs ranged from 74 nt (PgmiR35) to 336 nt (PgmiR20). This length range is almost similar to novel miRNA precursors of Japanese apricot [68]. Among 10 miRNAs, 6 had a 5' arm and 4 had a 3' arm. The stem loop structures of predicted novel miRNA candidates were drawn from the precursor sequences by using TurboFold (Fig. 4) [45].

Table 3. Predicted novel miRNAs in pomegranate

miRNA ID	Count	Seq (5p)	Seq (3p)
PgmiR08	4807	-	TCAAGTGATGATTGACGAGATC
PgmiR09	852	AGGCCCCACTGACCGTCGGAT	-
PgmiR14	358	-	TTTGATTCGAGGAATAAAGGC
PgmiR19	245	CTGTTTGGATTGCAGGTTATG	-
PgmiR20	102	-	TTAGATGACCATCAACAAACA
PgmiR22	1615	GGAATGGTTGTCTGGCTCGAGG	-
PgmiR23	323	CAGGAAGAGCAGTGAGCACGCAA	-
PgmiR25	115	GAAGCTGACGAGGGAGAGTGG	-
PgmiR31	255	-	TACTAGCTGTAGGGATATTGC
PgmiR35	1350	AATTGGACGGAAAAGACAGGG	-

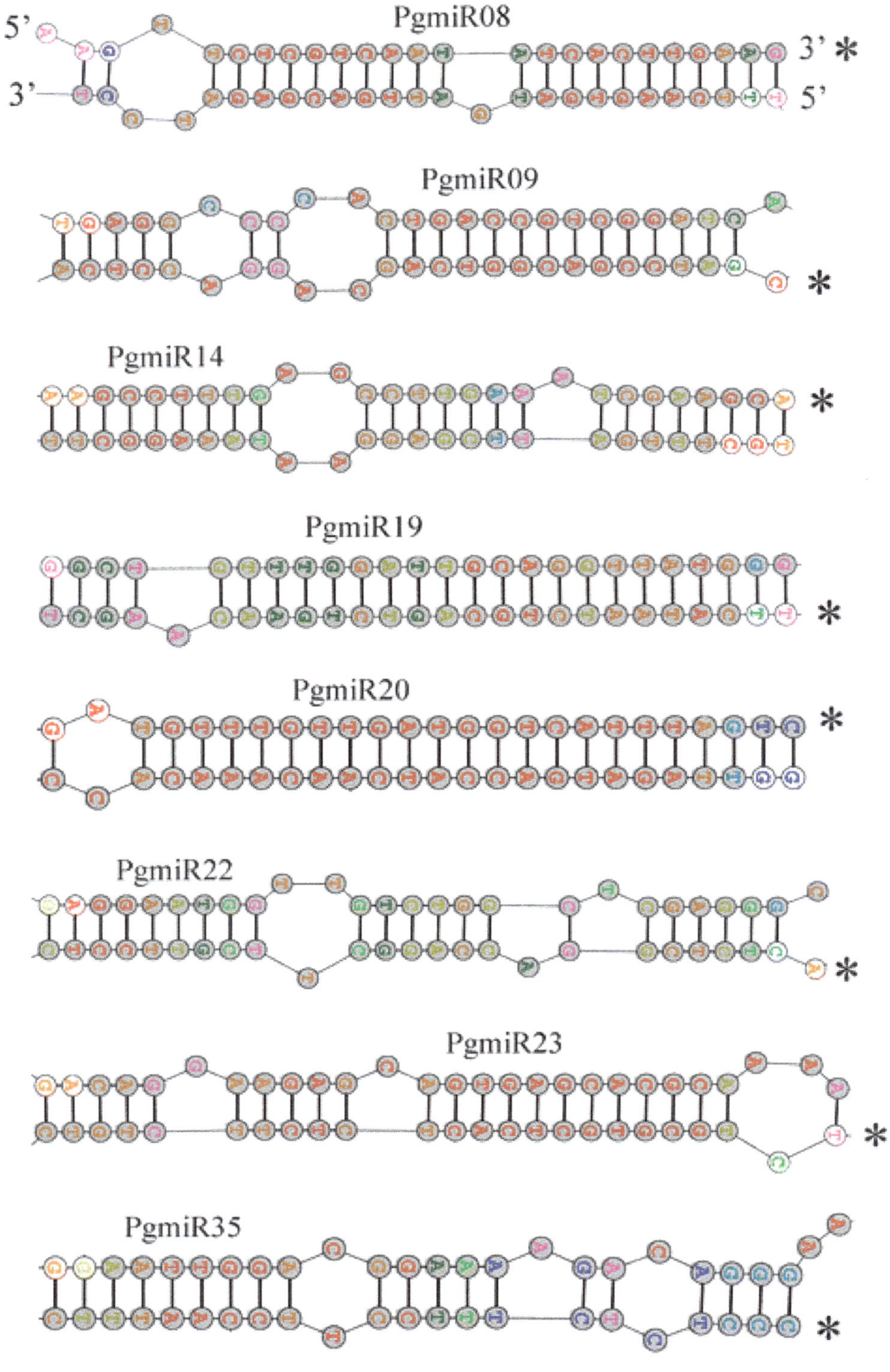

Figure 4. Predicted precursor structures of novel miRNAs found in pomegranate. The predicted fold-back structures of few selected miRNAs precursors from novel miRNA pool based on minimum folding free energy. The regions of miRNA are shaded with grey color. The miRNA guide strand is marked with asterisk.

Novel miRNA prediction was summarized as the base bias on the first position from the 5' end and base bias on each position (Fig. 5). With the exception of 18 nt for 18- to 26-nt small RNAs, the base bias at the first position from the 5' end had a strong preference for U but not G. Nucleotides A and U predominately occupied the first position base bias for the 18- and 20-nt small RNAs, respectively, which agreed with the base bias results for *Acipenser schrenckii* [69].None of the miRNAs in this range showed a G or C preference (Fig. 5a; Table S4). Even though nucleotide U was preferred more than 80 % of the time as a first base for 20- to 23-nt mRNAs, base biases for 21- to 23-nt novel miRNAs showed a pattern of U followed by A, C and G in the pomegranate library. For nucleotide bias at each position of 24-nt mRNAs, overall, nucleotide A was the most prevalent (37.7 %), followed by G (30.3 %), C (17.0 %), and U (15.0 %). The proportion with U at the first and second positions was 96.9 % and 56.9 %, respectively, which was similar to golden-thread orchid [70]. The predominant positions of bases in 24-nt sRNA tags from position 1 to 24 were UUGACAGAAGAUAGAGAGCACAGU (Fig. 5b; Table S5).

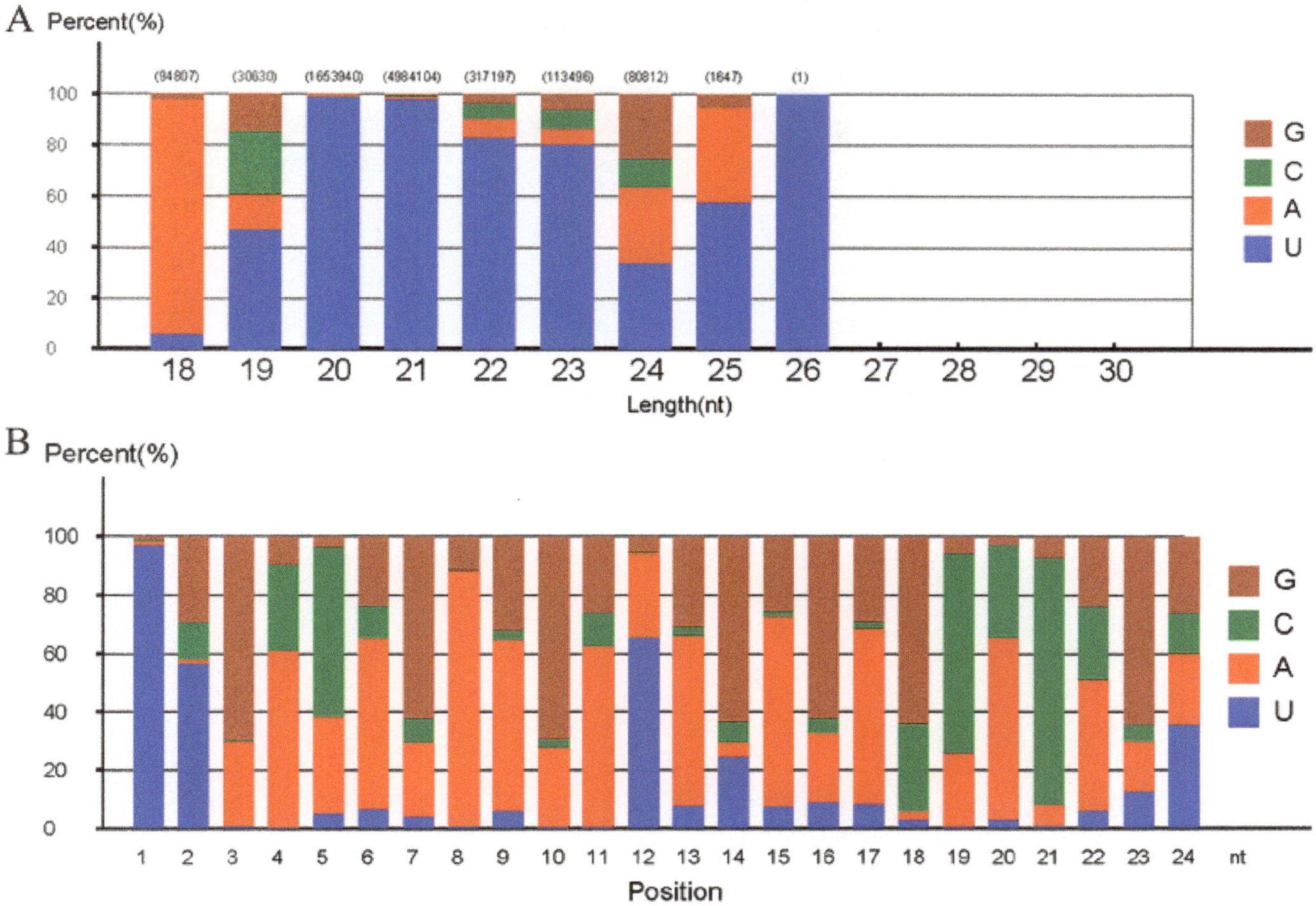

Figure. 5. miRNA variants and their nucleotide bias position. **a** First nucleotide bias for the first position of 18- to 26-nt miRNAs. Nucleotide U predominates. **b** MiRNA nucleotide bias for each position of 24-nt miRNAs.

Validation of High-throughput RNA-sequencing in Different Tissues

To elucidate the potential roles of miRNAs in pomegranate fruit development, we profiled the expression levels of known and novel miRNAs. miRNAs have wide expression in plant tissues and play multiple key regulatory roles in physiological and developmental processes [71]. Most miRNAs in plants regulate developmental processes by destroying their target mRNAs because the target gene has complete complementarity with miRNA [72]. We used stem-loop RT-qPCR with unique primer sets to validate the expression pattern of highly expressed miRNA families (PgmiR156, PgmiR157, PgmiR159, PgmiR160, PgmiR172, and PgmiR319) and their variants (PgmiR156a, PgmiR156g, PgmiR157b, PgmiR157c, and PgmiR159b) in leaves, male and female flowers, and different fruit developmental stages of pomegranate (Fig. 6). This method could confirm the existence of pomegranate miRNAs and also detect the expression of miRNAs in various tissues. We found a differential expression pattern across tissues. The miRNAs PgmiR156, PgmiR156a, PgmiR159a, PgmiR159b, PgmiR160b, and mPgiR319b were highly upregulated during later stages of fruit development; that of PgmiR172 was high in female flowers, then gradually decreased to a lower level with fruit maturity. Other family members were ubiquitously expressed in leaves and other tissues including fruits. Additionally, we validated a few novel miRNAs with high count reads from sequencing (Fig. 7). Their expression pattern differed among tissues. Novel miRNAs PgmiR08 and PgmiR22 showed high expression during the early developmental stage of fruit that receded toward the final maturity stage. Interestingly, PgmiR19 appeared to express only during fruit developmental stage and not in leaf and flower. These differentially expressed miRNAs may regulate different targets during fruit development and ripening.

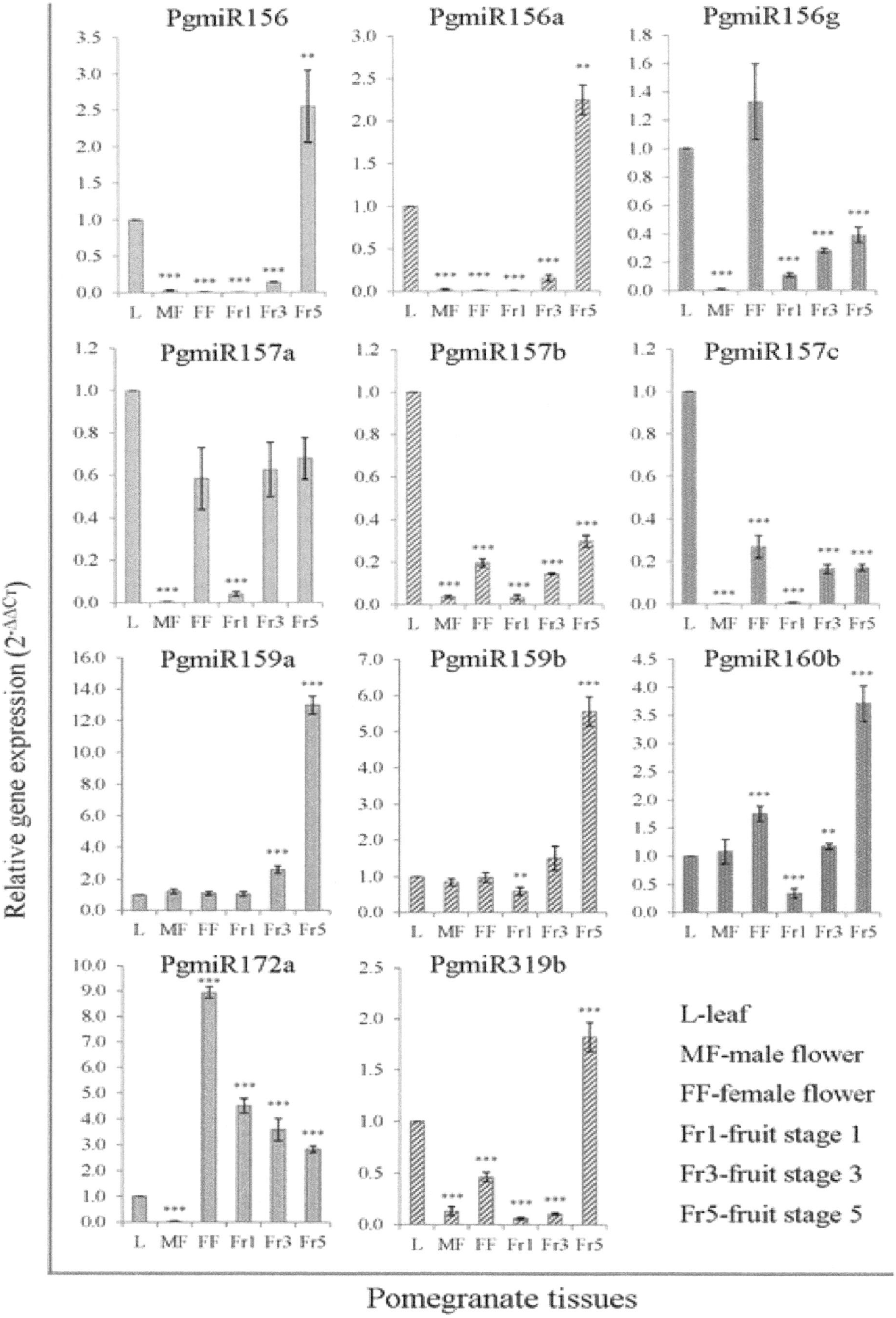

Figure 6. Stem-loop RT-qPCR validation of highly expressed known miRNAs and their variants in different tissues. Relative quantity is based on the expression of the reference gene 5.8 s ribosomal RNA. X-axis indicates different tissues and Y-axis the expression of miRNA relative to that in leaf tissue. Data are mean ± SD from three biological replicates. **,$P<0.01$; ***, $P<0.001$ by Student *t* test. Bar values higher or lower compared to leaf tissue indicates upregulation or downregulation, respectively.

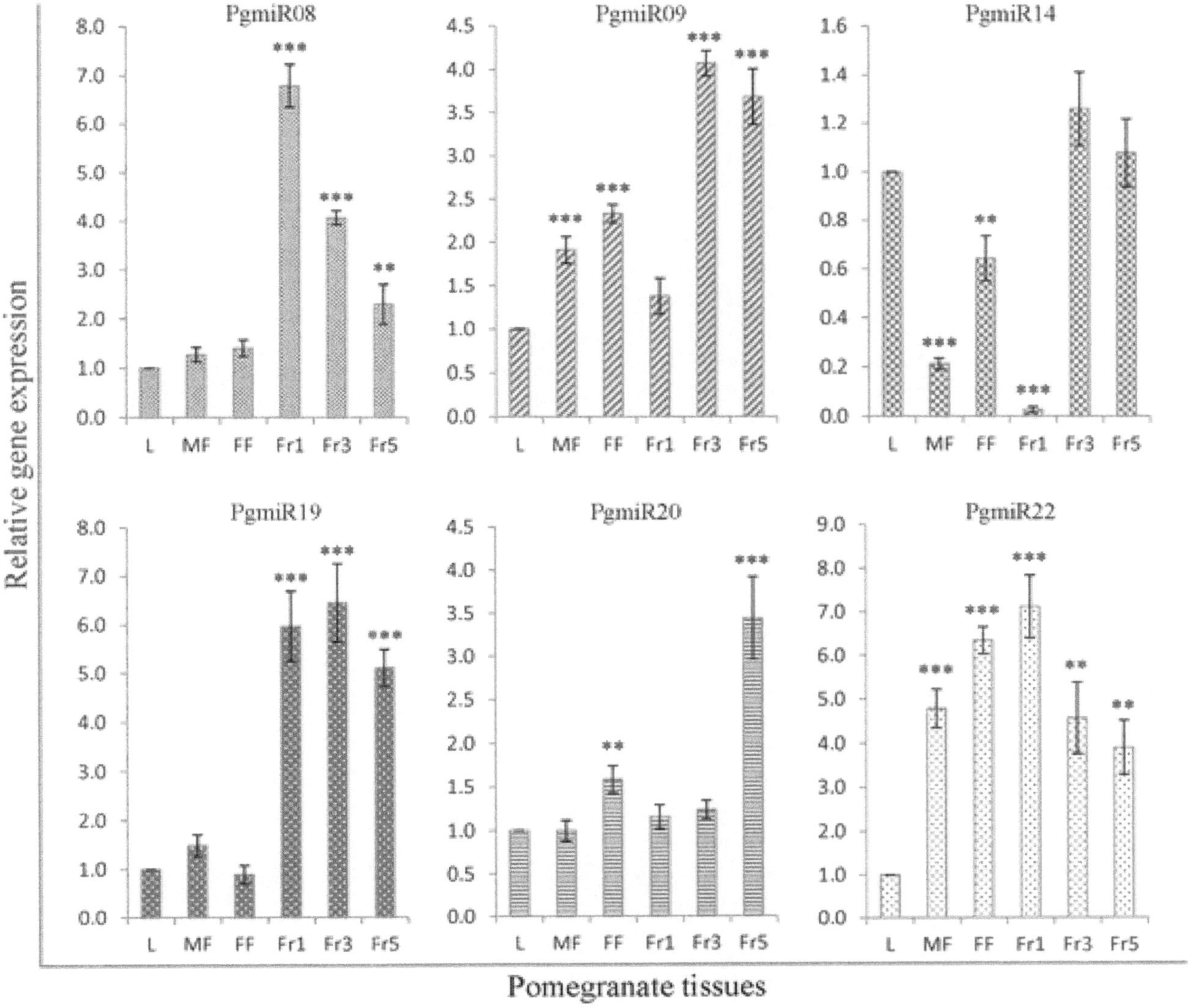

Figure 7. Stem-loop RT-qPCR validation of novel miRNAs in various pomegranate tissues. Relative quantity is based on expression of the reference gene 5.8 s ribosomal RNA. X-axis indicates different tissues (L, leaf; MF, male flower; FF, female flower; Fr1, fruit stage 1; Fr3, fruit stage 3; Fr5, fruit stage 5) and Y-axis indicates the expression of miRNA relative to that in leaf tissue. Data are mean±SD from three biological replicates. **, $P<0.01$; ***, $P<0.001$ by Student *t* test. Bar values higher or lower compared to leaf tissue indicates upregulation or downregulation, respectively.

Prediction of miRNA Target Genes, Gene Ontology (GO) and KEGG Pathway Analysis

Most of the targets of miRNA are conserved across several plants including Arabidopsis, rice, poplar, and wheat [73, 74, 75, 76]. Majority of them are various transcription factors including SQUAMOSA promoter binding protein-like (SPB/SPL) (miR156), NAM (miR164), MYB (miR159, miR172, miR319) that regulate plant development and phytohormone signaling [77]. SPL is one of the miR156 targets in Arabidopsis [78], with expression inversely related to that of miR156. SPL, which shows abundant expression in the absence of miR156 expression in early stages of fruit development, may destabilize the MYB-bHLH-WD40 complex to repress the anthocyanin biosynthetic pathway and further accumulation [64]. Keeping this hypothesis in mind, with increased SPL expression being a negative regulator of anthocyanin accumulation, the anthocyanin content in pomegranate might be still under the detectable level with increased flavonol quantity in early aril developmental stages. However, SPL expression may be decreased during later stages of maturity to accumulate anthocyanin with increased PgmiR156 expression. Although this conclusion is premature without quantifying SPL accumulation in different fruit stages, the increased expression of PgmiR156/PgmiR157 we observed might have a positive effect on increasing anthocyanin and proanthocynidin or tannin levels in mature pomegranate.

To better understand the functions of identified novel miRNAs in pomegranate, we predicted putative candidate genes by using bioinformatic analyses [79, 80]. A total of 288 target genes were identified for ten novel miRNAs and gene ontology with annotation details have been found. Consistent with previous reports, most of the novel miRNA targets belong to plant-specific transcription factors, (AP2, MYB, ARF, GRAS, PHD, and bZIP), followed by regulators of metabolic processes (protein kinases, LRR kinase, RLKs, etc.) and hormone signaling. In addition, there are several other targets whose functions are largely unknown. The targets of PgmiR08 ARFs, bHLH, SecY protein, TIR1 F-box, and auxin signaling F-box2 (AFB2) are shown to be involved in root and fruit development, anthocyanin accumulation as well as in abiotic stress. In contrast to climacteric fruits (apple, banana, tomato), notably little is known about the hormonal control of ripening in non-climacteric fruits such as pomegranate, grape, and strawberry [81] and it has been proven that even ethylene levels or respiration was considerably low during ripening of non-climacteric fruit [82]. That could be one possibility that we did not find any major ethylene pathway candidates in our target identification. Anthocyanin biosynthesis is a branch of the flavonoid pathway and genes involved in anthocyanin biosynthesis and regulation have been discovered and studied in several fruits, such as bHLH in apple [83], and MYB and bHLH in peach [84]. To support this notion, ARF10 plays key role in anthocyanin biosynthesis of pomegrante. The GO shows that MYB transcription factor, the target of PgmiR14, PgmiR22 and PgmiR31 is involved in multiple hormone signaling including gibberellic acid, ethylene and salicylic acid during fruit development and ripening [85].

In addition, GRAS transcription factor (PgmiR25), and nuclear transcription factor Y (PgmiR22), copper transporter (PgmiR09), disease resistance protein TIR-NBS-LRR, and LRR protein kinase (PgmiR31) are the targets of few novel miRNAs. Recently, genes coding for GRAS transcription factors were identified as targets of miRNAs during fruit development and ripening of tomato [86] and grapevine [87]. Moreover, F-box family proteins play vital roles in the signal transduction pathways of different hormones [88] and 166 F-box genes were identified during maturation and fruit ripening in apple [89]. Group of F-box genes targeted by PgmiR08 and PgmiR20 might participate mostly in auxin signaling pathway towards fruit ripening. During fruit development, synthesized sucrose in the leaf is transported to sink tissues such as fruits where it is directly used for metabolism or translocated into storage tissues for the synthesis of major storage products through carbohydrate metabolism [90]. Mutants of sucrose transporters (SUT) exclusively affected tomato fruit and seed development [91]. SUT2, the target of PgmiR31 and the key player in sucrose:hydrogen symporter activity, might be a key player in normal fruit development. The seed development is part of fruit maturity and ripening, and the development of both occurs simultaneously. In pomegranate, seeds which are inside the arils are surrounded by juice. A nuclear transcription factor Y subunit A-1 (NF-YA1) targeted by novel miRNAs PgmiR22 and PgmiR23, and a bZIP transcription factor targeted by PgmiR31 seem to involve in seed maturation and dormancy in the arils of pomegranate fruits.

Above all, tissue integrity and cation binding to the cell wall during fruit senescence is very important, and pectin methylesterase (PME) activity modifies tissue integrity in ripening tomato [92]. As an ubiquitous plant enzyme, PME catalyzes the deesterification of galactosyluronate methyl esters of pectin to their free carboxylic groups, and has been suggested to cause transesterification to uronoyl-sugar crosslinks [93]. PME has been implicated in various processes in ripening fruits including textural changes, formation of abscission zones and cell wall growth, maturation, and extensibility. Alongside, invertases may involve in the long-distance transport of sucrose and take part in phloem loading and unloading [94]. From our transcriptome and GO analysis, we believe that plant invertase/pectin methylesterase inhibitor could be targeted by PgmiR31 to aid in fruit ripening process. Overall, the known and unknown targets of novel microRNAs participate in pomegranate fruit development and further ripening process.

To evaluate the potential functions of the miRNA target genes, GO categories were assigned to all of the predicted genes, which resulted in three unique categories: cellular component, molecular function, and biological processes (Fig. 8). In the cellular component, the major categories were "cell," followed by "intracellular part" and "organelle". In the molecular function category, the major categories were "binding" and "catalytic activity." For biological process, the "cellular" and "metabolic processes" were the most abundant categories. Metabolic processes are the key active process in fruit development [66], and cellular processes and metabolic processes were the top two GO categories within biological processes.

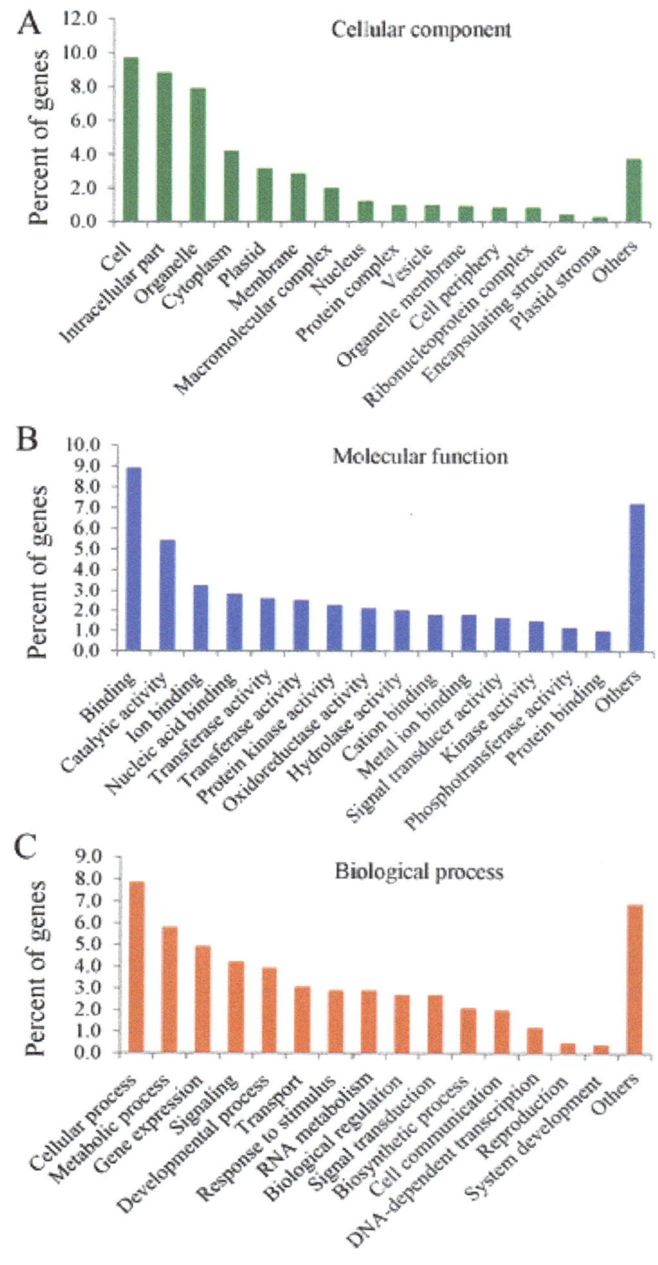

Figure 8. Gene ontology categories for miRNA targets in pomegranate. Target genes were classified into the categories cellular component (**a**), molecular function (**b**), and biological processes (**c**). Values in the Y axis are the percentage of target genes in different functional categories.

To further evaluate the completeness of the miRNA transcriptome and benefits of the annotation of the target candidates of known and novel miRNAs, all annotated sequences from poplar were identified by KEGG pathway groups. A total of 629 candidates from multiple KO pathways were identified according to *P*-value and *Q*-value from the KEGG database (Fig. 9). We were able to enrich 41 miRNA families together targeting those candidate genes in 107 major pathways related to metabolism of starch synthesis, amino acid synthesis, protein synthesis, plant-pathogen interaction, and hormone signal transduction, etc. In addition, biosynthesis of secondary metabolites, and fructose and mannose metabolism pathways which are important for fruit maturity also existed. To support the participation of KEGG pathways in pomegranate fruit development, the previous evidence shows that edible part of the pomegranate arils contain 10 % total sugars comprising fructose and glucose, ascorbic acid, citric acid, bioactive compounds such as phenolics and flavonoids, principally anthocyanins [56]. Specifically, novel miRNAs were found to be involved in different steps of multiple pathways, including ascorbate metabolism (conversion of L-ascorbate to L-dehydroascorbate), fatty acid metabolism, carbon fixation (change of ribose 5-phosphate to ribulose 5-phosphate), and RNA transport (by regulating members involved in nuclear pore complex and exon junction complex). More importantly, novel miRNA members participated in plant hormone transduction pathways such as auxin (regulating genes *TIR1*, *ARF* and *SAUR*), cytokinin (*CRE1* and *A-ARR*), gibberellin (*DELLA*), abscisic acid (*PP2C*), brassinosteroid (*BAK1/BRI1* and *BZR1/2*), and jasmonic acid (*JAZ* and*MYC2*). These key hormone related pathways may participate in synthesis of various phytocompounds in mature pomegranate fruit as the gene ontology suggested. KEGG pathway analysis showed 14 candidates in fructose and mannose metabolism, 1 in carbon fixation, 23 members in biosynthesis of secondary metabolites, and 4 candidates for the sucrose and starch metabolism. Altogether, the pomegranate fruit quality is largely impacted by the composition of sugar and acid, which is one of the most significant fruit development characteristics.

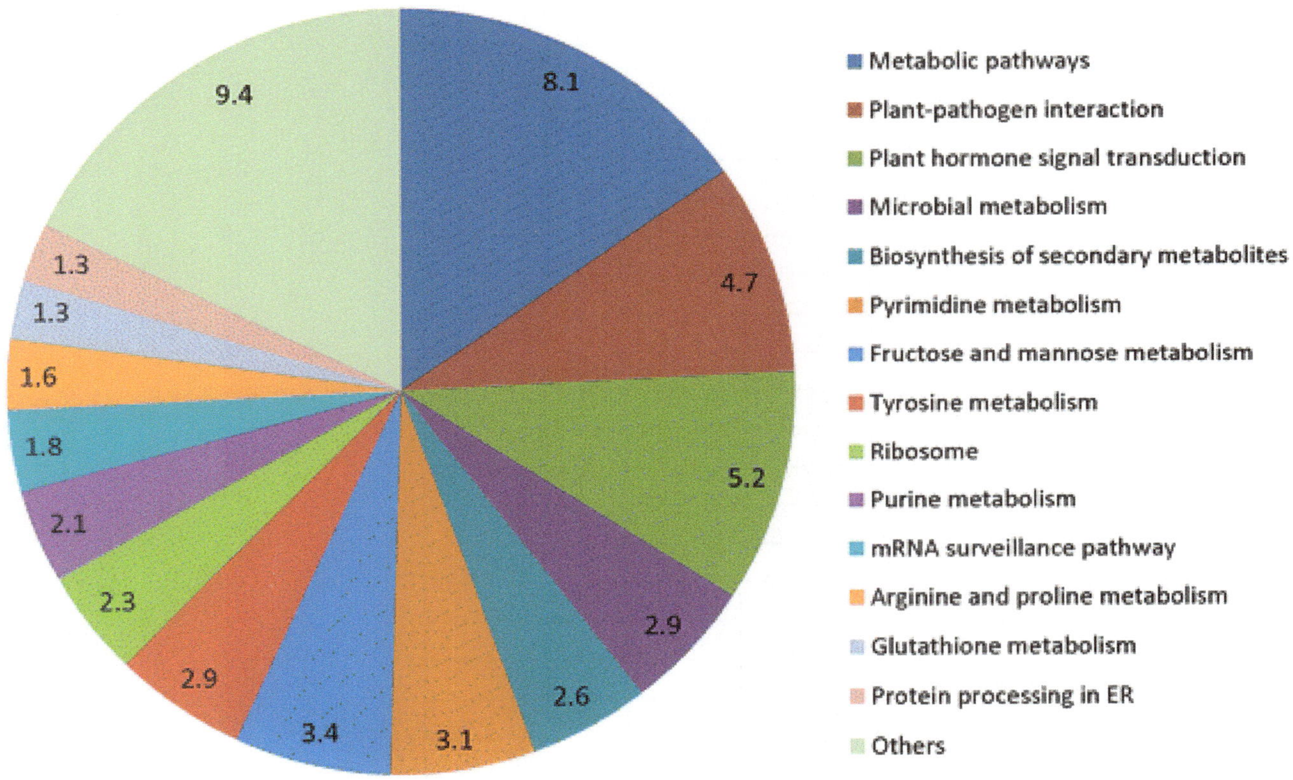

Figure 9. Annotation of miRNA targets based on KEGG orthology. Values are the percentage of target genes in different functional categories.

CONCLUSIONS

We used small RNA-sequencing of pomegranate with Illumina Hiseq2000 sequencing and identified 10 novel miRNAs. We reveal the differential expression of a few predominately expressed miRNAs and their variants in different developmental stages of fruit. This is the first report to investigate sRNAs in pomegranate, with a large number found as known and novel miRNAs. By searching the poplar genome, 288 putative target genes were predicted for the 10 novel miRNAs and then annotated by using GO and KEGG databases to explore their putative functions in different metabolic pathways. We revealed several fruit development pathways including sugar and acid, and plant hormone signaling. This identification of novel miRNAs in pomegranate will be valuable for further understanding the functions and regulatory mechanisms of miRNAs in other related plant species.

ACKNOWLEDGEMENTS

This project was supported by the USDA-NIFA grant (Proposal no. 2013-04023) and additionally supported by the Indian Council of Agricultural Research (ICAR). The authors thank the National Agricultural Innovation Project, ICAR, for sponsoring visits of Nripendra V. Singh and Ramajayam Devarajan. We are also grateful to Guru Jagadeeswaran for his suggestions to improve the quality of the manuscript significantly.

ADDITIONAL FILES

Table S1. Length distribution and frequency of small RNA tags. (XLSX 10 kb)

Length	Frequency
Nucleotide length	Read counts
3	22
4	20
5	19
6	26
7	36
8	47
9	41
10	111
11	76
12	97
13	273
14	952
15	9118
16	30244
17	74669
18	174566
19	282619
20	2064425
21	6512694
22	1880236
23	2665983
24	14918653
25	593675
26	242711
27	210081
28	95688
29	38676
30	8355
31	2102
32	759
33	138
34	26
35	11
36	3
37	2
38	5
39	2
40	1
41	1
42	20
43	4
44	27

Table S2. Total counts of variants of known miRNAs. (XLSX 11 kb). miRNA family with variants and abundance

miRNA family name used for qPCR	Length	Counts	Phase/ offset	Mismatch	Variant sequence	Reference sequence
PgmiR156 (csi-miR156)	20	1539285	0	0	TGACAGAAGAGAGTGAGCAC	TGACAGAAGAGAGTGAGCAC
PgmiR156a (bna-miR156a)	21	11593	0	0	TGACAGAAGAGAGTGAGCACA	TGACAGAAGAGAGTGAGCAC
PgmiR156g (ath-miR156g)	21	5901	0	1	CGACAGAAGAGAGTGAGCACA	TGACAGAAGAGAGTGAGCAC
miR156	20	4997	0	1	TGACAGAAGAGAGTGAGCAA	TGACAGAAGAGAGTGAGCAC
miR156	20	4969	0	1	TGACAGAAGAGAGTGAGCAT	TGACAGAAGAGAGTGAGCAC
miR156	19	4937	0	0	TGACAGAAGAGAGTGAGCA	TGACAGAAGAGAGTGAGCAC
miR156	21	4370	-1	0	TTGACAGAAGAGAGTGAGCAC	TGACAGAAGAGAGTGAGCAC
miR156	21	2884	0	1	TGACAGGAGAGAGTGAGCACA	TGACAGAAGAGAGTGAGCAC
miR156	20	2351	0	1	CGACAGAAGAGAGTGAGCAC	TGACAGAAGAGAGTGAGCAC
miR156	21	2135	-1	0	GTGACAGAAGAGAGTGAGCAC	TGACAGAAGAGAGTGAGCAC
miR156	20	2079	0	1	TGATAGAAGAGAGTGAGCAC	TGACAGAAGAGAGTGAGCAC
PgmiR157a (ath-miR157a)	21	3852427	0	0	TTGACAGAAGATAGAGAGCAC	TTGACAGAAGATAGAGAGCAC
miR157	20	13887	1	0	TGACAGAAGATAGAGAGCAC	TTGACAGAAGATAGAGAGCAC
PgmiR157b (ath-miR157b)	22	10081	0	0	TTGACAGAAGATAGAGAGCACA	TTGACAGAAGATAGAGAGCAC
miR157	21	9130	0	1	TTGACAGAAGATAGAGAGCAT	TTGACAGAAGATAGAGAGCAC
miR157	21	8651	0	1	TTGACAGAAGATAGAGAGCAG	TTGACAGAAGATAGAGAGCAC
miR157	20	6265	0	0	TTGACAGAAGATAGAGAGCA	TTGACAGAAGATAGAGAGCAC
miR157	21	5655	0	1	TCGACAGAAGATAGAGAGCAC	TTGACAGAAGATAGAGAGCAC
miR157	21	5353	0	1	TTGATAGAAGATAGAGAGCAC	TTGACAGAAGATAGAGAGCAC
miR157	22	4807	0	0	TTGACAGAAGATAGAGAGCACT	TTGACAGAAGATAGAGAGCAC

miR157	21	4121	0	1	TTGACGGAAGATAGAGAGCAC	TTGACAGAAGATAGAGAGCAC
miR157	21	3314	0	1	TTGACAGAAGATGGAGAGCAC	TTGACAGAAGATAGAGAGCAC
miR157	21	3257	0	1	TTGGCAGAAGATAGAGAGCAC	TTGACAGAAGATAGAGAGCAC
miR157	21	3097	0	1	ATGACAGAAGATAGAGAGCAC	TTGACAGAAGATAGAGAGCAC
miR157	21	3058	0	1	CTGACAGAAGATAGAGAGCAC	TTGACAGAAGATAGAGAGCAC
miR157	21	2990	0	1	TTGACAGAAGATAGAGAGCAA	TTGACAGAAGATAGAGAGCAC
miR157	21	2888	0	1	TTGACAGAAGATATAGAGCAC	TTGACAGAAGATAGAGAGCAC
miR157	21	2688	0	1	TTGACAGGAGATAGAGAGCAC	TTGACAGAAGATAGAGAGCAC
miR157	21	2590	0	1	TTAACAGAAGATAGAGAGCAC	TTGACAGAAGATAGAGAGCAC
PgmiR157c (ath-miR157c)	21	1931	0	1	GTGACAGAAGATAGAGAGCAC	TTGACAGAAGATAGAGAGCAC
miR157	21	1689	0	1	TTGACAGAAGAAAGAGAGCAC	TTGACAGAAGATAGAGAGCAC
miR157	21	1680	0	1	TAGACAGAAGATAGAGAGCAC	TTGACAGAAGATAGAGAGCAC
miR157	21	1575	0	1	TTGACTGAAGATAGAGAGCAC	TTGACAGAAGATAGAGAGCAC
PgmiR159a (ptc-miR159a)	21	3017	0	0	TTTGGATTGAAGGGAGCTCTA	TTTGGATTGAAGGGAGCTCTA
PgmiR159b (ptc-miR159b)	20	248	0	2	TTTGGATTGAAGGGAGCTTA	TTTGGATTGAAGGGAGCTCTA
miR159	22	106	0	0	TTTGGATTGAAGGGAGCTCTAA	TTTGGATTGAAGGGAGCTCTA
PgmiR160b (ptc-miR160b)	21	4146	0	0	GCGTATGAGGAGCCATGCATA	GCGTATGAGGAGCCATGCATA
miR160	22	180	-1	0	GGCGTATGAGGAGCCATGCATA	GCGTATGAGGAGCCATGCATA
PgmiR172a (ptc-miR172a)	21	1471	0	0	AGAATCTTGATGATGCTGCAT	AGAATCTTGATGATGCTGCAT
PgmiR319b (ppe-miR319b)	22	1461	0	0	TAGCTGCCGACTCATTCATCCA	TAGCTGCCGACTCATTCATCCA
miR319	21	157	1	2	AGCTGCCGACTCATTCACTCA	TAGCTGCCGACTCATTCATCCA
miR319	21	109	-1	2	TTAGCTGCCGACTCATTCACT	TAGCTGCCGACTCATTCATCCA

Table S3. List of novel miRNAs, their precursors with hairpin structure sequences. (XLSX 10 kb). Novel miRNA with their precursors forming hairpin structures

Id	mfe	Id(5p)	Id(3p)	count	count(5p)	count(3p)	seq	struct	seq(5p)	seq(3p)
PgmiR08	-51.84	-	Pg-SH-RNA-m0008-3p	4807	-	4807	CGACCAATGAAGTTCTCGTCAATATCACTTG AAGCAGTACGTC GCATCGGGCCTCTTCCCTG TCGGGCATGAGCTGGAG CAGCTTGGAAATTTTCAAGT GATGATTGACGAGATCCTCATGGGTC	.((((.((((. (.((((((((((((((((((((......((.((. ((.(((.((..(((....)))...))))).)))) .))........))))))))))).)))))))))).).)))).))))	-	TCAAGTGATG ATTGACGAGATC
PgmiR09	-54.3	Pg-SH-RNA-m0009-5p	-	852	852	-	ACGGTTCCTG AGGCCCCACTGACCGTCGGATCAGTGTGAA TTCCACCGATC CGACG GTC AGCAGG ACCTCATGAACCGTC	((((((.(((((.((..((((((((((((((. (((.(....)))).)))))))))))))))..)).))))).))))))).	AGGCCCCAC TGACCGTCGGAT	-
PgmiR14	-48.8	-	Pg-SH-RNA-m0014-3p	358	-	358	TTCGATTGAAGCCTTTGAGCCTT GAAATCGAAGCACAAAGCCTTCGA TGAAGGCTCTATGCTTTGATTCGA GGAATAAAGGCTTCGGTTAAA	...(((((((((((((..((((((. (((((((.(.(((((((...))))))).).)))))))))))))))..)))))))))))))...	-	TTTGATTCGA GGAATAAAGGC
PgmiR19	-67.9	Pg-SH-RNA-m0019-5p	-	245	245	-	TCTTATGGGGCTGTTTG GATTGCAGGTTATGGGAG GGGGTGGGGATCATGAAAGGG GTTTATGCTCCCGACCCCTTTT CATAATCTGCAGTCTGAACAAGCTGTTATGGGGA	((((((((((((((..((((((((((((((.. (((((((.((((.((((((.....)))))).)))).))))))..))))))))))))))..))).)))).)))))))).	CTGTTTGGATTG CAGGTTATG	-
PgmiR20	-73.1	-	Pg-SH-RNA-m0020-3p	102	-	102	GCCATGGAGATG TTTGTTGATGGTCATTTAGTCGCCGGCGCTC TTTCTCCATGGGGCATGCTT TGGGATCA TTCGAGCTCATCTT AGCTTATGATCGATCGAT TCGTATGCGATATGGCTGGGGT TCCATATGGGTTAGATGACCATCAACAAACACCTCCACCCG	....(((((.((((((((((((((((((((.. (...((.(((((....(((((..((((((.. ((((((((((..((((((......)))))))))))..))))...))))))..)))))..))))).))......). .)))))))))))))))))))).))))))....	-	TTAGATGACCA TCAACAAACA
PgmiR22	-63.6	Pg-SH-RNA-m0022-5p	-	1615	1615	-	TAGGGTGAGAGGAATGGTTGTCTGGCTCGAGGGCCA TTGAGGGGTGCCGATCAGTTATTCAATTG ATTAATTGAAGCCCAAACCCTAAATGACTCTCGG ACCAGGCTTCGTTCCTCCCACCCAA	..(((((.(((((((((..((((((..((((((. ((((.((((..((...(((((((.((....)).))))))).)).....)))).)))).))))))..)))))) .)))))))))).))))))..	GGAATGGTTGT CTGGCTCGAGG	-
PgmiR23	-50.1	Pg-SH-RNA-m0023-5p	-	323	323	-	GGGAATCTGACAGGAAGAGCAGTGAGCACGCAAAA GCAAAGGTATAGCAC ATC TATACAATTGGTCCTCTGC GTGCTCACTCTCTTCTGTCTAATTTCCT	((((((..(((((.(((((.((((((((((((.. (.(((..((((((.....)))))).)))).).....)))))))))))))))))))))))..)))))).	CAGGAAGAGCAG TGAGCACGCAA	-
PgmiR25	-62.6	Pg-SH-RNA-m0025-5p	-	115	115	-	TTCGGGGCCAGAAGCT GACGAGGGAGAGTGGATAAACGGTGTCTTTT GCAATCCATAGTGATCAAATGGT GTCGTTAGTTTTTC TCTCGGCAA TCTCCTTCAAAGCCATGCCATTAAAGGCATTGGTT CCTCACTCTCCCTCAA GGGCTTCTGATCTCTGAGC	(((((((.((((((((...((((((((((((..... ((((((((((.(((.........(((......... (((((..((.....))..)))))........)))......)))....))))))))))).....)))))))))))).. .))))))))..))))))).	GAAGCTGACGA GGGAGAGTGG	-
PgmiR31	-43.5	-	Pg-SH-RNA-m0031-3p	255	-	255	CAGAACAAATTG CAATATCTATT CAGCTGGTAGGC CTAAGCATTCTACTT ATGAATGTATAGCAC TACTACTAGCTG TAGGGATATTGCTGTTTGATCT	.(((.(((((.((((((((((...((((((((((.. (((.(((((((........))))))).))).....))))))))))...)))))))))).))))).)))	-	TACTAGCTGT AGGGATATTGC
PgmiR35	-31.94	Pg-SH-RNA-m0035-5p	-	1350	1350	-	GAGTTGAAGGAATTGGACGGAAAAGACAGGGAAACA A AAAGGCCCCCTCTCTTTCCTTCCAATTTCTTCAATCT	..((((((((((((((.(((((.((.((((. ....)))).))))))).))))))))))))))))..	AATTGGACGGA AAAGACAGGG	-
List is given only for novel miRNAs showing more than 100 counts										

Table S4. Novel miRNAs with the first nucleotide bias for 18- to 26-nt small RNAs. (XLSX 8 kb). mir_novel_first_base

len	A	U	C	G		U %	Predominant
18	87124	5581	84	2018	94807	91.8962	A
19	4324	14400	7383	4523	30630	47.0127	U
20	1909	1644003	515	7513	1653940	99.3992	U
21	47009	4907138	16888	13069	4984104	98.4558	U
22	23430	263773	20012	9982	317197	83.1575	U
23	6648	91583	8552	6713	113496	80.6927	U
24	24070	27747	8824	20171	80812	34.3352	U
25	606	955	0	86	1647	57.9842	U
26	0	1	0	0	1	100	U

Table S5. miRNA nucleotide bias at each position of 24-nt small RNAs. (XLSX 9 kb). mir_novel_base_bias

pos	A	U	C	G	total	high %
1	103066	6934244	54791	57448	7149549	U
2	101176	4073425	887017	2087931	7149549	U
3	2071446	68041	50101	4959961	7149549	G
4	4334646	30199	2105297	679407	7149549	A
5	2377206	372253	4160757	239333	7149549	C
6	4218313	474582	769508	1687146	7149549	A
7	1804972	306590	597976	4440011	7149549	G
8	6276202	67903	20929	784515	7149549	A
9	4218909	440498	207814	2282328	7149549	A
10	1939906	63164	191561	4954918	7149549	G
11	4432480	47751	813860	1855458	7149549	A
12	2030225	4707516	32507	379301	7149549	U
13	4191743	556057	201476	2200273	7149549	A
14	328105	1782721	525786	4512937	7149549	G
15	4663496	547949	126225	1811879	7149549	A
16	1736488	637333	325787	4449941	7149549	G
17	4307681	601115	178551	2062202	7149549	A
18	203711	224859	2151010	4569969	7149549	G
19	1795881	51949	4891074	410645	7149549	C
20	4493038	217280	2251867	187364	7149549	A
21	398056	54647	4650774	392132	5495609	C
22	230294	31000	128951	121260	511505	A
23	32930	25224	11668	124486	194308	G
24	19678	29045	11161	20928	80812	U
	56309648	22345345	25346448	45271773		
	37.7	15.0	17.0	30.3		

Table S6. Primers used in this study for stem-loop RT-qPCR. (XLSX 10 kb). miRNA sequence, their variants, and primers used in this study

Conserved miRNA variants	**miRNA sequence (5' to 3' direction)**
PgmiR156	TGACAGAAGAGAGTGAGCAC
PgmiR156a	TGACAGAAGAGAGTGAGCACA
PgmiR156g	CGACAGAAGAGAGTGAGCACA
PgmiR157a	TTGACAGAAGATAGAGAGCAC
PgmiR157b	TTGACAGAAGATAGAGAGCACA
PgmiR157c	GTGACAGAAGATAGAGAGCAC
PgmiR159a	TTTGGATTGAAGGGAGCTCTA
PgmiR159b	TTTGGATTGAAGGGAGCTTA
PgmiR160b	GCGTATGAGGAGCCATGCATA
PgmiR172a	AGAATCTTGATGATGCTGCAT
PgmiR319b	TAGCTGCCGACTCATTCATCCA
Novel miRNAs	
PgmiR08	TCAAGTGATGATTGACGAGATC
PgmiR09	AGGCCCCACTGACCGTCGGAT
PgmiR14	TTTGATTCGAGGAATAAAGGC
PgmiR19	CTGTTTGGATTGCAGGTTATG
PgmiR20	TTAGATGACCATCAACAAACA
PgmiR22	GGAATGGTTGTCTGGCTCGAGG
Primer ID	**Stem loop reverse transcriptase primers**
PgmiR156-RT	GTCGTATCCAGTGCAGGGTCCGAGGTATTCGCACTGGATACGAC**GTGCTC**
PgmiR156a-RT	GTCGTATCCAGTGCAGGGTCCGAGGTATTCGCACTGGATACGAC**TGTGCT**
PgmiR156g-RT	GTCGTATCCAGTGCAGGGTCCGAGGTATTCGCACTGGATACGAC**TGTGCT**
PgmiR157a-RT	GTCGTATCCAGTGCAGGGTCCGAGGTATTCGCACTGGATACGAC**GTGCTC**
PgmiR157b-RT	GTCGTATCCAGTGCAGGGTCCGAGGTATTCGCACTGGATACGAC**TGTGCT**
PgmiR157c-RT	GTCGTATCCAGTGCAGGGTCCGAGGTATTCGCACTGGATACGAC**GTGCTC**
PgmiR159a-RT	GTCGTATCCAGTGCAGGGTCCGAGGTATTCGCACTGGATACGAC**TAGAGC**
PgmiR159b-RT	GTCGTATCCAGTGCAGGGTCCGAGGTATTCGCACTGGATACGAC**TAAGCT**
PgmiR160b-RT	GTCGTATCCAGTGCAGGGTCCGAGGTATTCGCACTGGATACGAC**TATGCA**

PgmiR172a-RT	GTCGTATCCAGTGCAGGGTCCGAGGTATTCGCACTGGATACGAC**ATGCAG**
PgmiR319b-RT	GTCGTATCCAGTGCAGGGTCCGAGGTATTCGCACTGGATACGAC**TGGATG**
PgmiR08-RT	GTCGTATCCAGTGCAGGGTCCGAGGTATTCGCACTGGATACGAC**GATCTC**
PgmiR09-RT	GTCGTATCCAGTGCAGGGTCCGAGGTATTCGCACTGGATACGAC**ATCCGA**
PgmiR14-RT	GTCGTATCCAGTGCAGGGTCCGAGGTATTCGCACTGGATACGAC**GCCTTT**
PgmiR19-RT	GTCGTATCCAGTGCAGGGTCCGAGGTATTCGCACTGGATACGAC**CATAAC**
PgmiR20-RT	GTCGTATCCAGTGCAGGGTCCGAGGTATTCGCACTGGATACGAC**TGTTTG**
PgmiR22-RT	GTCGTATCCAGTGCAGGGTCCGAGGTATTCGCACTGGATACGAC**CCTCGA**
Primer ID	**Forward primers**
PgmiR156-F	GCGGCGTGACAGAAGAGAGT
PgmiR156a-F	GCGGCGTGACAGAAGAGAGTG
PgmiR156g-F	GCGGCGCGACAGAAGAGAGTG
PgmiR157a-F	GCGGCGTTGACAGAAGATAGA
PgmiR157b-F	GCGGCGTTGACAGAAGATAGAG
PgmiR157c-F	GCGTCTGTTGACAGAAGATAGA
PgmiR159a-F	GCGGCGTTTGGATTGAAGGGA
PgmiR159b-F	GCGGCGTTTGGATTGAAGGG
PgmiR160b-F	GCGGCGGCGTATGAGGAGCCA
PgmiR172a-F	GCGGCGAGAATCTTGATGATG
PgmiR319b-F	GCGGCGTAGCTGCCGACTCATT
PgmiR08-F	GCGGCGTCAAGTGATGATTGAC
PgmiR09-F	GCGGCGAGGCCCCACTGACCG
PgmiR14-F	GCGGCGTTTGATTCGAGGAAT
PgmiR19-F	GCGGCGCTGTTTGGATTGCAG
PgmiR20-F	GCGGCGTTAGATGACCATCAA
PgmiR22-F	GCGGCGGGAATGGTTGTCTGGC
Primer ID	**Universal reverse primer**
	GTGCAGGGTCCGAGGT

REFERENCES

1. Shulman Y, Fainberstein L, Lavee S. Pomegranate fruit development andmaturation. J Hortic Sci Biotechnol. 1984;59:265–74.
2. Holland D, Hatib K, Bar-Ya'akov I. Pomegranate: Botany, Horticulture, Breeding. In: Horticultural Reviews. Hoboken, NJ 07030-5774, USA: Wiley; 2009. p.127–191.
3. Rana J, Pradheep K, Verma VD. Naturally occurring wild relatives of temperate fruits in Western Himalayan region of India: an analysis. Biodivers Conserv. 2007;16(14):3963–91.
4. Singh NV, Abburi VL, Ramajayam D, Kumar R, Chandra R, Sharma KK, Sharma J, Babu KD, Pal RK, Mundewadikar DM, et al. Genetic diversity and association mapping of bacterial blight and other horticulturally important traits with microsatellite markers in pomegranate from India. Mol. Genet. Genomics 2015;290(4):1393-402. doi: 10.1007/s00438-015-1003-0.
5. da Silva JAT, Rana TS, Narzary D, Verma N, Meshram DT, Ranade SA. Pomegranate biology and biotechnology: a review. Sci Hortic. 2013;160:85–107.
6. Levin GM: Pomegranate roads: A Soviet botanist's exile from Eden: Pomegranate Roads; 2006
7. Morton JF. Fruits of warm climates. Miami: JF Morton; 1987. ISBN: 0-9610184-1-0.
8. Malviya S, Jha A, Hettiarachchy N. Antioxidant and antibacterial potential of pomegranate peel extracts. J Food Sci Technol. 2014;51(12):4132–7.
9. Wang JY, Zhu C, Qian TW, Guo H, Wang DD, Zhang F, Yin X. Extracts of black bean peel and pomegranate peel ameliorate oxidative stress-induced hyperglycemia in mice. Exp Ther Med. 2015;9(1):43–8.
10. Sreekumar S, Sithul H, Muraleedharan P, Azeez JM, Sreeharshan S. Pomegranate fruit as a rich source of biologically active compounds. BioMed Res Int. 2014;2014:686921.
11. Ahmed MM, Samir E-SA, El-Shehawi AM, Alkafafy ME. Anti-obesity effects of taif and Egyptian pomegranates: molecular study. Biosci Biotechnol Biochem. 2015;79(4):598–609.
12. Aslan A, Can Mİ, Boydak D. Anti-oxidant effects of pomegranate juice on *Saccharomyces cerevisiae* cell growth. Afr J Tradit Complement Altern Med. 2014;11(4):14–8.
13. Bellesia A, Verzelloni E, Tagliazucchi D. Pomegranate ellagitannins inhibit α-glucosidase activity in vitro and reduce starch digestibility under simulated gastro-intestinal conditions. Int J Food Sci Nutr. 2015;66(1):85–92.
14. Tanaka Y, Sasaki N, Ohmiya A. Biosynthesis of plant pigments: anthocyanins, betalains and carotenoids. Plant J. 2008;54(4):733–49.
15. Koes R, Verweij W, Quattrocchio F. Flavonoids: a colorful model for the regulation and evolution of biochemical pathways. Trends Plant Sci. 2005;10(5):236–42.
16. de Pascual-Teresa S, Sanchez-Ballesta MT. Anthocyanins: from plant to health. Phytochem Rev. 2008;7(2):281–99.
17. Tzulker R, Glazer I, Bar-Ilan I, Holland D, Aviram M, Amir R. Antioxidant activity, polyphenol content, and related compounds in different fruit juices and homogenates prepared from 29 different pomegranate accessions. J Agric Food Chem. 2007;55(23):9559–70.
18. Winkel-Shirley B. Flavonoid biosynthesis. A colorful model for genetics, biochemistry, cell biology, and biotechnology. Plant Physiol. 2001;126(2):485–93.
19. Nesi N, Jond C, Debeaujon I, Caboche M, Lepiniec L. The Arabidopsis *TT2* gene encodes an R2R3 MYB domain protein that acts as a key determinant for proanthocyanidin accumulation in developing seed. Plant Cell. 2001;13(9):2099–114.
20. Pelletier MK, Murrell JR, Shirley BW. Characterization of flavonol synthase and leucoanthocyanidin dioxygenase genes in Arabidopsis (further evidence for differential regulation of "early" and"late" genes). Plant Physiol. 1997;113(4):1437–45.
21. Shirley BW, Kubasek WL, Storz G, Bruggemann E, Koornneef M, Ausubel FM, Goodman HM. Analysis of Arabidopsis mutants deficient in flavonoid biosynthesis. Plant J. 1995;8(5):659–71.
22. Ben-Simhon Z, Judeinstein S, Nadler-Hassar T, Trainin T, Bar-Ya'akov I, Borochov-Neori H, Holland D. A pomegranate (*Punica granatum* L.) WD40-repeat gene is a functional homologue of Arabidopsis *TTG1* and is involved in the regulation of anthocyanin biosynthesis during pomegranate fruit development. Planta. 2011;234(5):865–81.
23. Zhao X, Yuan Z, Feng L, Fang Y: Cloning and expression of anthocyanin biosynthetic genes in red and white pomegranate. J Plant Res 2015;128(4):687–96. doi: 10.1007/s10265-015-0717-8.
24. Rouholamin S, Zahedi B, Nazarian-Firouzabadi F, Saei A. Expression analysis of anthocyanin biosynthesis key regulatory genes involved in pomegranate (*Punica granatum* L.). Sci Hortic. 2015;186:84–8.
25. Hamilton AJ, Baulcombe DC. A species of small antisense RNA in posttranscriptional gene silencing in plants. Science. 1999;286(5441):950–2.
26. Lima JC, Loss-Morais G, Margis R. MicroRNAs play critical roles during plant development and in response to abiotic stresses. Genet Mol Biol. 2012;35(4):1069–77.
27. Huntzinger E, Izaurralde E. Gene silencing by microRNAs: contributions of translational repression and mRNA decay. Nat Rev Genet. 2011;12(2):99–110.
28. Axtell MJ, Bowman JL. Evolution of plant microRNAs and their targets. Trends Plant Sci. 2008;13(7):343–9.
29. Jagadeeswaran G, Nimmakayala P, Zheng Y, Gowdu K, Reddy UK, Sunkar R. Characterization of the small RNA component of leaves and fruits from four different cucurbit species. BMC Genomics. 2012;13(1):1–13.
30. Manohar S, Jagadeeswaran G, Nimmakayala P, Tomason Y, Almeida A, Sunkar R, Levi A, Reddy UK. Dynamic regulation of novel and conserved miRNAs across

various tissues of diverse cucurbit species. Plant Mol Biol Report. 2012;31(2):335–43.

31. Martínez G, Forment J, Llave C, Pallás V, Gómez G. High-throughput sequencing, characterization and detection of new and conserved cucumber miRNAs. PLoS One. 2011;6(5):e19523.

32. Mica E, Piccolo V, Delledonne M, Ferrarini A, Pezzotti M, Casati C, Del Fabbro C, Valle G, Policriti A, Morgante M. High throughput approaches reveal splicing of primary microRNA transcripts and tissue specific expression of mature microRNAs in *Vitis vinifera*. BMC Genomics. 2009;10(1):558.

33. Schreiber AW, Shi B-J, Huang C-Y, Langridge P, Baumann U. Discovery of barley miRNAs through deep sequencing of short reads. BMC Genomics. 2011;12(1):129.

34. Xia R, Zhu H, An Y, Beers EP, Liu Z. Apple miRNAs and tasiRNAs with novel regulatory networks. Genome Biol. 2012;13(6):R47.

35. Jia X, Shen J, Liu H, Li F, Ding N, Gao C, Pattanaik S, Patra B, Li R, Yuan L. Small tandem target mimic-mediated blockage of microRNA858 induces anthocyanin accumulation in tomato. Planta. 2015;242(1):283–93.

36. He H, Liang G, Li Y, Wang F, Yu D. Two young MicroRNAs originating from target duplication mediate nitrogen starvation adaptation via regulation of glucosinolate synthesis in *Arabidopsis thaliana*. Plant Physiol. 2014;164(2):853–65.

37. Xue W, Wang Z, Du M, Liu Y, Liu J-Y. Genome-wide analysis of small RNAs reveals eight fiber elongation-related and 257 novel microRNAs in elongating cotton fiber cells. BMC Genomics. 2013;14(1):629.

38. Saminathan T, Nimmakayala P, Manohar S, Malkaram S, Almeida A, Cantrell R, Tomason Y, Abburi L, Rahman MA, Vajja VG et al. Differential gene expression and alternative splicing between diploid and tetraploid watermelon. J Exp Bot. 2015;66(5):1369–85.

39. Zarei A, Zamani Z, Mousavi A, Fatahi R, Alavijeh MK, Dehsara B, Salami SA. An effective protocol for isolation of high-quality RNA from pomegranate seeds. Asian Aust J Plant Sci Biotechnol. 2012;6:32–7.

40. Wang T, Chen L, Zhao M, Tian Q, Zhang W-H. Identification of drought-responsive microRNAs in *Medicago truncatula* by genome-wide high-throughput sequencing. BMC Genomics. 2011;12(1):367.

41. Li R, Yu C, Li Y, Lam T-W, Yiu S-M, Kristiansen K, Wang J. SOAP2: an improved ultrafast tool for short read alignment. Bioinformatics. 2009;25(15):1966–7.

42. Kulcheski FR, de Oliveira LF, Molina LG, Almerao MP, Rodrigues FA, Marcolino J, Barbosa JF, Stolf-Moreira R, Nepomuceno AL, Marcelino-Guimaraes FC et al. Identification of novel soybean microRNAs involved in abiotic and biotic stresses. BMC Genomics. 2011;12:307.

43. Meyers BC, Axtell MJ, Bartel B, Bartel DP, Baulcombe D, Bowman JL, Cao X, Carrington JC, Chen X, Green PJ et al. Criteria for annotation of plant microRNAs. Plant Cell. 2008;20(12):3186–90.

44. Kozomara A, Griffiths-Jones S. miRBase: integrating microRNA annotation and deep-sequencing data. Nucleic Acids Res 2010:gkq1027.

45. Harmanci AO, Sharma G, Mathews DH. TurboFold: iterative probabilistic estimation of secondary structures for multiple RNA sequences. BMC Biochem. 2011;12:108.

46. Allen E, Xie Z, Gustafson AM, Carrington JC. microRNA-directed phasing during trans-acting siRNA biogenesis in plants. Cell. 2005;121(2):207–21.

47. Schwab R, Palatnik JF, Riester M, Schommer C, Schmid M, Weigel D. Specific effects of microRNAs on the plant transcriptome. Dev Cell. 2005;8(4):517–27.

48. Ashburner M, Ball CA, Blake JA, Botstein D, Butler H, Cherry JM, Davis AP, Dolinski K, Dwight SS, Eppig JT. Gene ontology: tool for the unification of biology. Nat Genet. 2000;25(1):25–9.

49. Kanehisa M, Goto S. KEGG: kyoto encyclopedia of genes and genomes. Nucleic Acids Res. 2000;28(1):27–30.

50. Livak KJ, Schmittgen TD. Analysis of relative gene expression data using real-time quantitative PCR and the $2^{-\Delta\Delta CT}$method. Methods. 2001;25(4):402–8.

51. Heber D. Multitargeted therapy of cancer by ellagitannins. Cancer Lett. 2008;269(2):262–8.

52. Ismail T, Sestili P, Akhtar S. Pomegranate peel and fruit extracts: a review of potential anti-inflammatory and anti-infective effects. J Ethnopharmacol. 2012;143(2):397–405.

53. Ben-Arie R, Segal N, Guelfat-Reich S. The maturation and ripening of the 'Wonderful' pomegranate. J. Am. Soc. Hortic. Sci. 1984

54. Gil MI, García-Viguera C, Artés F, Tomás-Barberán FA. Changes in pomegranate juice pigmentation during ripening. J Sci Food Agric. 1995;68(1):77–81.

55. Miguel MG, Neves MA, Antunes MD. Pomegranate (*Punica granatum* L.): a medicinal plant with myriad biological properties-a short review. J Med Plants Res. 2010;4:2836–47.

56. Viuda-Martos M, Fernández-López J, Pérez-Álvarez J. Pomegranate and its many functional components as related to human health: a review. Compr Rev Food Sci Food Saf. 2010;9(6):635–54.

57. Elfalleh W, Hannachi H, Tlili N, Yahia Y, Nasri N, Ferchichi A. Total phenolic contents and antioxidant activities of pomegranate peel, seed, leaf and flower. J Med Plants Res. 2012;6:4724–30.

58. Fahlgren N, Howell MD, Kasschau KD, Chapman EJ, Sullivan CM, Cumbie JS, Givan SA, Law TF, Grant SR, Dangl JL. High-throughput sequencing of Arabidopsis microRNAs: evidence for frequent birth and death of MIRNA genes. PLoS One. 2007;2(2):e219.

59. Li D, Wang L, Liu X, Cui D, Chen T, Zhang H, Jiang C, Xu C, Li P, Li S. Deep sequencing of maize small RNAs reveals a diverse set of microRNA in dry and imbibed seeds. PLoS One. 2013;8(1):e55107.

60. Moxon S, Jing R, Szittya G, Schwach F, Pilcher RLR, Moulton V, Dalmay T. Deep sequencing of tomato short

RNAs identifies microRNAs targeting genes involved in fruit ripening. Genome Res. 2008;18(10):1602–9.

61. Song C, Wang C, Zhang C, Korir NK, Yu H, Ma Z, Fang J. Deep sequencing discovery of novel and conserved microRNAs in trifoliate orange (*Citrus trifoliata*). BMC Genomics. 2010;11(1):431.

62. Kozomara A, Griffiths-Jones S. miRBase: annotating high confidence microRNAs using deep sequencing data. Nucleic Acids Res 2013:gkt1181.

63. Wu G, Poethig RS. Temporal regulation of shoot development in *Arabidopsis thaliana* by miR156 and its target *SPL3*. Development. 2006;133(18):3539–47.

64. Gou JY, Felippes FF, Liu CJ, Weigel D, Wang JW. Negative regulation of anthocyanin biosynthesis in Arabidopsis by a miR156-targeted SPL transcription factor. Plant Cell. 2011;23(4):1512–22.

65. Xie K, Shen J, Hou X, Yao J, Li X, Xiao J, Xiong L. Gradual increase of miR156 regulates temporal expression changes of numerous genes during leaf development in rice. Plant Physiol. 2012;158(3):1382–94.

66. Wu J, Wang D, Liu Y, Wang L, Qiao X, Zhang S. Identification of miRNAs involved in pear fruit development and quality. BMC Genomics. 2014;15(1):953.

67. Zhao C-Z, Xia H, Frazier TP, Yao Y-Y, Bi Y-P, Li A-Q, Li M-J, Li C-S, Zhang B-H, Wang X-J. Deep sequencing identifies novel and conserved microRNAs in peanuts (*Arachis hypogaea* L.). BMC Plant Biol. 2010;10(1):3.

68. Gao Z, Shi T, Luo X, Zhang Z, Zhuang W, Wang L. High-throughput sequencing of small RNAs and analysis of differentially expressed microRNAs associated with pistil development in Japanese apricot. BMC Genomics. 2012;13(1):371.

69. Yuan L, Zhang X, Li L, Jiang H, Chen J. High-throughput sequencing of microRNA transcriptome and expression assay in the sturgeon, *Acipenser schrenckii*. PloS one. 2014;9(12):e115251.

70. Li X, Jin F, Jin L, Jackson A, Ma X, Shu X, Wu D, Jin G. Characterization and comparative profiling of the small RNA transcriptomes in two phases of flowering in *Cymbidium ensifolium*. BMC Genomics. 2015;16(1):622.

71. He L, Hannon GJ. MicroRNAs: small RNAs with a big role in gene regulation. Nat Rev Genet. 2004;5(7):522–31.

72. Hake S. MicroRNAs: a role in plant development. Curr Biol. 2003;13(21):R851–2.

73. Adai A, Johnson C, Mlotshwa S, Archer-Evans S, Manocha V, Vance V, Sundaresan V. Computational prediction of miRNAs in *Arabidopsis thaliana*. Genome Res. 2005;15(1):78–91.

74. Jin W, Li N, Zhang B, Wu F, Li W, Guo A, Deng Z. Identification and verification of microRNA in wheat (*Triticum aestivum*). J Plant Res. 2008;121(3):351–5.

75. Lu S, Sun Y-H, Chiang VL. Stress-responsive microRNAs in *Populus*. Plant J. 2008;55(1):131–51.

76. Sunkar R, Girke T, Zhu J-K. Identification and characterization of endogenous small interfering RNAs from rice. Nucleic Acids Res. 2005;33(14):4443–54.

77. Liu Q, Chen Y-Q. Insights into the mechanism of plant development: interactions of miRNAs pathway with phytohormone response. Biochem Biophys Res Commun. 2009;384(1):1–5.

78. Wang J-W, Czech B, Weigel D. miR156-regulated SPL transcription factors define an endogenous flowering pathway in*Arabidopsis thaliana*. Cell. 2009;138(4):738–49.

79. Lai EC, Tomancak P, Williams RW, Rubin GM. Computational identification of Drosophila microRNA genes. Genome Biol. 2003;4(7):R42.

80. Rhoades MW, Reinhart BJ, Lim LP, Burge CB, Bartel B, Bartel DP. Prediction of plant microRNA targets. Cell. 2002;110(4):513–20.

81. Adams LS, Seeram NP, Aggarwal BB, Takada Y, Sand D, Heber D. Pomegranate juice, total pomegranate ellagitannins, and punicalagin suppress inflammatory cell signaling in colon cancer cells. J Agric Food Chem. 2006;54(3):980–5.

82. Seymour GB, Taylor JE, Tucker GA. Biochemistry of fruit ripening: Dordrecht, Netherlands: Springer Science & Business Media; 2012.

83. Feng F, Li M, Ma F, Cheng L. Phenylpropanoid metabolites and expression of key genes involved in anthocyanin biosynthesis in the shaded peel of apple fruit in response to sun exposure. Plant Physiol Biochem. 2013;69:54–61.

84. Rahim MA, Busatto N, Trainotti L. Regulation of anthocyanin biosynthesis in peach fruits. Planta. 2014;240(5):913–29.

85. Kumar R, Khurana A, Sharma AK. Role of plant hormones and their interplay in development and ripening of fleshy fruits. J Exp Bot. 2014;65(16):4561–75.

86. Karlova R, van Haarst JC, Maliepaard C, van de Geest H, Bovy AG, Lammers M, Angenent GC, de Maagd RA. Identification of microRNA targets in tomato fruit development using high-throughput sequencing and degradome analysis. J Exp Bot. 2013;64(7):1863–78.

87. Grimplet J, Agudelo-Romero P, Teixeira RT, Martinez-Zapater JM, Fortes AM. Structural and functional analysis of the GRAS gene family in grapevine indicates a role of GRAS proteins in the control of development and stress responses. Frontiers Plant Sci. 2016;7:353.

88. Wang X, Kong H, Ma H. F-box proteins regulate ethylene signaling and more. Genes Dev. 2009;23(4):391–6.

89. Cui H-R, Zhang Z-R, Xu J-N, Wang X-Y. Genome-wide characterization and analysis of F-box protein-encoding genes in the*Malus domestica* genome. Mol Gen Genomics. 2015;290(4):1435–46.

90. Rae AL, Perroux JM, Grof CP. Sucrose partitioning between vascular bundles and storage parenchyma in the sugarcane stem: a potential role for the ShSUT1 sucrose transporter. Planta. 2005;220(6):817–25.

91. Hackel A, Schauer N, Carrari F, Fernie AR, Grimm B, Kühn C. Sucrose transporter *LeSUT1* and *LeSUT2* inhibition

affects tomato fruit development in different ways. Plant J. 2006;45(2):180–92.

92. Tieman DM, Handa AK. Reduction in pectin methylesterase activity modifies tissue integrity and cation levels in ripening tomato (*Lycopersicon esculentum* Mill.) fruits. Plant Physiol. 1994;106(2):429–36.

93. Fry SC. Cross-linking of matrix polymers in the growing cell walls of angiosperms. Annu Rev Plant Physiol. 1986;37(1):165–86.

94. Eschrich W. Free space invertase, its possible role in phloem unloading. Ber Deut Bot Ges. 1980;93(1):363–78.

CHAPTER 13

EXPLORING MIRNAS INVOLVED IN BLUE/UV-A LIGHT RESPONSE IN BRASSICA RAPA REVEALS SPECIAL REGULATORY MODE DURING SEEDLING DEVELOPMENT

Bo Zhou[1,2], Pengzhen Fan[1], Yuhua Li[1,2], Haifang Yan[1], and Qijiang Xu[1,2]

[1]College of Life Science, Northeast Forestry University, 26 Hexing Road, Harbin 150040, China.

[2]State Key Laboratory of Tree Genetics and Breeding, Northeast Forestry University, 26 Hexing Road, Harbin 150040, China.

ABSTRACT

Background

Growth, development, and pigment synthesis in Brassica rapa subsp. rapa cv. Tsuda, a popular vegetable crop, are influenced by light. Although microRNAs (miRNAs) have vital roles in the metabolic processes and abiotic stress responses of plants, whether miRNAs play a role in anthocyanin biosynthesis and development of Tsuda seedlings exposed to light is unknown.

Results

Seventeen conserved and 226 novel miRNAs differed at least 2-fold in response to blue and UV-A light compared with levels after a dark treatment. Real time PCR showed that BrmiR159, BrmiRC0191, BrmiRC0460, BrmiRC0323, BrmiRC0418, BrmiRC0005 were blue light-induced and northern blot revealed that the transcription level of BrmiR167 did not differ significantly among seedlings treated with dark, blue or UV-light. BrmiR156 and BrmiR157 were present in the greatest amount (number of reads) and among their 8 putative targets in the SPL gene family, only SPL9 (Bra004674) and SPL15 (Bra003305) increased in expression after blue or UV-A exposure. In addition, miR157-guided cleavage of target SPL9 mRNAs (Bra004674, Bra016891) and SPL15 mRNAs (Bra003305, Bra014599) took place 10 or 11 bases from the 5′ ends of the binding region in the miR157 sequence.

Conclusions

A set of miRNAs and their targets involved in the regulation of the light-induced photomorphogenic phenotype in seedlings of Brassica rapa was identified, providing new insights into blue and UV-A light-responsive miRNAs in seedlings of Tsuda and evidence of multiple targets for the miRNAs and their diverse roles in plant development.

Citation: Zhou B, Fan P, Li Y, Yan H, and Xu Q (2016). Exploring miRNAs involved in blue/UV-A light response in Brassica rapa reveals special regulatory mode during seedling development. BMC Plant Biology, 16:111. DOI 10.1186/s12870-016-0799-z

BACKGROUND

MicroRNAs (miRNAs) are noncoding, regulatory RNAs approximately 21 nt long that negatively regulate gene expression at the transcriptional and post-transcriptional levels via post-transcriptional cleavage of mRNA, inhibition of translation and RNA-dependent RNA polymerase (RdRP)-mediated second-strand synthesis, and *trans*-acting small interfering RNAs (ta-siRNAs) initiated by miRNA and miRNA-dependent DNA methylation [1–4]. Initially discovered in *C. elegans* [5], they have now been widely reported in plants and animals [6, 7], and some miRNAs or miRNA-like RNAs have even been reported in viruses and fungi [8,9]. In recent years, increasing evidence has demonstrated that miRNAs are involved in the regulation of many biological and metabolic processes such as development, signal transduction, metabolism and response to environmental signals in plants [10–12].

Environmental signals such as ultraviolet (UV) and blue regions of the light spectrum are important for plants to control a wide range of processes during growth and development [13, 14]. At least two classes of photoreceptors, mainly UV-B receptors and cryptochromes, absorb UV and blue light, respectively, in plants [15–18]. Blue and UV-A light-mediated photomorphogenic responses include de-etiolation, phototropism, stomatal opening, and anthocyanin accumulation [19], which are controlled by light-responsive genes. In particular, blue and UV-A light induce transcription of the genes involved in anthocyanin biosynthesis of *Brassica rapa* subsp. *rapa* [20]. Blue and UV-B light have also been reported to mediate miRNAs that regulate genes in maize and *Arabidopsis* [21, 22]. For example, in *Arabidopsis*, Cryptochrome 1 (CRY1) and Cryptochrome 2 (CRY2) mediate the expression of miR172 after blue light stimulation in a CONSTANS (CO)-independent manner to regulate photoperiodic flowering time [22]. Four miRNAs (miR160, miR165/166, miR167 and miR393) also respond to UV-B and might be involved in auxin signaling pathways of *Arabidopsis*[23]. Similarly, miRNA396 exposed to UV-B radiation in leaves of *Arabidopsis* is upregulated, leading to a decrease in transcripts of *GROWTH-REGULATING FACTOR* genes (*GRFs*) which result in inhibition of cell proliferation and leaf growth [24, 25]. Moreover, miR408 is coordinately regulated by SQUAMOSA PROMOTER BINDING PROTEIN-LIKE7 (SBP/SPL7) and ELONGATED HYPOCOTYL5 (HY5) in *Arabidopsis*, and the transcription levels of miR408 and its target genes are also changed in response to light and copper [26]. In maize, miR164, miR165, miR166, and miR398 are up-regulated, while miR156, miR171, miR172, miR396, and miR529 are down-regulated under UV-B radiation treatment. Furthermore, both miR156 and miR529 are decreased and their several targets (SBP transcripts) are increased after 8h UV-B in maize leaves [21].

To date, high-throughput sequencing has revealed novel microRNAs in different organs at different stages of development of *B. rapa* subsp. *pekinensis* cv. Chiifu [27], in seedlings of *B. rapa* subsp. *chinensis* stressed by heat [28] and in skotomorphogenic seedlings of *B. rapa* subsp. *rapa* cv. Tsuda [29]. In addition, many light signal transduction factors and genes involved in light-induced anthocyanin biosynthesis have been extensively studied in *Brassica rapa* [20, 23], but the miRNAs that are responsive to blue and UV-A light have not yet been systematically identified and characterized at the genome level.

Turnip (*Brassica rapa* subsp. *rapa*; Brassicaceae) is an important and popular cruciferous root vegetable. The turnip cv. Tsuda, a purple top cultivar, accumulates the anthocyanidin pelargonidin in swollen hypocotyls and in the upper mid-section of hypocotyls of the seedling after exposure to sunlight and blue and UV-A light [20, 30]. Because of the health-promoting role of anthocyanins, this red turnip cultivar Tsuda is very desirable, and its blue and UV-A light-induced anthocyanin biosynthesis is especially suitable to study the role of miRNAs in regulating anthocyanin production in response to various wavelengths of light.

To identify miRNAs involved in light-induced anthocyanin biosynthesis in *Brassica rapa*, high-throughput sequencing technology was used to obtain differentially expressed conserved and novel miRNAs in response to blue light, UV-A and dark treatment. A set of miRNAs and their targets were found to be involved in the regulation of light-induced photomorphogenic phenotype in seedlings of *Brassica rapa*. The most abundant miRNA156/157 could negatively regulate the transcript level of their targets SPL9 and SPL15 in the anthocyanin biosynthetic seedlings under blue light and UV-A induction. The identification of blue and UV-A light-responsive miRNAs could aid in understanding the mechanisms underlying plant response to light-induced anthocyanin biosynthesis.

RESULTS

High-throughput Analysis of small RNAs Responsive to Blue and UV-A Light

A small RNA (sRNA) library obtained from seedlings of turnip grown in the dark for 4 days as control and induced by blue light or UV-A for 1 day after growing in the dark for 3 days as treatment were sequenced using the Solexa system. A total of 16,859,441, 16,110,664 and 13,974,772 clean reads were obtained from the dark, blue and UV-A small RNA library, respectively. Among the clean reads, small RNA sequences varied widely in length (from 10 to 44 nt); the number of 20–24-nt sequences significantly outnumbered the shorter or longer sequences. The number of these 20–24-nt small RNAs accounted for 84.37 % of the total sequences for the dark treatment, 69.42 % for blue and 66.77 % for UV-A. The blue and the UV-A light treatments generated 337,524 and 298,261 unique sRNAs, respectively, fewer than in the dark treatment (359,531 unique sRNAs). Compared with sRNAs in the dark treatment, 28.17 % of the sRNAs were specific to blue light, while 24.88 % of the sRNAs were specific to UV-A (Fig. 1). Moreover, 63.51 %, 62.13 % and 56.75 % unique sequences were obtained from dark-, blue- and UV-A-treated seedlings, respectively, and they mapped completely onto the *Brassica* A genome from *B. rapa* (Chiifu-401) (http://brassicadb.org/brad/). Also, 7,033,801 (52.81 %), 5,365,249 (42.15 %) and 4,042,699 (36.57 %) small RNA reads from the dark, blue and UV-A small RNA library, respectively, were annotated as miRNAs. The rest of the sequences were found to be other types of RNA, including unannotated RNA, tRNA, rRNA, small nuclear RNA (snRNA), small nucleolar RNA (snoRNA) and intron or exon sequences. The numbers and proportions of different categories of small RNAs are shown in Table 1. The results revealed that the proportion of known miRNAs in blue- or UV-A-treated seedlings was lower than in the dark-grown seedlings.

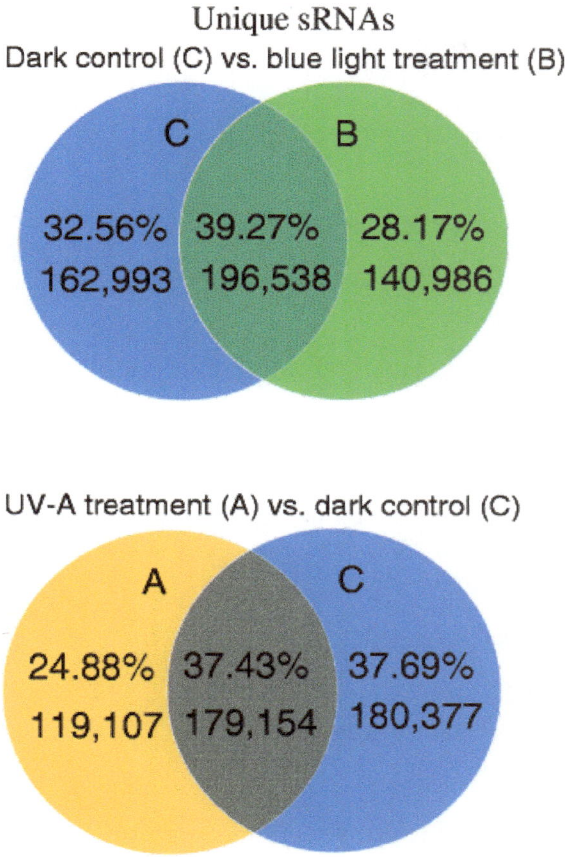

Figure 1. Venn diagram of unique sRNAs in sRNA library for Tsuda seedlings after blue light or UV-A treatment in comparison with the dark treatment. Left, dark-control specific sRNAs (C) vs. blue-light specific (B); right, dark-control specific sRNAs vs. UV-A specific (A).

Table 1. Distribution of small RNA reads among different categories in the sRNA library after dark, Blue light or UV-A treatment of *Brassica rapa* subsup. rapa cv. Tsuda

Category	Dark treatment				Blue light treatment				UV-A light treatment			
	Unique sRNAs	Percent (%)	Total sRNAs	Percent (%)	Unique sRNAs	Percent (%)	Total sRNAs	Percent (%)	Unique sRNAs	Percent (%)	Total sRNAs	Percent (%)
Total	359,531	100 %	13,319,035	100 %	337,524	100%	12,727,958	100 %	298,261	100 %	1,105,3984	100 %
exon_ antisense	7952	2.21 %	151,963	1.14 %	7847	2.32%	117,444	0.92 %	7742	2.60 %	167,459	1.51 %
exon_ sense	7916	2.20 %	94,268	0.71 %	7545	2.24%	100,660	0.79 %	7259	2.43 %	79,121	0.72 %
intron_ antisense	12,312	3.42 %	183,025	1.37 %	11,345	3.36%	148,510	1.17 %	9167	3.07 %	146,973	1.33 %
intron_ sense	19,413	5.40 %	274,427	2.06 %	17,941	5.32%	250,571	1.97 %	13,782	4.62 %	183,113	1.66 %
miRNA	2,349	0.65 %	7,033,801	52.81 %	2,365	0.70%	5,365,249	42.15 %	2,395	0.80 %	4,042,699	36.57 %
rRNA	20,611	5.73 %	1,235,211	9.27 %	22,614	6.70%	864,612	6.79 %	35,121	11.78 %	1,412,822	12.78 %
snRNA	311	0.09 %	2,640	0.02 %	383	0.11%	3,134	0.02 %	394	0.13 %	3,668	0.03 %
snoRNA	189	0.05 %	1546	0.01 %	457	0.14%	7999	0.06 %	324	0.11 %	3431	0.03 %
tRNA	5715	1.59 %	1,137,537	8.54 %	10,857	3.22%	2,791,679	21.93 %	11,743	3.94 %	1,945,982	17.60 %
unann	282,763	78.65 %	3,204,617	24.06 %	256,170	75.90%	3,078,100	24.18 %	210,334	70.52 %	3,068,716	27.76 %

Identification of Conserved and Novel Candidate miRNAs in Seedlings of *B. Rapa* Subsp. *rapa* cv. Tsuda after Different Light Treatments

All unique small RNA sequences were aligned with the currently known miRNAs in the miRNA database miRBase (Release 20) and then screened to determine their hairpin structures. In total, 138 precursors belonging to 54 conserved miRNA families in the dark-responsive miRNAs, 140 precursors belonging to 54 conserved miRNA families in the UV-A-responsive miRNA library, and 129 precursors belonging to 53 conserved miRNA families in the blue-light-responsive miRNA library were separately identified as homologs of known miRNA families from diverse plant species. Among these families, miR156, miR157, miR167, miR168 were abundant, accounting for more than 90 % of the total conserved and novel miRNA reads in dark-, blue-light- and UV-A-responsive sRNA libraries.

To uncover novel miRNAs from *B. rapa*, all unannotated sRNAs were analyzed using a bioinformatic approach. According to criteria for annotating novel plant miRNAs [31], 70 miRNA sequences (including their miRNA*s) from 126 loci were identified. Among these miRNAs, 10 novel miRNAs (BrmiRC0035, BrmiRC0091, BrmiRC0099, BrmiRC0144, BrmiRC0153, BrmiRC0191, BrmiRC0211, BrmiRC0250, BrmiRC0415, BrmiRC0460) were found in all three sRNA libraries. Only four novel miRNAs (BrmiRC0311, BrmiRC784, BrmiRC799, BrmiRC839) were detected in blue-light- and UV-A-specific-responsive sRNA libraries.

Differential Expression of miRNAs in Response to Dark, Blue and UV-A Light

Based on the results of high-throughput sequencing, miR5175 and miR3630 were only detected in the UV-A sRNA library, miR5300 was only found in the blue sRNA library, and miR1535 and miR5138 were identified only in the dark sRNA library. By normalizing the number of miRNA reads (on the basis of transcripts per million, TPM) in the library, the relative abundance of miR391, miR1439, miR2111, miR2911, miR2916, miR5083 in the UV-A sRNA library was 2 times higher than in the dark sRNA library, whereas the abundance of miR396 and miR1885 was the opposite. In the blue-light-responsive sRNA library, aside from miR391, miR2111 and miR5083, miR1511 was more abundant than in the dark sRNA library, whereas miR5072 and miR5139 were less abundant. However, the transcription level of most miRNAs did not obviously differ in the seedlings among the dark, blue light and UV-A light treatments. Among the novel miRNAs, 23 candidate miRNAs were downregulated, and 7 were upregulated more than 2-fold after UV-A treatment. In particular, BrmiRC0305 and BrmiRC0491 were upregulated more than 4-fold after UV-A. Moreover, 9 candidate miRNAs were downregulated, and 10 candidate miRNAs were upregulated more than 2-fold after blue light. There were also 110 candidate miRNAs specific to the UV-A-light-induced library, 67 candidate miRNAs specific to blue-light-induced library and 226 specific to the dark-treatment library (Fig. 2).

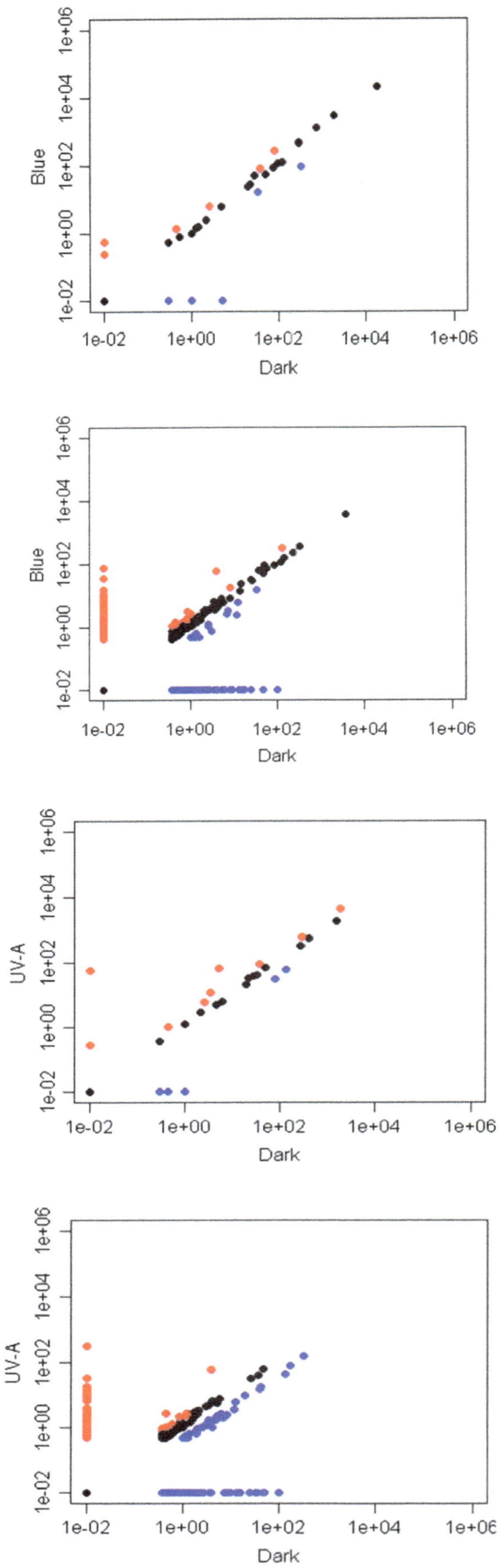

Figure 2. Differential expression of miRNAs in Tsuda seedings after dark, blue light or UV-A treatment. Red dots denote the ratio of log $2(y_{treatment}/x_{control}) \geq 1$; blue dots denote the ratio of log $2\ (y_{treatment}/x_{control}) \leq -1$.

Expression Analysis of Potential miRNAs Responsive to Dark, Blue and UV-A Light by qRT-PCR and RNA Blot

To confirm the original high-throughput sequencing results, we selected three conserved miRNAs (BrmiR156 with more than 3 million reads, BrmiR167 with more than 100,000 reads, and BrmiR159 with more than 2000 reads) as examples to analyze their expression using an RNA blot. In addition, five novel candidate miRNAs and BrmiR159 were analyzed by real time PCR. The RNA blot and real time PCR results demonstrated that all tested miRNAs were expressed in seedlings in the dark, blue light and UV-A treatments and that the transcription levels of the different miRNAs varied (Figs. 3, 4). Transcription of most of the selected miRNAs (BrmiR159, BrmiRC0191, BrmiRC0460, BrmiRC0323, BrmiRC0418, BrmiRC0005) was higher after the blue light treatment than after the dark or the UV-A light treatment (Fig. 3). Northern blot showed that the transcription level of BrmiR167 was not significantly different among the seedlings from the dark, blue and UV-light induction, while the expression of BrmiR156 was slightly inhibited by blue and UV-A light (Fig. 4). When the relative expression levels of candidate microRNAs shown in RNA blots (gray level ratios between the blot results of microRNAs and the U6) or by RT-PCR were compared with the normalized reads of microRNAs from the high-throughput sequencing library of the dark, blue light and UV-A light induced seedlings, all selected miRNAs had the same expression profiles as in the original high-throughput sequencing results (Fig. 3, 4).

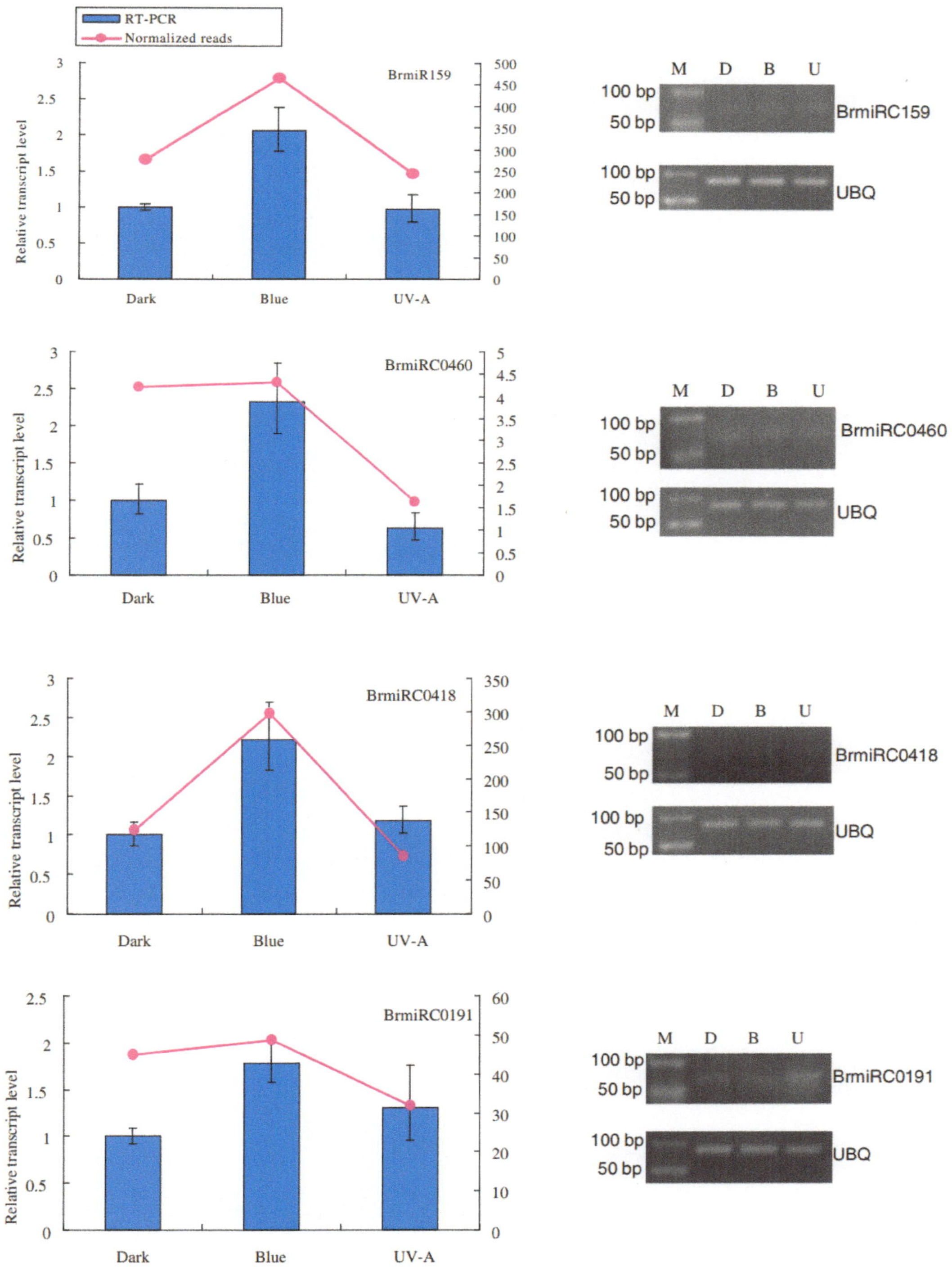

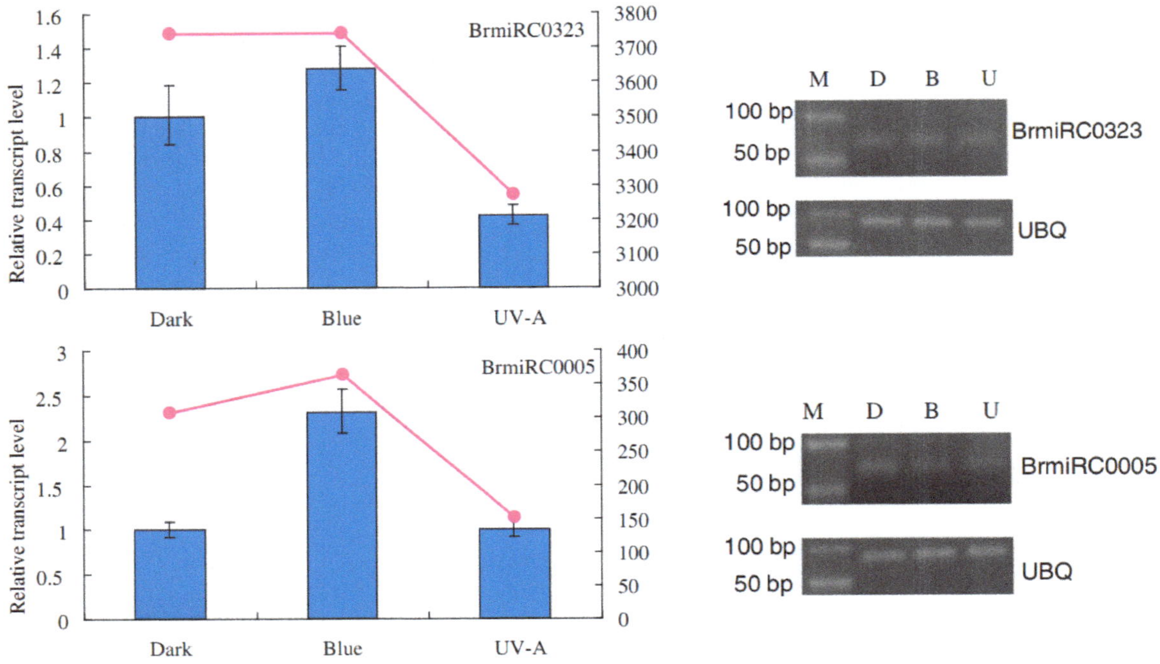

Figure 3. Transcript expression of conserved and novel miRNA predicted from Tsuda after treatment with dark, blue light or UV-A. Relative expression ($2^{-\Delta\Delta CT}$) of conserved BrmiR159 and predicted BrmiRC0191, BrmiRC0460, BrmiRC0323, BrmiRC0418, BrmiRC0005 was analyzed by real time PCR, and the qRT-PCR products were separated by electrophoresis. The UBQ gene was used as the internal control, and seedlings in the dark control were used as the calibrator. Lanes: M, size marker (10 bp DNA Ladder, MBI); D, dark; B, blue light; A, UV-A. Each bar indicates the mean ± SE of triplicate assays after correcting for template quantity relative to the UBQ gene. The dotted line indicates the normalized reads for the corresponding miRNAs from high-throughput sequencing.

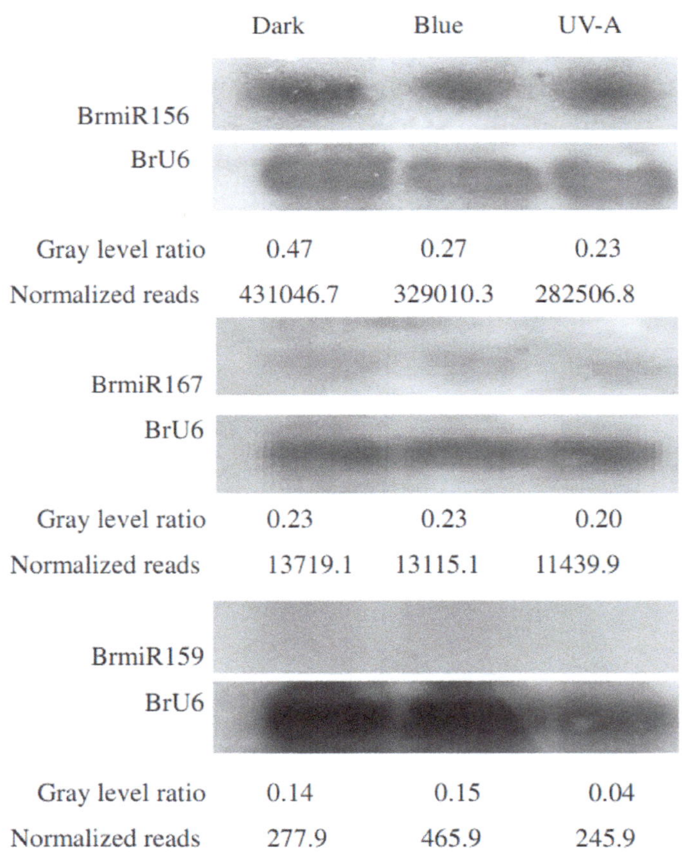

Figure 4. Northern blot analysis of BrmiR156, BrmiR167 and BrmiR159 in seedlings of Tsuda after exposure to dark, blue light or UV-A. 5′-Digoxigenin-labeled DNA oligonucleotide with complementary sequence to miRNA was used as the probe; BrU6 was used as the control. The gray level ratio between miRNAs and BrU6 was used to compare relative expression to normalized reads.

Prediction of miRNA Target Genes

The putative target genes for all identified miRNAs from the seedlings in the different light treatments were predicted by using psRNA Target program. In total, 270 potential targets genes were found for 51 conserved miRNAs and 3589 for 802 novel miRNAs. Many of the miRNAs had multiple targets, indicating the diverse regulatory roles of these miRNAs. Certainly, the targets that are common to more than one miRNA, such as Bra026884 (target of miR161 and miR400), Bra000791 (target of BrmiRC0930 and BrmiRC0965), Bra000965 (target of BrmiRC0598 and BrmiRC0776), Bra001160 (target of BrmiRC0149 and BrmiRC0801), also have different miRNA-binding sites, indicative of the complex regulatory network of miRNAs. Most of these putative target genes encode transcription factors such as SPL (targets of miR156, miR157 and miR1088), Auxin response factor (ARF) (targets of miR160 and miR167), Myeloblastosis (MYB) (targets of miR159 and miR828), Basic leucine zipper (bZIP) (BrmiRC0651, and BrmiRC0940), AGAMOUS-like (AGL) (targets of miR824 and BrmiRC0607), WRKY (targets of BrmiRC0049, BrmiRC0149, BrmiRC0177 and BrmiRC0181) and APETALA2 (AP2) (targets of miR172), which have known or putative functions in a wide variety of biological processes. Interestingly, miR164, BrmiRC0117, BrmiRC0178, BrmiRC0601, BrmiRC0982 targeted genes that encode NAM, ATAF1, 2, and CUC2 (NAC) proteins, and miR1885, BrmiRC0149, BrmiRC0177, BrmiRC0311, BrmiRC0801, BrmiRC0883 targeted genes encoding a disease resistance protein ((Toll Interleukin-1 Receptor) TIR-(Nucleotide-Binding Site) NBS- (Leucine-Rich Repeat) LRR class). Genes targeted by miR161, miR400 and BrmiRC0797 encode a pentatricopeptide (PPR) repeat-containing protein. In addition, some miRNAs targeted genes that are involved in light signal transduction and morphogenic responses. For example, miR2111 was predicted to target a cop8 (constitutive photomorphogenic) homolog, miR399 was predicted to target SPA2 (SPA1-RELATED 2), BrmiRC0406 was predicted to target CRY2, and BrmiRC0826 was predicted to target phytochrome and flowering time regulatory protein 1.

Analysis of miR156, miR157 and Their Targets after Blue Light and UV-A Exposure

In the high-throughput sequencing analyses of the blue light and UV-A induction compared with the dark treatment, miR156 and miR157 had about 1.5-fold change in their transcript numbers. RT-PCR also showed a significant change in expression of miR157 after UV-A treatment compared with the dark treatment (two-tailed *t*-test, *p*-value is 0.0259, $t=3.132$, $df=5$), but under blue light treatment compared with the dark treatment the expression did not differ significantly ($P<0.05$) (p-value is 0.129, $t=1.817$, $df=5$).

We then explored miR156 and miR157 and their putative targets in more detail. The 13 target genes of miR156 and miR157 belong to 7 members of the *SPL* gene family and have highly conserved sequences in the binding site (Table 2 and Fig. 5). RT-PCR showed that the expression of predicted targets *SPL9*(Bra004674) and *SPL15* (Bra003305) induced by blue light and UV-A was approximately 3 times higher than in the dark treatment. The transcript level of the other detected targets, *SPL10* (Bra010949 and Bra030041),*SPL2* (Bra027478 and Bra033671), *SPL6* (Bra038324) and *SPL13* (Bra022766), did not differ significantly among seedlings grown in the dark, blue or UV-A light (Fig. 6). The miR157-guided cleavage of target *SPL9*mRNAs (Bra004674, Bra016891) and *SPL15* mRNAs (Bra003305, Bra014599) was detected as expected. The cleavage sites were 10 or 11 bases from the 5′ ends of the binding region in the miR157 sequences (Fig. 7).

Table 2. Predicted binding site of miR156 and miR157 on sequences of candidate target genes in SPL family

Candidate gene	Start region number in mRNA sequence	mRNA sequence of predicted binding site area	Amino acid sequence of corresponding binding site area	Stop region number in amino sequence
Bra027478 (SPL2)	922	GAUGGUGCUCUCUCUCUUCUGUCAAAU	QDLDGALSLLSNSTAW	316
Bra033671 (SPL2)	841	GAUGGUGCUCUCUCUCUUCUGUCAAAU	QDLDGALSLLSNSTPW	293
Bra004363 (SPL6)	631	ACUUGUGCUUGCUCUCUUCUGUCAGCU	PRSTCACSLLSAQSQQ	223
Bra038324 (SPL6)	673	ACUUGUGCUCUCUCUCUUCUGUCAGCU	PRSTCALSLLSAQSQQ	237
Bra016891 (SPL9)	736	AACUGUGCUCUCUCUCUUCUGUCAAAU	GDSNCALSLLSNPHQ	257

Bra004674 (SPL9)	739	AACUGUGCUCUCUCUCUUCUGUCAAAC	SVTNCALSLLSNPHQP	259
Bra010949 (SPL10)	991	UACAGUGCUCUCUCUCUUCUGUCAACG	QDFYSALSLLSTSSDS	343
Bra030041 (SPL10)	1015	UACAGUGCUCUCUCUCUUCUGUCAACG	QDFYSALSLLSTSSDS	351
Bra030040 (SPL11)	1067	CACCGUGCUCUCUCUCUUCUGUCAACU	QEFHRALSLLSTSSSD	369
Bra032822 (SPL11)	994	CACCGUGCUUUCUCUCUUCUGUCAACU	QDFHRAFSLLSTSSGP	344
Bra022766 (SPL13)	844	GAUUGCGCUCUCUCUCUUCUGUCAUCA	HDSDCALSLLSSSSSH	294
Bra003305 (SPL15)	562	AGCUGUGCUCUCUCUCUUCUGUCAAAC	TDSSCALSLLSNYTNP	201
Bra014599 (SPL15)	550	AGCUGUGUUCUCUCUCUUCUGUCAAAC	TDSSCVLSLLSNSNTT	196

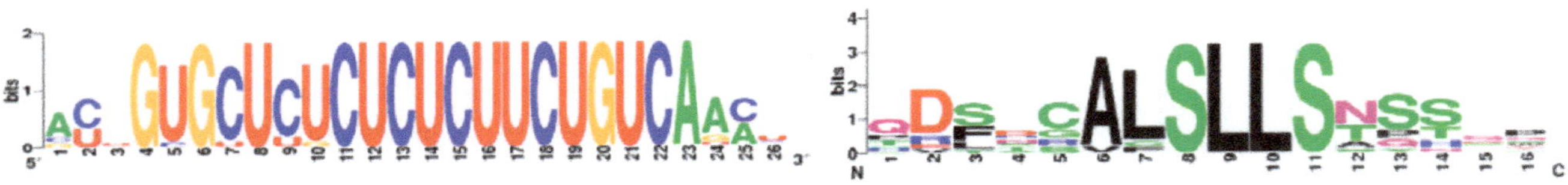

Figure 5. Conserved sequence analysis of predicted binding site for miR156 and miR157 in target mRNA and amino sequence. Multiple sequence alignment was done using WebLogo (http://weblogo.berkeley.edu/logo.cgi).

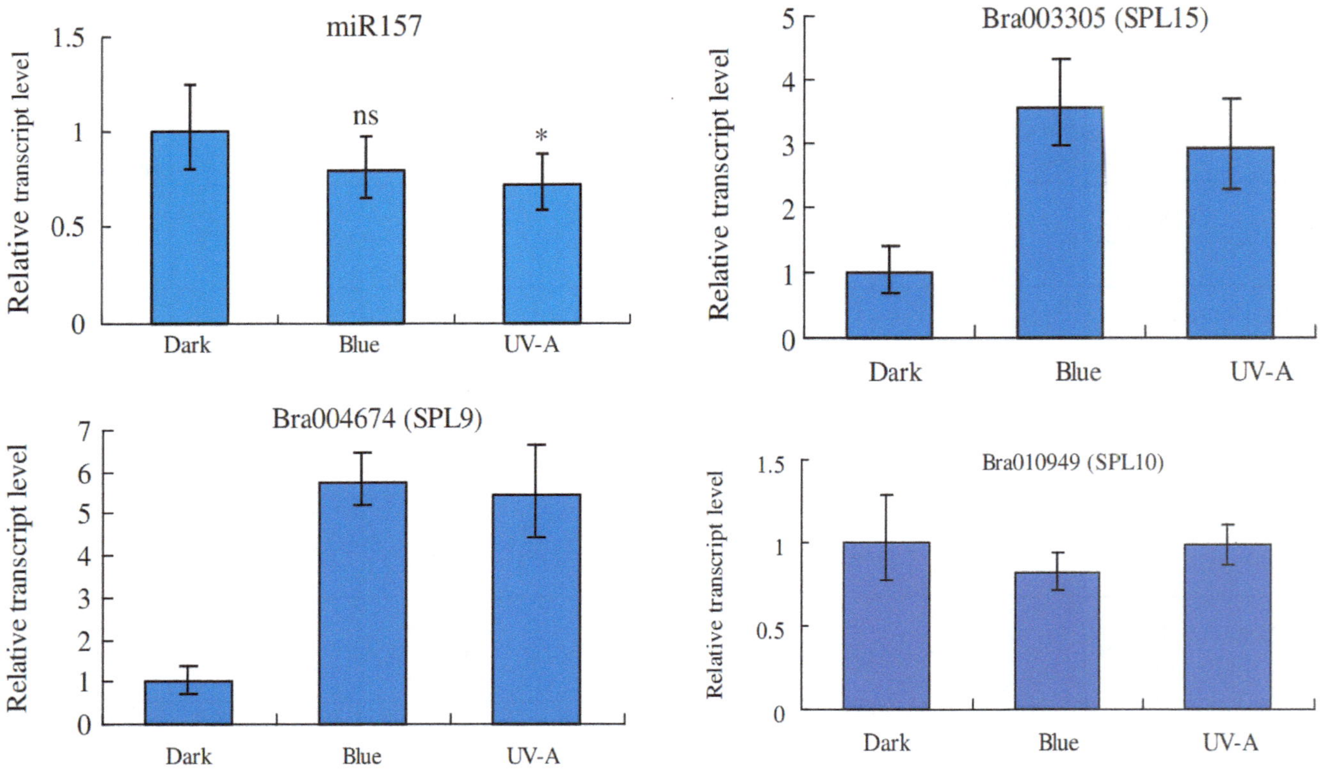

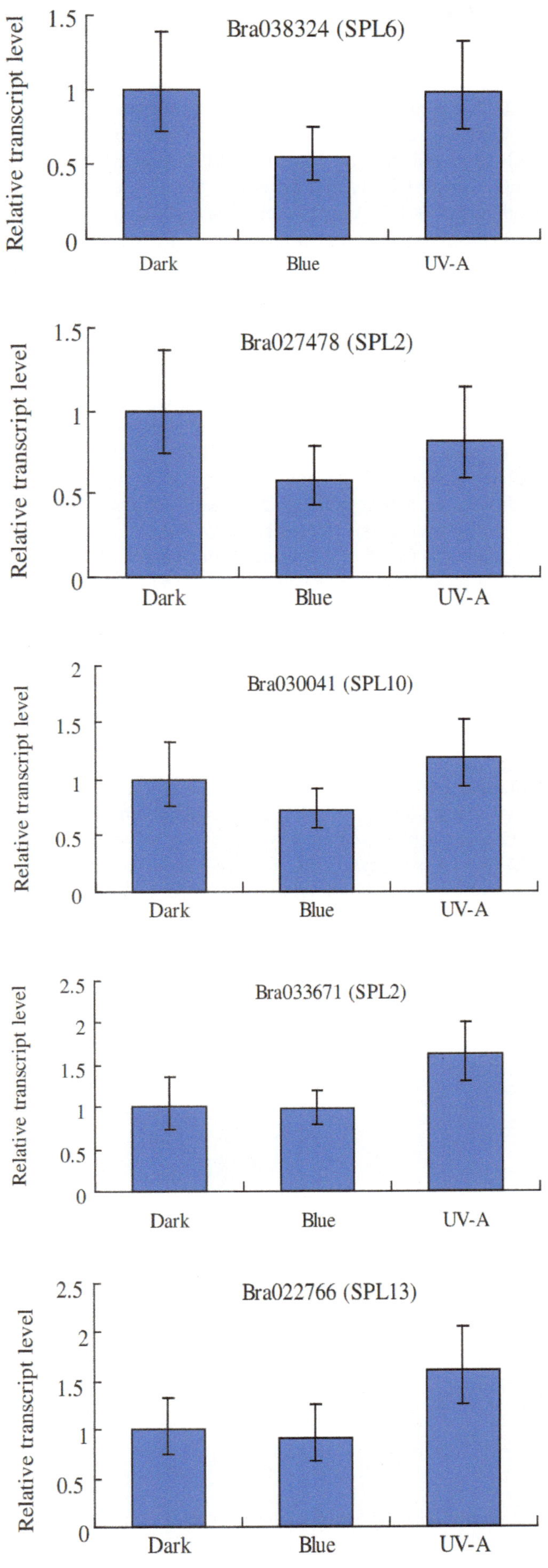

Figure 6. Mean relative transcript level of BrmiR157 and its potential target BrSPL genes in seedlings of Tsuda exposed to dark, blue light or UV-A. Relative expression ($2^{-\Delta\Delta CT}$) of conserved BrmiR157 and predicted Bra004674 (SPL9), Bra003305 (SPL15), Bra010949 (SPL10), Bra038324 (SPL6), Bra027478 (SPL2), Bra030041 (SPL10), Bra033671 (SPL2), Bra022766 (SPL13) were analyzed by real time PCR. The UBQ gene served as the internal control to correct for template quantity; seedlings in the dark control served as the calibrator. Bar indicates ± SE of triplicate assays. Statistically significant differences between blue light treatment and dark, UV-A treatment and dark are shown (*t*-test, $*P < 0.05$; ns, not significant difference).

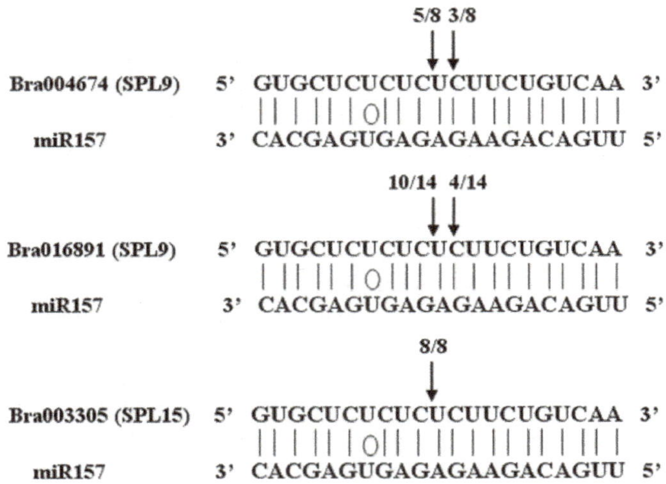

Figure 7. Sites of BrmiR157-mediated cleavage on target BrSPL9 and BrSPL15 mRNAs. Positions of dominant 5′-RACE products of BrSPL9 (Bra004674, Bra016891) and BrSPL15 (Bra003305) mRNA are indicated by a vertical arrow in the binding region.

DISCUSSION

In this study, high-throughput sequencing was used to obtain more knowledge about gene regulation by miRNAs induced by blue light and UV-A, which induce anthocyanin biosynthesis in seedlings of the Tsuda cultivar of *Brassica rapa* subsp. *rapa*. By comparing the miRNAs and the miRNA precursors in *Arabidopsis*with the sRNA library of blue- and UV-A-treated seedlings of Tsuda turnip, we were able to identify the most common conserved miRNA families in the cultivar. Certainly, many Tsuda turnip-specific miRNAs were also obtained, but the number of reads was limited (from 5 to 50,000), and most were not over 1000. This result is consistent with our previous research on the dark-treated seedlings [29].

As technology has developed, a number of miRNAs have been identified in plants, and their roles in regulation have begun to be elucidated. For example, higher levels of pigments accumulated in transgenic*Arabidopsis* which over-expressed HY5-regulated miR408, resulting in a phenotype opposite of the lower pigment levels caused by a mutation in *hy5* [32]. MiR172 mediates photoperiodic flowering independent of*CONSTANS* in *Arabidopsis* [22, 33]. High-throughput sequencing of *Brassica rapa* subsp. *chinensis* showed that miR398, miR399, miR827, miR5716 and miR1885 were downregulated under heat stress, while miR156, miR5714, miR5718 and miR5726 significantly increased [28]. In our light-induced RNA library, BrmiR391, BrmiR2111, BrmiR5083 and BrmiRC0132, BrmiRC0448, BrmiRC0491 were induced by both blue light and UV-A. Furthermore, miR156/157, miR159/319, miR160, miR165/166, miR167, miR169, miR170/171, miR172, miR393, miR398 and miR401, which are responsive to UV-B in *Arabidopsis* [34], were all detected in light-treated seedlings of *Brassica rapa*, but most had no obvious difference in transcription level. Only miR156/157, miR398 induced by UV-A, miR159, miR319 induced by blue light, miR172 induced by blue and UV-A light showed over 1.5-fold change in expression compared with the levels in dark-treated seedlings. In the analysis of the predicted targets, the BrmiR391 target was annotated as flavin adenine dinucleotide (FAD) binding, the BrmiR2111 target as a COP8 (constitutive photomorphogenic) homolog, the BrmiR5083 target as a potassium transporter family protein, the BrmiRC0132 target as a cytochrome family protein, the BrmiRC0448 target MYB domain protein 16, the BrmiRC0491 target as potassium transporter 11. Most of these targets are involved in photomorphogenesis and signal transduction [35–38].

Interestingly, miR156 and miR157 had the most reads in our detected sRNA libraries. Although the fold change in transcript levels decreased about 1.5 fold in the blue light and in UV-A treatment compared with the dark treatment, the read numbers for miR156 and miR157 were over 100,000 less in the Tsuda seedlings grown in blue or UV-A light than in the dark. Additionally, the RNA blot and the RT-PCR also showed little downregulation of miR156 or miR157 in blue light and UV-A light. The downregulation expression pattern of miR156 in light-induced anthocyanin synthesis seedlings of turnip is consistent with that in *Arabidopsis*under exogenous sugar induction [39, 40]. The level of miR156 is also reduced when the plant grows from the juvenile phase to adult phase and the expression of miR156 dwindles, ultimately leading to the reproductive phase transition [41–43]. The targets of miR156 and miR157, SPL transcription factor families, play important roles in regulating diverse aspects of plant growth and development including the vegetative phase and floral transition in *Arabidopsis* [44, 45]. Moreover, miR156 is also positively regulated by SPLs, and a negative feedback loop regulates the expression of its targets [45]. Our results also showed that the miR156 and miR157 targets, SPL9 and SPL15, were specifically upregulated by blue light and UV-A light in Tsuda seedlings. The difference of expression between miR156/157 and SPL9, SPL15 suggests that miR156 or miR157 may target SPL9 and SPL15 more effectively than other SPLs and may be involved in photomorphogenesis induced by blue light and UV-A light during the seedling development of Tsuda. A regulatory relationship may exist between miRNA 156, miRNA157 and their targets SPL9, SPL15 although we cannot exclude the possibility that SPL9 and SPL15 are upregulated by other blue or UV-A light responsive transcription factors.

The SPL genes targeted by miR156 can be grouped into four categories: SPL9/SPL15, SPL6/SPL13, SPL11/SPL10/SPL2 and SPL3/SPL4/SPL5 [46]. SPL9 and SPL15 regulate shoot maturation and leaf initiation [47]. Overexpression of miR156 in switchgrass (*Panicum virgatum*) [48], *Arabidopsis* [49] and*Brassica napus* [50] led to reduced SPL15 expression, increased shoot branching or enhanced carotenoid content [48–50]. On the other hand, reduced miR156 activity in *Arabidopsis* results in a high level of SPL9 and negatively regulates anthocyanin accumulation by competing with TT8 in binding to anthocyanin-specific R2R3-MYBs to disrupt the stabilization of the MYB-bHLH-WD40 (MBW) complex [51]. The activity of MBW transcription factor complexes is essential to activate anthocyanin biosynthesis genes and thus for anthocyanin pigments to accumulate [52]. In this regard, the stability of the MBW complex might be increasingly impaired as the level of miR156 and miR157 transcripts decreases in seedlings of turnip exposed to blue light and UV-A light. The *Arabidopsis* mutant *sk156* expresses miR156b at a higher level than in the wild type, and miR156b-induced *SPL15* suppression is partially responsible for the increased carotenoid abundance in seeds and altered morphology of the mutant adult plant [49]. Moreover, photosynthates promote the vegetative phase change by repressing the expression of miR156 in*Arabidopsis* [39]. Our results also showed that when dark-grown seedlings were exposed to blue or UV-A light, the transcript level of miR156 and miR157 decreased, thus reducing the suppression of their targets*SPL9* and *SPL15*; the seedlings had the photomorphogenic phenotype such as anthocyanin biosynthesis, short hypocotyls, and cotyledon expansion (Fig. 8).

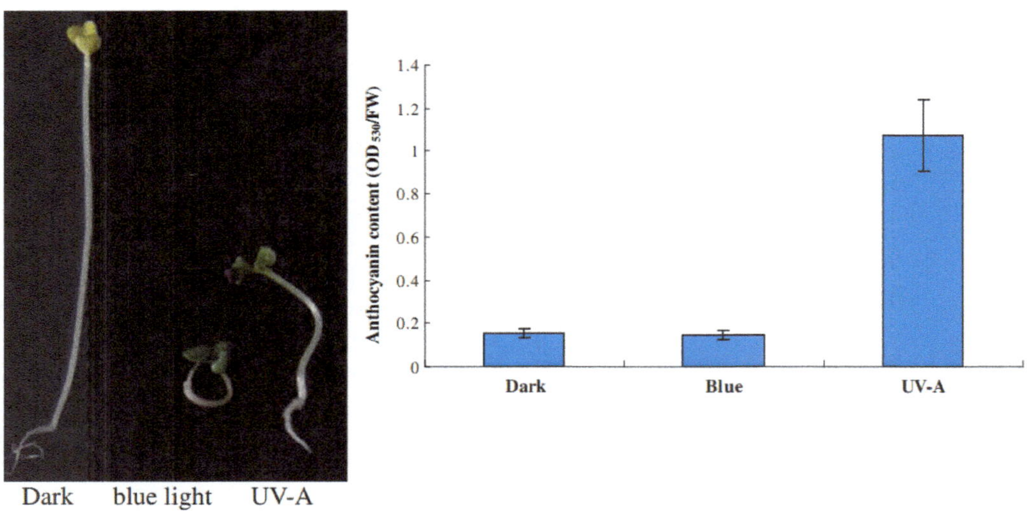

Figure 8. Phenotype of *Brassica rapa* subsp. *rapa* cv. Tsuda seedling and anthocyanin content after dark, blue light or UV-A treatment. Seedlings were exposed to blue light at 10 W m^{-2}, UV-A at 3 W m^{-2}, or to no light (dark). Anthocyanin was extracted from fresh whole seedlings, and the concentration was determined using absorbance at OD_{530} per gram of fresh mass. Vertical bars indicate ± SE ($n = 8$).

On the basis of our data and previous studies [20, 23, 30, 48–51], we can deduce that blue light/UV-A activates light signal transduction genes, some of which positively regulate anthocyanin biosynthesis in seedlings of *Brassica rapa* subsp. *rapa* cv. Tsuda. When positive anthocyanin biosynthesis regulators such as R2R3-MYB, basic helix-loop-helix (bHLH), WD40 [53] express increasingly by light induction, then negative feedback involving miRNAs such as miR156, miR157 and their targets SPL9 and SPL15 are also activated to negatively regulate the anthocyanin biosynthetic pathway and maintain a new balance for the transcription of the biosynthetic genes in seedlings of Tsuda turnip. Thus, light induces the maximum expression of these regulated genes. Previous studies on Tsuda seedlings also showed that the expression of chalcone synthase gene (*BrCHS*), dihydroflavonol 4-reductase gene (*BrDFR*) and Production of Anthocyanin Pigment 1 gene (*BrPAP1 or MYB75*), which are activated by MBW in blue light and UV-A, declined to low levels after they reached a peak [20].

CONCLUSIONS

Light-responsed miRNAs were identified by Solexa sequencing of the sRNA library from photomorphogenic seedlings of *Brassica rapa* subsp. *rapa* cv. Tsuda exposed to blue light and UV-A. A qRT-PCR and northern blot analysis confirmed the expression profiles of a subset of conserved and novel miRNAs that were identified by high-throughput sequencing from the seedlings grown in the dark, blue light, or UV-A. Many putative targets for these miRNAs were predicted to be involved in plant growth, development and response to light. Transcripts of miR156 and miR157 were the most abundant in the sRNA library, and their respective targets SPL9 and SPL15 were shown to vary greatly in transcription level during seedlings photomorphogenesis from dark to blue-light- or UV-A-light-exposed. These results provide new insights into blue and UV-A light-responsive miRNAs in the seedlings of turnip and supply evidence for miR156/157-guided cleavage of target SPL9 (Bra004674, Bra016891) and SPL15 (Bra003305, Bra014599).

METHODS

Plant Materials and Light Treatments

Seeds of the turnip *Brassica rapa* L. subsp. *rapa* cv. Tsuda which originated from Dr. Kawabata Saneyuki's laboratory (University of Tokyo) were sown in a row on wet filter paper. The seedlings were grown in the dark at 25 °C for 3 days until they were approximately 4 cm tall. For light treatments, these dark-grown seedlings were then irradiated with blue light (470 nm light-emitting diode LED, NSPB5205, Nichia) at 10 W m^{-2} or UV-A light (UV-A fluorescent lamp, FL10BLB, Toshiba, filtered through soda-lime glass plates, peak at 350 nm) at 3 W m^{-2} for 24 h. For the dark control, seedlings were grown in the dark. After a 24-h irradiation, whole seedlings (ground parts and roots) were collected and ground in liquid nitrogen to isolate total RNA using TRNzol-A$^+$ reagent according to the manufacturer's instructions (TIANGEN, Biotech, Beijing, China). Anthocyanin was estimated as described previously [23] by calculating the ratio of OD_{530} to gram fresh mass.

Small RNA Isolation and Sequencing

A turnip seedling RNA library was constructed according to previously described methods [54]. In brief, small RNAs 18–30 nt long were separated from total RNA on a 15 % denaturing polyacrylamide gel and purified. The isolated small RNAs were then ligated to a 5′ adapter and a 3′ adapter and reverse transcribed into cDNA and subsequently amplified by PCR. After purification, small RNA libraries were sequenced directly using Solexa sequencing technology by Beijing Genomics Institute (BGI) (Shenzhen, Guangdong, China).

High-throughput Sequencing Data Analysis and miRNA Identification

Clean reads were obtained by removing all low-quality reads, adapter reads, and contaminant reads. The high-quality sRNA sequences were then used to analyze length distribution and mapped to *Brassica rapa*genome sequences (http://brassicadb.org/brad/index.php) using the program SOAP [55]. Noncoding RNAs, e.g., rRNAs, tRNAs, snRNAs (small nuclear RNA) and snoRNAs (small nucleolar RNA), in the NCBI GenBank (http://www.ncbi.nlm.nih.gov/genbank/) and Rfam (http://www.sanger.ac.uk/Software/Rfam) databases were eliminated [56, 57]. The conserved miRNAs were annotated by sequence alignment using the database miRBase20.0 (http://www.mirbase.org/index.shtml) with 0- to 3-base mismatches, with gaps counted as mismatches [58]. Potential novel miRNAs were identified using Mireap (http://sourceforge.net/projects/mireap/) and mfold (http://unafold.rna.albany.edu/?q=mfold) to predict whether they can be folded into a hairpin secondary structure [59].

Prediction of Potential Target Genes for *Brassica Rapa* miRNAs

The potential targets of conserved and novel miRNAs were predicted by the psRNA Target program (http://plantgrn.noble.org/psRNATarget/) using default parameters. *Brassica rapa*, de novo scaffolds assembly v1.1 (2011–08–30) cds, (http://brassicadb.org/brad/) was used as the genomic library for the target search. The functional annotations of these target genes were then analyzed using the Phytozome v7.0 transcript database (http://www.phytozome.net/).

Small RNA Blot Analysis

Small RNA blots were analyzed as described previously [60] by probing the target small RNAs separated in a 4 % agarose gel using a 5′-digoxigenin-labeled DNA oligonucleotide probe complementary to the miRNA sequence. U6 snRNA was detected as a control and the sequence of the probe (BrU6) was 5′TTCTCGATTTGTGCGTGTCATC CTTGCGCAGGGGCCATGCTAATCTTCTC, with a digoxigenin molecule at the 5′ end. The relative expression of the detected miRNAs was calculated as the gray level ratio of images between miRNAs and BrU6.

Reverse Transcription (RT) and Real Time PCR Analysis of miRNAs

Total RNA from turnip seedlings exposed to blue or UV-A light or kept in the dark and stem-loop primers for miRNAs and a reverse primer for the housekeeping gene were used for multiplex cDNA synthesis. [61–63]. The cDNA samples were then used for real-time PCR using miRNA-specific primers.

Quantitative real-time PCR was carried out on an ABI 7500 real-time system (Applied Biosystems, California, USA) with the POWER SYBR GREEN PCR Master Mix (Applied Biosystems). The RT-PCR assays for the miRNAs were done in triplicate for each cDNA sample. The PCR thermal cycling parameters were 95 °C for 10 min followed by 40 cycles of 95 °C for 15 s and 60 °C for 1 min. A comparative CT (ΔΔCT) method was used to calculate the relative amounts of the transcripts [64], then the RT-PCR products were separated by gel electrophoresis. Amplification of a fragment of the UBQ gene was chosen as an internal control, and the seedlings in the dark control were used as a calibrator. All experiments were done with three biological replicates and three technical replicates. The reaction specificities were tested with melt gradient-dissociation curves, then electrophoresis.

5′ RNA Ligase-mediated RACE (RLM-5 RACE) PCR

Total RNA (3 μg) from light-treated seedlings was mixed equally and purified using PolyA Tract mRNA Isolation System IV (Promega, Madison, WI, USA), according to the manufacturer's instructions. The 5′-Full RACE Kit (TaKaRa, Dalian, China) was used without the alkaline phosphatase and tobacco acid pyrophosphatase steps to detect the cleavage sites of the miRNA-targeted genes. According to the manufacturer's instructions, the 5′ RACE adapter was directly ligated to mRNA. The nested 5′ RACE Outer/Inner primer and three gene-specific nested primers were used for rapid amplification of cDNA ends. The PCR products were purified in a 2 % agarose gel, cloned into the pCRII vector (Invitrogen, Carlsbad, CA, USA), and at least 15 independent clones were sequenced.

ACKNOWLEDGEMENTS

We are grateful to Dr. Beth E. Hazen for improving the English of this manuscript.

REFERENCES

1. Bologna NG, Voinnet O. The diversity, biogenesis, and activities of endogenous silencing small RNAs in Arabidopsis. Annu Rev Plant Biol. 2014; 65: 473–503.
2. Khraiwesh B, Arif MA, Seumel GI, Ossowski S, Weigel D, Reski R, Frank W. Transcriptional control of gene expression by microRNAs. Cell. 2010; 140(1): 111–22.
3. Voinnet O. Origin, biogenesis, and activity of plant microRNAs. Cell. 2009; 136(4): 669–87.
4. Li YF, Zheng Y, Addo-Quaye C, Zhang L, Saini A, Jagadeeswaran G, Axtell MJ, Zhang W, Sunkar R. Transcriptome-wide identification of microRNA targets in rice. Plant J. 2010; 62(5): 742–59.
5. Lee RC, Feinbaum RL, Ambros V. The C. elegans heterochronic gene lin-4 encodes small RNAs with antisense complementarity to lin-14. Cell. 1993; 75(5): 843–54
6. Zhang BH, Pan XP, Cobb GP, Anderson TA. Plant microRNA: A small regulatory molecule with big impact. Dev Biol. 2006; 289(1): 3–16.
7. Ambros V. The functions of animal microRNAs. Nature. 2004; 431(7006): 350–5.
8. Lee HC, Li L, Gu W, Xue Z, Crosthwaite SK, Pertsemlidis A, Lewis ZA, Freitag M, Selker EU, Mello CC, et al. Diverse pathways generate microRNA-like RNAs and Dicer-independent small interfering RNAs in fungi. Mol Cell. 2010; 38(6): 803–14.
9. Cullen BR. Viruses and microRNAs: RISCy interactions with serious consequences. Genes Dev. 2011; 25(18): 1881–94.
10. Jones-Rhoades MW, Bartel DP, Bartel B. MicroRNAS and their regulatory roles in plants. Annu Rev Plant Biol. 2006; 57: 19–53.
11. Zhang B, Wang Q, Pan X. MicroRNAs and their regulatory roles in animals and plants. J Cell Physiol. 2007; 210(2): 279–89.
12. Sun G. MicroRNAs and their diverse functions in plants. Plant Mol Biol. 2012; 80(1): 17–36.
13. Briggs WR, Christie JM. Phototropins 1 and 2: versatile plant blue-light receptors. Trends Plant Sci. 2002; 7(5): 204–10.
14. Jenkins GI, Christie JM, Fuglevand G, Long JC, Jackson JA. Plant responses to UV and blue light: biochemical and genetic approaches. Plant Sci. 1995; 112(2): 117–38.
15. Hayes S, Velanis CN, Jenkins GI, Franklin KA. UV-B detected by the UVR8 photoreceptor antagonizes auxin signaling and plant shade avoidance. Proc Natl Acad Sci U S A. 2014; 111(32): 11894–9.
16. Jenkins GI. The UV-B photoreceptor UVR8: from structure to physiology. Plant Cell. 2014; 26(1): 21–37.
17. Cashmore AR, Jarillo JA, Wu YJ, Liu D. Cryptochromes: blue light receptors for plants and animals. Science. 1999; 284(5415): 760–5.
18. Lin C, Shalitin D. Cryptochrome structure and signal transduction. Annu Rev Plant Biol. 2003; 54: 469–96.
19. Lin C. Blue light receptors and signal transduction. Plant Cell. 2002; 14(Suppl): S207–225.
20. Wang Y, Zhou B, Sun M, Li Y, Kawabata S. UV-A light induces anthocyanin biosynthesis in a manner distinct from synergistic blue+UV-B light and UV-A/blue light responses in different parts of the hypocotyls in turnip seedlings. Plant Cell Physiol. 2012; 53(8): 1470–80.
21. Casati P: Analysis of UV-B regulated miRNAs and their targets in maize leaves. Plant Signal Behav 2013, 8(10): doi: 10 4161/psb 26758
22. Jung JH, Seo YH, Seo PJ, Reyes JL, Yun J, Chua NH, Park CM. The GIGANTEA-regulated microRNA172 mediates photoperiodic flowering independent of CONSTANS in Arabidopsis. Plant Cell. 2007; 19(9): 2736–48.
23. Zhou B, Li Y, Xu Z, Yan H, Homma S, Kawabata S. Ultraviolet A-specific induction of anthocyanin biosynthesis in the swollen hypocotyls of turnip (Brassica rapa). J Exp Bot. 2007; 58(7): 1771–81.
24. Baucher M, Moussawi J, Vandeputte OM, Monteyne D, Mol A, Perez-Morga D, El Jaziri M. A role for the

miR396/GRF network in specification of organ type during flower development, as supported by ectopic expression of Populus trichocarpa miR396c in transgenic tobacco. Plant Biol (Stuttg). 2013; 15(5): 892–8.

25. Casadevall R, Rodriguez RE, Debernardi JM, Palatnik JF, Casati P. Repression of growth regulating factors by the microRNA396 inhibits cell proliferation by UV-B radiation in Arabidopsis leaves. Plant Cell. 2013; 25(9): 3570–83.
26. Zhang H, Zhao X, Li J, Cai H, Deng XW, Li L. MicroRNA408 is critical for the HY5-SPL7 gene network that mediates the coordinated response to light and copper. Plant Cell. 2014; 26(12): 4933–53.
27. Kim B, Yu HJ, Park SG, Shin JY, Oh M, Kim N, Mun JH. Identification and profiling of novel microRNAs in the Brassica rapa genome based on small RNA deep sequencing. BMC Plant Biol. 2012; 12: 218.
28. Yu X, Wang H, Lu Y, de Ruiter M, Cariaso M, Prins M, van Tunen A, He Y. Identification of conserved and novel microRNAs that are responsive to heat stress in Brassica rapa. J Exp Bot. 2012; 63(2): 1025–38.
29. Zhou B, Fan P, Li Y. High-throughput sequence analysis of small RNAs in skotomorphogenic seedlings of Brassica rapa ssp. rapa. Gene. 2014; 548(1): 68–74.
30. Zhou B, Wang Y, Zhan Y, Li Y, Kawabata S. Chalcone synthase family genes have redundant roles in anthocyanin biosynthesis and in response to blue/UV-A light in turnip (Brassica rapa; Brassicaceae). Am J Bot. 2013; 100(12): 2458–67.
31. Meyers BC, Axtell MJ, Bartel B, Bartel DP, Baulcombe D, Bowman JL, Cao X, Carrington JC, Chen X, Green PJ, et al. Criteria for annotation of plant MicroRNAs. Plant Cell. 2008; 20(12): 3186–90.
32. Zhang H, He H, Wang X, Yang X, Li L, Deng XW. Genome-wide mapping of the HY5-mediated gene networks in Arabidopsis that involve both transcriptional and post-transcriptional regulation. Plant J. 2011; 65(3): 346–58.
33. Wang JW. Regulation of flowering time by the miR156-mediated age pathway. J Exp Bot. 2014; 65(17): 4723–30.
34. Zhou X, Wang G, Zhang W. UV-B responsive microRNA genes in Arabidopsis thaliana. Mol Syst Biol. 2007; 3: 103.
35. Montgomery BL. Sensing the light: photoreceptive systems and signal transduction in cyanobacteria. Mol Microbiol. 2007; 64(1): 16–27.
36. Wei N, Kwok SF, von Arnim AG, Lee A, McNellis TW, Piekos B, Deng XW. Arabidopsis COP8, COP10, and COP11 genes are involved in repression of photomorphogenic development in darkness. Plant Cell. 1994; 6(5): 629–43.
37. Nakamura M, Satoh T, Tanaka S, Mochizuki N, Yokota T, Nagatani A. Activation of the cytochrome P450 gene, CYP72C1, reduces the levels of active brassinosteroids in vivo. J Exp Bot. 2005; 56(413): 833–40.
38. Zhang H, Yin W, Xia X. Shaker-like potassium channels in Populus, regulated by the CBL-CIPK signal transduction pathway, increase tolerance to low-K+ stress. Plant Cell Rep. 2010; 29(9): 1007–12.
39. Yang L, Xu M, Koo Y, He J, Poethig RS. Sugar promotes vegetative phase change in Arabidopsis thaliana by repressing the expression of MIR156A and MIR156C. Elife. 2013; 2: e00260.
40. Yu S, Cao L, Zhou CM, Zhang TQ, Lian H, Sun Y, Wu J, Huang J, Wang G, Wang JW. Sugar is an endogenous cue for juvenile-to-adult phase transition in plants. Elife. 2013; 2: e00269.
41. Wang H. The miR156/SPL Module, a Regulatory Hub and Versatile Toolbox, Gears up Crops for Enhanced Agronomic Traits. Mol Plant. 2015; 8(5): 677–88.
42. Chen X, Zhang Z, Liu D, Zhang K, Li A, Mao L. SQUAMOSA promoter-binding protein-like transcription factors: star players for plant growth and development. J Integr Plant Biol. 2010; 52(11): 946–51.
43. Xie K, Shen J, Hou X, Yao J, Li X, Xiao J, Xiong L. Gradual increase of miR156 regulates temporal expression changes of numerous genes during leaf development in rice. Plant Physiol. 2012; 158(3): 1382–94.
44. Wang JW, Czech B, Weigel D. miR156-regulated SPL transcription factors define an endogenous flowering pathway in Arabidopsis thaliana. Cell. 2009; 138(4): 738–49.
45. Wu G, Park MY, Conway SR, Wang JW, Weigel D, Poethig RS. The sequential action of miR156 and miR172 regulates developmental timing in Arabidopsis. Cell. 2009; 138(4): 750–9.
46. Guo AY, Zhu QH, Gu X, Ge S, Yang J, Luo J. Genome-wide identification and evolutionary analysis of the plant specific SBP-box transcription factor family. Gene. 2008; 418(1–2): 1–8.
47. Schwarz S, Grande AV, Bujdoso N, Saedler H, Huijser P. The microRNA regulated SBP-box genes SPL9 and SPL15 control shoot maturation in Arabidopsis. Plant Mol Biol. 2008; 67(1–2): 183–95.
48. Fu C, Sunkar R, Zhou C, Shen H, Zhang JY, Matts J, Wolf J, Mann DG, Stewart Jr CN, Tang Y, et al. Overexpression of miR156 in switchgrass (Panicum virgatum L.) results in various morphological alterations and leads to improved biomass production. Plant Biotechnol J. 2012; 10(4): 443–52.
49. Wei S, Gruber MY, Yu B, Gao MJ, Khachatourians GG, Hegedus DD, Parkin IA, Hannoufa A. Arabidopsis mutant sk156 reveals complex regulation of SPL15 in a miR156-controlled gene network. BMC Plant Biol. 2012; 12: 169.
50. Wei S, Yu B, Gruber MY, Khachatourians GG, Hegedus DD, Hannoufa A. Enhanced seed carotenoid levels and branching in transgenic Brassica napus expressing the Arabidopsis miR156b gene. J Agric Food Chem. 2010; 58(17): 9572–8.
51. Gou JY, Felippes FF, Liu CJ, Weigel D, Wang JW. Negative regulation of anthocyanin biosynthesis in Arabidopsis by a miR156-targeted SPL transcription factor. Plant Cell. 2011; 23(4): 1512–22.

52. Petroni K, Tonelli C. Recent advances on the regulation of anthocyanin synthesis in reproductive organs. Plant Sci. 2011; 181(3): 219–29.

53. Guo N, Cheng F, Wu J, Liu B, Zheng S, Liang J, Wang X. Anthocyanin biosynthetic genes in Brassica rapa. BMC Genomics. 2014; 15: 426.

54. Song C, Wang C, Zhang C, Korir NK, Yu H, Ma Z, Fang J. Deep sequencing discovery of novel and conserved microRNAs in trifoliate orange (Citrus trifoliata). BMC Genomics. 2010; 11: 431.

55. Li R, Li Y, Kristiansen K, Wang J. SOAP: short oligonucleotide alignment program. Bioinformatics. 2008; 24(5): 713–4.

56. Altschul SF, Gish W, Miller W, Myers EW, Lipman DJ. Basic local alignment search tool. J Mol Biol. 1990; 215(3): 403–10.

57. Griffiths-Jones S. miRBase: the microRNA sequence database. Methods Mol Biol. 2006; 342: 129–38. Griffiths-Jones S, Saini HK, van Dongen S, Enright AJ. miRBase: tools for microRNA genomics. Nucleic Acids Res. 2008; 36(Database issue): D154–158.

58. Zuker M. Mfold web server for nucleic acid folding and hybridization prediction. Nucleic Acids Res. 2003; 31(13): 3406–15.

59. Fan P, ZHOU B, Li Y. A Safer Method for Northern Blot Analysis of MicroRNAs in Turnip *Brassica rapa*'Tsuda'. Fenzi Zhiwu Yuzhong (online) (Molecular Plant Breeding). 2013; 11(28): 1206–10.

60. Chen C, Ridzon DA, Broomer AJ, Zhou Z, Lee DH, Nguyen JT, Barbisin M, Xu NL, Mahuvakar VR, Andersen MR, et al. Real-time quantification of microRNAs by stem-loop RT-PCR. Nucleic Acids Res. 2005; 33(20): e179.

61. Gao ZH, Wei JH, Yang Y, Zhang Z, Xiong HY, Zhao WT. Identification of conserved and novel microRNAs in Aquilaria sinensis based on small RNA sequencing and transcriptome sequence data. Gene. 2012; 505(1): 167–75.

62. Lao K, Xu NL, Yeung V, Chen C, Livak KJ, Straus NA. Multiplexing RT-PCR for the detection of multiple miRNA species in small samples. Biochem Biophys Res Commun. 2006; 343(1): 85–9. Livak KJ, Schmittgen TD. Analysis of relative gene expression data using real-time quantitative PCR and the 2(−Delta Delta C(T)) Method. Methods. 2001; 25(4): 402–8.

CHAPTER 14

DIFFERENTIAL GENE EXPRESSION ANALYSIS OF 'GRANNY SMITH' APPLE (MALUS DOMESTICA BORKH.) DURING FRUIT SKIN COLORATION

X.J. Zhang[a], L.X. Wang[a], Y.L. Liu[a], X.X. Chen[b], Y.Z. Yang[a], and Z.Y. Zhao[a,c]

[a]College of Horticulture, Northwest A&F University, Yangling, Shaanxi 712100, PR China

[b]College of Natural Resources and Environment, Northwest A&F University, Yangling, Shaanxi 712100, PR China

[c]Apple E&T Research Centre of Shaanxi Province, Yangling, Shaanxi 712100, PR China

ABSTRACT

'Granny Smith' apples growing under normal sunlight develop green skin, whereas the peel turns red due to anthocyanin accumulation after the removal of a bagging treatment. Two anthocyanins, Cyanidin 3-O-galactoside (cy3-gal) and Cyanidin 3-O-arabinoside (cy3-ara), were detected in the red 'Granny Smith' apple peels, and cy3-gal was determined to be chiefly responsible for the red color. The content of cy3-gal was more than 98% of the total anthocyanin in the red 'Granny Smith' peels. To better understand the molecular basis of anthocyanin biosynthesis in 'Granny Smith' apples, we performed a quantitative real-time PCR (qRT-PCR) analysis of anthocyanin biosynthetic genes (MdCHS, MdF3H, MdDFR, MdANS, MdUFGT, and MdMYB1). Our results indicate that the expression of these genes (except MdCHS) was associated with increased anthocyanin accumulation in the skin of 'Granny Smith' apples. Four selected genes obtained from the 'Granny Smith' skin cDNA library, phytoene synthase (PSY), WD40 repeat protein, polygalacturonase (PG), and galactosidase (GAL), were also confirmed by qRT-PCR. We found that these genes were differently expressed during 'Granny Smith' apple skin coloration, suggesting that they are directly or indirectly involved in pigment accumulation. In conclusion, anthocyanin biosynthesis in 'Granny Smith' apples is the result of interactions between multiple enzymes in the anthocyanin biosynthesis pathway, and the coloring mechanism of 'Granny Smith' apples may be similar to that of red-skinned cultivars.

INTRODUCTION

Apple (*Malus domestica* Borkh.) is one of the most important fruit crops in temperate regions of the world. Apple color is an important factor when considering consumer appeal, with red-colored fruits being more popular (King and Cliff, 2002). Apple skin color is thought to be determined by the interaction of anthocyanin molecules, a class of flavonoids that is responsible for the red skin color (Lancaster and Dougall, 1992), with other compounds, such as chlorophyll and carotenoids. There are more anthocyanins in red-skinned apple cultivars than in non-red-skinned cultivars (Honda et al., 2002).

In recent years, significant progress has been made in understanding the genetic regulation of anthocyanin biosynthesis in apples. The genes encoding anthocyanin biosynthesis enzymes have been isolated and studied (Honda et al., 2002 and Kim et al., 2003). The transcriptional levels of these genes are positively correlated with anthocyanin accumulation (Honda et al., 2002 and Ben-Yehudah et al., 2005), and the MYB-bHLH-WD40 regulatory protein complex controls their expression (Koes et al., 2005 and Allan et al., 2008). In

Citation: X.J. Zhang, L.X. Wang, Y.L. Liu, X.X. Chen, Y.Z. Yang, Z.Y. Zhao, Differential gene expression analysis of 'Granny Smith' apple (Malus domestica Borkh.) during fruit skin coloration, South African Journal of Botany, Volume 88, September 2013, Pages 125-131, ISSN 0254-6299, http://dx.doi.org/10.1016/j.sajb.2013.07.009.

apples, MYB transcription factors (*MdMYB1*, *MdMYBA*, and *MdMYB10*) and bHLH transcription factors (*MdbHLH3* and *MdbHLH33*) have been isolated and characterized (Takos et al., 2006, Ban et al., 2007a and Espley et al., 2007). In addition to genetic factors, external factors, such as light, temperature, mineral nutrition, and orchard management practices, also affect apple anthocyanin biosynthesis, with light being the most important and essential factor (Saure, 1990), and the transcription of *MdMYB1* and several anthocyanin biosynthetic genes is regulated by light (Takos et al., 2006, Ban et al., 2007a, Ju et al., 1997 and Ubi et al., 2006). Recent studies have proven that bagging treatment is an effective practice to study the effect of light on fruit anthocyanin synthesis (Ju, 1998, Zhang et al., 2011 and Wei et al., 2011). It is also a useful technique for studying anthocyanin biosynthesis and gene expression in apples (Ju, 1998, Takos et al., 2006 and Ban et al., 2007a).

'Granny Smith' apples display a green background with a pink blush in exposed areas (Reay, 1999). We found that, during the fruit ripening phase, bagged fruits that are re-exposed to sunlight accumulate significant levels of anthocyanin in the skin (Fig. 1). Previous research commonly used red-skinned apple cultivars as experimental materials to elucidate the molecular regulation of anthocyanin biosynthesis (Takos et al., 2006 and Xu et al., 2012). However, informations regarding the genetic regulation of anthocyanin biosynthesis for non-red-skinned cultivars are rare. The present study was aimed to understand the molecular mechanisms of color variability in red 'Granny Smith' apple skin, by comparison with the non-bagged fruits in terms of anthocyanin content and transcription of anthocyanin biosynthetic genes. In addition, in order to identify potential color-related genes of 'Granny Smith', a light-specific subtractive cDNA library was constructed using SSH technology (Diatchenko et al., 1996), which has been widely used to identify differentially expressed genes (Ban et al., 2007b, Han et al., 2011, Zhang et al., 2011 and Zhou et al., 2012). Moreover, this cDNA library would provide a foundation for future cloning and functional analysis of genes related to 'Granny Smith' coloration.

Figure 1. The 'Granny Smith' apple. a. Control fruit (175 DAFB); b. bagged fruit (10 DABR).

MATERIALS AND METHODS

Plant Materials

The fruits used in this experiment were obtained from the orchard of the Northwest A&F University apple experimental station in the Baishui, Shaanxi Province in 2010. For the bagging treatment, each fruit was bagged with two-layer paper bags (Hong Tai, China) on May 25th, approximately 30 days after full bloom (DAFB). The bags were 13 × 16 cm, and the inner layer was made of red paper with a wax coating, while the outside layer was taupe. The bags were removed at 165 DAFB. The fruits in the bagged group were collected at 0, 2, 4, 6, 8, 10, and 15 days after the bag removal (DABR). The control group fruits were grown under normal sunlight conditions, and they were collected at 165, 167, 169, 171, 173, 175, and 180 DAFB. At each point in time, 10 individual fruits were collected from different locations in the tree canopy. The peels, including 1 mm of cortical tissue, were carefully peeled off with a knife, frozen immediately in liquid nitrogen and stored at − 80 °C until use.

Apple Skin SSH Library Construction

Total RNA from the skin tissues of 'Granny Smith' apple was isolated using a hot borate method (Yao et al., 2005). Isolation of mRNA was performed by Oligotex® mRNA Mini Kit (QIAGEN GmbH, Germany), and cDNA was synthesized by the SMART cDNA synthesis kit (Clontech, Palo Alto, CA, USA). The SSH library of the 'Granny Smith' fruit peel was constructed using the PCR-Select™ Subtractive Hybridization kit (Clontech, Palo Alto, CA, USA) according to the manufacturer's instructions, with the cDNA of the peel exposed to sunlight for 0 day as the driver and the mixed peel cDNA (2, 4, 6, 8 and 10 DABR) as the tester. The second PCR products, after subtraction, were recycled and linked into the pGEM-T-Easy Vector (Promega, USA) and then were transformed into*Escherichia coli* strain TOP10 (Tiangen Biotech (Beijing) Co., LTD) and screened on the LB solid medium plate containing IPTG + X-gal + Amp at 37 °C overnight. Subsequently, recombination rate was tested by blue–white screening.

Sequencing and BLAST Analysis

White colonies were isolated on selective media, Luria-Bertani ampicillin (LB-Amp) solid medium, and these clones were grown overnight in 100 μL LB-Amp medium at 37 °C, respectively. Two μL of bacterium culture for each colony was used to amplify the inserts with SP6 and T7 primers. The cDNA fragments were sequenced by Invitrogen Trading Co. (Shanghai). All the inserted sequences were queried for similarity in the NCBI database using the BLASTX sequence comparison software at the website (http://www.ncbi.nlm.nih.gov/BLAST).

Anthocyanin Extraction and Measurement

For the high-performance liquid chromatography (HPLC) analysis, the anthocyanins were extracted from 1 g of finely ground plant material in 1 mL 1% (v/v) HCl-methanol for 24 h at 4 °C on a rotating wheel in the dark. The samples were clarified by centrifugation at 13,000 ×g for 15 min at 4 °C, and then 1.5 mL of supernatant was transferred to an autosampler vial. The analysis was conducted using an HPLC PDA (Waters, USA), which was equipped with a model 1525 binary solvent delivery system (Waters, USA), an on-line degasser and a 2707 autosampler (Waters, USA). The data acquisition and processing were performed with Breeze software. For all the samples, the anthocyanin separations were performed using a 250 × 4.6 mm i.d., 5 μm C18 column (Diamonsil, China). Solvent A was 10% (v/v) formic acid, and solvent B was methanol. The gradient of solvent B was as follows: 0 min, 17%; 1 min, 17%; 10 min, 35%; 20 min, 37%; and 25 min, 100%. The gradient was run at a flow rate of 1 mL · min^{-1} at a column temperature of 40 °C, and a 5 μL sample was injected. Absorbance was measured at 530 nm. The standards were Cyanidin 3-O-galactoside, Cyanidin 3-O-glucoside (Sigma, USA), Cyanidin 3-O-arabinoside, and Cyanidin 3-O-rutinoside (Polyphenols Laboratories, Hanaveien). Anthocyanins were expressed as mg/100 g fresh wight (FW).

Analysis of Gene Expression Profile by qRT-PCR

Isolation of total RNA from the pericarp of 'Granny Smith' apples and first-strand cDNA synthesis were performed as described above. The transcript levels of *MdCHS*, *MdF3H*,*MdDFR*, *MdANS*, *MdUFGT*, *MdMYB1*, and four selected genes were analyzed by qRT-PCR using SYBR green master mix (Takara, Japan) and an IQ5 real-time PCR cycler (Bio-Rad Laboratories, USA) in a reaction volume of 25 μL. All the gene-specific primers were designed using the Primer Premier 5.0 program (Supplementary Table 1). According to the manufacturer›s instructions, the reaction mixtures were heated to 95 °C for 30 s, followed by 40 cycles at 95 °C for 10 s, 60 °C for 30 s and 72 °C for 30 s, followed by 72 °C for 7 min. A melting curve was generated for each sample at the end of each run to ensure the purity of the amplified products. The raw data were analyzed with Bio-Rad IQ5 software.

Statistical Analysis

Statistical analysis was conducted with the SPSS 17.0. ANOVA was used for analysis of anthocyanin content and gene expression levels. The significance of the differences between means was estimated by Duncan test ($P \leq 0.05$). Experimental data were the average of three replications ± s.e.

RESULTS

Construction of the Subtracted cDNA Library

A subtracted cDNA library was constructed to identify the molecular determinants of 'Granny Smith' apple skin color. A total of 1000 randomly chosen white transformants were selected, and more than 90% of the colonies produced a single band. The insert sizes ranged from 200 to 800 bp.

One hundred and sixteen high-quality ESTs were obtained after removing the low-quality and repetitive sequences. Of these genes, 73 had high homology to genes of known function (Supplementary Table 2). The annotation of the genes obtained from the 'Granny Smith' apple skin SSH library was predicted by BLAST, and the genes were sorted into functional categories based on BLAST analysis. The largest category was metabolism-related genes, with a total of 36 ESTs. This group included genes with homology to phytoene synthase, cysteine protease inhibitor, isoflavone reductase, pyruvate decarboxylase, sarcosine oxidase, etc. The second largest category was stress/defense related genes, totaling 13 ESTs. Five transcription factors were also present in this cDNA library, including BZIP domain class transcription factor, WD-repeat protein, C3HL domain class transcription factor, Ttg1-like protein, and MYB domain class transcription factor. The functional descriptions of the other three gene categories are listed in Supplementary Table 2. The remaining 43 ESTs had no match in GenBank, suggesting that they represent unique 'Granny Smith' genes or are from nonconserved regions of known genes. Putative functions of all ESTs were annotated based on the highest scoring matches (e-values lower than $10e^{-4}$) by BLAST in NCBI.

Changes of Anthocyanin Content during Fruit Coloration

The anthocyanin composition and concentration of 'Granny Smith' apple peels were assessed by HPLC. The control group fruits have green skin during fruit maturation, and anthocyanins were barely detected. Two anthocyanins, cy3-gal and cy3-ara, were detected in the red 'Granny Smith' apple peels in our study (Fig. 2). Cy3-gal is chiefly responsible for the red color, ranging from 0 to 30.92 mg/100 g FW during 'Granny Smith' coloration (Fig. 3b). Cy3-ara was present in smaller amounts, ranging from 0 to 0.43 mg/100 g FW (Fig. 3a). The content of cy3-gal was greater than 98% of the total anthocyanin content of the red 'Granny Smith' peels. There was very little anthocyanin biosynthesis in the skin tissues of bagged fruits at 0 DABR, but it increased rapidly at 4 DABR, then accumulated progressively.

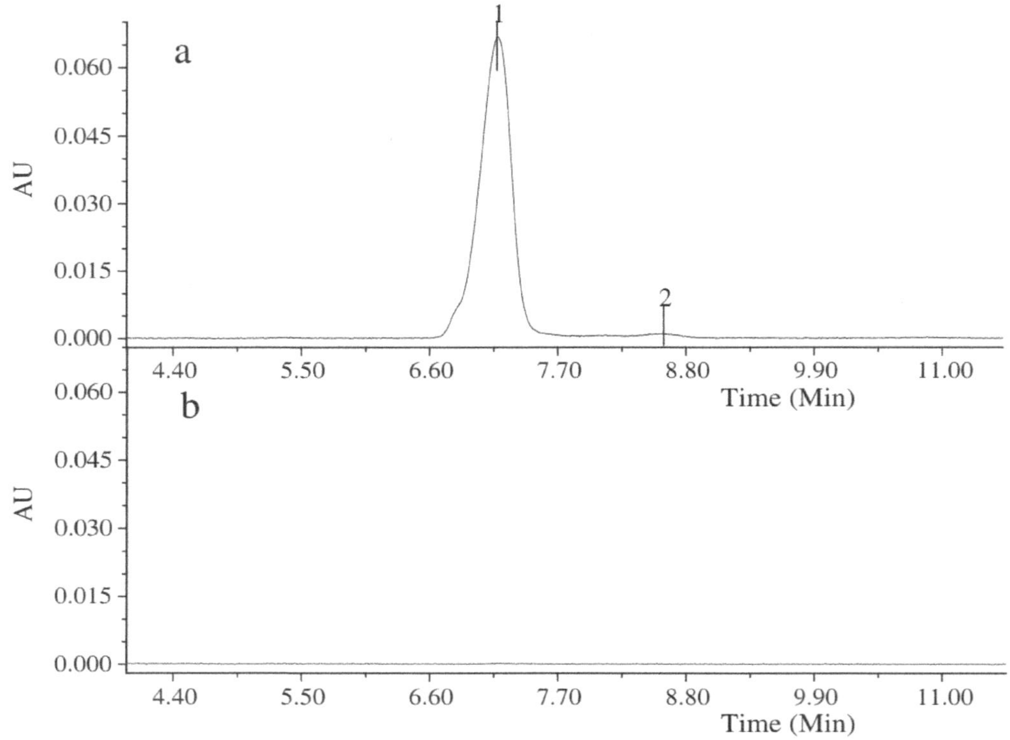

Figure 2. HPLC separation of anthocyanins in peels of 'Granny Smith' apple. a. Bagged fruit (10 DABR); b. Control fruit (175 DAFB). Peak 1: Cyanidin 3-O-galactoside; Peak 2: Cyanidin 3-O-arabinoside.

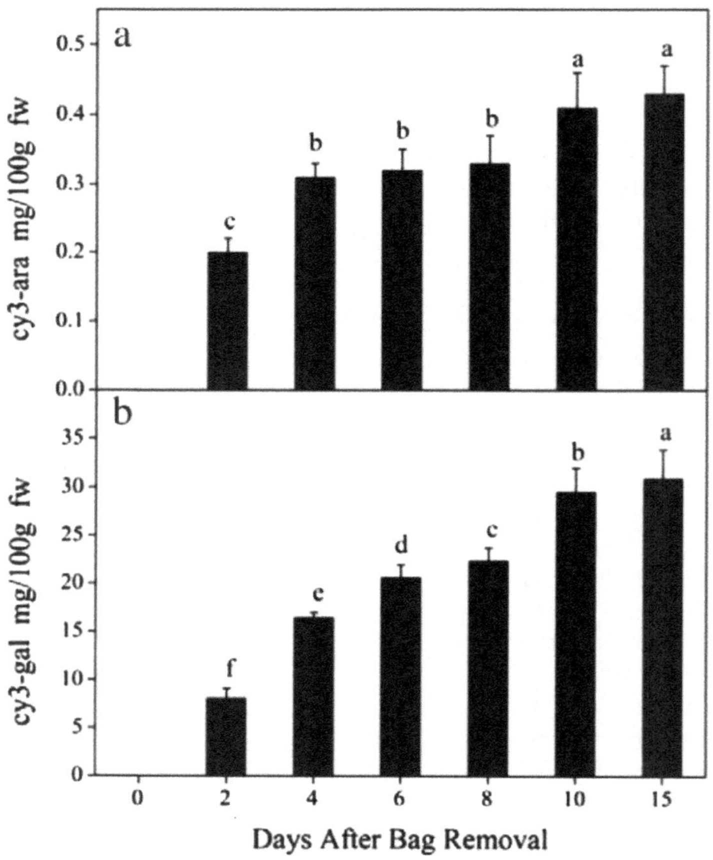

Figure 3. Contents of anthocyanins in red 'Granny Smith' apple peel after bag removal. a. Cyanidin 3-O-arabinoside; b. Cyanidin 3-O-galactoside. Data are means ± s.e. of three replications. Different letters above the bars indicate significant ($P < 0.05$) differences according to the Duncan test.

qRT-PCR for Anthocyanin Biosynthetic Genes

qRT-PCR was used to measure the relative expression of anthocyanin related genes in 'Granny Smith' apples. In the bagged group, the transcript levels of all six genes were very low at 0 DABR, and they increased rapidly in the skin tissues exposed to sunlight (Fig. 4). *MdF3H* transcripts rapidly increased, reaching a maximum level at 2 DABR, then decreasing to a lower level at 10 DABR, and then increased again. *MdCHS* and *MdDFR*transcripts increased, reaching a maximum level of expression at 4 DABR, then gradually decreased. The transcripts of the downstream genes (*MdANS* and *MdUFGT*) in the anthocyanin pathway and the regulation the *MdMYB1* gene reached maximum expression levels at 6, 2, and 2 DABR, respectively (Fig. 4). In the control group, the expressions levels of all six genes (expect *MdCHS*) were significantly lower than those in the bagged group. Among these genes, *MdCHS* and *MdUFGT* transcripts peaked at 169 DAFB, whereas the others showed no major fluctuations during fruit coloration.

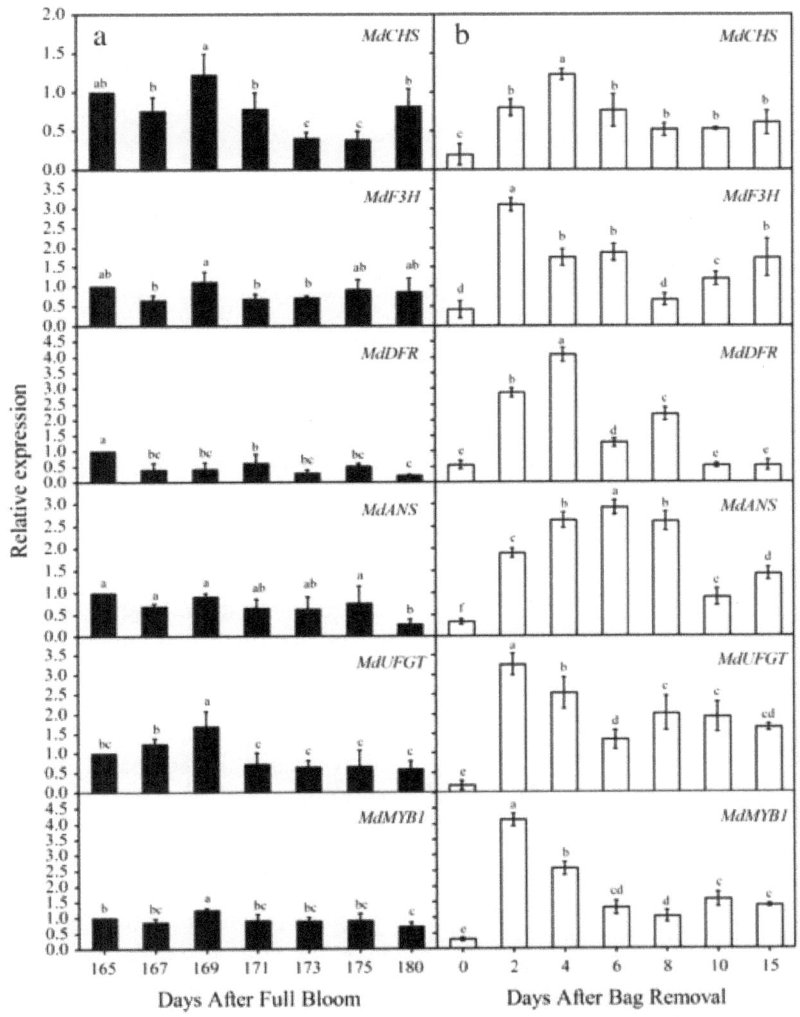

Figure 4. Expression profiling of 'Granny Smith' anthocyanin biosynthesis genes during fruit color change. The genes amplified included: *MdCHS* (AB074485), *MdF3H* (AB074486), *MdDFR* (AF117268), *MdANS* (AB074487), *MdUFGT* (AF117267) and *MdMYB1* (DQ886414). Each column height indicates the relative mRNA abundance, relative to 165 DAFB which was set to 1. All qRT-PCR reactions were normalized using the Ct value corresponding to *MdActin* (AB638619). a. Control group; b. Bagged group. Data are means ± s.e. of three replications. Different letters above the bars indicate significant ($P < 0.05$) differences according to the Duncan test.

qRT-PCR for Selected Genes

Four selected genes were detected using gene-specific primers. All of these genes were differentially expressed in 'Granny Smith' apple skin during fruit coloration (Fig. 5). In the bagged group, transcript levels of WD40 gene were up-regulated and reached the peak at 4 DABR, then gradually decreased, similar to those in the control group. Expression of PSY gene was up-regulated from 0 DABR, and reached a peak at 10 DABR in bagged fruits, however, it was down-regulated in the control group. The GAL gene transcript level rapidly increased, reaching a maximum level of expression at 4 DABR, then decreased in the bagged group. However, in the control group, GAL levels appeared to fluctuate, as the levels were high at 169 DAFB, decreased dramatically at 171 DAFB, and then increased steadily up to 180 DAFB. The transcript levels of the PG gene increased steadily in the control groups up to 169 DAFB, then rapidly decreased. In the bagged groups, PG expression followed a similar pattern to that of the GAL gene in the control groups (Fig. 5).

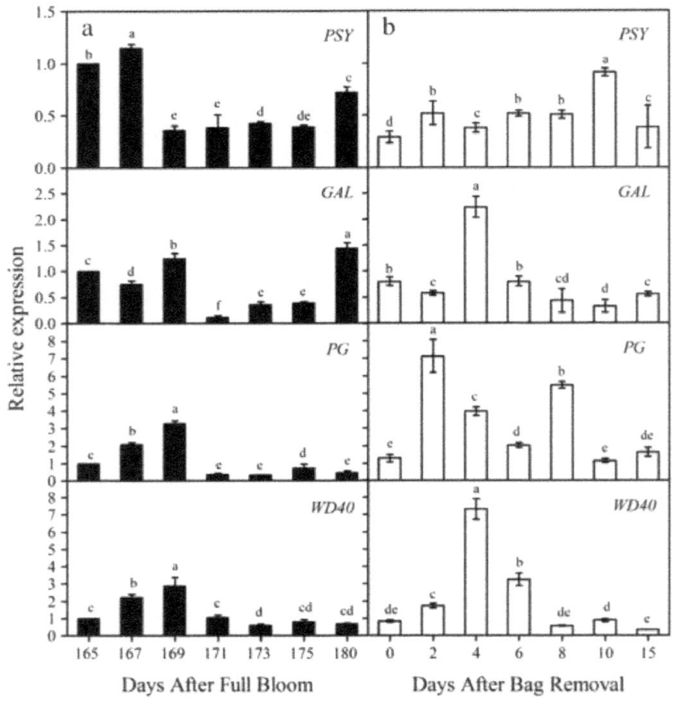

Figure 5. Expression profiling of four selected genes from the SSH library of 'Granny Smith'. The genes amplified included: *PSY* (JK750344); *GAL* (JK973900); *PG* (JK750375) and *WD40* (JK974070). Each column height indicates the relative mRNA abundance, relative to 165 DAFB which was set to 1. All qRT-PCR reactions were normalized using the Ct value corresponding to *MdActin* (AB638619). a. Control group; b. Bagged group. Data are means ± s.e. of three replications. Different letters above the bars indicate significant ($P < 0.05$) differences according to the Duncan test.

DISCUSSION

To study the bagging effect on anthocyanin accumulation in 'Granny Smith' apple skin, the anthocyanin content and the expression level of its regulatory genes were compared between bagged and non-bagged fruit. Previous research has shown that the HPLC chromatograms of red-skinned apple extracts revealed three anthocyanins (cy3-gal, cy3-glu, and cy3-ara), and cy3-gal is the most abundant anthocyanin present in red cultivars (Kondo et al., 2002 and Liu et al., 2013). In the present work, only two anthocyanins, cy3-gal and cy3-ara, were detected in red 'Granny Smith' apple peels (Fig. 2), whereas anthocyanins were barely detected in non-bagged fruit. These results indicate that the bagged fruits had a much greater potential to accumulate anthocyanin than the control group. Correspondingly, the transcription of the six biosynthetic genes significantly increased in the bagged fruit. Moreover, the differences in four late biosynthetic genes were more obvious than the upstream gene, *MdCHS*, which had a similar expression level in both the bagged group and the control group (Fig. 4). Among these genes,*MdUFGT* is the most important gene for anthocyanin biosynthesis in apples (Ju et al., 1997). Our results showed that the transcript levels of *MdUFGT* in red skin, bagged 'Granny Smith' fruits were higher than those of the control fruits, and our previous study revealed that only UFGT activity significantly correlated with anthocyanin accumulation in 'Granny Smith' apples (Liu et al., 2013), suggesting that UFGT may be an important factor in anthocyanin biosynthesis in 'Granny Smith' apples.

It is well known that anthocyanin biosynthetic genes were regulated by MYB-bHLH-WD40 protein complex, which was influenced by external environment factors (Koes et al., 2005 and Allan et al., 2008), and light is known to be the most important external factor regulating anthocyanin synthesis (Lancaster and Dougall, 1992 and Saure, 1990). In our research, the transcription of *MdMYB1* and biosynthetic genes was low in dark-grown fruits but increased dramatically after bag removal (Fig. 4). Moreover, anthocyanin content and the transcript levels of *MdMYB1* and four late genes in the anthocyanin pathway were dramatically elevated in the re-exposed fruits compared to the control group. Here, the transcription of *MdMYB1* was positively correlated with the transcript levels of biosynthetic genes. Similar results were reported by Takos et al. (2006) and Xu et al. (2012). Taken together, our results indicate that anthocyanin accumulation in red peels of 'Granny Smith' apples is the result of interactions between multiple key enzymes in the anthocyanin pathway, similar to red-skinned cultivars.

We also examined the genotype of 'Granny Smith' apples using two alleles markers (Cheng et al., 1996 and Takos et al., 2006), and our results showed that 'Granny Smith' apple has the same genotype as some red-skinned cultivars (date not shown). This finding supported that green-skinned cultivars may contain the 'red' gene (Ju et al., 1997), and the expression of anthocyanin biosynthetic genes in green-skinned apples might be blocked in some way (Kim et al., 2003), which can be reversed by bagging treatment. Moreover, other researchers have suggested that there may be a delayed ripening response that prevents pigment synthesis and accumulation in green-skinned apples (Cheng et al., 1996). All of these indicated that anthocyanin accumulation in 'Granny Smith' apple is not an accidental phenomenon, and the coloring mechanism of 'Granny Smith' apples may be either similar or identical to that of red-skinned cultivars.

To better understand the effect of bagging treatment on 'Granny Smith' apples anthocyanin biosynthesis, a subtracted cDNA library was constructed to identify the molecular determinants of 'Granny Smith' apple skin color. The transcription of four selected genes (PSY, WD40, PG, and GAL), which were involved in apple coloration or softening, was determined by qRT-PCR. The results showed that these selected genes were differentially expressed in 'Granny Smith' skin during fruit coloration (Fig. 5), suggesting that they are directly or indirectly involved in pigment accumulation.

Carotenoid color ranges from yellow to red to orange and plays an important role in many fruits coloration (Bartley and Scolnik, 1995). PSY is a key regulatory gene in the carotenoid biosynthesis pathway; accordingly, this gene has been a potential target of genetic manipulation in plants to enhance carotenoid levels (Busch et al., 2002 and Kato et al., 2004). PSY genes have been isolated and analyzed from numerous higher plants (Busch et al., 2002, Kato et al., 2004 and Qin et al., 2011), but little is known about their functions and potentials to improve carotenoid levels in rosaceous plants, such as apple, pear, and peach. In this study, PSY gene was up-regulated in bagged fruits (Fig. 5) during fruit coloration, suggesting that the PSY gene plays an important role in the regulation of apple skin color.

Another color-related gene, WD40 repeat protein, combines with MYB and bHLH to regulate anthocyanin biosynthesis (Morita et al., 2006 and Gonzalez et al., 2008). In latter reports, WD40 repeat protein was shown to be involved in anthocyanin biosynthesis regulation (Stommel et al., 2009, Matus et al., 2010 and Zohar et al., 2011). The *MdTTG1* gene encoding an apple WD40 repeat protein has been reported that it could regulate anthocyanin accumulation in transgenic *Arabidopsis* (Brueggemann et al., 2010). In our study, the transcription of WD40 gene was low in the dark-grown fruits, but increased dramatically after bag removal and was up-regulated in bagged fruits skins during fruit coloration (Fig. 5). The data suggest that this gene is responsive to light and involved in anthocyanin biosynthesis in apples.

Fruit coloration and softening are two important indexes to evaluate fruit maturity, and fruit softening is associated with changes in the composition of the cell wall (Fischer and Bennett, 1991 and Barnavon et al., 2001). Recent work suggests that the changes in the cell wall result from the action of cell wall-modifying enzymes, such as polygalacturonase (PG) and galactosidase (GAL) (Tateishi et al., 2007, Liu et al., 2011 and Zheng et al., 2012). In our study, the expression of PG and GAL during fruit ripening was increased during 'Granny Smith' fruit coloration, similar to previous studies (Wu et al., 1993, Ross, 1994 and Mann et al., 2008). Moreover, the expression levels of PG and GAL were higher in fruit maturity prophase, and they were also higher in the bagged fruits than the control group (Fig. 5).

CONCLUSIONS

'Granny Smith' skin rapidly becomes red when the fruits are re-exposed to sunlight after bag removal. Two anthocyanins, cy3-gal and cy3-ara, were detected in the red 'Granny Smith' apple peels, and cy3-gal was chiefly responsible for red color. Expression analysis of anthocyanin biosynthetic genes (*MdCHS*, *MdF3H*, *MdDFR*, *MdANS*, and *MdUFGT*) showed that all of the selected genes (except *MdCHS*) in the anthocyanin pathway were markedly elevated in the re-exposed fruits, and their expression was highly correlated with anthocyanin accumulation. *MdMYB1* was found to regulate these structural genes in anthocyanin biosynthesis pathway. To gain a better understanding of the molecular regulation of anthocyanin biosynthesis in 'Granny Smith', we constructed a subtractive cDNA library and obtained 73 unigenes that are mainly involved in metabolism and disease/defense. The expression changes of four selected genes (PSY, WD40, PG, and GAL) were confirmed using qRT-PCR, and the results showed that these genes are directly or indirectly involved in 'Granny Smith' coloration during fruit maturation. Taken together, anthocyanin accumulation in 'Granny Smith' apples is the result of interactions between multiple key enzymes in the anthocyanin pathway, and the coloring mechanism of 'Granny Smith' may be similar or identical to that of red-skinned cultivars.

ACKNOWLEDGMENTS

This research was supported by the earmarked fund for Modern Agro-industry Technology Research System, China (CARS-28). We are all extremely grateful to Dr. Zahoor Ahmad Bhat, who devoted much time and effort to revise the manuscript for us.

APPENDIX A. SUPPLEMENTARY DATA

Supplementary Table 1: Primers used for qRT-PCR to verify the expression pattern of differentially expressed genes.

Gene	Forward primer sequence	Reverse primer sequence	Genbank ID
Actin	TGTGTTTCCTAGTATTGTTGGTCGC	GCCAGATCTTCTCCATGTCATCC	AB638619
MdCHS	ACCCAGGTGGCCCTGCTATTCT	TCCCTCCCCGGTCGTTTTGA	AB074485
MdF3H	AGTGCCCTCAGCCCGACCTAAC	ATCCATGTCTTCCCATCATCCC	AB074486
MdDFR	AAATACCCCGAATA-CAATATACCCAC	TGAACTCAAACCCTATCTCCCT-CAAC	AF117268
MdANS	GATTGGATGAAGGGAGGCTG-GAGA	GGCTGAGGGCATTTTGGGTAG-TAG	AB074487
MdUFGT	GCCCTTCCAAACACTCTCTTCTC	CCCGTCAGCCACATCGTACACC	AF117267
MdMYB1	ACCAATGTGATAAGACCT-CAGCCCC	CATTTCCATTCTTTGTTGACGAC-GA	DQ886414

WD40	CCCTGACTGCCTCTTCTCCC	GCGATTCGGTGGTGCTGGTG	JK974070
PSY	TGAAGTCTGTGCTGAGTATGC-CAAA	TGTATGAAGCATTAGGTCCATC-CACT	JK750344
GAL	TGACTGCCAAGGAAGTCTAC-CATATAC	ACACGAAATACCCTGTTAGAGT-GAGC	JK973900
PG	AAACATAAGAGGAACGAGT-GCTTCC	AACCCTAGGTTTTAACATC-TAGGGGA	JK750375

Supplementary Table 2: Identification of EST clones from 'Granny Smith' apple skin SSH library.

Clone	Acc. no.	Annotation	Identity (%)	*E*-value
Metabolism				
Md-31	JK750344	Phytoene synthase protein (PSY) mRNA [*Fragaria ananassa*]	88%	9e-97
Md-12	JK750326	Isoflavone reductase related protein [*Pyrus communis*]	95%	2e-59
Md-1	JK750315	Cysteine protease inhibitor mRNA [*Pyrus communis*]	94%	1e-100
Md-71	JK750383	Cinnamoyl CoA reductase (CCR7) [*Populus trichocarpa*]	81%	0.85
Md-94	JK750405	Aspartic proteinase 2 mRNA [*Castanea mollissima*]	87%	4e-58
Md-46	JK750359	Plasma membrane intrinsic protein [*Pyrus communis*]	94%	3e-46
Md-22	JK750335	beta Tubulin like protein [*Pyrus pyrifolia*]	97%	8e-99
Md-173	JK973965	Dehydroascorbate reductase (DHAR) [*Malus domestica*]	99%	0.0
Md-88	JK750399	Acyl-CoA oxidase [*Ricinus communis*]	82%	2e-54
Md-50	JK750363	Epoxide hydrolase2 (eh2 gene) [*Prunus persica*]	84%	3e-64
Md-23	JK750336	Glucose-6-phosphate precursor [*Helianthus annuus*]	85%	2e-67
Md-79	JK750391	Gibberellic acid receptor mRNA [*Gossypium hirsutum*]	80%	2e-70
Md-14	JK750328	Pdbcs-L14 putative lipoxigenase [*Prunus dulcis*]	90%	6e-40
Md-20	JK750333	beta-1,3-Glucanase [*Fragaria ananassa*]	82%	2e-26
Md-211	JK974003	Alcohol acyl transferase (AAT1) mRNA [*Malus domestica*]	99%	3e-88
Md-77	JK750389	60S ribosomal protein [*Vitis vinifera*]	88%	4e-48
Md-139	JK973931	beta Tubulin like protein [*Pyrus pyrifoliavar*]	97%	9e-99
Md-25	JK750338	Pyruvate decarboxylase (PDC) mRNA [*Prunus armeniaca*]	87%	3e-74
Md-13	JK750327	Sarcosine oxidase (SOX1) [*Populus trichocarpa*]	82%	6e-16
Md-236	JK974028	Galactokinase, putative, mRNA [*Ricinus communis*]	83%	3e-18
Md-37	JK750350	Ethylene receptor (ETR1) mRNA [*Malus domestica*]	98%	3e-75
Md-41	JK750354	PpGAL4 mRNA for beta-D-galactosidase [*Pyrus pyrifolia*]	96%	2e-111

Md-44	JK750357	Dehydrin2 mRNA [*Eriobotrya japonica*]	95%	3e-123
Md-179	JK973971	Glutathione S-transferase mRNA [*Medicago sativa*]	79%	3e-48
Md-59	JK750371	Xyloglucan endotransglucosylase/hydrolase [*Malus domestica*]	88%	3e-43
Md-55	JK750367	Chlorophyll a/b-binding protein mRNA [*Pyrus communis*]	97%	5e-85
Md-34	JK750347	40S ribosomal protein S20-2 mRNA [*Medicago truncatula*]	85%	1e-33
Md-63	JK750375	PcPG1 mRNA for polygalacturonase-1 [*Pyrus communis*]	97%	9e-102
Md-64	JK750376	Chloroplast diphosphate synthase [*Populus trichocarpa*]	89%	3e-105
Md-66	JK750378	Cetyl-CoA acetyltransferase, mitochondrial [*Ricinus communis*]	86%	4e-89
Md-108	JK973903	3-Ketoacyl-CoA reductase 1 mRNA [*Gossypium hirsutum*]	86%	5e-46
Md-271	JK974063	2,4-Dienoyl-CoA reductase mRNA [*Jatropha curcas*]	75%	2e-27
Md-255	JK974047	Poly phenoloxidase 5 precursor, mRNA [*Pyrus bretschneideri*]	93%	6e-32
Md-248	JK974040	Plastid alpha-amylase (AMY10) mRNA [*Malus domestica*]	98%	2e-76
Md-193	JK973985	Thiazole biosynthetic enzyme mRNA [*A.glutinosa*]	86%	0.0
Md-53	JK750365	2,4-Dienoyl-CoA reductase mRNA [*Jatropha curcas*]	74%	2e-27
Stress/defense				
Md-72	JK750384	Genes for F-box proteins, S ribonuclease [*Pyrus pyrifolia*]	90%	2e-23
Md-73	JK750385	alpha-Farnesene synthase (AFS) mRNA [*Pyrus pyrifolia*]	99%	1e-153
Md-65	JK750377	PHL1 photolyase (PhL) gene [*Pyrus pyrifolia*]	98%	7e-39
Md-57	JK750369	Defender against cell death 1 [*Malus domestica*]	99%	5e-98
Md-58	JK750360	Abscisic stress ripening protein [*Pyrus pyrifolia*]	92%	1e-09
Md-76	JK750388	Desiccation-related protein [*Glycine max*]	84%	3e-06
Md-67	JK750379	Heat shock protein 70-3 (HSP70-3) gene [*Nicotiana tabacum*]	76%	7e-10
Md-68	JK750370	Heat shock protein binding protein [*Ricinus communis*]	90%	1e-46
Md-27	JK750340	Thaumatin-like protein (tl gene) [*Malus domestica*]	97%	3e-144
Md-26	JK750339	Metallothionein-like protein(AMT2) [*Malus domestica*]	80%	1e-104
Md-35	JK750348	Universal stress protein [*Vitis vinifera*]	88%	6e-49
Md-62	JK750374	Dehydration-responsive protein RD22 [*Prunus persica*]	81%	1e-96
Md-113	JK973908	Metallothionein-like protein mRNA [*Malus domestica*]	98%	6e-144
Transcription factor				

Md-7	JK750321	BZIP domain class transcription factor [*Malus domestica*]	100%	7e-74
Md-69	JK750381	WD-repeat protein [*Ricinus communis*]	88%	5e-17
Md-24	JK750337	C3HL domain class transcription factor [*Malus domestica*]	99%	3e-170
Md-279	JK974070	Ttg1-like protein mRNA [*Malus domestica*]	100%	0.0
Md-280	JK974071	MYB domain class transcription factor [*Malus domestica*]	99%	7e-120
Signal transduction				
Md-4	JK750318	Glucose-6-phosphate dehydrogenase [*Populus suaveolens*]	85%	2e-93
Md-86	JK750397	Cinnamyl alcohol dehydrogenase mRNA [*Eriobotrya japonica*]	96%	3e-144
Md-80	JK750392	Anx1 (Anx1) mRNA [*Gossypium hirsutum*]	78%	5e-64
Md-5	JK750319	Cinnamyl alcohol dehydrogenase 1 [*Eriobotrya japonica*]	94%	4e-45
Md-61	JK750373	Glyceraldehyde 3-phosphate dehydrogenase [*Fragaria ananassa*]	93%	5e-67
Transporters				
Md-30	JK750343	Major allergen and lipid transfer protein [*Malus domestica*]	94%	3e-152
Md-83	JK750394	Nitrate transporter mRNA [*Malus hupehensis*]	99%	1e-80
Md-60	JK750372	White-brown-complex ABC transporter [*Populus trichocarpa*]	80%	7e-47
Md-92	JK750403	Cholesterol transport protein [*Populus trichocarpa*]	84%	4e-97
Md-190	JK973982	Pdbcs-L30 putative lipid transfer protein [*Prunus dulcis*]	91%	0.008
Md-272	JK974064	Sorbitol transporter (SOT6) mRNA [*Malus domestica*]	100%	4e-43
Md-122	JK973917	Histone acetyl transferase [*Populus trichocarpa*]	84%	2e-33
Md-200	JK973992	Isolate 14A3 auxin-repressed protein (arp) [*Pyrus pyrifolia*]	90%	8e-50
Others				
Md-54	JK750366	S6PDH mRNA [*Malus domestica*]	98%	2e-77
Md-2	JK750316	Elongation factor 1 alpha (EF-1alpha) [*Malus domestica*]	84%	5e-48
Md-8	JK750322	DNA-binding protein (T231) mRNA [*Nicotiana tabacum*]	80%	1e-13
Md-74	JK750386	Histone H2B (HTB11) mRNA [*Arabidopsis thaliana*]	89%	4e-38
Md-208	JK974000	Nucleosome/chromatin assembly factor [*Populus trichocarpa*]	83%	1e-18
Md-49	JK750362	Chloroplast DNA [*Pyrus pyrifolia*]	100%	6e-57

REFERENCES

1. Allan, A.C., Hellens, R.P., Laing, W.A., 2008. MYB transcription factors that color our fruit. Trends in Plant Science 13, 99–102.
2. Ban, Y., Honda, C., Hatsuyama, Y., Igarashi, M., Bessho, H., 2007a. Isolation and functional analysis of a MYB transcription factor gene that is a key regulator for the development of red coloration in apple skin. Plant & Cell Physiology 48, 958–970.
3. Ban, Y., Honda, C., Bessho, H., Pang, X.M., Moriguchi, T., 2007b. Suppression subtractive hybridization identifies genes induced in response to UV-B irradiation in apple skin: isolation of a putative UDP-glucose 4-epimerase. Journal of Experimental Botany 58, 182–1834.
4. Barnavon, I., Doco, T., Terrier, N., Ageorges, A., Romieu, C., Pellerin, P., 2001. Involvement of pectin methylesterase during the ripening of grape berries: partial cDNA isolation, transcript expression and changes in the degree of methyl-esterification of cell wall. Phytochemistry 58, 693–701.
5. Bartley, G.E., Scolnik, P.A., 1995. Plant carotenoids: pigments for photoprotection, visual attraction, and human health. The Plant Cell 7, 1027–1038.
6. Ben-Yehudah, G., Korchinsky, R., Redel, G., Ovadya, R., Oren-Shamir, M., Cohen, Y., 2005. Color accumulation patterns and the anthocyanin biosynthetic pathway in 'Red Delicious' apple variants. The Journal of Horticultural Science and Biotechnology 80, 187–192.
7. Brueggemann, J., Weisshaar, B., Sagasser, M., 2010. A WD40-repeat gene from Malus × domestica is a functional homologue of Arabidopsis thaliana TRANSPARENT TESTA GLABRA1. Plant Cell Reports 29, 285–294.
8. Busch, M., Seuter, A., Hain, R., 2002. Functional analysis of the early steps of carotenoid biosynthesis in tobacco. Plant Physiology 128, 439–453.
9. Cheng, F.S., Weeden, N.F., Brown, S.K., 1996. Identification of co-dominant RAPD markers tightly linked to fruit skin colour in apple. Theoretical and Applied Genetics 93, 222–227.
10. Diatchenko, L., Lau, Y.F., Campbell, A.P., Chenchik, A., Moqadam, F., Huang, B., Lukyanov, S., Lukyanov, K., Gurskaya, N., Sverdlov, E.D., Siebert, P.D., 1996. Suppression subtractive hybridization: a method for generating differentially regulated or tissue-specific cDNA probes and libraries. Proceedings of the National Academy of Sciences of the United States of America 93, 6025–6030.
11. Espley, R.V., Hellens, R.P., Putterill, J., Stevenson, D.E., Kutty-Amma, S., Allan, A.C., 2007. Red coloration in apple fruit is due to the activity of the MYB transcription factor, MdMYB10. The Plant Journal 49, 414–427.
12. Fischer, R.L., Bennett, A.B., 1991. Role of cell wall hydrolases in fruit ripening. Annual Review of Plant Physiology and Plant Molecular Biology 42, 675–703.
13. Gonzalez, A., Zhao, M., Leavitt, J.M., Lloyd, A.M., 2008. Regulation of the anthocyanin biosynthetic pathway by the TTG1/bHLH/Myb transcriptional complex in Arabidopsis seedlings. The Plant Journal 53, 814–827.
14. Han, S.E., Lee, H.E., Heo, S., Sung, S.K., Kim, W.T., Kim, D., 2011. Isolation and characterization of genes expressed differently in mature fruits of 'Redfield' and 'Greensleeves' apples. Horticulture, Environment and Biotechnology 52, 413–421.
15. Honda, C., Kotoda, N., Wada, M., Kondo, S., Kobayashi, S., Soejima, J., Zhang, Z., Tsuda, T., Moriguchi, T., 2002. Anthocyanin biosynthetic genes are coordinately expressed during red coloration in apple skin. Plant Physiology and Biochemistry 40, 955–962.
16. Ju, Z.G., 1998. Fruit bagging, a useful method for studying anthocyanin synthesis and gene expression in apple. Scientia Horticulturae 77, 155–164.
17. Ju, Z.G., Yuan, Y.B., Liu, C.L., Wang, Y.Z., Tian, X.P., 1997. Dihydroflavonol reductase activity and anthocyanin accumulation in 'Delicious', 'Golden Delicious' and 'Indo' apples. Scientia Horticulturae 70, 31–43.
18. Kato, M., Ikoma, Y., Matsumoto, H., Sugiura, M., Hyodo, H., Yano, M., 2004. Accumulation of carotenoids and expression of carotenoid biosynthetic genes during maturation in citrus fruit. Plant Physiology 134, 824–837.
19. Kim, S.-H., Lee, J.R., Hong, S.T., Yoo, Y.K., An, G., Kim, S.R., 2003. Molecular cloning and analysis of anthocyanin biosynthesis genes preferentially expressed in apple skin. Plant Science 165, 403–413.
20. King, M.C., Cliff, M.A., 2002. Development of a model for prediction of consumer liking from visual attributes of new and established apple cultivars. Journal American Pomological Society 56, 223–229.
21. Koes, R., Verweij, W., Quattrocchio, F., 2005. A colorful model for the regulation and evolution of biochemical pathways. Trends in Plant Science 10, 236–242.
22. Kondo, S., Maeda, M., Kobayashi, S., Honda, C., 2002. Expression of anthocyanin biosynthetic genes in Malus sylvestris L. 'Mutsu' non-red apples. The Journal of Horticultural Science and Biotechnology 77, 718–723.
23. Lancaster, J.E., Dougall, D.K., 1992. Regulation of skin color in apples. Critical Reviews in Plant Sciences 10, 487–502.
24. Liu, Y.Z., Dong, T., Lei, Y., Deng, X.X., Gu, Q.Q., 2011. Isolation of a polygalacturonase gene from citrus sinensis fruit and its expression relative to fruit mastication trait, fruit development, and calcium or boron treatments. Plant Molecular Biology Reporter 29, 51–59.
25. Liu, Y.L., Che, F., Wang, L.X., Meng, R., Zhang, X.J., Zhao, Z.Y., 2013. Fruit coloration and anthocyanin biosynthesis after bag removal in non-red and red apples (Malus × domestica Borkh.). Molecules 18, 1549–1563.
26. Mann, H.S., Alton, J.J., Kim, S.H., Tong, B.S., 2008. Differential expression of cell-wallmodifying genes and novel cDNAs in apple fruit during storage. Journal of the American Society for Horticultural Science 133, 152–157.
27. Matus, J.T., Poupin, M.J., Canon, P., Bordeu, E., Alcalde, J.A., Arce-Johnson, P., 2010. Isolation of WDR and

bHLH genes related to flavonoid synthesis in grapevine (Vitis vinifera L.). Plant Molecular Biology 72, 607–620.

28. Morita, Y., Saitoh, M., Hoshino, A., Nitasaka, E., Iida, S., 2006. Isolation of cDNAs for R2R3- MYB, bHLH and WDR transcriptional regulators and identification of c and ca mutations conferring white flowers in the Japanese morning glory. Plant & Cell Physiology 47, 457–470.

29. Qin, X.Q., Coku, A., Inoue, K., Tian, L., 2011. Expression, subcellular localization, and cisregulatory structure of duplicated phytoene synthase genes in melon (Cucumis melo L.). Planta 234, 737–748.

30. Reay, P.F., 1999. The role of low temperatures in the development of the red blush on apple fruit ('Granny Smith'). Scientia Horticulturae 79, 113–119.

31. Ross, G.A., 1994. A cDNA clone for endopolygalacturonase from apple. Plant Physiology 105, 1437–1438.

32. Saure, M.C., 1990. External control of anthocyanin formation in apple. Scientia Horticulturae 42, 181–218.

33. Stommel, J.R., Lightbourn, G.J., Winkel, B.S., Griesbach, R.J., 2009. Transcription factor families regulate the anthocyanin biosynthetic pathway in Capsicum annuum. Journal of the American Society for Horticultural Science 134, 244–251.

34. Takos, A.M., Jaffé, F.W., Jacob, S.R., Bogs, J., Robinson, S.P., Walker, A.R., 2006. Light-induced expression of a MYB gene regulates anthocyanin biosynthesis in red apples. Plant Physiology 142, 1216–1232.

35. Tateishi, A., Shiba, H., Ogihara, J., Isobe, K., Nomura, K., Watanabe, K., Inoue, H., 2007. Differential expression and ethylene regulation of β-galactosidase genes and isozymes isolated from avocado (Persea americana Mill.) fruit. Postharvest Biology and Technology 45, 56–65.

36. Ubi, B.E., Honda, C., Bessho, H., Kondo, S., Wada, M., Kobayashi, S., Moriguchi, T., 2006. Expression analysis of anthocyanin biosynthetic genes in apple skin: effect of UV-B and temperature. Plant Science 170, 571–578.

37. Wei, Y.Z., Hu, F.C., Hu, G.B., Li, X.J., Huang, X.M., Wang, H.C., 2011. Differential expression of anthocyanin biosynthetic genes in relation to anthocyanin accumulation in the pericarp of Litchi Chinensis Sonn. PloS One 6, 1–10.

38. Wu, Q.D., Szakács-Dobozi, M., Hemmat, M., Hrazdina, C., 1993. Endopolygalacturonase in apples (Malus domestica) and its expression during fruit ripening. Plant Physiology 102, 219–225.

39. Xu, Y.T., Feng, S.Q., Jiao, Q.Q., Liu, C.C., Zhang, W.W., Chen, W.Y., Chen, X.S., 2012. Comparison of MdMYB1 sequences and expression of anthocyanin biosynthetic and regulatory genes between Malus domestica Borkh. cultivar 'Ralls' and its blushed sport. Euphytica 185, 157–170.

40. Yao, Y.X., Zhao, L.L., Hao, Y.J., Zhai, H., 2005. Effective extraction of total RNA in apple flesh with improved hot borate method. Journal of Fruit Science 22, 737–740.

41. Zhang, X.D., Allan, A.C., Yi, Q., Chen, L.M., Li, K.Z., Shu, Q., Su, J., 2011. Differential gene expression analysis of Yunnan red pear, Pyrus Pyrifolia, during fruit skin coloration. Plant Molecular Biology Reporter 29, 305–314.

42. Zheng, X.L., Jing, G.X., Liu, Y., Jiang, T.J., Jiang, Y.M., Li, J.R., 2012. Expression of expansin gene, MiExpA1, and activity of galactosidase and polygalacturonase in mango fruit as affected by oxalic acid during storage at room temperature. Food Chemistry 132, 849–854.

43. Zhou, Q., Gao, H., Wang, M., Xu, Y., Guo, Y.Z., Wan, Y.Z., Zhao, Z.Y., 2012. Characterization of defense-related genes in the 'Qinguan' apple in response to Marssonina coronaria. South African Journal of Botany 80, 36–43.

44. Zohar, B.S., Sylvie, J., Talia, N.H., 2011. A pomegranate (Punica granatum L.) WD40-repeat gene is a functional homologue of Arabidopsis TTG1 and is involved in the regulation of anthocyanin biosynthesis during pomegranate fruit development. Planta 234, 865–881.

CHAPTER 15

IMPACT OF DROUGHT, HEAT, AND THEIR COMBINATION ON CHLOROPHYLL FLUORESCENCE AND YIELD OF WILD BARLEY (HORDEUM SPONTANEUM)

Christoph Jedmowski[1], Ahmed Ashoub[1,2], Osama Momtaz[2], and Wolfgang Brüggemann[1,3]

[1]Department of Ecology, Evolution and Diversity, University of Frankfurt, Max von Laue Strasse 13, 60438 Frankfurt, Germany

[2]Agricultural Genetic Engineering Research Institute (AGERI), ARC, Giza 12619, Egypt

[3]Biodiversity and Climate Research Center, Senckenberganlage 25, 60325 Frankfurt, Germany

ABSTRACT

The impact of (long-term) drought acclimation and (short-term) heat stress and their combination on fast chlorophyll fluorescence induction curves (OJIP) and grain yield was tested using pot-grown plants of wild barley (Hordeum spontaneum) originating from Northern Egypt. Concerning agronomic traits, the main effect of drought was decreased biomass accumulation and grain yield, while heat specifically affected floral development. The treatments caused specific inhibitions of photosystem II (PSII) functionality. While heat stressed plants showed a reduction of maximum quantum efficiency of PSII (φP_0), an indication of effects on oxygen evolving complex (OEC) functionality, and the connectivity of PSII units, these features were entirely missing in drought acclimated plants. Drought caused a reduction of the Performance Index (PI_{abs}) and of the relative amplitude of the IP-phase of the OJIP induction curve (ΔV_{IP}). Individuals suffering from a combination of drought and heat showed a better ability to recover photosynthetic electron transport after the relief of stress in comparison to heat stressed plants. However, this improved capacity to recover was not accompanied by an increased grain yield. Thus, we conclude that chlorophyll fluorescence measurements provide valuable physiological data; however, their use in agronomic studies for the prediction of agronomic traits should be done with some precaution.

INTRODUCTION

During the forthcoming decades, global agriculture will have to face dramatic climatic changes [1, 2]. Water availability will be a limiting factor for agricultural activities in arid and semiarid regions. Even areas that do not undergo water availability problems so far are expected to experience prolonged periods of drought. Moreover, increased atmosphere temperature and heat wave incidences are expected to negatively impact crop yield and food security [3].

Citation: Christoph Jedmowski, Ahmed Ashoub, Osama Momtaz, and Wolfgang Brüggemann, "Impact of Drought, Heat, and Their Combination on Chlorophyll Fluorescence and Yield of Wild Barley (Hordeum spontaneum),"Journal of Botany, vol. 2015, Article ID 120868, 9 pages, 2015. doi:10.1155/2015/120868

Drought and heat are two main factors limiting plant productivity, and in the natural environment these two abiotic stresses often occur simultaneously. Thus, understanding the impact of heat and drought on plant development and productivity is particularly valuable. For our study, we used wild barley (Hordeum spontaneum) the progenitor of domesticated H. vulgare L. H. spontaneum is self-pollinating annual grass, interfertile with cultivated barley, and mainly distributed in the Mediterranean and Irano-Turanian regions [4], the "Fertile Crescent," and eastward to Central Asia [5]. It is considered a beneficial source for the inbreeding of tolerance genes against biotic and abiotic stresses into cultivated barley [6]. Suzuki et al. [7] summarized the specific impact of drought and heat on growth and development of cereals. Typical consequences of drought are reduced plant height, a decrease of the total number of spikes, and a reduced seed weight, whereas heat stress (HS) affects pollen development, resulting in a lowered number of seeds per plant [8]. In addition, drought and heat also have a specific impact on photosynthesis, a key process of plant metabolism. Reduced water availability triggers stomatal closure, minimizing water loss via transpiration and, on the other hand, leading to a lowered CO_2 partial pressure inside the leaf. In addition to this stomatal limitation of photosynthesis, nonstomatal effects (e.g., reduced activities of Calvin Benson cycle enzymes) occur under more severe drought stress conditions [9]. Mild HS (approx. 35°C) affects photosynthesis by a decrease of Ribulose 1,5-bisphosphate carboxylase/oxygenase (Rubisco) activity [10], caused by the heat susceptibility of Rubisco activase [11]. At higher temperatures, HS inhibits the photosynthetic electron transport by disturbing photosystem II (PSII) functionality. In particular, the structural integrity of the PSII associated oxygen evolving complex (OEC) is affected [12].

Measurements of chlorophyll fluorescence are reliable tools for the detection of disturbances of photosynthesis caused by environmental stresses [13]. The analysis of fast chlorophyll fluorescence induction curves (OJIP) according to Strasser et al. [14, 15] provides quick information about the functionality of the chloroplast electron transport chain. It was shown that OJIP induction curves are sensitive to a plethora of biotic and abiotic stresses [16–18].

The aims of this study are to determine the impact of heat, drought, and their combination on the floral development and the photosynthetic performance of wild barley and to evaluate the potential and the limitations of chlorophyll fluorescence measurements in estimating plant productivity under stress conditions.

MATERIALS AND METHODS

Plant Material and Growth Conditions

H. spontaneum from the northern coastal region of Egypt (Wadi Habies 31°24′00″N/27°03″00E) was used in this study. Seeds were germinated on wet filter paper for 2 days. Four seedlings were transferred into a 16 × 16 × 20 (depth) cm^3 pot filled with 1 kg of a commercial peat soil (type C700, Stender Erden, Schermbeck, Germany) and defined as one biological replicate. Seedlings were grown in a climate chamber under 16 h photoperiod at a light intensity of 400 $\mu mol\, m^{-2} s^{-1}$, temperatures of 26°C (light)/24°C (dark), and air humidity of 70%. Volumetric soil water content (SWC) was kept at 30% using a Theta Probe ML2x sensor combined with a HH2 moisture meter (Delta-T Devices Ltd., Cambridge, UK). Control (CTR) plants were kept under the conditions mentioned above throughout the experiment.

Stress Treatments

Plants subjected to drought treatment were left unwatered at the stage of 2 leaves for three weeks, until soil water potential (Ψ) reached −2.5 MPa, as measured using a Psypro psychrometer (Wescor, Utah, USA). This long-term drought treatment will be termed "drought acclimation" (DA), sensu Lambers et al. [19] to distinguish it from short-term stress treatments, like the applied heat stress (HS). For HS treatment, plants at the stage of four leaves were transferred into a transparent incubator with temperature control (Certomat H,Sartorius, Göttingen, Germany) under a light intensity of 200 $\mu mol\, m^{-2} s^{-1}$. Temperature was increased gradually by 1°C per 20 min from 26°C until 42°C and plants were incubated for 1.5 h at this temperature in the light and further 0.5 h in the dark. Leaf temperature was checked using an infrared thermometer. Air humidity was maintained by supplying water containers to the center of the hood. For combined stresses (CS), plants at the end of DA were subjected to the same heat treatment regime but omitting the presence of air humidity supplies. The third leaf of one plant per pot was used to measure relative leaf water content (RWC) as described by Smart and Bingham [20]. Following stress treatments, plants were recovered by rewatering to 5% SWC in the case of DA and CS or to 30% SWC for HS and returned to the growing conditions described in the previous section.

Chlorophyll Fluorescence Measurements

Fast chlorophyll fluorescence induction curves (OJIP) were measured with a Pocket Plant efficiency analyzer (PEA, Hansatech, King's Lynn, UK), following 1 h of dark adaptation. Measurements were conducted at the end of the DA, HS, and CS treatments and again after 24 h of recovery.

Table 1: Definitions and explanation of the JIP-test parameters used in this study. See Strasser et al. [15] for further information about the theory of the JIP-test and the derivation of the parameters.

	Fluorescence parameters
F_0	Minimum fluorescence intensity (at t = 10 μs)
F_M	Maximum fluorescence intensity
V_J	Relative variable fluorescence at 2 ms; $V_J = (F_{2\,ms} - F_0)/(F_M - F_0)$
V_I	Relative variable fluorescence at 30 ms; $V_I = (F_{30\,ms} - F_0)/(F_M - F_0)$
ΔV_{IP}	Relative amplitude of the IP-phase of the OJIP induction curve; $\Delta V_{IP} = 1 - V_I$
	Quantum efficiencies, flux ratios
φP_0	Quantum yield of the Q_A reduction; $\varphi P_0 = (1 - F_0)/F_M$
Ψ_0	Probability that a trapped exciton is used for electron transport beyond Q_A; $\Psi_0 = 1 - V_J$
RC/ABS	Density of active RCs on absorption basis
	Performance Index
PI_{abs}	Performance Index on absorption basis $PI_{abs} = [RC/ABS]\cdot[\varphi P_0/(1 - \varphi P_0)]\cdot[\Psi_0/(1 - \Psi_0)]$

Agronomic Measurements

Plants exposed to DA and CS were rewatered to 5% SWC (($\Psi = -1$ MPa) and kept under this water supply until seed development was completed. CTR and HS plants were kept at a SWC of 30%. The number of spikes, the relative number of empty spikes (spikes without any seed), and the number of seeds per replicate were counted. In addition, seeds were weighed immediately after harvesting and the weight of one thousand seeds was calculated. All plants of a single pot were considered as one biological replicate.

Statistical Analysis

Ten measurements per treatment were used to calculate the presented (averaged) OJIP curves, and ten biological replicates were used to carry out the yield analysis measurements. One-way analysis of variance (ANOVA) with Tukey's post hoc test was used to assess the significance of the differences among treatments at a significance level of $p < 0.05$. Comparison of two datasets was conducted using Student's t-test. All datasets shown are presented as the means + standard deviation (SD). Significant changes were marked with one ($p < 0.05$), two ($p < 0.01$), or three ($p < 0.001$) asterisks.

RESULTS

Plant Water Status

DA and CS treatment reduced the RWC of H. spontaneum plants significantly to 80.3% and 73.1%, respectively, in comparison to the CTR group (96.1%), whereas under HS the RWC remained stable (94.8%,). The observed difference between DA and CS was not statistically significant ($p = 0.27$).

Chlorophyll Fluorescence

All treatments had a specific effect on chlorophyll fluorescence parameters. DA resulted in a decrease of PI_{abs}, which was partly reversed during recovery (Figure 1(a)). In contrast, the maximum quantum yield was only slightly affected by DA treatment (Figure 1(b)). PI_{abs} (−75%) and φP_0 (−13%) behaved in a similar way to HS and CS plants. However, significant differences between these two groups were observed after recovery, as only CS plants showed increased PI_{abs} and φP_0 values after returning them to standard growing conditions. The fraction of active (Q_A-reducing) reaction centers on absorption base (RC/ABS) was not altered significantly by DA treatment (Figure 1(c)), while HS and CS caused a reduction to 70% of the CTR group. Again, recovery revealed differences between HS and CS plants. The relative amplitude of the IP-phase of the OJIP transient (ΔV_{IP}) was significantly reduced by DA treatment (Figure 1(d)). Recovery measurements showed a slight increase of ΔV_{IP}, but still the mean value was lower than under well-watered conditions. Under HS and CS, ΔV_{IP} increased by 15%, and this effect persisted after recovery of the HS plants. In case of the recovered CS plants, ΔV_{IP} dropped to 75% of the CTR value, similar to the DA recovery plants. Drought had no effect on the relative variable fluorescence of the O-J and the O-K phase of the OJIP induction curves (Figures 2(a) and 2(d)). Again, HS and CS showed an identical reaction under stress conditions (Figures 2(b), 2(c), 2(e), and 2(f)), exhibiting a positive maximum of ΔV_{OJ} (at 300–400 μs) and ΔV_{OK} (at 100–200 μs). These maxima persisted in plants recovering from HS but decreased (ΔV_{OJ}) or disappeared (ΔV_{OK}) in CS plants after recovery

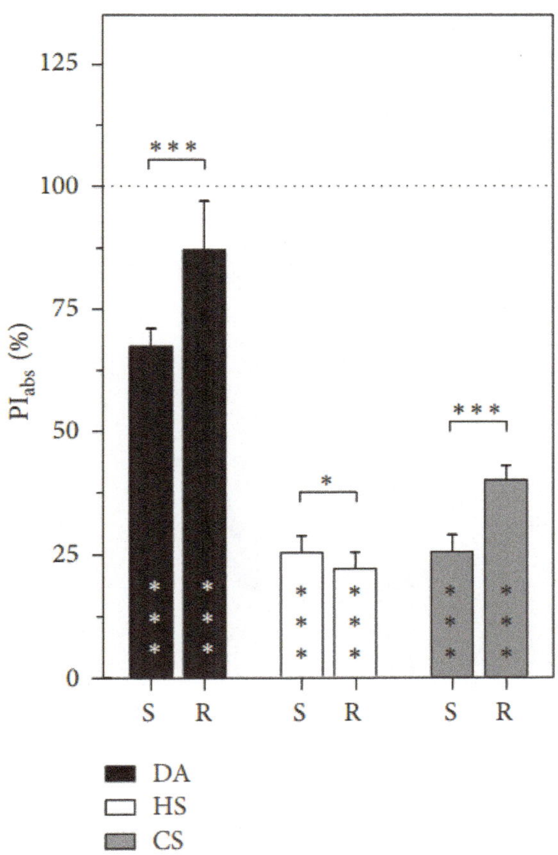

(a)

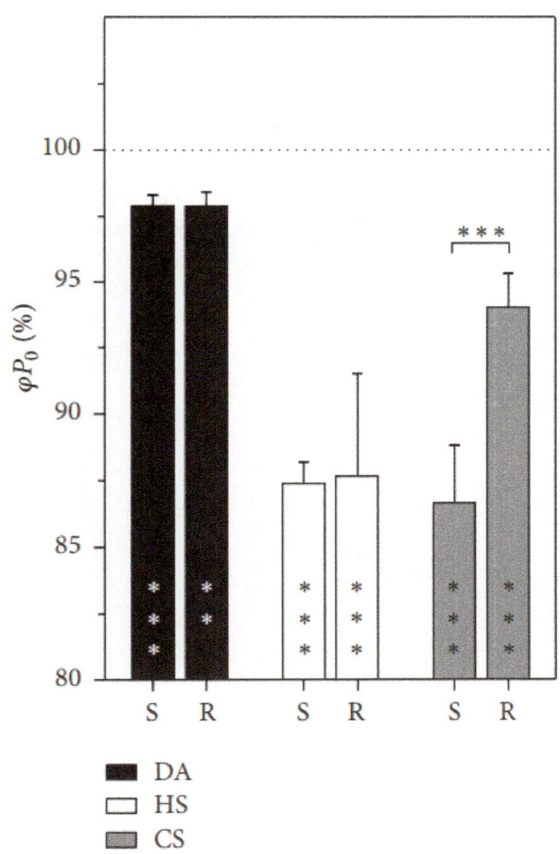

(b)

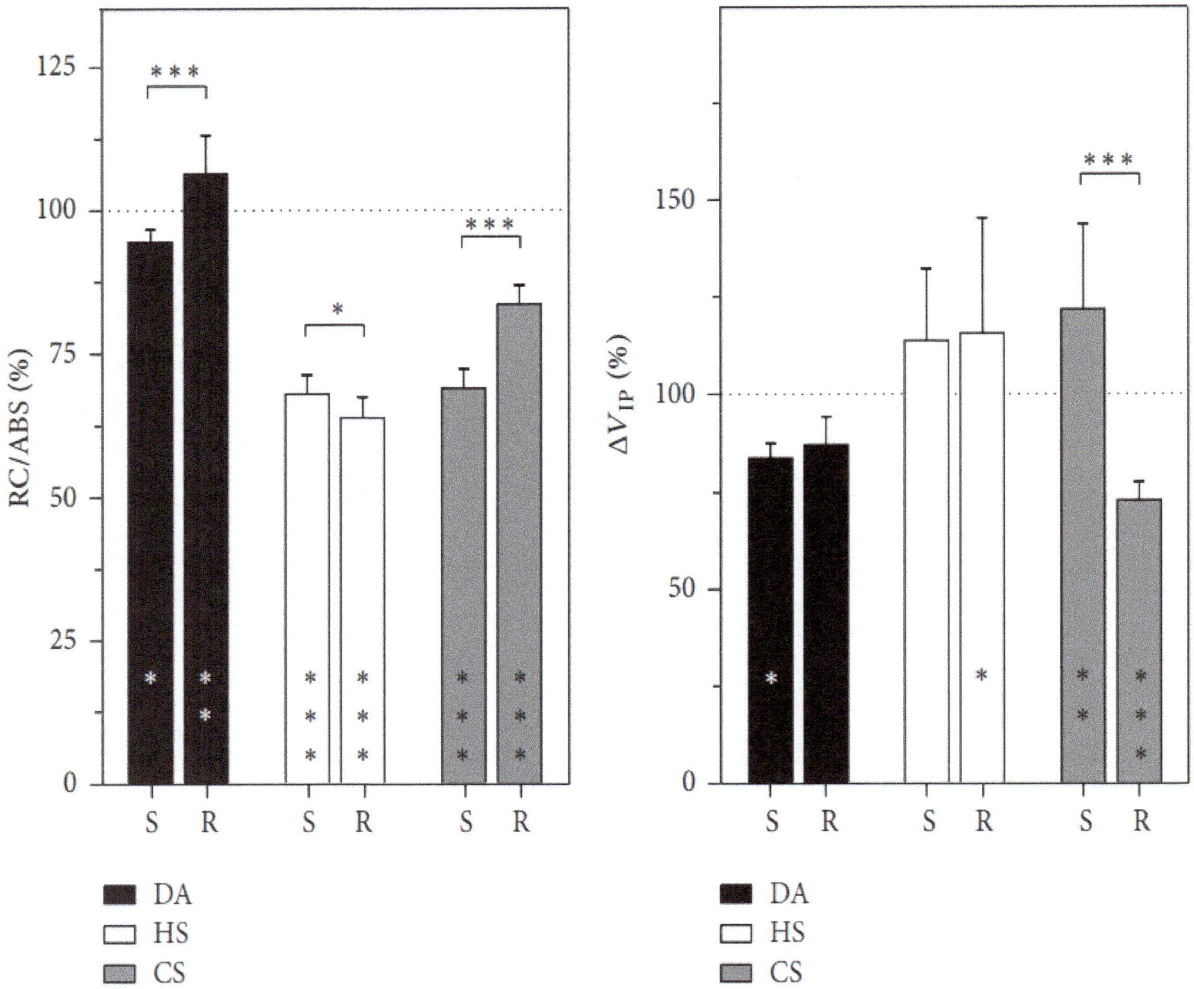

(c) (d)

Figure 1: Effect of drought, heat, or their combination on Performance Index (PI_{abs}, a), the maximum quantum yield of PSII (φP_0, b), the relative fraction of active reaction centers (RCs/ABS, c), and the relative amplitude of the IP-phase of the OJIP induction curve (ΔV_{IP}, d). Values are presented relative to the unstressed control group; DA = drought acclimated, HS = heat stress, CS = combined stresses, S = stress, and R = recovery; mean + SD, $n = 10$. Asterisks inside the columns represent significance in comparison to control values (=100%), asterisks above the connecting brackets indicate significance between stress and recovery within one treatment. Significant changes were marked with one ($p < 0.05$), two ($p < 0.01$), or three ($p < 0.001$) asterisks.

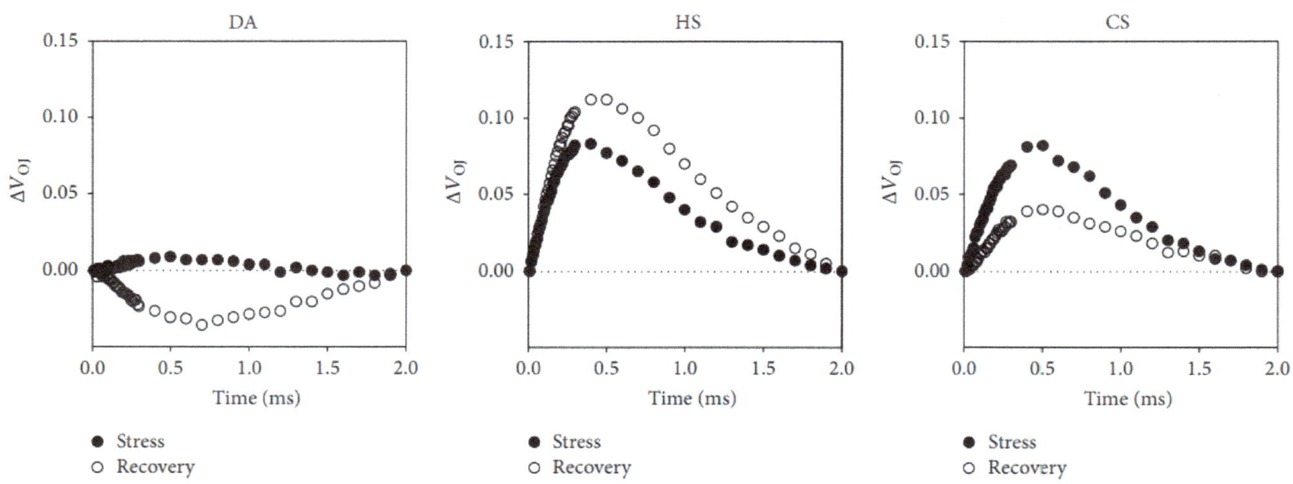

(a) (b) (c)

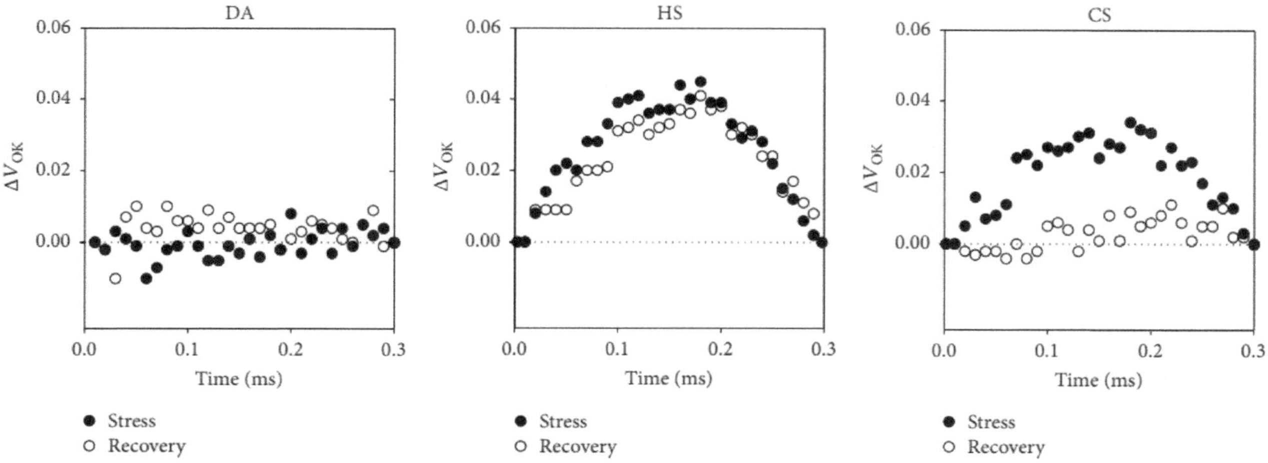

(d) (e) (f)

Figure 2: Differences of the relative variable fluorescence of O-J-normalised (ΔV_{OJ}, a–c) and O-K-normalised (ΔV_{OK}, d–f) induction curves of wild barley plants suffering drought, heat, or their combination and after 24 hours of recovery: $\Delta V_{OJ}(t) = [(F_t - F_0)_{DA/HS/CS}/(F_J - F_0)_{DA/HS/CS}] - [(F_t - F_0)_{CTR}/(F_J - F_0)_{CTR}]$, $\Delta V_{OK}(t) = [(F_t - F_0)_{DA/HS/CS}/(F_K - F_0)_{DA/HS/CS}] - [(F_t - F_0)_{CTR}/(F_J - F_0)_{CTR}]$; CTR = control, DA = drought acclimated, HS = heat stress, and CS = combined stresses; mean, $n = 10$.

Agronomic Measurements

Drought led to a decrease of the number of spikes per replicate (5.6, Figure 3(a)). CTR plants generated 14.3 spikes on average. CS plants behaved in a similar way to DS plants, exhibiting 6.8 spikes per replicate. HS had no effect on the total number of spikes. In contrast, the number (6.1, Figure 3(b)) and the frequency (52%, Figure 3(c)) of empty spikes were increased compared to the CTR group (2.7, 20.1%). DA and CS had no significant effect on the total and relative number of empty spikes. The number of seeds per replicate (Figure4(a)) was reduced under DA (31.7) and CS (25.2) conditions. Under HS, the amount of seeds was also lowered from approximately 80 to 65 per replicate, yet the difference to the CTR plants (87) was not statistically significant. Seed weight was only slightly affected under the applied stresses (Figure 4(b)). Under DA and CS treatment, plants produced smaller seeds (DA: 32.8 g/1000 seeds; CS: 33.9 g/1000 seeds) compared to the CTR group (38.1 g).

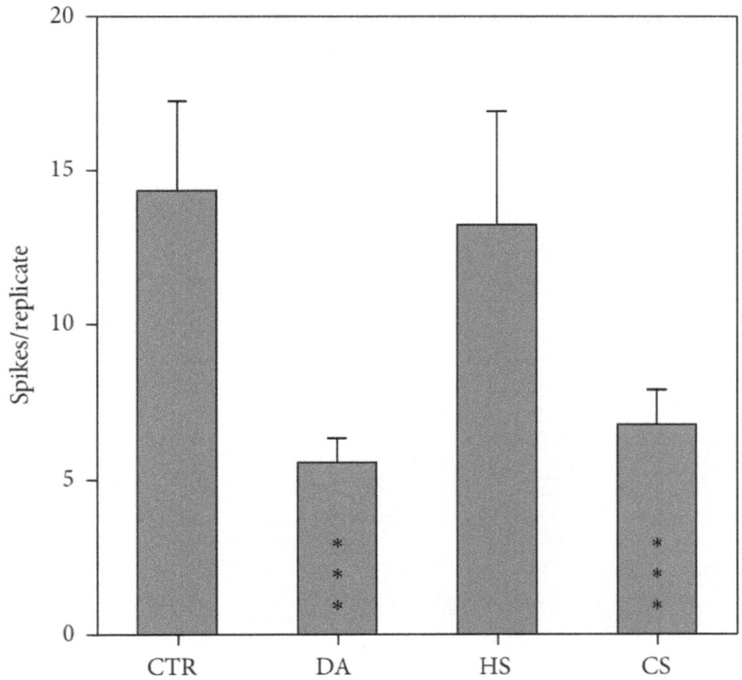

(a)

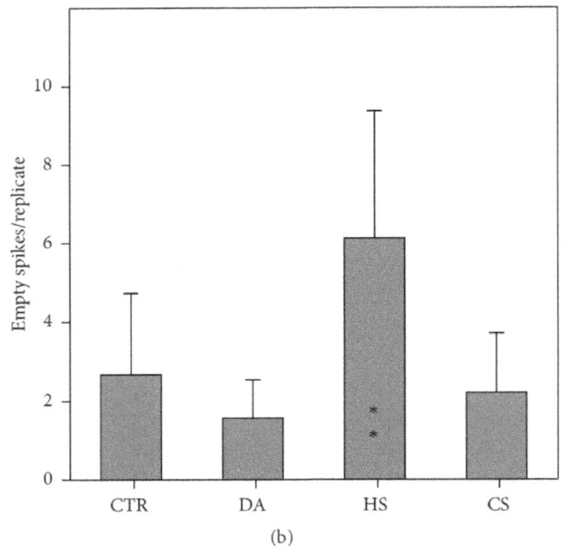

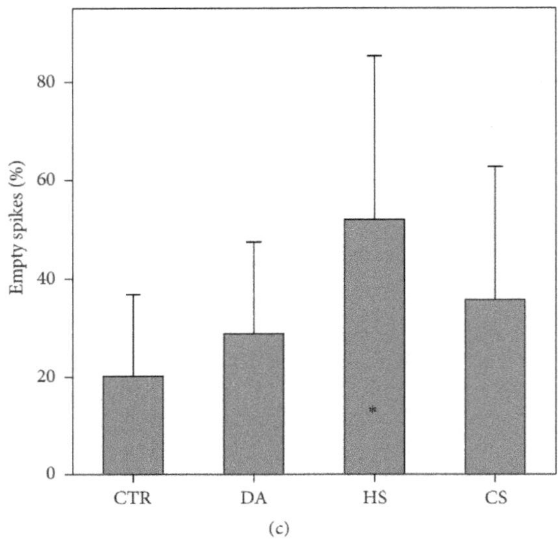

Figure 3: Number of spikes (a), number of empty spikes (b), and relative fraction of empty spikes (c) under control conditions and after drought acclimation, heat, or combined stresses treatment; CTR = control, DA = drought acclimated, HS = heat stress, and CS = combined stresses; mean + SD, $n = 10$. Asterisks inside the columns represent significance in comparison to control values. Significant changes were marked with one ($p < 0.05$), two ($p < 0.01$), or three ($p < 0.001$) asterisks.

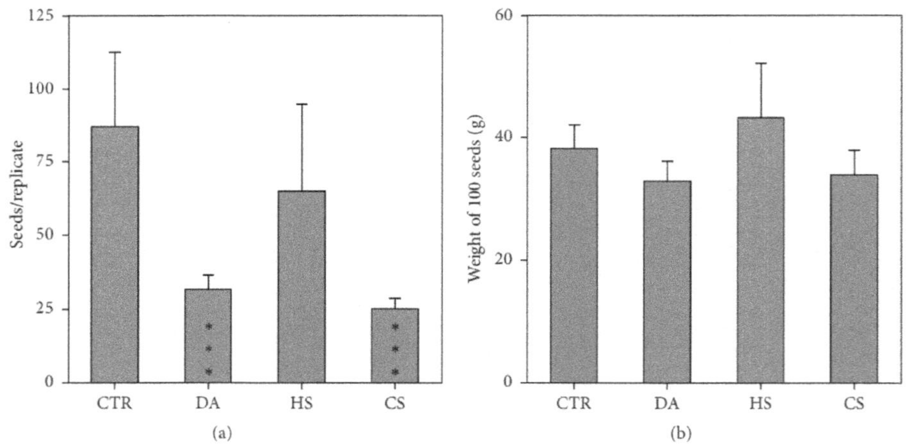

Figure 4: Number of seeds per replicate (a) and weight of 1000 seeds (b) under control conditions and after drought acclimation, heat, or combined stresses treatment; CTR = control, DA = drought acclimated, HS = heat stress, and CS = combined stresses; mean + SD, $n = 10$. Asterisks inside the columns represent significance in comparison to control values. Significant changes were marked with one ($p < 0.05$), two ($p < 0.01$), or three ($p < 0.001$) asterisks.

DISCUSSION

Chlorophyll Fluorescence

In our study, all applied treatments affected the functionality of PSII electron transport (Figures 1 and 2). The JIP-test was used in several studies aiming for the quantification of the impact of environmental stresses on photosynthesis. Oukarroum et al. [22] and Jedmowski et al. [21] evaluated the functionality of the photosynthetic electron transport chain functionality of H. vulgare varieties under drought stress and during recovery. In these studies, the PI_{abs} revealed differences between varieties as a function of drought and rewatering. Additionally, it was shown for wild barley genotypes under drought that the PI_{abs}—measured on dark adapted leaves—correlates well with maximum photosynthesis rates measured under saturating light conditions [18]. Thus, the lowered PI_{abs} observed in our study supports the theory that the reduced number of spikes and seeds per replicate under DA and CS conditions was caused by reduced carbon assimilation. Drought only had a minor effect on the maximum quantum efficiency of PSII. However, the observed change was statistically significant due to the very small standard deviation of φP_0. This finding confirms the results of several previous studies, reporting that φP_0 is relatively insensitive to decreased leaf water contents (e.g., [23, 24]).

As in the case of PI_{abs}, the change of φP_0 was highly similar for HS and CS plants. The decrease of φP_0 was mainly caused by a decrease of F_M and not by an increase of F_0 (data not shown), indicating that leaf temperature did not exceed the critical temperature for PSII [25], which is associated with a steep increase of F_0. Interestingly, PI_{abs} and φP_0 only recovered (partially) in the case of CS, while HS plants did not show increased values after 24 hours under standard growing conditions.

HS and CS reduced the fraction of Q_A-reducing RCs. Non-Q_A-reducing RCs are thought to act as "quenching sinks," dissipating the absorbed light energy exclusively as heat [15]. A reduced RC/ABS was already observed in other HS studies for soybean [26] and for wild barley [27]. Increased heat dissipation under environmental stress can be interpreted as a protective mechanism, as it reduces the risk of overexcitation of PSII under conditions of diminished linear electron transport. Like for the parameters PI_{abs} and φP_0, RC/ABS partially recovered in CS plants while there was no recovery of active RCs in HS plants.

Oukarroum et al. [28] reported for varieties of cultivated barley that the IP-phase of the OJIP transient is sensitive to drought treatment. Reduced ΔV_{IP} under drought stress was correlated with a decrease of the relative PI_{abs}, expressed as the so-called drought factor index (DFI). Additionally, it was shown for wild barley under progressive drought that during the early stages of drought stress the differential relative variable fluorescence around the I-step reacts more sensitively than the preceding J-step [18]. Our results for DA plants confirm the sensitivity of ΔV_{IP} and show that a recovery period of 24 h is not sufficient for restoring the initial kinetics of the IP-phase. The amplitude of IP-phase was assigned to be caused by a reduction of PSI content by correlating ΔV_{IP} with the relative amount of the PSI psaD protein as well as with the maximum amplitude of 820 nm transmission [29]. Accordingly, the authors suggested the use of ΔV_{IP} as a semiquantitative measure of changes of the PSI content.

While DA decreased the amplitude of the IP-phase, HS and CS treatment increased ΔV_{IP}. Since ΔV_{IP} is associated with the ratio of functional PSI to functional PSII units rather than with the absolute content of PSI [29], this is not to be interpreted as an increase in PSI units. It is rather to be seen as the result of the higher tolerance of PSI over PSII to HS [30]. Zushi et al. [31] confirmed this finding using the JIPtest analysis for tomato leaves under HS. CS plants showed the same behavior as HS plants immediately after the HS. However, after recovery, PSII functionality has been restored largely (cf. Figure 1(b)), and now the (persisting) effect of the DA on the PSI to PSII ratio became evident.

The fluorescence rise during the first 2 ms in the normalised OJIP induction curve offers additional information on developing limitations on the donor side of PSII, when visualized in the ΔV_{OJ} curves. HS and CS initiated the appearance of a so-called K-band at 300–400 μs (Figures 2(b) and 2(c)). The K-band is a characteristic feature of heat induced damage of PSII structure and was observed in several studies [17, 32]. While, in principle, changes in the earliest phase of the fluorescence induction curve can also be due to differences in antenna size or in the plastoquinone pool redox state, the general interpretation of such effects under heat stress is to attribute it to damage to the electron donor side of PSII, for example, by structural damage of the OEC [33, 34]. Under DA treatment, no K-band developed (Figure 2(a)), in contrast to observations of Oukarroum et al. [22] and Jedmowski et al. [21] for barley under severe drought stress. An increased relative fluorescence in the OK-phase (ΔV_{OK}, "L-band") developed under HS and CS conditions, which was absent under DA treatment. The appearance of the Lband is thought to be associated with decreased energetic connectivity of the PSII units [35], leading to fluorescence emission from those antenna chlorophylls, which are unable to transfer their excitation energy into the RC. While the K- and L-band under CS treatment were largely reversible (Figures 2(c) and 2(f)), the effect of HS alone pertained over 24 hours (Figures 2(b) and 2(e)).

The persistence of the peaks in the ΔV_{OJ} and ΔV_{OK} curves in case of the HS plants is remarkable. The recovery processes of PSII after a heat pulse treatment (20–40 s at 50°C) were studied by Tóth et al. [36] using barley seedlings. Here, a K-band developed after heat treatment and disappeared entirely after 4 h of recovery. The authors proposed a light-dependent mechanism including the breakdown of damaged PSII RCs after 4 hours and the de novo protein synthesis of PSII units within 24 h, leading to restoral of the initial OJIP kinetics. It is difficult to judge why our plants did not show the same behavior, as they had a light period during the 24 hours of recovery. In difference to Tóth et al. [36], we did not use a short heat pulse, but a longer treatment at a lower temperature, including an acclimation phase (gradual increase of temperature). In consequence, during our treatment, plants had the possibility to adjust and regulate their metabolism and gene expression, for example, by the production of heat shock proteins [37]. It might be speculated that under our experimental conditions different mechanisms of regulation of PSII induced by the acclimation to high temperatures may be responsible for the apparently different results. The results of the chlorophyll fluorescence measurements clearly suggest the induction of (a) protective mechanism(s) during DA, resulting in a better PSII recovery of CS plants from heat treatment in comparison to the HS group, while the acute effect of heat was highly similar for HS and CS plants.

Agronomic Measurements

The effect of the stresses on photosynthesis and growth processes is best integrated in the final yield of the plants, given as number of seeds per replicate and seed weight (Figures 4(a) and 4(b)). The continuous and moderate drought treatment resulted in a significant decrease of the number of spikes and seeds per replicate as a consequence of the continuously lower overall carbon assimilation. The same effect was also found under CS, most likely as a result of the drought component. HS, in contrast, affected the production of fertile spikes significantly; however, in DA plants, this was not the case. In wheat plants, increasing temperatures above 30°C during the transition from the vegetative to the reproductive stage and during floret formation resulted in complete sterility [38]. Additionally, floral abnormalities induced by HS lead to spikelet sterility in rice production [39]. In principle, there are two possible explanations for the smaller proportion of empty spikes in the CS group. Firstly, CS plants may have induced a protection mechanism during DA treatment, leading to a "cross-resistance" of floral development to increased temperatures. Secondly, the DA treatment may have shifted the onset or the speed of flowering. If this would be the case, the applied heat treatment would have affected CS and HS plants at different stages of floral development. Guan et al. [40] described an acceleration of heading for rice cultivars under drought stress, which can be interpreted as a drought escape strategy ensuring flower development before a critical degree of plant desiccation is reached. On the other hand, Rollins et al. [41] did not observe any effect of drought stress in two cultivated barley genotypes concerning the time to reach maturity. Since the number of spikes per replicate was the same for DA and CS plants (and lower than for CTR plants), we suggest that this feature was determined before the heat treatment. On the other hand, seed weight was not significantly affected by any of the stresses. Rollins et al. [41] found long-term HS (one week at 36°/32°C day/night temperatures) and CS to significantly affect seed weight, while drought alone had no effect on this parameter in the case of H. vulgare. We conclude that the drought component of our CS treatment dominated the (short-term) heat component in our experiment, and consequently the seed weight was not affected in our case.

CONCLUDING REMARKS AND OUTLOOK

Measurements of fast chlorophyll fluorescence induction curves revealed a characteristic pattern of inhibitions of the photosynthetic electron transport. In the case of long-term drought treatment, fluorescence data could be used for estimating productivity, as the reduced yield was directly related to photosynthetic limitations. For HS, disturbances of floral development were the decisive feature determining productivity. In this case, chlorophyll fluorescence data is uncoupled from agronomic traits. This is also obvious for plants suffering CS, as their advantage over well-watered plants in restoring the photosynthetic electron transport after heat stress had no impact on productivity; that is, their grain yield was irreversibly determined during the short-term heat treatment. However, under conditions of a long-term high growth temperature, as applied, for example, in the experiments performed by Rollins et al. [41], a negative impact of heat on photosynthesis may again lead to a coupling of fluorescence parameters and agronomic traits. This should be clarified in future experiments. In summary, we conclude that chlorophyll fluorescence is a useful tool assisting in estimating crop performance under environmental stresses, though its meaningful use is restricted to situations in which carbon fixation is the dominating cause of yield limitation.

ACKNOWLEDGMENTS

This study was funded by the German Federal Ministry of Education and Research, Project no. 01DH14007, and the Science and Technology Development Fund, Egypt, Project no. 5034, in the framework of the German Egyptian Research Fund (GERF).

REFERENCES

1. D. S. Battisti and R. L. Naylor, "Historical warnings of future food insecurity with unprecedented seasonal heat," Science, vol. 323, no. 5911, pp. 240–244, 2009.
2. G. C. Nelson, D. van der Mensbrugghe, H. Ahammad et al., "Agriculture and climate change in global scenarios: why don't the models agree," Agricultural Economics, vol. 45, no. 1, pp. 85–101, 2014.
3. IPCC, "Summary for policymakers," in Climate Change 2013: The Physical Science Basis. Contribution of Working Group I to the Fifth Assessment Report of the Intergovernmental Panel on Climate Change, T. F. Stocker, D. Qin, G. K. Plattner et al., Eds., Cambridge University Press, Cambridge, UK, 2013.
4. J. R. Harlan and D. Zohary, "Distribution of wild wheats and barley," Science, vol. 153, no. 3740, pp. 1074–1080, 1966.
5. T.-M. Choo, "Genetic resources of tibetan barley in China," Crop Science, vol. 42, no. 5, pp. 1759–1760, 2002.
6. R. P. Ellis, B. P. Forster, L. L. Handley et al., "Wild barley: a source of genes for crop improvement in the 21st century?" Journal of Experimental Botany, vol. 51, no. 342, pp. 9–17, 2000.
7. N. Suzuki, R. M. Rivero, V. Shulaev, E. Blumwald, and R. Mittler, "Abiotic and biotic stress combinations,"New Phytologist, vol. 203, no. 1, pp. 32–43, 2014.
8. P. V. V. Prasad, S. R. Pisipati, I. Momčilović, and Z. Ristic, "Independent and combined effects of high temperature and drought stress during grain filling on plant yield and chloroplast EF-Tu expression in spring wheat," Journal of Agronomy and Crop Science, vol. 197, no. 6, pp. 430–441, 2011.
9. D. W. Lawlor and W. Tezara, "Causes of decreased photosynthetic rate and metabolic capacity in water-deficient leaf cells: a critical evaluation of mechanisms and integration of processes," Annals of Botany, vol. 103, no. 4, pp. 561–579, 2009.

10. U. Feller, S. J. Crafts-Brandner, and M. E. Salvucci, "Moderately high temperatures inhibit ribulose-1,5-bisphosphate carboxylase/oxygenase (Rubisco) activase-mediated activation of Rubisco," Plant Physiology, vol. 116, no. 2, pp. 539–546, 1998.

11. S. J. Crafts-Brandner and M. E. Salvucci, "Rubisco activase constrains the photosynthetic potential of leaves at high temperature and CO_2," Proceedings of the National Academy of Sciences of the United States of America, vol. 97, no. 24, pp. 13430–13435, 2000.

12. D. Nash, M. Miyao, and N. Murata, "Heat inactivation of oxygen evolution in Photosystem II particles and its acceleration by chloride depletion and exogenous manganese," Biochimica et Biophysica Acta, vol. 807, no. 2, pp. 127–133, 1985.

13. M. Brestič and M. Živčák, "PSII fluorescence techniques for measurement of drought and high temperature stress signal in crop plants: protocols and applications," in Molecular Stress Physiology of Plants, G. R. Rout and A. B. Das, Eds., pp. 87–131, Springer, New Delhi, India, 2013.

14. R. J. Strasser, A. Srivastava, and M. Tsimilli-Michael, "The fluorescence transient as a tool to characterize and screen photosynthetic samples," in Probing Photosynthesis: Mechanisms, Regulation and Adaption, M. Yunus, U. Pathre, and P. Mohanty, Eds., pp. 445–483, Taylor & Francis, London, UK, 2000.

15. R. J. Strasser, M. Tsimilli-Michael, and A. Srivastava, "Analysis of the chlorophyll a fluorescence transient," in Chlorophyll a Fluorescence: A Signature of Photosynthesis, G. C. Papageorgiou and Govindjee, Eds., vol. 19 of Advances in Photosynthesis and Respiration, pp. 321–362, Springer, Dordrecht, The Netherlands, 2004.

16. A. J. Strauss, G. H. J. Krüger, R. J. Strasser, and P. D. R. van Heerden, "Ranking of dark chilling tolerance in soybean genotypes probed by the chlorophyll a fluorescence transient O-J-I-P," Environmental and Experimental Botany, vol. 56, no. 2, pp. 147–157, 2006.

17. S. Mathur, A. Jajoo, P. Mehta, and S. Bharti, "Analysis of elevated temperature-induced inhibition of photosystem II using chlorophyll a fluorescence induction kinetics in wheat leaves (Triticum aestivum),"Plant Biology, vol. 13, no. 1, pp. 1–6, 2011.

18. C. Jedmowski, S. Bayramov, and W. Brüggemann, "Comparative analysis of drought stress effects on photosynthesis of Eurasian and North African genotypes of wild barley," Photosynthetica, vol. 52, no. 4, pp. 564–573, 2014.

19. H. Lambers, F. S. Chapin III, and T. L. Pons, "Assumptions and approaches," in Plant Physiological Ecology, H. Lambers, F. S. Chapin, and T. L. Pons, Eds., pp. 1–9, Springer, New York, NY, USA, 1998.

20. R. E. Smart and G. E. Bingham, "Rapid estimates of relative water content," Plant Physiology, vol. 53, no. 2, pp. 258–260, 1974.

21. C. Jedmowski, A. Ashoub, and W. Brüggemann, "Reactions of Egyptian landraces of Hordeum vulgareand Sorghum bicolor to drought stress, evaluated by the OJIP fluorescence transient analysis," Acta Physiologiae Plantarum, vol. 35, no. 2, pp. 345–354, 2013.

22. A. Oukarroum, S. E. Madidi, G. Schansker, and R. J. Strasser, "Probing the responses of barley cultivars (Hordeum vulgare L.) by chlorophyll a fluorescence OLKJIP under drought stress and re-watering,"Environmental and Experimental Botany, vol. 60, no. 3, pp. 438–446, 2007.

23. D. Christen, S. Schönmann, M. Jermini, R. J. Strasser, and G. Défago, "Characterization and early detection of grapevine (Vitis vinifera) stress responses to esca disease by in situ chlorophyll fluorescence and comparison with drought stress," Environmental and Experimental Botany, vol. 60, no. 3, pp. 504–514, 2007.

24. M. Živčák, M. Brestič, K. Olšovská, and P. Slámka, "Performance index as a sensitive indicator of water stress in Triticum aestivum L," Plant, Soil and Environment, vol. 54, no. 4, pp. 133–139, 2008.

25. U. Schreiber and J. A. Berry, "Heat-induced changes of chlorophyll fluorescence in intact leaves correlated with damage of the photosynthetic apparatus," Planta, vol. 136, no. 3, pp. 233–238, 1977.

26. J. A. De Ronde, W. A. Cress, G. H. J. Krüger, R. J. Strasser, and J. Van Staden, "Photosynthetic response of transgenic soybean plants, containing an Arabidopsis P5CR gene, during heat and drought stress,"Journal of Plant Physiology, vol. 161, no. 11, pp. 1211–1224, 2004.

27. C. Jedmowski and W. Brüggemann, "Imaging of fast chlorophyll fluorescence induction curve (OJIP) parameters, applied in a screening study with wild barley (Hordeum spontaneum) genotypes under heat stress," Journal of Photochemistry and Photobiology B: Biology, vol. 151, pp. 153–160, 2015.

28. A. Oukarroum, G. Schansker, and R. J. Strasser, "Drought stress effects on photosystem i content and photosystem II thermotolerance analyzed using Chl a fluorescence kinetics in barley varieties differing in their drought tolerance," Physiologia Plantarum, vol. 137, no. 2, pp. 188–189, 2009.

29. M. G. Ceppi, A. Oukarroum, N. Çiçek, R. J. Strasser, and G. Schansker, "The IP amplitude of the fluorescence rise OJIP is sensitive to changes in the photosystem I content of leaves: a study on plants exposed to magnesium and sulfate deficiencies, drought stress and salt stress," Physiologia Plantarum, vol. 144, no. 3, pp. 277–288, 2012.

30. M. Havaux, H. Greppin, and R. J. Strasser, "Functioning of photosystems I and II in pea leaves exposed to heat stress in the presence or absence of light. Analysis using in-vivo fluorescence, absorbance, oxygen and photoacoustic measurements," Planta, vol. 186, no. 1, pp. 88–98, 1991.

31. K. Zushi, S. Kajiwara, and N. Matsuzoe, "Chlorophyll a fluorescence OJIP transient as a tool to characterize and evaluate response to heat and chilling stress in tomato leaf and fruit," Scientia Horticulturae, vol. 148, pp. 39–46, 2012.

32. A. Oukarroum, S. El Madidi, and R. J. Strasser, "Exogenous glycine betaine and proline play a protective

role in heat-stressed barley leaves (Hordeum vulgare L.): A chlorophyll a fluorescence study," Plant Biosystems, vol. 146, no. 4, pp. 1037–1043, 2012.

33. S. Z. Tóth, G. Schansker, G. Garab, and R. J. Strasser, "Photosynthetic electron transport activity in heat-treated barley leaves: the role of internal alternative electron donors to photosystem II," Biochimica et Biophysica Acta, vol. 1767, no. 4, pp. 295–305, 2007.

34. G. P. Wang, F. Li, J. Zhang, M. R. Zhao, Z. Hui, and W. Wang, "Overaccumulation of glycine betaine enhances tolerance of the photosynthetic apparatus to drought and heat stress in wheat," Photosynthetica, vol. 48, no. 1, pp. 30–41, 2010.

35. R. J. Strasser and A. D. Stirbet, "Estimation of the energetic connectivity of PS II centres in plants using the fluorescence rise O–J–I–P: fitting of experimental data to three different PS II models," Mathematics and Computers in Simulation, vol. 56, no. 4-5, pp. 451–461, 2001.

36. S. Z. Tóth, G. Schansker, J. Kissimon, L. Kovács, G. Garab, and R. J. Strasser, "Biophysical studies of photosystem II-related recovery processes after a heat pulse in barley seedlings (Hordeum vulgare L.),"Journal of Plant Physiology, vol. 162, no. 2, pp. 181–194, 2005.

37. A. Ashoub, M. Baeumlisberger, M. Neupaertl, M. Karas, and W. Brüggemann, "Characterization of common and distinctive adjustments of wild barley leaf proteome under drought acclimation, heat stress and their combination," Plant Molecular Biology, vol. 87, no. 4-5, pp. 459–471, 2015.

38. H. S. Saini and D. Aspinall, "Abnormal sporogenesis in wheat (Triticum aestivum L.) induced by short periods of high temperature," Annals of Botany, vol. 49, no. 6, pp. 835–846, 1982.

39. Y. Takeoka, K. Hiroi, H. Kitano, and T. Wada, "Pistil hyperplasia in rice spikelets as affected by heat stress," Sexual Plant Reproduction, vol. 4, no. 1, pp. 39–43, 1991.

40. Y. S. Guan, R. Serraj, S. H. Liu et al., "Simultaneously improving yield under drought stress and non-stress conditions: a case study of rice (Oryza sativa L.)," Journal of Experimental Botany, vol. 61, no. 15, pp. 4145–4156, 2010.

41. J. A. Rollins, E. Habte, S. E. Templer, T. Colby, J. Schmidt, and M. von Korff, "Leaf proteome alterations in the context of physiological and morphological responses to drought and heat stress in barley (Hordeum vulgare L.)," Journal of Experimental Botany, vol. 64, no. 11, pp. 3201–3212, 2013.

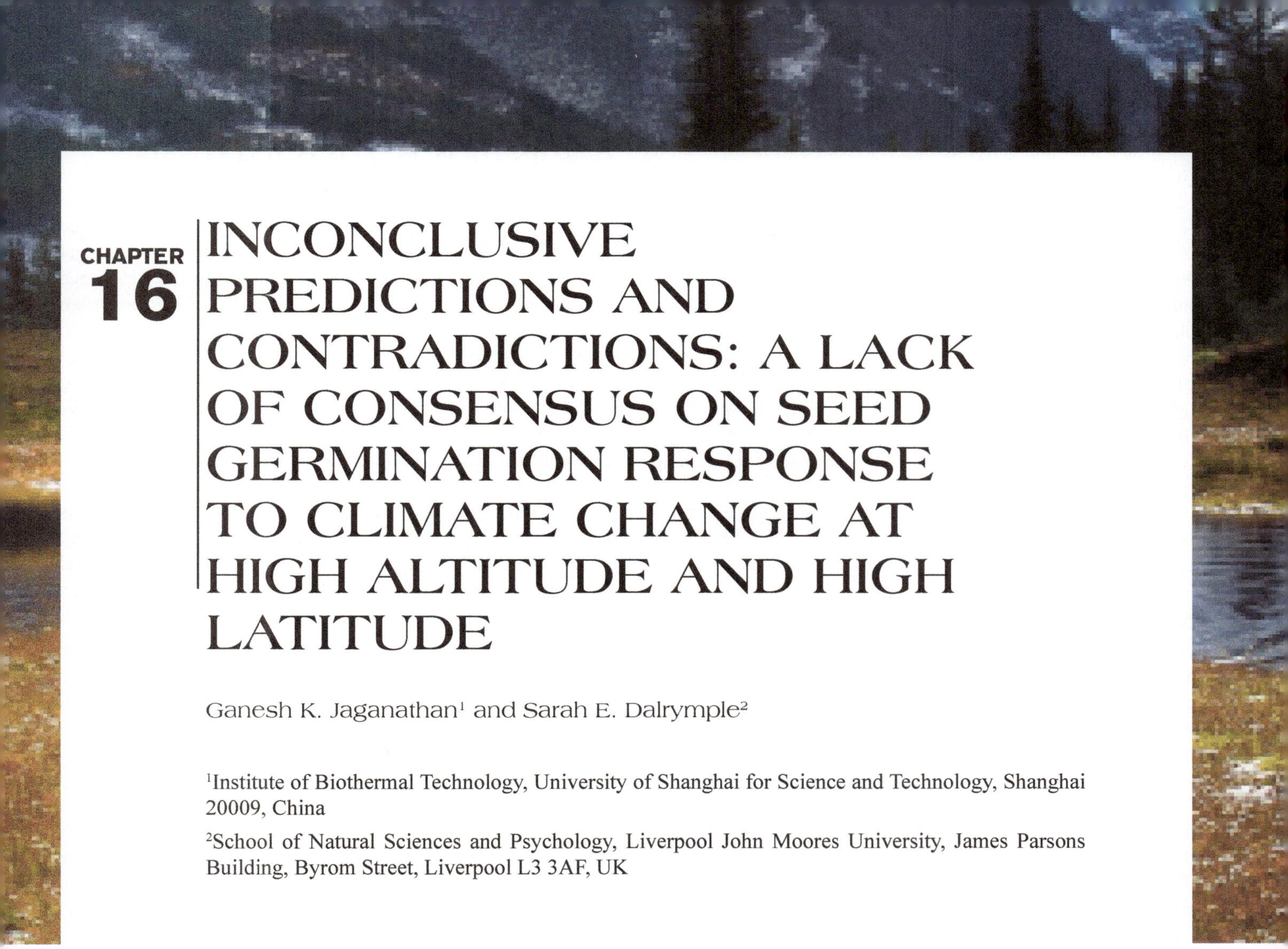

CHAPTER 16

INCONCLUSIVE PREDICTIONS AND CONTRADICTIONS: A LACK OF CONSENSUS ON SEED GERMINATION RESPONSE TO CLIMATE CHANGE AT HIGH ALTITUDE AND HIGH LATITUDE

Ganesh K. Jaganathan[1] and Sarah E. Dalrymple[2]

[1]Institute of Biothermal Technology, University of Shanghai for Science and Technology, Shanghai 20009, China

[2]School of Natural Sciences and Psychology, Liverpool John Moores University, James Parsons Building, Byrom Street, Liverpool L3 3AF, UK

ABSTRACT

Climate change directly affects arctic-alpine plants and acute responses to increased temperatures may be seen in their reproductive fitness and germination ability. However, uncertainties prevail in predicting whether a future warmer climate favors or hampers seed germination in high latitude and high altitude soils and seed germination research in such systems has not been able to provide generalizable patterns of response. The available literature on this subject has been conducted at various locations contributing to difficulties in predicting the response of arctic-alpine seeds to climate change. Here, we show that discrepancies in seed collection, dormancy breaking treatments, and germination conditions found in the published literature are possible reasons for our inability to draw large scale conclusions. We explore how these factors influence the results and highlight the fact that many of the previous investigations have reported the effects of warmer temperature, rather than a warmer climate and all the associated complex environmental interactions, on seed germination. We recommend that long-term monitoring of seed response to treatments that mimic the present and future alpine climate is likely to produce more ecologically meaningful insights and suggest several practical steps that researchers can take that would facilitate greater coherence between studies.

INTRODUCTION

Germination is one the most critical events in the life-cycle of many plants around the world [1]. Germination usually begins immediately in nondormant seeds after suitable temperature and moisture are available [1, 2]. However, dormant seeds delay germination and often use the unsuitable climatic conditions to break dormancy and germinate during the growing season [3]. By producing seeds that can synchronize the timing of germination to the most favorable period of the year, plants can bet-hedge against unpredictably variable environments [4, 5]. Such a highly complex trait has been acting as a selective evolutionary pressure resulting from climatic variation over millions of years [5]. As a consequence, some species find a particular habitat more suitable, whilst other habitats become detrimental for seed germination and continued plant growth. The suitable environment allows the persistence of those adapted species to germinate, grow, and reproduce, thus maintaining species abundance. However, this adaptation has been undergoing a rapid transformation in recent years, predominantly due to the effects of global warming bringing significant changes to climatic conditions [6, 7]. It has been estimated that the earth's surface temperature

Citation: Ganesh K. Jaganathan and Sarah E. Dalrymple, "Inconclusive Predictions and Contradictions: A Lack of Consensus on Seed Germination Response to Climate Change at High Altitude and High Latitude," Journal of Botany, vol. 2016, Article ID 6973808, 13 pages, 2016. doi:10.1155/2016/6973808

has risen by 0.6°C over the past 100 years and the projections for future warming suggested an alarming 2°C in this century [8, 9]. These discussions have stimulated a plethora of studies focusing on the germination capacity of seeds at ecosystem levels [10, 11] or of different life-forms [8]. It is becoming clear that the effects of global warming are vegetation specific [12, 13]. For example, in the tropics and fire-prone environments, the increased summer temperatures are likely to result in long and high summer temperature along with more frequent or high intensity fire, altering the germination and persistence of many species [14, 15]. In contrast, the temperate regions might experience more nonseasonal rainfall and prolonged drought, which may favor only some of the present species to persist [16]. Consequently, understanding and predicting how germination phenology changes in warmer climatic conditions is of imperative scientific and practical importance, because such information is helpful in devising appropriate conservation actions.

Although our understanding of how seeds germinate in the warmer world is not comprehensive across various ecosystems, we focus on the results available on alpine and arctic ecosystems. This by no means denotes that other ecosystems are less important; however, the effects of global warming have become more apparent in alpine regions with upward migration of plants being witnessed already [17–19], prompting a need to review the literature and use these lessons learned to plan conservation protocols. In addition, high altitude and high latitude regions are highly fragile ecosystems that are extremely susceptible to even a slight increase in temperature. Thus, these ecosystems are regarded as a "reliable indicator" for predicting the effects of climate change worldwide [19]. Secondly, unlike other ecosystems, global warming results in extinctions of arctic-alpine species, as these cold-adapted species migrate to higher altitudes and latitudes in search of diminishing suitable climate space [9, 17]. Thirdly, we know little about the process of germination even under the current climate, as it has been long thought that alpine plants typically select vegetative reproduction, although there is growing consensus that seeds represent key survival mechanisms of alpine species [20].

Given that the appropriate conditions required to complete germination are extremely narrow for arctic-alpine species [21–23], many mechanisms have been evolved to synchronize the timing of germination with the short-growing season, so that the seedling establishment and plant growth are maximized [21, 22, 24]. There are a small but increasing number of studies addressing the possible effects of warming on germination ecology of alpine species [25–32]. However, the results of these studies have led to uncertainties in answering the important question: does a warmer alpine climate favor or hamper the germination of seeds? If one attempts to generalize seed germination response under warmer conditions, a great deal of heterogeneity results from evaluating data from different methods employed in each study. The overarching aim of this contribution is to outline methods used in studying the germination response of alpine/arctic seeds to increasing temperatures and provide a crucial reevaluation of their merits. In particular, we attempt to show that there are some inconsistencies in methods employed by previous studies affecting our ability to predict the real-life effects.

THEORETICAL PREDICTIONS FOR GERMINATION RESPONSE TO CLIMATE CHANGE

Climate change affects the seed germination and persistence of many seed bearing species across various environments, in an ecosystem-specific way [33–36]. This complex interface of species with the prevailing climatic condition often guides species to develop dormancy, which will be broken with a particular set of conditions following seed dispersal but before growing season [3, 5]. In particular, the changing climate is affecting arctic-alpine species in numerous ways [37] and this effect is the result of many complex interactions between the different biological, hydrological, climatic, and edaphic components of the surrounding ecosystems [38]. Consequently, it is useful to untangle the different mechanisms affecting seeds in order to understand how climate change might alter germination success and whether the experimental investigations analysed later in this review adequately address all the critical aspects of climate change.

The key component of climate change is an increase in atmospheric temperature and changes of +3.3°C in alpine regions have been predicted by the end of this century [39], potentially affecting seed development, persistence, and germination. This average masks a seasonal differential of +2.7°C in the spring as opposed to +3.8°C in the summers by 2098 and the coincidence of peak temperatures with crucial stages of dormancy breaking and germination has become the focus of experimental manipulations, some of which are reviewed later. Further, the increased temperatures have led to greater variability and less predictability in arctic-alpine weather patterns [38, 40], meaning that windows for suitable germination may become more or less frequent and will certainly become more irregular in their occurrence.

The indirect yet severe impacts of atmospheric warming on seeds of arctic-alpines include loss of early snow cover and late accumulation of snow in the autumn, fall of snow as precipitation, thin snow, increased windiness and resultant desiccation, high nocturnal temperature, and low diurnal temperature range. The surface reservoirs of frozen water, snow beds and ice sheets, are predicted to continue the general trend of recession [41, 42] that has already been recorded in arctic [38] and alpine [43] regions. Snow beds are of particular importance to seeds due to their capacity for ameliorating the harsh winter weather by insulating the soil surface below [44]. Dense snow cover ensures seeds overwinter in a "safe-repository" and the constant temperature experienced under snow cover, which is close to 0°C, provides the right conditions for cold-stratification, breaking dormancy, and allowing seeds to germinate in the growing season [24, 45, 46]. Variation in snow thickness and lesser duration of snow cover may mean that the seeds maintain dormancy during the growing season. Seeds maintaining dormancy may be added to the "soil seed bank" and germinate in the future year or lose viability during persistence, thus affecting the dynamics of plant communities. Another outcome of snow bed loss is that colder temperatures are experienced by dispersed seeds and therefore elevated seed death due to ice formation in the internal tissues results. However, a third possible response is that earlier snow melt and later snow accumulation may also facilitate greater seed production in species that are normally restricted in size and, therefore, seed set due to a very short-growing season [47]. The accentuated seasonal variation together with markedly different temperatures is posing additional problems to the alpine seeds. Overall, the whole life-cycle of reproductive phase of plants from seed production to germination occurs in a warmer temperature [48–51]. The variation in maternal environment during seed maturation would have profound effect in determining the continued existence of a particular species [52]. Seeds mature with different levels of dormancy based on the location in which they mature, a phenomenon that could allow the seeds to judge duration required for dormancy break [53–55]. For example, seeds maturing at higher altitudes have deep PD and require longer duration of cold-stratification [24]. This is an adaptation as at higher altitudes snow cover is long and decrease in altitude correlates with lesser snow cover duration. However, climate change may alter both the temperature at which the seeds mature and the conditions at which these seeds persist in soil, as the length of snow cover will be greatly altered. The interaction between these factors affecting seed germination is less understood.

Soil temperatures of thawing permafrost regions are predicted to warm by 2–4°C [38, 56] and this has particularly significant implications for seeds because this effectively deepens the soil horizon within which the seed may be dispersed. This may lead to greater burial depths and more infrequent exposure to conditions required for both dormancy breaking and germination [24] possibly leading to higher mortality of seeds over the long term.

Soil moisture has not been monitored in arctic regions but the surface water balance (which is closely correlated with soil moisture being the balance between precipitation and PET) has been shown to be in decline indicating drying of soils and this pattern is predicted to continue in alpine regions [39]. Water cycle changes in high latitudes and high altitudes also include increased groundwater infiltration where permafrost has thawed and ceases to be a physical barrier to water movement from the surface and surface flows such as rivers and streams that are predicted to increase with resultant greater erosion of soils particularly in the winter when there are more likely to be ungerminated viable seeds within these soils [57]. However, the general pattern of river and stream flow is difficult to predict because the aforementioned increase of water infiltration to underground aquifers might represent a diversion from surface flows and increased evapotranspiration may both result in reduced flow rates.

It is not yet known to what extent these varying factors will affect seed germination and it will undoubtedly be the case that the interplay between different influences may shift to favor net germination increase or decrease at very small scales. For this reason, it is imperative that experimental investigations simulate conditions without confounding their own results and are meticulous about reporting the conditions under which they germinate seeds or arctic-alpine species.

EMPIRICAL STUDIES OF WARMING EFFECTS ON SEED GERMINATION

To identify any generalized trend of seed germination in high altitudes during warmer climate, we searched the ISI Web of Knowledge database for studies addressing the effect of warming temperature on seed germination in alpine/arctic ecosystem using key terms "alpine seed germination and climate change" or "arctic seed germination and climate change". This search resulted in 48 papers. We screened all the papers to identify the studies intended to determine the effects of warmer temperature on seed germination pattern. To be included in this work, seeds must be germinated at a warmer temperature and information should be presented about how these seeds were collected and any dormancy breaking treatments applied. With these as the criteria for selecting studies, we created a list by entering the species name and other parameters reported in Table 1 from 9 studies that met the criteria [25–31, 58, 59]. Briefly, each row contains information about how a particular species used in the original study was treated and multiple entries were made if the same species is treated under two different conditions. In this way, we were able to identify 191 cases belonging to 92 taxa with some overlap amongst nine studies [25–31, 58, 59] which subjected seeds collected from alpine/arctic areas to a warming treatment in comparison to a control. In some cases, this control was a result of seeds being stored outside and subjected to natural conditions; in others, the control and warming treatments were created in cultivation to simulate predicted climatic warming.

Table 1: Summary of methodologies of studies subjecting alpine seeds to warming treatments. In the dataset, each combination of reference, taxa, and treatment was represented by a different row totaling 191 combinations (see methods sections for details). The final parameter summarizes the impact of warming treatments using a simple description of difference in germination compared to the control by taxa; "no effect" represents a nonsignificant response ($p > 0.05$) and "no measure of statistical significance" means that authors did not attempt to report the outcomes of warming treatments using analyses of means and variances.

Parameter	Categories/type of response	n	Unit
Reference/taxa/treatment combinations		191	Rows
Reference		9	Studies
Species		92	Taxa
Seed collection method	From plant	26	Rows
	Soil excavation	90	Rows
	Seeds purchased	3	Rows
	Not described	72	Rows
Transport (from field to experimental facility)	Ambient temperature	90	Rows
	Not described	101	Rows

Manipulation of postharvest storage conditions	Dry	150	Rows
	Cold (temperature maintained at c. 5°C)	6	Rows
	Not described	35	Rows
Postharvest storage time	2 days	27	Rows
	1 week	5	Rows
	2 weeks	46	Rows
	2 h–3 wk	6	Rows
	6 weeks	90	Rows
	Not described	17	Rows
Postharvest storage temperature	Room temperature	51	Rows
	5°C	6	Rows
	15°C	90	Rows
	Not described	44	Rows
Light conditions during postharvest storage	Dark	6	Rows
	Not described	185	Rows
Dormancy breaking treatment	Cold-stratification (given under snow cover and different temperature equated to winter conditions and collected after overwintering)	137	Rows
	Assumed to be stratified during postharvest storage	9	Rows
	Cold-stratification plus gibberellic acid	45	Rows
Cold-stratification duration	4 weeks	5	Rows
	12 weeks	3	Rows
	20 weeks	46	Rows
	32 weeks/8 months	12	Rows
	34 weeks	6	Rows
	39 weeks	16	Rows
	10 months	90	Rows
	Not applicable	1	Rows
	Not described	12	Rows
Cold-stratification temperature	0.5°C	26	Rows
	−7 to 0°C	16	Rows
	1–12 wks at 21/15°C, 12–32 wks at 0.5°C	3	Rows
	20 weeks at 2°C, 12 weeks warm stratification	1	Rows
	Beneath snow for approximately 24 weeks and then stored at 2°C	8	Rows
	1–4 wks (0.5°C), 5–8 wks (−5°C), 9–12 wks (−10°C), 13–16 wks (−5°C), and 17–20 wks (0.5°C)	23	Rows
	−20°C	5	Rows
	Field conditions	96	Rows
	Not applicable	1	Rows
	Not described	12	Rows
Light conditions during cold-stratification	Dark	84	Rows
	Natural conditions at midlatitudes	90	Rows
	Not described	5	Rows
	Not applicable	12	Rows
Stage at which temperature increase was imposed	Seed maturation	5	Rows
	Dormancy breaking	46	Rows
	Germination	124	Rows
	Both (dormancy breaking and germination)	16	Rows

Temperature increase compared to control conditions (°C)	1–3°C	5	Rows
	1	2	Rows
	2.4°C	8	Rows
	2.5°C	46	Rows
	2°C	3	Rows
	3°C	11	Rows
	4°C	101	Rows
	6–10°C	6	Rows
	8°C	3	Rows
	5/0°C	2	Rows
	10/4°C	2	Rows
	15/9°C	2	Rows
Germination medium	Agar	16	Rows
	Excavated soil	101	Rows
	Commercial soil	46	Rows
	Filter paper	11	Rows
	Sown in field	2	Rows
	Sown in field (removed natural soil and used commercial soil)	9	Rows
	Soil + perlite	6	Rows
Light conditions during germination	Natural + additional 4 hrs illumination during winter	90	Rows
	Natural	11	Rows
	24 h light	5	Rows
	Simulated field conditions	85	Rows
Length of germination/seedling monitoring time	35 days	5	Rows
	8 weeks	8	Rows
	12 weeks	6	Rows
	13 weeks	57	Rows
	16 weeks	8	Rows
	24 weeks	6	Rows
	30 weeks	90	Rows
	1 year	9	Rows
	4 years	2	Rows
Postgermination viability tests	Embryo inspection	3	Rows
	Not described	188	Rows
Overall effect of warming on proportion/number of seeds germinating	Increase	31	Rows
	Decrease	19	Rows
	No effect	51	Rows
	No measure of statistical significance	90	Rows

[i]Row used here describes a species entered with all its details including seed collection, transport, postharvest storage, postharvest storage conditions, postharvest storage duration, dormancy breaking, cold-stratification conditions, cold-stratification duration, and germination conditions. In the case of same species treated with different conditions, multiple entries were made. Thus, a species can be represented by more than one row. Each unit given as row indicates how many rows correspond to each category. Thus, each unit given as row indicates a particular set of treatment-taxa combination.

[ii]Graae et al. [27] storedseeds of Cerastium alpinum L. ssp. lanatum (Lam.) Asch. and Graebn, Polygonum viviparum L., and Saxifraga cernua L. at room temperature in paper bags in dry state. But the moisture content of the seeds was not given. They assumed the seeds do not require any further stratification to germinate.

[iii]Müller et al. [58] collected soil core containing seeds roughly about 10 months after the seeds had dispersed. They handled the seeds by splitting them in to two groups: (1) one receiving gibberellic acid treatment after 10 months in soil, presumably cold-stratified in the field (45 cases); (2) one receiving no gibberellic acid treatment meaning the seeds had just undergone cold-stratification (45 cases). Please note that since both the groups underwent cold-stratification, data reporting cold-stratification time and duration (in subsequent rows) describes the cold-stratification time and duration even in gibberellic acid treated seeds.

[iv]Some studies only applied warming during germination phase or cold-stratification stage. Few studies applied warming during both cold-stratification and germination phases.

We were able to include only five studies in their entirety; all the combination of treatments applied to different species listed in these five studies met our review requirements: 6 rows [27], 6 rows [31], 46 rows [28], 16 rows [26], and 90 rows [25]. In some other studies, only part of the investigation made it into our analysis as treatments were omitted because they could not be satisfactorily compared to the control. For example, Müller et al. [58] subjected seeds to four treatments of which only one is included in this summary, due to the lack of control in other treatments. Also, many species used in various treatments from this study showed no germination and based on their final germination percentages and comparison of species across various treatments only six species were included in the present work (6 rows). Shevtsova et al. [30] subjected their warming treatments using a Free Air Temperature Increase system to increase the soil temperature by 3°C over different combinations of subsequent time periods. Only the results from warming over the longest time period were included in this analysis (11 rows). In Klady et al. [29], seeds collected from five sites (5 rows) were tested, each having different soil characteristics; we consider germination was altered if there was a significant difference at least in one site. Hobbie and Chapin [59] performed their experiments with five species, but our analysis included only two species (2 rows) that germinated more than 5%.

DIVERSITY OF METHODS USED IN WARMING STUDIES AFFECTING OUR CAPACITY TO DRAW GENERALIZED CONCLUSIONS

Across the studies, a great diversity of methods exists in the way these seeds were handled from maturation to germination (Table 1). We discuss each of these in turn.

Seed Collection

The first factor to be considered is the seed collection method. Seeds may be collected directly from the plant, from the soil surface, or as part of a soil sample using coring or digging out whole turfs. The choice of collection technique has implications for the conclusions that can be drawn, but even within one of these methods, the potential for heterogeneity is great and prevents comparison of results from different studies.

Most of the studies collected fresh seeds that matured under current climatic conditions from soil or directly from plants at the end of maturation drying [26–28, 30, 31]. Seeds of most arctic and alpine species are shed during late summer or autumn and due to the requirement for dormancy breaking, which is satisfied during overwintering, these seeds do not germinate until spring [24, 54]. After overwintering, these seeds come out of dormancy and acquire the ability to germinate at wide range of temperature ranging from as low as 5 to 30°C [22, 24]. However, germination is not an assured event every spring because of the unpredictability in the climatic conditions. As a result, many species are known to enter "soil seed banks" so that these seeds germinate in a later year having favorable conditions. Thus, simple collection of seeds from the plant or soil surface is expected to provide more information about germinating seeds from the "transient seed bank" (i.e., seeds live in soil for less than 5 years) but overlooks how seeds which entered the "persistent seed bank" (i.e., seeds live in soil for more than 5 years) are affected by global warming. Thus, an understanding of germination from recently shed seeds and seeds present in seed banks is necessary. To provide more information on the latter situation, Hoyle et al. [25] dug out 10 cm deep soil samples containing seeds and germinated them in glasshouse conditions to understand the effects of a warming climate on Australian alpine seeds persisted in soil. Therefore, this soil core contained seeds from both "transient" and "persistent" soil banks with the potential that recently shed seeds will become part of the persistent seed bank.

Overall, these methods of collecting seeds from soil or soil core matured under current environmental conditions ignore the fact that maternal environmental conditions in which the seeds mature will be warmer in the future climate. There is clear evidence to show that seeds matured in one microsite may have different physiology compared with other microsites and year [60, 61]. Seeds of arctic-alpine species maturing under warmer conditions are likely to have different physiological characteristics including changes in dormancy status and seed size, than seeds matured under current conditions [50, 58, 62–64]. Interestingly, evidence of rapid climate change affecting the flowering phenology and the duration of seed maturation have been documented in other ecosystems [65–67]. The only investigation that included seeds matured in warmer climatic conditions from an alpine environment is the work by Klady et al. [29]. However, the postmaturation experiments included in their analysis unfortunately do not reflect that of the real-world future climatic conditions (see below).

Postharvest Storage

In many cases, seeds after collection were used almost immediately in the experiment (Table 1). Since the physiology of seeds might change during postharvest storage, the immediate use of seeds in the experiments provides some confidence that seeds have similar characteristics if they had persisted in soil. However, Hoyle et al. [25] air-dried and stored seeds for 6 weeks before incorporating their specimen in subsequent analysis (Table 1). It is therefore not known if the dormancy state of the seeds had altered by this process. Other cases recorded from all the 9 studies that reported to have stored seeds for a few weeks (see Table 1) might also have similar limitations. We believe that this will have been done to avoid seed decay but recommend that, whenever possible, direct transfer to experimental conditions is undertaken.

Dormancy Breaking

Other important factors that merit discussion are the conditions in which dormancy breaking was applied. Given the growing awareness that (1) seed collection time and/or methods influences the dormancy status of seeds and (2) methods used under ex situ conditions may not represent prevailing in situ conditions, ecologists are increasingly becoming conscious of employing ecologically meaningful dormancy breaking treatments. In their natural high latitude and high altitude environments, seeds are kept under snow cover in a fully imbibed state with constant temperature close to 0°C. Thus, amongst several dormancy alleviation methods developed (e.g., cold-dry stratification and GA_3 application), cold-wet stratification is regarded as ecologically relevant [28, 68]. According to C. C. Baskin and J. M. Baskin [22], dormancy breaking with cold-stratification occurs only in imbibed seeds;

if the seeds are completely dry (physiologically <40% RH) the dormancy may not be broken. Liu et al. [68] tested the best dormancy breaking treatments in 489 Eastern Tibet Plateau alpine species and found cold-wet conditions had resulted in better germination compared with dry warm or dry cold conditions.

In certain situations, seeds were allowed to overwinter in the field under current environmental conditions and subsequently germinated in glass house at a warmer temperature inferred from earlier literature [25, 27]. Many compelling new attempts have shown interest in studying the reverse, whereby the effects of various possible cold-stratification conditions that are predicted to result from a warming future climate are compared with the current winter conditions [26, 28]. It has been assumed that these methods would inform us about the effects of a warmer climate on dormancy breaking. These approaches might need some degree of caution because of the variation in cold-stratification conditions applied in those studies. Of particular note is the variation in temperatures and duration of cold-stratification: from −20°C to 0.5°C and 12 weeks to 10 months (Table 1). In the latter case, this was represented by the only soil excavation study [25]; soils were collected in the summer prior to seed set to ensure that seeds had been subjected to natural overwintering conditions. Of course, the 10 months between seed set and soil collection would constitute a time period longer than, and temperatures warmer than, that conventionally thought of as describing the cold-stratification period. On the other hand, Graae et al. [27] assumed seeds stored dry did not require any further stratification and thus did not specially apply any dormancy breaking treatment.

The variation in cold-stratification conditions used in these studies might be interpreted as an attempt by the study authors to equate experimental conditions to the location-specific temperature variation predicted to prevail in the locations from which seeds were collected. However, this seems not to be the case as it does not explain the use of one constant temperature such as −20°C [29], which does not reflect the real-world situation. There is a great deal of evidence to show that seeds under snow cover experience a constant and favorable temperature for dormancy break due to the insulating effects [45, 69]. Although global warming is predicted in some scenarios, to reduce the snow thickness and also the snow cover duration [70], even under these situations, seeds are likely to be exposed to very low temperatures only for a short period of time. Furthermore, predictions for snow cover and duration are likely to very greatly. In addition to these difficulties, it is extremely challenging, if not impossible, to suggest whether optimal cold-stratification conditions might be met, as the temperature experienced by seeds in the soil is affected by not only the depth of the seeds but also the location. For example, at lower elevations, the seeds experience warmer soil temperatures than the same species at higher elevations. One meaningful approach would be to use data loggers to monitor the temperature and use these temperatures in the glass house (see recommendation).

Another approach to break dormancy in the seed samples collected from alpine/arctic locations is the use of gibberellic acid (GA_3) (Table 1). Hoyle et al. [25] applied GA_3 to the seed samples that had undergone 10 months of cold-stratification in soil (45 rows in Table 1). GA_3 treatment has important effects on dormancy alleviation, especially if the seeds have physiological dormancy (PD) [71]. GA_3 application in dormancy breaking studies estimates the level of PD, that is, nondeep, intermediate, and deep [22]. Hoyle et al. [25] showed that 17 out of 45 species only germinated if the dormancy is broken by applying between $5\,mg\,L^{-1}$ and $1\,g\,L^{-1}$ GA_3 to their Australian alpine soil sample. This result indicates even after cold-stratification many species remained still dormant, which were broken with GA_3. Although recurring studies are showing PD is a more common form of dormancy found in alpine species, it is not the only form of dormancy prevalent in alpine seeds. Soil samples dug out from random plots are also likely to contain seeds with physical, morphological, morphophysiological, and combinational dormancy of various level [22]. Thus, breaking dormancy with GA_3 following cold-stratification in the field eventually underestimates the prevalence of other forms of dormancy in their sample. For example, species having an impermeable seed coat, that is, physical dormancy (PY), are less likely to break dormancy when treated with GA_3. Further, they did not report any information about the nongerminated seeds at the time of terminating experiment. Thus, it is not known whether a proportion of seeds failed to overcome dormancy with GA_3 treatment and remain nongerminated.

Germination

One of the fundamental problems with germinating multiple species ex situ is selecting an optimal temperature that is appropriate for all species. In general, the effects of global warming were predicted based on one constant temperature or, sometimes, an alternating temperature regime to mimic diurnal fluctuations to promote germination and compared with a warmer temperature inferred from literature [25]. Criticism of using one temperature to understand the alpine seed behavior has a long history (see Körner [21] for detailed discussion). It is often portrayed that freshly matured alpine seeds germinate well at higher temperature than at a lower temperature [22]. Consequently, many studies have attempted to germinate seeds at various combinations of higher temperature cycles, for example, 35°/25°C, 30°/20°C, or 25°/15°C (e.g., [72]).

However, the situation in the natural environment is entirely different from this simplistic "one" temperature treatment. Temperature increases gradually following snow melt bringing growing conditions for some of the early germinating species [21, 23]. In contrast, late germinating or high-temperature requiring species remainquiescent (sensu [73]) in the soil until the appropriate temperature for germination is attained. These differential temperature requirements are even apparent within genus; after stratification Carex remota resulted in more than 80% germination between 12 and 33°C, whereas C. pendula germinated to 80% only at a higher temperature ranging between 24 and 33°C, although few C. remota and C. pendula germinated at temperatures as low as 6°C and 9°C, respectively [74]. Similarly, Marchand and Roach [75] showed germination of Juncus trifidus occurs only at relatively low temperatures between 12 and 23°C, in contrast to the higher temperature (15 and 28°C) requirement in Potentilla tridentata. In species such as Arenaria groenlandica, germination occurred at a range of temperatures between 12 and 28°C, although the percentage of germinated seeds differs with temperature.

In a warmer alpine climate, the decreased snow cover duration may not satisfy the stratification requirements evolved in seeds adapted at current environmental conditions. Similarly, late snow arrival could also shift seed germination time of some species, for example, Geum, Veronica, and Oxyria, to autumn (i.e., immediately after dispersal), instead of germinating after snow melt in spring as they do currently, probably also due to the seeds maturing in warmer conditions resulting in nondormant or conditionally dormant seeds [26]. Longer snow-free growing seasons expected from warmer climate in the future might increase the window of opportunity for germination, but the success of seedling establishment may be increased or decreased depending largely upon the location and species [30, 31, 59, 76, 77].

In addition, further limitations are incurred by the great range of temperature increases imposed after stratification during germination ranging from 1°C to diurnal peaks of 15°C above the control (Table 1). This range of temperatures is not a problem per se; the response of species to various warming treatments might be illuminating if the earlier treatment of seeds was more consistent. However, the potential for affecting the response of seeds prior to any experimentally controlled treatment is great given the variety of conditions applied during transport, postharvest storage, and pretreatment phases (Table 1). The apparent control of a particular temperature providing conditions for germination and the impact of warmer temperature occurring during the whole growing season affecting germination timing and emergence are also documented in other environments [13, 78–80]. Thus, we emphasize that the response of seed germination in the future climate is more complex and increased research should shed more light on this subject.

Postgermination Viability Testing

A major problem with testing germination of many species at one temperature is that species requiring temperatures outside the given condition generally do not germinate. It is therefore apparently incorrect to conclude seed germination failure as nonviability [81, 82], but rather that the appropriate temperature for germination has not been possibly provided. Evaluating the nongerminated seeds after the germination tests by "tetrazolium test" has alternatively been proved useful in identifying viable but nongerminated seeds [83, 84]. Whilst one cannot ignore the fact that it may not be possible to test the viability when experiments are performed in natural environment, we strongly emphasize whenever possible that future studies must test the nongerminated seeds at the end of germination experiments using tetrazolium test or other viability tests, as this approach could potentially serve as a reasonable way to identify the viable seeds that may require additional dormancy breaking treatments before they germinate. Thus, it is very surprising that 188 cases did not use any postviability assessments (Table 1).

Variation in Germination Response across Species Incorporated in Different Studies

The final parameter in Table 1 reports the impact of warming treatments on the proportion or number of seeds germinating. Thirty-one cases were identified where the impact of warming increased germination, 19 cases showed a decrease in germination after warming, and 51 showed no statistically significant difference to the control (Table 1). There are also inconsistencies in the time of application of warming. Only five cases applied warming treatment during seed maturation. But 46 cases were identified that applied the warming during stratification/dormancy breaking whilst 124 applied warming during the germination phase. The remaining 16 cases looked at warming during both the stratification and the germination phases (Table 1).

It should be noted that many taxa were included in several studies and only seven genera were represented by just one case each in the dataset. This serves to highlight our point that species response to warming appears to be inconsistent, but this conclusion is confounded by the diversity of methods employed and species response. For example, Dryas octopetala is represented four times in our dataset, in all cases germinated in soil; however, germination responds differentially to warming; at low increased temperatures (2.5°C) no effect could be discerned [28] whilst at 3°C the response was negative (i.e., less germination). When the warming treatment varied 6–10°C above the control, the response was positive with Dryas octopetala showing increased germination [30]. Besides the variation in temperature used in the original studies possibly to determine the effects of different warming temperature on seed germination, another possible reason for the lack of consistent results in this species might also be due to the variation in cold-stratification duration and temperature applied to break dormancy, c.34 weeks in field [58] compared to 24 months under snow cover [30] and 0.5°C for 20 wks or 1–4 wks (0.5°C), 5–8 wks (−5°C), 13–16 wks (−10°C), and 17–20 wks (0.5°C) [28]. Similarly, Vaccinium uliginosum was reported to germinate higher [27], lower [28], or at the same level [28, 30] at an increased temperature. Although the difference in increased temperature range used in their studies appears to be the reason for such variation in germination response, the disparity of cold-stratification treatments employed in those studies also plays an important role contributing to the mixed results.

Similar conflicting patterns also exist in numerous other species. Germinating Betula pubescens ssp.czerepanovii at 2.5°C above current temperature level did not have any effect on germination [28], but when the temperature rose 3°C above current level the final germination percentage was significantly reduced [30]. However, accepting this result is at least partly perplexed by the dormancy breaking duration and conditions employed in those studies. Milbau et al. [28] used sequential dormancy breaking conditions: 1–4 wks (0.5°C), 5–8 wks (−5°C), 13–16 wks (−10°C), and 17–20 wks (0.5°C), whereas Shevtsova et al. [30] stored seeds beneath snow cover for approximately 24 weeks and then stored at 2°C (for an unknown period of time) before germinating them. For Empetrum hermaphroditum, a 2.5°C increase in temperature has no significant effect on germination [28], but when temperature increases 10/4, 15/9, and 5/0°C (day/night) compared to control the germination significantly improved [27]. One is tempted to conjecture that breaking dormancy in E. hermaphroditum by cold-stratification and incubating the seeds at higher temperature may result in higher germination percentage, but what effect global warming could have on dormancy breaking phase was not addressed by Graae et al. [27]. Seeds of Epilobium angustifolium germinated to a lesser percentage both when kept under snow cover [30] and after sequential dormancy breaking temperatures [28], but applying cold-stratification at a constant temperature of 0.5°C improved germination when tested at 2.5°C higher than current temperature, explicitly indicating the way by which cold-stratification is applied determines the final germination percentage. Similar inconsistent patterns were also found in Vaccinium uliginosum, as more germination was reported by Graae et al. [27], but lesser germination percentage was reported by Milbau et al. [28].

Because of these conflicting patterns, we agree with the summaries of various authors included here that response to warming is a species-specific impact which cannot be generalized across functional type or regional flora. The nine studies included in this analysis used seeds maturing in different locations: Alaska [59], Australia [25], Canada [58], Greenland [27], Sweden [28, 30], and Italy [26], thus possibly having different initial dormancy and germination ability as reviewed by Jaganathan et al. [24], leading to a difference in the germination pattern. However, we also add that the variable response we identified even using coarse measures of increased or decreased germination within species shows that the species response is highly dependent on the experimental set-up. This fact undermines any attempt to extrapolate experimental responses to the real world and makes predictions difficult, if not impossible, using existing data. Thus, any future attempts to make real predictions of alpine seed germination should consider the fact that, in the warmer world, seeds will be exposed to higher temperatures from flowering to germination. Indeed the whole life-cycle of arctic-alpine plants will be happening at a higher temperature. Thus, using season-specific temperature change (predicted in warmer arctic and alpine soils) to break dormancy and germination of seeds matured in a warmer conditions would provide more realistic insights.

Some general recommendations can be made based on the lessons learned from previous studies. These are common guidelines, but it is evident that studies are increasing without paying particular attention to these issues. Thus, we believe that following these recommendations might help better predict the effects of global warming on high latitude and high altitude ecosystems.

- Efforts must be taken to grow plants in higher temperatures and collect the seeds that matured in a warmer climate or establish permanent plots (see Elkin et al. [9] for detailed information on how to establish long-term plots) and monitor the process of seed maturation and germination in field conditions. ITEX has already taken steps to monitor the changes in seed characteristics grown in a warmer temperature by growing plants in warmer conditions; we suggest this approach should also be extended to germination and seedling phase.
- In case of planning experiments ex situ, seeds must be collected from both plants and freshly shed seeds from the uppermost soil surface (most likely mid-to-late autumn). Although seeds of most alpine plants are small and collecting them at the time of maturity is complex, most of the studies reviewed here collected seeds at the time of maturity indicating collection of freshly matured seeds is possible. In addition, complementary studies must be performed by collecting the soil core and determining the depths to which the targeted species has been buried. Replication should aim to represent the spatial heterogeneity of the system using the scale of plant community patches as a guide. Researchers should be aware that an appropriate system of randomly located plots (using a stratified sampling procedure if necessary) with a set number of cores from within each plot would constitute proper replication.
- Transport seeds and soil samples containing seeds as quickly as possible to conditions that mimic natural environmental changes (i.e., the current conditions or control and the warming treatment(s), both with appropriate photoperiods) in which seeds will be allowed to experience the natural conditions; that is, "temperature control" chambers capable of lowering temperatures to subzero level. However, prior to transporting the seeds, soil from cores should be pooled and mixed by plot and kept moist at levels that mimic natural conditions. Whenever possible, moist soil must be used as germinating medium. This also helps us to monitor the process of germination until seedling stage.
- It is recommended to split the samples collected both from plants and from soil core into various subsamples that are treated to different temperature regimes in parallel. That is to say, one chamber must be designed to match with the temperature of current alpine climate. Other(s) should replicate soil temperature according to recent projections relevant to the location of seed collection. It is well-known that maternal effect may determine the physiological state of seeds, thus affecting the level of dormancy and germination temperature in the freshly collected seeds [24, 85]. To avoid this problem, we suggest using a data logger to measure the soil temperature of seed collected site and follow this temperature in the chamber. Thus, seeds experience field conditions but in the chamber. If this is done over long time periods, that is, 1 to 2 years, the pattern of germination and dormancy breaking under the current environmental conditions can be determined. Using the other subsamples in additional chambers, warming treatments can be applied, that is, current temperature determined from data loggers + temperature replicating the projections under warmer conditions, for example, 2°C. In this way, seeds experience warmer climate from dormancy breaking to germination.
- Seeds need to be monitored regularly to ensure that germination does not go unnoticed. Seedlings should be identified as soon as possible and removed or grown in a different medium until identification is possible.
- When the long-term experiment seems to be finished, the remaining seeds should be sieved from the soil and checked for viability, for example, using the tetrazolium test. Further, we suggest using tetrazolium test in freshly collected seeds, after dormancy breaking step (if applied separately) and possibly at the end of germination to compare the number of seeds deemed viable.

ACKNOWLEDGMENTS

This work would not have been made possible without the help of Ms. Xiao Qun, who spent hours in checking the database created. The authors thank all researchers who provided valuable comments on earlier draft of this paper.

REFERENCES

1. M. Fenner and K. Thompson, The Ecology of Seeds, Cambridge University Press, 2005.
2. R. L. Long, M. J. Gorecki, M. Renton et al., "The ecophysiology of seed persistence: a mechanistic view of the journey to germination or demise," Biological Reviews, vol. 90, no. 1, pp. 31–59, 2015.
3. W. E. Finch-Savage and G. Leubner-Metzger, "Seed dormancy and the control of germination," New Phytologist, vol. 171, no. 3, pp. 501–523, 2006.
4. D. L. Venable, "Bet hedging in a guild of desert annuals," Ecology, vol. 88, no. 5, pp. 1086–1090, 2007.
5. C. G. Willis, C. C. Baskin, J. M. Baskin et al., "The evolution of seed dormancy: environmental cues, evolutionary hubs, and diversification of the seed plants," New Phytologist, vol. 203, no. 1, pp. 300–309, 2014.
6. J. Plue, P. De Frenne, K. Acharya et al., "Climatic control of forest herb seed banks along a latitudinal gradient," Global Ecology and Biogeography, vol. 22, no. 10, pp. 1106–1117, 2013.
7. A. Cochrane, C. J. Yates, G. L. Hoyle, and A. B. Nicotra, "Will among-population variation in seed traits improve the chance of species persistence under climate change?" Global Ecology and Biogeography, vol. 24, no. 1, pp. 12–24, 2015.
8. G.-R. Walther, E. Post, P. Convey et al., "Ecological responses to recent climate change," Nature, vol. 416, no. 6879, pp. 389–395, 2002.
9. C. Elkin, A. G. Gutiérrez, S. Leuzinger et al., "A 2 °C warmer world is not safe for ecosystem services in the European Alps," Global Change Biology, vol. 19, no. 6, pp. 1827–1840, 2013.
10. B. Luna, B. Pérez, I. Torres, and J. M. Moreno, "Effects of incubation temperature on seed germination of mediterranean plants with different geographical distribution ranges," Folia Geobotanica, vol. 47, no. 1, pp. 17–27, 2012.
11. P. De Frenne, B. J. Graae, J. Brunet et al., "The response of forest plant regeneration to temperature variation along a latitudinal gradient," Annals of Botany, vol. 109, no. 5, pp. 1037–1046, 2012.
12. L. R. Iverson, M. W. Schwartz, and A. M. Prasad, "How fast and far might tree species migrate in the eastern United States due to climate change?" Global Ecology and Biogeography, vol. 13, no. 3, pp. 209–219, 2004.
13. R. P. Neilson, L. F. Pitelka, A. M. Solomon et al., "Forecasting regional to global plant migration in response to climate change," BioScience, vol. 55, no. 9, pp. 749–759, 2005.
14. M. K. J. Ooi, T. D. Auld, and A. J. Denham, "Climate change and bet-hedging: interactions between increased soil temperatures and seed bank persistence," Global Change Biology, vol. 15, no. 10, pp. 2375–2386, 2009.
15. F. Giorgi and P. Lionello, "Climate change projections for the Mediterranean region," Global and Planetary Change, vol. 63, no. 2-3, pp. 90–104, 2008.
16. J. Kreyling, "Winter climate change: a critical factor for temperate vegetation performance," Ecology, vol. 91, no. 7, pp. 1939–1948, 2010.
17. J.-P. Theurillat and A. Guisan, "Potential impact of climate change on vegetation in the European Alps: a review," Climatic Change, vol. 50, no. 1-2, pp. 77–109, 2001.
18. G.-R. Walther, S. Beißner, and C. A. Burga, "Trends in the upward shift of alpine plants," Journal of Vegetation Science, vol. 16, no. 5, pp. 541–548, 2005.
19. S. Dullinger, T. Dirnböck, and G. Grabherr, "Modelling climate change-driven treeline shifts: relative effects of temperature increase, dispersal and invasibility," Journal of Ecology, vol. 92, no. 2, pp. 241–252, 2004.
20. T. A. Forbis, "Seedling demography in an alpine ecosystem," American Journal of Botany, vol. 90, no. 8, pp. 1197–1206, 2003.
21. C. Körner, Alpine Plant Life: Functional Plant Ecology of High Mountain Ecosystems, Springer, 2003.
22. C. C. Baskin and J. M. Baskin, Seeds: Ecology, Biogeography, and Evolution of Dormancy and Germination, Elsevier, San Diego, Calif, USA, 2nd edition, 2014.
23. W. D. Billings and H. A. Mooney, "The ecology of arctic and alpine plants," Biological Reviews, vol. 43, no. 4, pp. 481–529, 1968.
24. G. K. Jaganathan, S. E. Dalrymple, and B. Liu, "Towards an understanding of factors controlling seed bank composition and longevity in the alpine environment," The Botanical Review, vol. 81, no. 1, pp. 70–103, 2015.
25. G. L. Hoyle, S. E. Venn, K. J. Steadman et al., "Soil warming increases plant species richness but decreases germination from the alpine soil seed bank," Global Change Biology, vol. 19, no. 5, pp. 1549–1561, 2013.
26. A. Mondoni, G. Rossi, S. Orsenigo, and R. J. Probert, "Climate warming could shift the timing of seed germination in alpine plants," Annals of Botany, vol. 110, no. 1, pp. 155–164, 2012.
27. B. J. Graae, I. G. Alsos, and R. Ejrnaes, "The impact of temperature regimes on development, dormancy breaking and germination of dwarf shrub seeds from arctic, alpine and boreal sites," Plant Ecology, vol. 198, no. 2, pp. 275–284, 2008.
28. A. Milbau, B. J. Graae, A. Shevtsova, and I. Nijs, "Effects of a warmer climate on seed germination in the subarctic," Annals of Botany, vol. 104, no. 2, pp. 287–296, 2009.
29. R. A. Klady, G. H. R. Henry, and V. Lemay, "Changes in high arctic tundra plant reproduction in response to long-term experimental warming," Global Change Biology, vol. 17, no. 4, pp. 1611–1624, 2011.
30. A. Shevtsova, B. J. Graae, T. Jochum et al., "Critical periods for impact of climate warming on early seedling

establishment in subarctic tundra," Global Change Biology, vol. 15, no. 11, pp. 2662–2680, 2009.

31. B. J. Graae, R. Ejrnæs, F. L. Marchand et al., "The effect of an early-season short-term heat pulse on plant recruitment in the Arctic," Polar Biology, vol. 32, no. 8, pp. 1117–1126, 2009.

32. M. Drescher and S. C. Thomas, "Snow cover manipulations alter survival of early life stages of cold-temperate tree species," Oikos, vol. 122, no. 4, pp. 541–554, 2013.

33. C. M. D›Antonio and P. M. Vitousek, "Biological invasions by exotic grasses, the grass/fire cycle, and global change," Annual Review of Ecology and Systematics, vol. 23, no. 1, pp. 63–87, 1992.

34. D. J. Kriticos, R. W. Sutherst, J. R. Brown, S. W. Adkins, and G. F. Maywald, "Climate change and the potential distribution of an invasive alien plant: Acacia nilotica ssp. indica in Australia," Journal of Applied Ecology, vol. 40, no. 1, pp. 111–124, 2003.

35. R. Condit, S. P. Hubbell, and R. B. Foster, "Changes in tree species abundance in a neotropical forest: impact of climate change," Journal of Tropical Ecology, vol. 12, no. 2, pp. 231–256, 1996.

36. I. Ibáñez, J. S. Clark, S. LaDeau, and J. Hille Ris Lambers, "Exploiting temporal variability to understand tree recruitment response to climate change," Ecological Monographs, vol. 77, no. 2, pp. 163–177, 2007.

37. V. F. Briceño, G. L. Hoyle, and A. B. Nicotra, "Seeds at risk: how will a changing alpine climate affect regeneration from seeds in alpine areas?" Alpine Botany, vol. 125, no. 2, pp. 59–68, 2015.

38. L. D. Hinzman, N. D. Bettez, W. R. Bolton et al., "Evidence and implications of recent climate change in Northern Alaska and other Arctic regions," Climatic Change, vol. 72, no. 3, pp. 251–298, 2005.

39. A. Gobiet, S. Kotlarski, M. Beniston, G. Heinrich, J. Rajczak, and M. Stoffel, "21st century climate change in the European Alps—a review," Science of the Total Environment, vol. 493, pp. 1138–1151, 2014.

40. G. Parolo and G. Rossi, "Upward migration of vascular plants following a climate warming trend in the Alps," Basic and Applied Ecology, vol. 9, no. 2, pp. 100–107, 2008.

41. J. P. Dedieu, A. Lessard-Fontaine, G. Ravazzani, E. Cremonese, G. Shalpykova, and M. Beniston, "Shifting mountain snow patterns in a changing climate from remote sensing retrieval," Science of the Total Environment, vol. 493, pp. 1267–1279, 2014.

42. M. Beniston, H. F. Diaz, and R. S. Bradley, "Climatic change at high elevation sites: an overview," Climatic Change, vol. 36, no. 3-4, pp. 233–251, 1997.

43. S. Chersich, K. Rejšek, V. Vranová, M. Bordoni, and C. Meisina, "Climate change impacts on the Alpine ecosystem: an overview with focus on the soil," Journal of Forest Science, vol. 61, no. 11, pp. 496–514, 2015.

44. R. G. Björk and U. Molau, "Ecology of alpine snowbeds and the impact of global change," Arctic, Antarctic, and Alpine Research, vol. 39, no. 1, pp. 34–43, 2007.

45. R. J. Williams, "Patterns of air temperature and accumulation of snow in subalpine heathlands and grasslands on the Bogong High Plains, Victoria," Australian Journal of Ecology, vol. 12, no. 2, pp. 153–163, 1987.

46. K. D. Sommerville, A. J. Martyn, and C. A. Offord, "Can seed characteristics or species distribution be used to predict the stratification requirements of herbs in the Australian Alps?" Botanical Journal of the Linnean Society, vol. 172, no. 2, pp. 187–204, 2013.

47. Y. Vitasse, S. Delzon, C. C. Bresson, R. Michalet, and A. Kremer, "Altitudinal differentiation in growth and phenology among populations of temperate-zone tree species growing in a common garden,"Canadian Journal of Forest Research, vol. 39, no. 7, pp. 1259–1269, 2009.

48. M. V. Price and N. M. Waser, "Effects of experimental warming on plant reproductive phenology in a subalpine meadow," Ecology, vol. 79, no. 4, pp. 1261–1271, 1998.

49. C.-H. A. Wahren, M. D. Walker, and M. S. Bret-Harte, "Vegetation responses in Alaskan arctic tundra after 8 years of a summer warming and winter snow manipulation experiment," Global Change Biology, vol. 11, no. 4, pp. 537–552, 2005.

50. C. Galen and M. L. Stanto, "Short-term responses of alpine buttercups to experimental manipulations of growing season length," Ecology, vol. 74, no. 4, pp. 1052–1058, 1993.

51. G. Kudo and A. S. Hirao, "Habitat-specific responses in the flowering phenology and seed set of alpine plants to climate variation: implications for global-change impacts," Population Ecology, vol. 48, no. 1, pp. 49–58, 2006.

52. G. Bernareggi, M. Carbognani, A. Petraglia, and A. Mondoni, "Climate warming could increase seed longevity of alpine snowbed plants," Alpine Botany, vol. 125, no. 2, pp. 69–78, 2015.

53. B. Liebst and J. Schneller, "Seed dormancy and germination behaviour in two Euphrasia species (Orobanchaceae) occurring in the Swiss Alps," Botanical Journal of the Linnean Society, vol. 156, no. 4, pp. 649–656, 2008.

54. E. Schwienbacher, J. A. Navarro-Cano, G. Neuner, and B. Erschbamer, "Seed dormancy in alpine species,"Flora: Morphology, Distribution, Functional Ecology of Plants, vol. 206, no. 10, pp. 845–856, 2011.

55. I. Wagner and A. M. Simons, "Intraspecific divergence in seed germination traits between high- and low-latitude populations of the arctic-alpine annual Koenigia islandica," Arctic, Antarctic, and Alpine Research, vol. 40, no. 1, pp. 233–239, 2008.

56. W. Haeberli and M. Beniston, "Climate change and its impacts on glaciers and permafrost in the Alps,"Ambio, vol. 27, no. 4, pp. 258–265, 1998.

57. E. J. Scherff, C. Galen, and M. L. Stanton, "Seed dispersal, seedling survival and habitat affinity in a snowbed plant: Limits to the distribution of the snow buttercup, Ranunculus adoneus," Oikos, vol. 69, no. 3, pp. 405–413, 1994.

58. E. Müller, E. J. Cooper, and I. G. Alsos, "Germinability of arctic plants is high in perceived optimal conditions but low in the field," Botany, vol. 89, no. 5, pp. 337–348, 2011.

59. S. E. Hobbie and F. S. Chapin, "An experimental test of limits to tree establishment in Arctic tundra,"Journal of Ecology, vol. 86, no. 3, pp. 449–461, 1998.

60. J. Beckstead, S. E. Meyer, and P. S. Allen, "Bromus tectorum seed germination: between-population and between-year variation," Canadian Journal of Botany, vol. 74, no. 6, pp. 875–882, 1996.

61. G. K. Jaganathan, "Influence of maternal environment in developing different levels of physical dormancy and its ecological significance," Plant Ecology, vol. 217, no. 1, pp. 71–79, 2016.

62. G. H. R. Henry and U. Molau, "Tundra plants and climate change: the International Tundra Experiment (ITEX)," Global Change Biology, vol. 3, no. 1, pp. 1–9, 1997.

63. J. Walck and K. Dixon, "Time to future-proof plants in storage," Nature, vol. 462, no. 7274, p. 721, 2009.

64. A. M. Arft, M. D. Walker, J. Gurevitch et al., "Responses of Tundra plants to experimental warming: meta-analysis of the international tundra experiment," Ecological Monographs, vol. 69, no. 4, pp. 491–511, 1999.

65. S. J. Franks, S. Sim, and A. E. Weis, "Rapid evolution of flowering time by an annual plant in response to a climate fluctuation," Proceedings of the National Academy of Sciences of the United States of America, vol. 104, no. 4, pp. 1278–1282, 2007.

66. Ø. Johnsen, C. G. Fossdal, N. Nagy, J. MØlmann, O. G. Dæhlen, and T. SkrØppa, "Climatic adaptation inPicea abies progenies is affected by the temperature during zygotic embryogenesis and seed maturation,"Plant, Cell & Environment, vol. 28, no. 9, pp. 1090–1102, 2005.

67. A. Hedhly, J. I. Hormaza, and M. Herrero, "Global warming and sexual plant reproduction," Trends in Plant Science, vol. 14, no. 1, pp. 30–36, 2009.

68. K. Liu, J. M. Baskin, C. C. Baskin et al., "Effect of storage conditions on germination of seeds of 489 species from high elevation grasslands of the eastern Tibet Plateau and some implications for climate change," American Journal of Botany, vol. 98, no. 1, pp. 12–19, 2011.

69. C. Lütz, Plants in Alpine Regions: Cell Physiology of Adaption and Survival Strategies, Springer, Vienna, Austria, 2011.

70. S. Wipf, V. Stoeckli, and P. Bebi, "Winter climate change in alpine tundra: plant responses to changes in snow depth and snowmelt timing," Climatic Change, vol. 94, no. 1-2, pp. 105–121, 2009.

71. S. Turner and D. Merritt, "Seed germination and dormancy," in Plant Germplasm Conservation in Australia: Stratergies and Guidelines for Developing, Managing an Utlizing Ex Situ Collection, C. A. Offord and P. F. Meagher, Eds., Australian Network for Plant Conservation, Canberra, Australia, 2009.

72. Y. Shimono and G. Kudo, "Comparisons of germination traits of alpine plants between fellfield and snowbed habitats," Ecological Research, vol. 20, no. 2, pp. 189–197, 2005.

73. A. J. Murdoch and R. H. Ellis, "Dormancy, viability and longevity," in Seeds: The Ecology of Regeneration in Plant Communities, M. Fenner, Ed., pp. 183–214, CABI Publishing, Wallingford, UK, 2000.

74. M. Brändel and W. Schütz, "Temperature effects on dormancy levels and germination in temperate forest sedges (Carex)," Plant Ecology, vol. 176, no. 2, pp. 245–261, 2005.

75. P. J. Marchand and D. A. Roach, "Reproductive strategies of pioneering alpine species: seed production, dispersal, and germination," Arctic and Alpine Research, vol. 12, no. 2, pp. 137–146, 1980.

76. R. Crawford, Plants at the Margin: Ecological Limits and Climate Change, Cambridge University Press, Cambridge, UK, 2008.

77. J. L. Walck, S. N. Hidayati, K. W. Dixon, K. Thompson, and P. Poschlod, "Climate change and plant regeneration from seed," Global Change Biology, vol. 17, no. 6, pp. 2145–2161, 2011.

78. N. M. Nadkarni and R. Solano, "Potential effects of climate change on canopy communities in a tropical cloud forest: an experimental approach," Oecologia, vol. 131, no. 4, pp. 580–586, 2002.

79. F. T. Short and H. A. Neckles, "The effects of global climate change on seagrasses," Aquatic Botany, vol. 63, no. 3-4, pp. 169–196, 1999.

80. M. B. Davis, R. G. Shaw, and J. R. Etterson, "Evolutionary responses to changing climate," Ecology, vol. 86, no. 7, pp. 1704–1714, 2005.

81. L. Bliss, "Seed germination in arctic and alpine species," Arctic, vol. 11, no. 3, pp. 180–188, 1958.

82. M. T. K. Arroyo, L. A. Cavieres, C. Castor, and A. M. Humaña, "Persistent soil seed bank and standing vegetation at a high alpine site in the central chilean andes," Oecologia, vol. 119, no. 1, pp. 126–132, 1999.

83. L. A. Cavieres and M. T. K. Arroyo, "Seed germination response to cold stratification period and thermal regime in Phacelia secunda (Hydrophyllaceae)—altitudinal variation in the Mediterranean Andes of central Chile," Plant Ecology, vol. 149, no. 1, pp. 1–8, 2000.

84. L. Giménez-Benavides, A. Escudero, and F. Pérez-García, "Seed germination of high mountain Mediterranean species: altitudinal, interpopulation and interannual variability," Ecological Research, vol. 20, no. 4, pp. 433–444, 2005.

85. S. E. Meyer, S. G. Kitchen, and S. L. Carlson, "Seed germination timing patterns in Intermountain Penstemon (Scrophulariaceae)," American Journal of Botany, vol. 82, no. 3, pp. 377–389, 1995.

CHAPTER 17

EFFECTS OF CLIMATE AND LAND USE ON HERBACEOUS SPECIES RICHNESS AND VEGETATION COMPOSITION IN WEST AFRICAN SAVANNA ECOSYSTEMS

Issouf Zerbo[1], Markus Bernhardt-Römermann[2], Oumarou Ouédraogo[1], Karen Hahn[3], and Adjima Thiombiano[1]

[1]Laboratory of Plant biology and Ecology, University of Ouagadougou, 03 BP 7021, Ouagadougou 03, Burkina Faso

[2]Institute of Ecology, Friedrich Schiller University Jena, Dornburger Strasse 159, 07743 Jena, Germany

[3]Department of Ecology and Geobotany, Institute for Ecology, Evolution and Diversity, J. W. Goethe University, Max-von-Laue-Strasse 13, 60438 Frankfurt am Main, Germany

ABSTRACT

West African Savanna ecosystems are undergoing severe changes in their vegetation composition due to the impact of human land use and changes in climatic conditions. This study aims to examine the effect of climate, land use, and their interaction on species richness and composition of West African herbaceous vegetation. Plot based vegetation sampling was done in Burkina Faso. Specific richness and diversity indices were used to determine the effect of land use, climate, and their interaction. An importance value was computed to determine herbaceous species dominating the communities. Frequency of species is used to examine their distribution pattern. The results showed that climate significantly influenced herbaceous specific richness more than land use. However, land use had a significant effect on herbaceous vegetation composition. Herbaceous species diversity changed with environmental conditions. The floristic composition of dominant species is driven by both climate and land use. The frequency of distribution demonstrated that herbaceous species occurrences were more influenced by the mixed effect of climate and land use than their separate effects. Occasional and rare species are the most important part of herbaceous vegetation. Thus heterogeneity of Savanna ecosystem and vulnerability of herbaceous species are high.

INTRODUCTION

West African Savanna ecosystems are undergoing severe changes in their vegetation composition and species cover due to the impact of human land use and changes in climatic conditions. Indeed, African Savannas represent a fundamental environment for people practicing stock breeding and agriculture [1] and collecting nontimber forest products (NTFP) for multiple purposes [2–4]. In recent decades, the most dramatic changes were increasing human population and consequently an increase in land use intensity often related to increased livestock density and the extension of agricultural lands, or a shortening of fallow periods [5, 6]. Climate change and human activities (land use, chopping, fire, and grazing) are leading to the degradation of these ecosystems [3, 7–13]. These changes threaten biodiversity and sustainable land use [6].

Citation: Issouf Zerbo, Markus Bernhardt-Römermann, Oumarou Ouédraogo, Karen Hahn, and Adjima Thiombiano, "Effects of Climate and Land Use on Herbaceous Species Richness and Vegetation Composition in West African Savanna Ecosystems," Journal of Botany, vol. 2016, Article ID 9523685, 11 pages, 2016. doi:10.1155/2016/9523685

Several studies predicted dramatic changes for West Africa vegetation for the next 50 years: for example, Savannas are predicted to expand in the next few centuries at the expense of tropical forests, mainly as a result of deforestation and human fires [14]. On the contrary, a tendency to greening is predicted because of the increase in precipitation in West Africa [15]. The main question is how climate and land use interact to affect West African herbaceous vegetation?

Many authors studied the effect of land use and climate on West African Savannas vegetation separately. Most of these studies showed that anthropogenic activities affect woody vegetation composition [16–20] and climate influences its species richness and stand structures [21–25]. Other studies revealed that land use and climate affect vegetation composition [12, 26–30]. Authors who studied the herbaceous vegetation demonstrated that grazing, fire, and selected cutting [5, 9, 10, 31, 32] influenced herbaceous biomass. Devineau and Fournier [33] pointed out that the characteristics of the top soil horizon are most important for herbaceous species installation. Among these studies, none focused implicitly on the impact of climate, land use, and their interaction on Savanna herbaceous species richness and composition. Given that herbaceous species are the most sensitive ecosystem components of Savanna ecosystems [25, 33] and climate change, it is essential to study the effect of the interaction of land use and climate.

The scope of this study is to determine the effect of climate, land use, and their interaction on herbaceous species richness and composition in Savanna vegetation. We assume that investigating these factors and their interaction may clarify which change may take place in West African Savanna herbaceous diversity. With this study we will test the following hypotheses:

- The specific richness or taxonomic richness of herbaceous vegetation is more driven by climate than by land use.
- The composition of herbaceous vegetation varies due to climate and land use conditions.
- The frequency of herbaceous species depends more on the interaction between climate and land use than on their individual effects.

METHODS

Study Sites

The study was conducted along a climatic gradient in three protected areas and their surroundings used by local people for multiple purposes (fallows, grazing, collecting of nonforest product, etc.). The three study sites are representative for the phytogeographical zoning of Burkina Faso described by Guinko [34] (Figure 1).

- The Sylvopatoral Reserve and Partial Fauna Reserve of Sahel: it is a protected area of IUCN category VI [35]. It is located in the Sahelian zone with a mean annual rainfall ranging between 300 mm and 600 mm. Mean annual temperature ranges from 23.5°C to 35°C.
- The Northern part of National W Biosphere Reserve: W Biosphere Reserve is a protected area of IUCN category II [35]. The study site is located in the North Sudanian zone. The mean annual rainfall ranges between 600 mm and 900 mm. Mean annual temperature ranges from 25.5°C to 33°C.
- The Southern part of Classified Forest and Partial Fauna Reserve of Pama: Pama Reserve is a protected area of IUCN category IV [35]. The study site is located in the South Sudanian zone. The mean annual rainfall ranges between 900 mm and 1200 mm. Mean annual temperature ranges from 25.9°C to 31.4°C.

We distinguished between two types of land use (communal area and protected area); in "protected areas" all human activities are strictly prohibited, while in "communal areas" all kinds of human land use activities are allowed (like grazing, agriculture, and harvest of nontimber forest products).

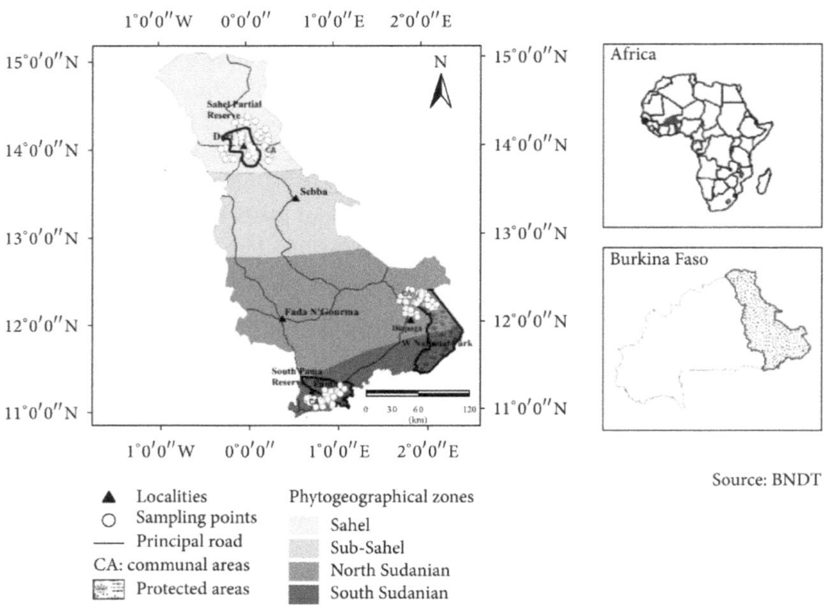

Figure 1: Study sites.

Data Collection

Per study site, two contrasting land use types were compared: protected and communal areas. Per land use type and study site, herbaceous vegetation composition was sampled during the rainy season (September to November) on plots of 10 m × 10 m. Additionally, we distinguished between four different habitats (dry, bowal, fresh, and wet). For each habitat 10 plots were sampled. Thus, vegetation sampling resulted in 240 plots (2 land use types × 4 habitats × 3 study areas × 10 repetitions). Per plot the percentage of cover occupied by each herbaceous species was visually estimated using the method of Braun-Blanquet [36] with the cover classes r (rare species): cover less than 1%, +: cover less than 1%, 1a: cover between 1 and 3%, 1b: cover between 3 and 5%, 2a: cover between 5 and 15%, 2b: cover between 15 and 25%, 3: cover between 25 and 50%, 4: cover between 50 and 75%, and 5: cover between 75 and 100%.

Data Analysis

Specific richness or taxonomic richness of an ecosystem is defined as the number of species or taxa that can be found there regardless of the number of individuals or the biomass of each taxon [37]. To examine the effect of climate gradient and land use on specific richness, we considered family richness, genera richness, and species richness as the number of families or genera or species occurring in a given climate zone or land use type. To estimate the effect of the explanatory variables climate and land use on specific richness, species richness, genera richness, and family richness (response variables) we used Generalized Linear Models with Poisson errors [21].

There are several numbers of diversity which differ greatly and make comparisons difficult and confusing, which hampers also a sound interpretation [38]. To describe the effect of the interaction of climate and land use on herbaceous vegetation composition, we used the series of diversity numbers and evenness presented by Hill [39], which is well indicated for ecological interpretation. Hill's family of diversity numbers are

- $N0 = S$, where S is the number of species in a plot;
- $N1 = e^{H'}$, where $H' = -\text{Sum}(P_i * \ln(P_i))$ is Shannon's index;
- $N2 = 1/D$, where $D = \text{Sum}(P_i^2)$ is Simpson's index, with P_i relative abundance of species in a plot.

$N0$ is the effective number of species in the sample regardless of their abundance. $N1$ measure the number of abundant species in the sample. $N2$ is the number of very abundant species.

Evenness (E) = H/ ln(S)

It is an expression of the balance in the distribution of individuals among the species. Its value approaches 0 when one species is highly dominant and 1 when all species have similar densities. E is independent from species numbers occurring on a plot.

Diversity indices were analyzed with the linear model ($\alpha = 0.05$) [21, 37] because they fulfilled the assumptions of normality and variance homogeneity. In all cases climate and land use conditions are treated as explanatory variables and diversity indices as response variables. Student's t-test at 5% of significance was used as host-post test to detect difference between combinations.

To evaluate β-diversity (similarity between climate-land use conditions), Sorensen similarity index based on species data abundance and presence/absence was used to evaluate similarity between each climate and land use condition. This index potentially varies between 0 and 1: a value close to 1 indicates greater similarity between plots and hence low beta-diversity [40].

The importance value of each species was calculated per climate and land use combination to identify dominant herbaceous species. The importance value index was considered as the sum of Relative Frequency and Relative Dominance according to the following parameters:

- Frequency = number of plots in which the species occurs/total plot number × 100.
- Relative Frequency = frequency of a species/sum of all frequencies × 100.
- Relative Dominance = total cover of a species/total cover of all species × 100.
- Importance value = Relative Frequency + Relative Dominance.

In order to determine the effect of climate, land use, and their interactions on the occurrence of herbaceous species, the frequency of each species was computed and compared [41]. An χ^2-test was used to observe the significance of the variation of species frequency [41]. In addition relative abundance of species and rank-frequency diagrams [37, 42] were used to determine herbaceous vegetation structure between climate-land use conditions. Frequent species are those that are widely distributed within all climate-land use conditions. Occasional and rare species are those that are confined to one type of climate-land use conditions or which are accidentally found in an environment:

- Frequency ≥ 50%: frequent species.
- 25 ≥ Frequency < 50%: fairly frequent species.
- 5 ≥ Frequency < 25%: occasional species.
- Frequency ≤ 5%: rare species.

All statistical analyses were done using R.3.0.2 [43].

RESULTS

Richness of Herbaceous Vegetation along Climatic and Land Use Gradient

The inventoried flora of the three sites included 374 species of which 307 are herbaceous species and 67 juveniles of woody species. The herbaceous species belong to 152 genera and 39 families. Overall, herbaceous vegetation is dominated by the families of Poaceae (31.8%), Fabaceae (13.2%), Cyperaceae (9.5%), and Malvaceae (6.4%).

Specific richness increased along the climatic gradient from the Sahel to the South Sudanian vegetation zone (Table 1). The number of families did not vary significantly between the three climatic zones (Table 1). On the contrary the number of genera and specific richness were significantly lower in the Sahel than in the Sudanian zones ($p < 0.05$). The same pattern was found overall and for communal area and protected area.

Table 1: Climate effect on specific richness of herbaceous vegetation (GLMs analysis).

		Sahel	North Sudanian	South Sudanian	*F*	*p*
Families	Overall	24	33	34	2.0800	0.3535
	Protected area	15	30	28	5.9071	0.0522
	Communal area	25	27	33	1.2019	0.5483
Genus	Overall	74	117	125	15.1200	0.0005*
	Protected area	31	100	102	49.2051	<0.0001*
	Communal area	74	95	105	7.6152	0.0403*
Species richness	Overall	130	198	208	21.2500	<0.0001*
	Protected area	41	147	161	87.9035	<0.0001*
	Communal area	128	150	168	7.4399	0.0459*

*Significant effect of the statistic test.

Analyzing the influence of land use, species, genera, and family richness did not vary significantly between protected and communal areas ($p < 0.05$) except for the Sahel where we observed higher diversities in the communal areas (Table 2).

Table 2: Land use effect on specific richness of herbaceous vegetation (GLMs analysis).

		Protected area	Communal area	*F*	*p*
Families	Overall	33	37	0.2287	0.6325
	Sahel	12	25	4.6665	0.0308*
	North Sudanian	30	27	0.1580	0.6910
	South Sudanian	28	33	0.4103	0.5218
Genus	Overall	125	145	1.4828	0.2233
	Sahel	31	74	18.1381	<0.0001*
	North Sudanian	100	95	0.1282	0.7203
	South Sudanian	102	105	0.0435	0.8348
Species richness	Overall	223	263	3.2959	0.0695
	Sahel	41	128	47.0106	<0.0001*
	North Sudanian	147	150	0.0303	0.8618
	South Sudanian	161	168	0.1489	0.6995

Significant effect of the statistic test.

Diversity and Similarity of Herbaceous Species between Climate and Land Use Conditions

The analysis of diversity of herbaceous vegetation by Hill numbers allowed underlining the variations of mean species numbers ($N0$), abundant species numbers, and very abundant species numbers between climate and land use types (Table 3).

Table 3: Herbaceous vegetation diversity indices along climate-land use conditions.

Climatic zone	Land use	Mean richness $N0$	exp(H') $N1$	$1/D$ $N2$	Evenness (E)
Sahel	Communal area	14.34 ± 4.71^{d}	8.23 ± 3.15^{d}	1.27 ± 0.23^{a}	0.77 ± 0.09^{c}
	Protected area	19.50 ± 4.74^{bc}	10.05 ± 3.12^{cd}	1.20 ± 0.06^{abc}	0.77 ± 0.05^{c}
North Sudanian	Communal area	21.67 ± 7.90^{b}	13.31 ± 5.43^{b}	1.15 ± 0.08^{bc}	0.83 ± 0.06^{b}
	Protected area	15.86 ± 5.90^{cd}	10.58 ± 4.14^{c}	1.19 ± 0.12^{b}	0.84 ± 0.06^{b}
South Sudanian	Communal area	25.58 ± 6.34^{a}	18.00 ± 4.60^{a}	1.09 ± 0.04^{cd}	0.89 ± 0.03^{a}
	Protected area	19.51 ± 6.35^{b}	14.63 ± 4.96^{b}	1.11 ± 0.04^{d}	0.90 ± 0.03^{a}

Superscript letters indicate significant differences according to Student's t-test. The numbers with the same letters indicate that there is no significant difference.

Mean species numbers by plot ($N0$) showed that the South Sudanian zone was more diverse than North Sudanian zone which was also more diverse than the Sahelian zone (d.f. = 5, $F = 28.56, p < 0.0001$). Besides, in each climate zone, there is a significant difference between communal and protected areas. Mean species number varies from 14.34 ± 4.71 in the Sahel communal area to 25.58 ± 6.34 species in the South Sudanian communal area.

The analysis of herbaceous vegetation by Exp(H') ($N1$) revealed that the number of abundant species increases significantly (d.f. = 5, $F = 28.65$, $p < 0.0001$) with the precipitation gradient from the Sahel to the South Sudanian zone. Besides, in the Sahelian zone, there are no significant differences between communal and protected areas. In contrast, in the North and South Sudanian zone, we observed that the number of abundant species in communal areas is significantly high compared to that in protected areas. The number of abundant species varied from 8.23 ± 3.15 species in the Sahel communal area to 18.00 ± 4.60 in the South Sudanian communal area.

By considering $1/D$ ($N2$), the number of very important species decreases significantly from the Sahel to the South Sudanian zone (d.f. = 5, $F = 12.14$, $p < 0.0001$). But in each climate zone there are no significant differences between communal and protected areas in terms of numbers of very abundant species. The number of very abundant species does not exceed two (1.09 ± 0.04 in the South Sudanian communal area to 1.27 ± 0.23 in the Sahelian communal area).

The evenness pattern increased significantly (d.f. = 5, $F = 28.56$, $p < 0.0001$) with the precipitation gradient from the Sahel to the South Sudanian zone. Evenness index is high in each climate and land use combination but does not differ significantly between communal and protected areas regardless of the climate zone. Its value ranges from 0.77 ± 0.09 in the Sahel to 0.90 ± 0.03 in the South Sudanian protected area.

The composition of herbaceous vegetation varied according to the climate and land use conditions. Sorensen coefficient of similarity reveals that similarity is low between all climate and land use conditions (Table 4). There was a higher difference between the Sahelian and Sudanian zones than between South Sudanian and North Sudanian zone. The difference between communal and protected area was higher in the Sahel and lower in the Sudanian zones.

Table 4: Sorensen similarities of climate and land use types.

Climatic zone	Land use	Sahel		North Sudanian		South Sudanian	
		Communal	**Protected**	**Communal**	**Protected**	**Communal**	**Protected**
Sahel	Communal	1					
	Protected	0.25	1				
North Sudanian	Communal	0.23	0.17	1			
	Protected	0.16	0.07	0.50	1		
South Sudanian	Communal	0.20	0.07	0.50	0.45	1	
	Protected	0.16	0.07	0.31	0.48	0.53	1

Dominant Herbaceous Species according to Climate-Land Use Conditions

In the Sahelian zone, four herbaceous species were found as new for the flora of Burkina Faso. They areDigitaria fuscescens (Presl) Henrard (Poaceae) on wet habitat; Trichoneura mollis (Kunth) Ekman (Poaceae) on sand dune; Sporobolus spicatus (Vahl) Kunth (Poaceae) on wet habitat, and Abutilon macropodum Guill. & Perr. (Malvaceae) on sand dune.

We identified the dominant herbaceous species per climatic zone and land use type on the basis of their importance value index (IVI ≥ 5) (Table 5). In the communal area of the Sahel, the dominant species wereZornia glochidiata, Schoenefeldia gracilis, Panicum laetum, Cassia obtusifolia, and Alysicarpus ovalifolius. The protected area was characterized by a dominance of Digitaria horizontalis, Zornia glochidiata, Eragrostis tremula, Alysicarpus ovalifolius, and Cenchrus biflorus. In the communal area of the North Sudanian zone,Microchloa indica, Loudetia togoensis, Pennisetum pedicellatum, Tripogon minimus, and Zornia glochidiata are the most dominant species. In the protected area, important species are Loudetia togoensis, Hyparrhenia involucrata, Andropogon pseudapricus, Microchloa indica, and Andropogon gayanus. In the South Sudanian communal area, Andropogon pseudapricus, Pennisetum pedicellatum, Microchloa indica, Tephrosia pedicellata,and Spermacoce stachydea are the most dominant species. In the protected area, the most important ones areAndropogon gayanus, Andropogon pseudapricus, Hyparrhenia involucrata, Chamaecrista pratensis, andSorghastrum bipennatum.

Table 5: Herbaceous species with an importance value ≥ 5 according to climate and land use conditions.

Climatic zone	**Sahel**		**North Sudanian**		**South Sudanian**	
Land use	**Communal area**	**Protected area**	**Communal area**	**Protected area**	**Communal area**	**Protected area**
Alysicarpus ovalifolius	6.24	9.70	4.64	2.82	6.16	2.99

Andropogon gayanus	3.72	—	—	7.54	3.37	11.64
Andropogon pseuda-pricus	—	—	—	8.37	9.91	10.09
Aristida adscensionis	5.44	—	—	—	—	—
Cassia obtusifolia	7.76	4.38	—	—	3.24	—
Cenchrus biflorus	3.19	9.79	—	—	—	—
Cenchrus prieurii	—	7.22	—	—	—	—
Ceratotheca sesa-moides	—	5.87	—	—	—	—
Chamaecrista pra-tensis	2.79	6.51	3.28	2.86	4.55	6.26
Corchorus tridens	4.03	6.87	3.55	—	—	—
Digitaria gayana	—	5.18	—	—	—	—
Digitaria horizon-talis	—	33.61	—	—	—	—
Eragrostis tremula	5.41	9.87	—	—	—	—
Hyparrhenia involu-crata	—	—	—	9.03	—	7.83
Indigofera bracteo-lata	—	—	—	—	3.77	5.38
Indigofera pilosa	—	7.37	—	—	—	—
Loudetia togoensis	—	—	9.19	12.72	—	3.11
Microchloa indica	—	—	9.51	8.05	7.33	—
Mitracarpus hirtus	—	—	5.08	—	—	—
Panicum laetum	8.49	—	—	—	—	—
Pennisetum pedicel-latum	—	—	9.09	5.17	8.16	4.42
Schoenefeldia gracilis	17.68	—	—	—	—	—
Setaria pumila	—	—	5.93	—	—	—
Sorghastrum bipen-natum	—	—	—	—	3.40	5.43
Spermacoce radiata	5.32	7.03	3.70	—	4.12	—
Spermacoce stachy-dea	—	—	3.90	3.03	6.71	5.40
Sporobolus pyrami-dalis	—	—	5.44	—	3.29	—
Stylosanthes erecta	—	6.82	—	—	—	—
Tephrosia bracteo-lata	—	—	—	2.63	—	5.11
Tephrosia pedicellata	—	4.79	5.78	3.06	7.25	—
Tripogon minimus	—	—	8.35	5.00	4.10	—
Zornia glochidiata	15.63	24.82	6.65	—	—	—

Impact of Climate, Land Use, and Their Interaction on Herbaceous Species Frequency

The occurrence of 23.6% of all species was affected by land use type and 59.8% by climate (Figure 2). Interestingly, for 84.5% of the species, the interaction between climate and land use was important.

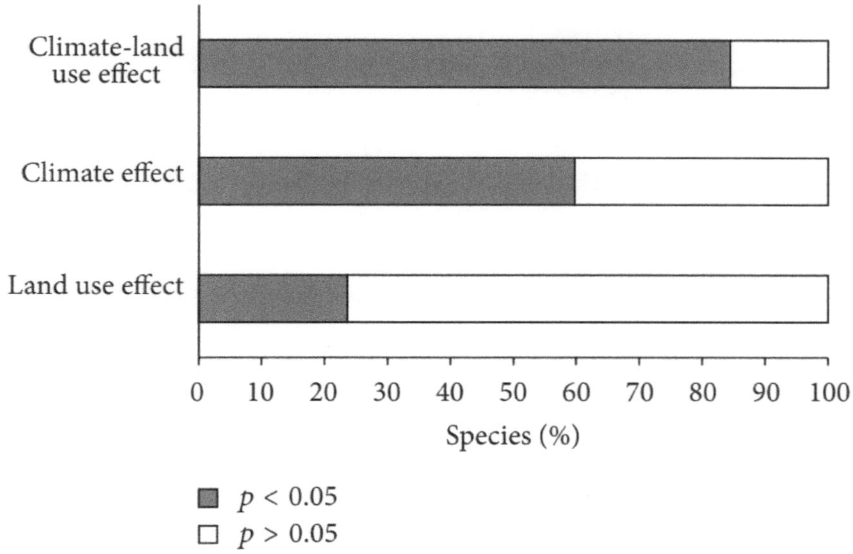

Figure 2: Percentage of herbaceous species according to their responses to land use, climate gradient, and their interaction ($p < 0.05$: significant responses; $p > 0.05$ no significant responses).

In all climate zone and land use combinations, herbaceous vegetation is dominated by occasionally occurring and rare species excepted in the protected area of the Sahel (Figure 3). The abundance of occasional species ranged from 41.6% (South Sudanian protected area) to 57.1% (North Sudanian protected area). That of rare species varied from 28.0% (North Sudanian communal area) to 45.3% (South Sudanian protected area). In the Sahel protected area, frequent species (51.2%) and occasional species are abundant (36.6%).

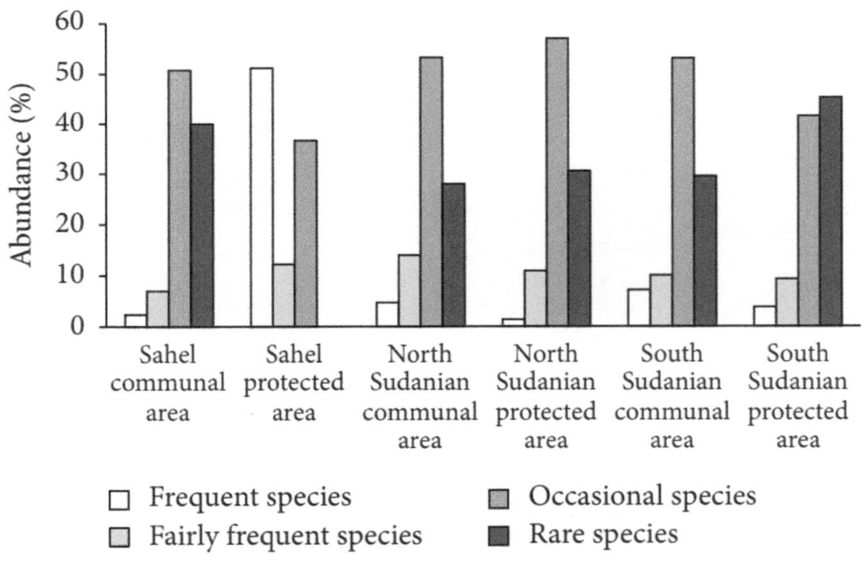

Figure 3: Abundance of herbaceous species according to their frequency classes.

The rank-frequency diagram allows distinguishing frequent species, fairly frequent species, occasional species, and rare species. Figure 4 illustrated only some herbaceous species among several.

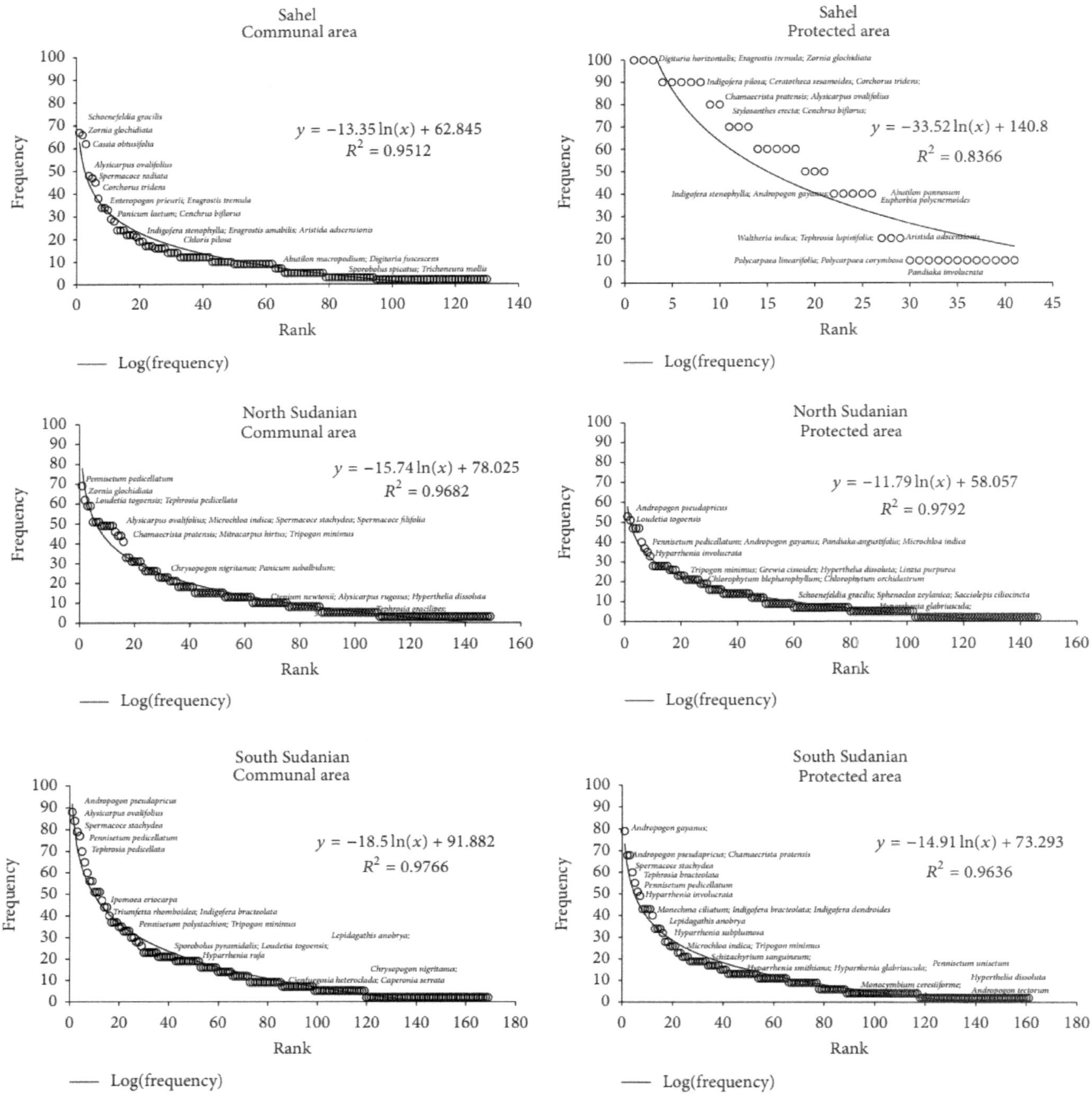

Figure 4: Rank-frequency diagram.

In the Sahel communal area, *Schoenefeldia gracilis*, *Zornia glochidiata, and Cassia obtusifolia* were the most frequent species (Figure 4). *Alysicarpus ovalifolius* and *Enteropogon prieurii* were found among fairly frequent species. Among occasional species were *Aristida adscensionis* and *Indigofera stenophylla*. *Trichoneura mollis*, *Sporobolus spicatus*, and *Digitaria fuscescens* are found among rare species. In the protected area of the same climatic zone *Digitaria horizontalis, Eragrostis tremula*, and *Cenchrus biflorus* were frequent. Fairly frequent species were represented by *Abutilon pannosum* and *Indigofera stenophylla*. Among occasional species were *Tephrosia lupinifolia*, *Aristida adscensionis*, and *Pandiaka involucrata*.

In the North Sudanian communal area, *Pennisetum pedicellatum*, *Loudetia togoensis*, and *Microchloa indicabelong* to the most frequent species. Among fairly frequent species were *Chamaecrista pratensis, Spermacoce stachydea*, and *Tripogon minimus*. *Sporobolus pyramidalis* and *Chrysopogon nigritanus* were among occasional species. Among rare species were *Alysicarpus rugosus*, *Ctenium newtonii*, and *Hyperthelia dissoluta*. In the protected area, only *Andropogon pseudapricus* and *Loudetia togoensis* were more frequent. *Andropogon gayanusand Hyparrhenia involucrata* were found among fairly frequent species. *Hyperthelia dissoluta, Linzia purpurea, Chlorophytum blepharophyllum*, and *Chlorophytum orchidastrum* were encountered among occasional species.

Hyparrhenia glabriuscula, Sacciolepis ciliocincta, and *Sphenoclea zeylanica* can be cited as rare species.

In the South Sudanian communal area, *Andropogon pseudapricus, Alysicarpus ovalifolius, Spermacoce stachydea*, and *Pennisetum pedicellatum* were more frequent. *Tripogon minimus* and *Pennisetum polystachion* were encountered among fairly frequent species. Occasional species were represented by species like *Sporobolus pyramidalis* and *Hyparrhenia rufa*. Among rare species were *Chrysopogon nigritanus, Cienfuegosia heteroclada*, and *Caperonia serrata*. In the protected area, most frequent species were *Andropogon gayanus, Andropogon pseudapricus*, and *Chamaecrista pratensis*. Among fairly frequent species were *Monechma ciliatum, Indigofera dendroides*, and *Lepidagathis anobrya*. Occasional species can be represented by *Hyparrhenia smithiana, Hyparrhenia glabriuscula*, and *Pennisetum unisetum* and rare species by *Andropogon tectorum* and *Monocymbium ceresiiforme*.

DISCUSSION

Effect of Climatic Gradient and Land Use on Savanna Herbaceous Species Richness

In Savanna ecosystems, herbaceous species richness increases with increasing annual precipitation. Species richness in the Sudanian zone is higher than in the Sahelian zone. These results are in accordance with Bognounou et al. [21] and Schmidt et al. [27], who observed, respectively, increasing of woody and overall species numbers with the increasing of rainfall and the duration of the rainy season. Species richness at landscape scale in West Africa is expected to rise with precipitation towards the South [44], a pattern in accordance with our results found for Burkina Faso [21, 27, 45].

The absence of a significant difference in species richness between the protected and the communal area in the Sudanian zone provides evidence that land use affects species richness of herbaceous vegetation less than climate. Devineau et al. [26] and Shackleton [46] found at the landscape scale, that herbaceous vegetation richness does not differ inside and outside protected areas.

The difference observed for the Sahel between species richness in communal and protected areas might be explained by the fact that this protected area was realized by fencing. Compared to the Sudanian zone of Burkina Faso, in the Sahel zone, the protected area (the Sylvopatoral Reserve and Partial Fauna Reserve of Sahel) was most affected by human pressure. To reduce the impact of human pressure in this protected area in Sahel zone the work was done in fenced area. Consequently, these protection issues probably might explain the observed difference in species richness between protected and communal areas.

The dominance of families of Poaceae, Fabaceae, Cyperaceae, and Malvaceae in all study areas reflects that these families are the most frequent in Savanna herbaceous vegetation in West Africa [1, 45, 47]. The Poaceae family is well represented, as grasses are the largest family of vascular plants in Savannas [45, 48–50].

Effect of Climate and Land Use on Herbaceous Vegetation Diversity and Composition

Hill's diversity numbers reveal that herbaceous vegetation was more diverse in the Sudanian zone than the Sahelian zone. The significant difference of mean species and abundant species numbers along the climate gradient shows that precipitation increase favors an increase of herbaceous species richness in Savanna ecosystem. The low values found for very abundant species show that the different climate and land use conditions were dominated by a group of species and the relative high evenness value underlines that dominant species have a regular distribution of their individuals [1, 24, 51, 52]. In fact, species diversity is often dominated by species evenness patterns [51].

The similarity in herbaceous species composition and abundance between climate and land use conditions is low. The low similarity in herbaceous species composition and abundance between climate-land use conditions reflects differences in climate conditions and land use regimes. In addition, the low to moderate level of similarity between climate and land use combinations indicates high beta-diversity and accentuates the importance of climate and land use in explaining herbaceous species diversity at larger spatial scale.

The difference in herbaceous vegetation composition is related to the dual effect of climate and land use conditions. Climate conditions explain the difference in species composition between different climate zones [21, 27]. Land use causes differences in species composition within a climate zone. Indeed, human disturbances influence composition and distribution of plant species [5]. Patterns of species richness and cover are determined by multiple environmental factors (climate change and human activities) [53]. Generally, local species richness and diversity of Savanna ecosystems are maintained by dynamic interactions between local colonization from species pools at larger spatial scales and local extinction due to competitive exclusion [11]. Thus the herbaceous vegetation of West African Savannas is dominated by a pool of species which vary according to the climatic zones and land use conditions. In the Sahel, Schoenefeldia gracilis, Zornia glochidiata, Panicum laetum, and Cassia obtusifolia (communal area) and Digitaria horizontalis and Zornia glochidiata(protected area) are the most dominant herbaceous species. According to Kadeba et al. [52], in the Sahel,Schoenefeldia gracilis is a dominant species of glacis and Cassia obtusifolia and Panicum laetum are dominant species in lowlands and Digitaria horizontalis and Zornia glochidiata on sand hills. The predominance of grass species such as Microchloa indica, Loudetia togoensis, Pennisetum pedicellatum, Tripogon minimus, andAndropogon pseudapricus in the communal areas of the Sudanian zone could be explained by grazing pressure which favors installation of annual or small size herbaceous species [50]. Protection allows establishment and occurrence of taller robust species such as Hyparrhenia involucrata, Andropogon gayanus, Andropogon pseudapricus, and Hyparrhenia subplumosa.

Effect of Climate and Land Use on Savanna Herbaceous Species Frequency of Distribution

The frequency of herbaceous species is more influenced by the interaction between climatic conditions and land use regimes than their separate effects. In fact, the distribution of herbaceous species is principally affected by climate, which determines geographical affinity [27], and land use, which affects the spatial heterogeneity of vegetation and influences ecosystem processes [1, 54]. This results in the fact that a species which is the most frequent under a given climate and land use condition becomes rare under other climate and land use conditions.

Overall, most of the listed species are classified as occasional or rare. The low percentage of frequent species and the high rate of occasional and rare species implies on the one hand a great heterogeneity among herbaceous plant communities of each climate and land use condition and on the other hand the vulnerability of these rare herbaceous species. These results are in concordance with the studies of Sambaré [42] who revealed that in Savanna plant communities the percentage of rare and occasional species is high. Indeed, the presence of several types of habitats (dry, bowal, fresh, and wet) within the different climate and land use conditions could favor the maintenance of distinct herbaceous vegetation communities. Some species could be exacting and therefore remain confined to particular environmental conditions. This is the case of Trichoneura mollis, Sporobolus spicatus, Abutilon macropodium, and Digitaria fuscescens, which are rare species only found in the communal Sahelian area. These species have not been found up to now in Burkina Faso; their presence is certainly linked to the transhumant pastoralism grazing. Indeed, their occurrence has been reported in Niger and other Sahelian countries [55]. Linzia purpurea, Pennisetum unisetum, Andropogon tectorum, andMonocymbium ceresiiforme are rare species encountered exclusively in protected areas of the Sudanian zone [56]. In fact, Pennisetum unisetum is a perennial grass species consumed during food storage period and it develops on alluvial soils. Monocymbium ceresiiforme (perennial grass) found on poor acid soils with fine gravel has magic and religious use for hunting protection and against evil spell. Andropogon tectorum (perennial grass) prefers ferruginous fresh soils under shade of woodland Savannas or riparian forests [55]. Linzia purpurea is forb which develops in woodland Savanna on detritus [56]. The specificity of these species for particular soils conditions makes them rare. Their absence in communal areas is certainly due to habitats destruction in these areas.

CONCLUSION

This study shows that specific richness ofherbaceous vegetation increases significantly with precipitation, and is, thus, mainly determined by climate, whereas the effect of land use is less important. Herbaceous vegetation diversity described by mean richness and the numbers of abundant species reveals that the Sudanian zone and the communal areas were, respectively, more diverse than the Sahelian zone and the protected areas. The number of very important species demonstrated that only few species are dominant and the evenness pattern showed that these dominant herbaceous species exhibit a regular distribution of their individuals within herbaceous communities. The composition of Savanna herbaceous species is dependent on climate and land use conditions. Land use affects more the composition of herbaceous vegetation than its richness. The occurrence of herbaceous species in Savanna ecosystem is more controlled by the interaction between climate and land use than by their separate effects. Occasional and rare species constitute the most important part of Savanna herbaceous vegetation diversity, characteristic of heterogeneity of ecosystems and vulnerability of herbaceous species.

ACKNOWLEDGMENTS

The authors thank the DFG (German Research Foundation, BE 4143/2-1, and HA 6789/1-1) and the Ministry of Secondary and Higher Education of Burkina Faso for financial support.

REFERENCES

1. J. Yoka, J. J. Loumeto, J. Djego, and J. Vouidibio, "Évaluation de la diversité floristique en herbacées des savanes de la cuvette congolaise (République du Congo)," Afrique Science, vol. 9, no. 2, pp. 110–123, 2013.
2. B. Belem, B. Marie, I. Nacoulma et al., "Use of non wood forest products by local people bordering the 'Parc National Kaboré Tambi', Burkina Faso," The Journal of Transdisciplinary Environmental Studies, vol. 6, no. 1, pp. 1–21, 2007.
3. T. D. Houehanou, A. E. Assogbadjo, R. G. Kakaï, M. Houinato, and B. Sinsin, "Valuation of local preferred uses and traditional ecological knowledge in relation to three multipurpose tree species in Benin (West Africa)," Forest Policy and Economics, vol. 13, no. 7, pp. 554–562, 2011.
4. I. Ouédraogo, B. M. I. Nacoulma, K. Hahn, and A. Thiombiano, "Assessing ecosystem services based on indigenous knowledge in south-eastern Burkina Faso (West Africa)," International Journal of Biodiversity Science, Ecosystems Services and Management, vol. 10, no. 4, pp. 313–321, 2014.
5. J. Alcamo, R. Schaldach, J. Koch, C. Kölking, D. Lapola, and J. Priess, "Evaluation of an integrated land use change model including a scenario analysis of land use change for continental Africa," Environmental Modelling and Software, vol. 26, no. 8, pp. 1017–1027, 2011.
6. I. Ouedraogo, M. Tigabu, P. Savadogo, H. Compaoré, P. C. Odén, and J. M. Ouadba, "Land cover change and its relation with population dynamics in Burkina Faso, West Africa," Land Degradation and Development, vol. 21, no. 5, pp. 453–462, 2010.
7. A. J. Dougill, E. D. G. Fraser, and M. S. Reed, "Anticipating vulnerability to climate change in dryland pastoral systems: using dynamic systems models for the Kalahari," Ecology and Society, vol. 15, no. 2, pp. 1–17, 2010.
8. P. Poschlod, J. P. Bakker, and S. Kahmen, "Changing land use and its impact on biodiversity," Basic and Applied Ecology, vol. 6, no. 2, pp. 93–98, 2005.
9. L. Sawadogo, D. Tiveau, and R. Nygård, "Influence of selective tree cutting, livestock and prescribed fire on herbaceous biomass in the savannah woodlands of Burkina Faso, West Africa," Agriculture, Ecosystems and Environment, vol. 105, no. 1-2, pp. 335–345, 2005.
10. P. Savadogo, L. Sawadogo, and D. Tiveau, "Effects of grazing intensity and prescribed fire on soil physical and

hydrological properties and pasture yield in the savanna woodlands of Burkina Faso," Agriculture, Ecosystems and Environment, vol. 118, no. 1–4, pp. 80–92, 2007.

11. P. Savadogo, M. Tigabu, L. Sawadogo, and P. C. Odén, "Examination of multiple disturbances effects on herbaceous vegetation communities in the Sudanian savanna-woodland of West Africa," Flora: Morphology, Distribution, Functional Ecology of Plants, vol. 204, no. 6, pp. 409–422, 2009.
12. L. G. Houessou, O. Teka, I. Toko, A. M. Lykke, and B. Sinsin, "Land use and land-cover change at 'W' biosphere reserve and its surroundings areas in Benin Republic (West Africa)," Environment and Natural Resources Research, vol. 3, no. 2, pp. 87–101, 2013.
13. G. Bocksberger, P. Daget, T. Jansen et al., "Poaceae distribution in West Africa: evolution and conservation," in Proceedings of the International Research on Food Security, Natural Resource Management and Rural Development Talk, Tropentag, October 2009, pp. 6–9, University of Hamburg.
14. A. Alexandre, M. Bouvet, and L. Abbadie, "The role of savannas in the terrestrial Si cycle: a case-study from Lamto, Ivory Coast," Global and Planetary Change, vol. 78, no. 3-4, pp. 162–169, 2011.
15. J. Heubes, I. Kühn, K. König, R. Wittig, G. Zizka, and K. Hahn, "Modelling biome shifts and tree cover change for 2050 in West Africa," Journal of Biogeography, vol. 38, no. 12, pp. 2248–2258, 2011.
16. S. Paré, M. Tigabu, P. Savadogo, P. C. Odén, and J. M. Ouadba, "Does designation of protected areas ensure conservation of tree diversity in the Sudanian dry forest of Burkina Faso?" African Journal of Ecology, vol. 48, no. 2, pp. 347–360, 2010.
17. S. A. Kaboré, M. Hien, D. Ouédraogo, T. R. E. Diallo, K. Hahn, and H. B. Nacro, "Use of ecosystem services of Sarcocephalus latifolius (Sm.) E.A.Bruce and induced effect of human pressure on the species in the Southwestern Region of Burkina Faso," Ethnobotany Research and Applications, vol. 12, pp. 561–570, 2014.
18. O. Ouédraogo, L. Bondé, J. I. Boussim, and A. Linstädter, "Caught in a human disturbance trap: responses of tropical savanna trees to increasing land-use pressure," Forest Ecology and Management, vol. 354, pp. 68–76, 2015.
19. A. C. Adomou, Vegetation patterns and environmental gradients in Benin. Implications for biogeography and conservation [Ph.D. thesis], Wageningen University, Wageningen, The Netherlands, 2005.
20. L. Bondé, O. Ouédraogo, F. Kagambèga, and J. I. Boussim, "Impact des gradients topographique et anthropique sur la diversité des formations ligneuses soudaniennes," Bois et Forêts des Tropiques, vol. 318, no. 4, pp. 15–26, 2013.
21. F. Bognounou, A. Thiombiano, P. Savadogo, I. J. Boussim, P. C. Oden, and S. Guinko, "Woody vegetation structure and composition at four sites along a latitudinal gradient in Western Burkina Faso," Boiset for Êtsdes Tropiques, vol. 300, no. 2, pp. 29–44, 2009.
22. S. S. H. Biaou, Tree recruitment in West African dry woodlands: the interactive effects of climate, soil, fire and grazing [Ph.D. thesis], Wageningen University, Wageningen, The Netherlands, 2009.
23. K. Adjonou, R. Bellefontaine, and K. Kokou, "Les forêts claires du Parc national Oti-Keran au Nord-Togo: Structure, dynamique et impacts des modifications climatiques récentes," Sécheresse, vol. 20, no. 1, pp. 1–10, 2009.
24. B. Yélémou, P. Savadogo, S. Traoré, J. Millogo-rasolodimby, and V. Hien, "Floristic diversity of Piliostigmaasssociations in relation to latitudinal gradient, soil and climate variables in Burkina Faso, West Africa,"Tropical Ecology, vol. 56, no. 1, pp. 57–76, 2015.
25. J.-L. Devineau, "To what extent does land-use affect relationships between the distribution of woody species and climatic change? A case study along an aridity gradient in western Burkina Faso," Plant Ecology, vol. 212, no. 6, pp. 959–973, 2011.
26. J.-L. Devineau, A. Fournier, and S. Nignan, "'Ordinary biodiversity' in western Burkina Faso (West Africa): what vegetation do the state forests conserve?" Biodiversity and Conservation, vol. 18, no. 8, pp. 2075–2099, 2009.
27. M. Schmidt, H. Kreft, A. Thiombiano, and G. Zizka, "Herbarium collections and field data-based plant diversity maps for Burkina Faso," Diversity and Distributions, vol. 11, no. 6, pp. 509–516, 2005.
28. M. Schmidt, K. König, and J. V. Müller, "Modelling species richness and life form composition in Sahelian Burkina Faso with remote sensing data," Journal of Arid Environments, vol. 72, no. 8, pp. 1506–1517, 2008.
29. M. Schmidt, T. Janßen, S. Dressler et al., "The West African vegetation database. Vegetation databases for the 21st century," Biodiversity & Ecology, vol. 4, pp. 105–110, 2012.
30. J. Heubes, M. Schmidt, B. Stuch et al., "The projected impact of climate and land use change on plant diversity: an example from West Africa," Journal of Arid Environments, vol. 96, pp. 48–54, 2013.
31. P. Savadogo, D. Zida, L. Sawadogo, D. Tiveau, M. Tigabu, and P. C. Odén, "Fuel and fire characteristics in savanna—woodland of West Africa in relation to grazing and dominant grass type," International Journal of Wildland Fire, vol. 16, no. 5, pp. 531–539, 2007.
32. P. Savadogo, D. Tiveau, L. Sawadogo, and M. Tigabu, "Herbaceous species responses to long-term effects of prescribed fire, grazing and selective tree cutting in the savanna-woodlands of West Africa,"Perspectives in Plant Ecology, Evolution and Systematics, vol. 10, no. 3, pp. 179–195, 2008.
33. J.-L. Devineau and A. Fournier, "Integrating environmental and sociological approaches to assess the ecology and diversity of herbaceous species in a Sudan-type savanna (Bondoukuy, western Burkina Faso)," Flora—Morphology, Distribution, Functional Ecology of Plants, vol. 202, no. 5, pp. 350–370, 2007.
34. S. Guinko, Végétation de la Haute-volta [dissertation thesis], L›Univerisité de Bordeaux III, 1984.

35. A. Thiombiano and D. Kampmann, Atlas de la Biodiversité de l›Afrique de l›Ouest, Tome II: Burkina Faso, Biodiversity Atlas of West Africa, Frankfurt am Main, Germany, 2010.
36. J. Braun-Blanquet, Plant Sociology. The Study of Plant Communities, Hafner Publishing Company, 1st edition, 1932.
37. R. Kindt and R. Coe, Tree Diversity Analysis. A Manual and Software for Common Statistical Methods for Ecological and Biodiversity Studies, World Agroforestry Centre (ICRAF), Nairobi, Kenya, 2005.
38. J. A. Ludwig and J. F. Reynolds, Statistical Ecology. A Primer in Methods and Computing, John Wiley & Sons, Hoboken, NJ, USA, 1988.
39. M. O. Hill, "Diversity and evenness: a unifying notation and its consequences," Ecology, vol. 54, no. 2, pp. 427–432, 1973.
40. A. E. Magurran, Measuring Biological Diversity, Blackwell, Malden, Mass, USA, 2004.
41. S. Godefroid and N. Koedam, "Distribution pattern of the flora in a peri-urban forest: an effect of the city-forest ecotone," Landscape and Urban Planning, vol. 65, no. 4, pp. 169–185, 2003.
42. O. Sambaré, Diversity, structure and phytosociology of riparian forests of Burkina Faso (West Africa) [Doctoral Dissertation], University of Ouagadougou, Ouagadougou, Burkina Faso, 2013.
43. R Development Core Team, R: A Language and Environment for Statistical Computing, R Foundation for Statistical Computing, Vienna, Austria, 2007.
44. A. C. Adomou, Vegetation patterns and environmental gradients in Benin: implications for biogeography and conservation [Ph.D. thesis], Wageningen Universiteit, Wageningen, The Netherlands, 2005.
45. T. Becker and J. V. Müller, "Floristic affinities, life-form spectra and habitat preferences of the vegetation of two semi-arid regions in Sahelian West and Southern Africa," Basic and Applied Dryland Research, vol. 1, no. 1, pp. 33–50, 2007.
46. C. M. Shackleton, "Comparison of plant diversity in protected and communal lands in the Bushbuckridge lowveld savanna, South Africa," Biological Conservation, vol. 94, no. 3, pp. 273–285, 2000.
47. M. Hejcman, A. A. Camara, M. Antonínová, V. Pavlu, T. Cerny, and A. T. Bâ, "Analysis of the herbaceous undergrowth of the woody savanna in the Fathala reserve, Delta du Saloum national park (Senegal),"Belgian Journal of Botany, vol. 138, no. 2, pp. 119–128, 2005.
48. E. P. S. Assédé, A. C. Adomou, and B. Sinsin, "Magnoliophyta, biosphere reserve of Pendjari, atacora province, Benin," Check List, vol. 8, no. 4, pp. 642–661, 2012.
49. A. D. Q. Agnew, "A field key to upland kenya grasses," Journal of East African Natural History, vol. 95, no. 1, pp. 1–83, 2006.
50. G. Bocksberger, Diversity and evolution of dominant African savanna grasses [Doctoral Dissertation], Johann Wolfgang Goethe University, Frankfurt, Germany, 2012.
51. M. C. Rutherford and L. W. Powrie, "Impacts of heavy grazing on plant species richness: a comparison across rangeland biomes of South Africa," South African Journal of Botany, vol. 87, pp. 146–156, 2013.
52. A. Kadeba, B. M. I. Nacoulma, A. Ouédraogo et al., "Land cover change and plants diversity in the Sahel: a case study from northern Burkina Faso," Annals of Forest Research, vol. 58, no. 1, pp. 109–123, 2015.
53. T. J. Stohlgren, A. J. Owen, and M. Lee, "Monitoring shifts in plant diversity in response to climate change: a method for landscapes," Biodiversity and Conservation, vol. 9, no. 1, pp. 65–86, 2000.
54. A. Hoshino, Y. Yoshihara, T. Sasaki et al., "Comparison of vegetation changes along grazing gradients with different numbers of livestock," Journal of Arid Environments, vol. 73, no. 6-7, pp. 687–690, 2009.
55. P. Poilecot, "Les poacées du Niger. Description-Illustration. Ecologie-Utilisations," Boissiera, vol. 56, pp. 1–766, 1999.
56. A. Thiombiano, M. Schmidt, S. Dressler, A. Ouédraogo, K. Hahn, and G. Zizka, "Catalogue des plantes vasculaires du Burkina Faso," Boissiera, vol. 65, p. 391, 2012.

CHAPTER 18

PHYSIOLOGICAL AND BIOCHEMICAL CHANGES IN MOTH BEAN (VIGNA ACONITIFOLIA L.) UNDER CADMIUM STRESS

Poornima D. Vijendra[1,2], Kavitha M. Huchappa[1], Roopa Lingappa[1], Giridhara Basappa[2], Sathisha G. Jayanna[3], and Vadlapudi Kumar[1,2]

[1]Department of Biochemistry, Kuvempu University, P.G. Centre, Shivagangothri, Davangere, Karnataka 577002, India

[2]Department of Biochemistry, Davangere University, Shivagangothri, Davangere, Karnataka 577002, India

[3]Department of Biochemistry, Kuvempu University, Shankaraghatta, Shivamogga District, Karnataka 577451, India

ABSTRACT

Moth bean (Vigna aconitifolia L.), a drought resistant legume, possesses high nutritional value. Cadmium (Cd) is a nonessential and the most toxic heavy metal in plants. The present study was to test the hypothesis of whether moth bean being a drought resistant legume can withstand the cadmium stress. Ten-day-old moth bean seedlings were subjected to cadmium stress and investigated for a period of 15 days every 3-day intervals. Cadmium quantification in moth bean tissues suggests root accumulation and translocation to aerial parts in a concentration dependent manner. Results of physiological and biochemical studies revealed that cadmium has affected the growth parameters like shoot and root lengths and tissue dry weights. Significant alternations in relative water content and cell membrane stability were observed in stressed seedlings. Similarly superoxide radical, lipoxygenase activity, membrane lipid peroxidation products, protein carbonyls, and reduced glutathione and nonprotein thiols were found increased in stressed seedlings compared to controls. However, hydrogen peroxide and ascorbic acid levels were not altered significantly in both stressed and control seedlings. Cadmium translocation ability from roots to aerial parts and elevated levels of nonenzymatic antioxidants in stressed seedlings suggest the cadmium stress withstanding ability of moth bean.

INTRODUCTION

Heavy metals are of great environmental and human health concerns due to their widespread occurrence, persistence in ecosystems, and toxic properties [1]. Agricultural crops differ widely in their tolerance to toxic metals. Tolerance of plants to heavy metals is genetically determined [2], so that identification of plant genotypes differing in resistance to heavy metals is a promising approach, not only for studying mechanisms protecting plants against toxic metals, but also for the selection of those adapted to production in presence of heavy metals in contaminated soils [3].

Cadmium (Cd) is highly toxic to all living organisms and is one of the most toxic heavy metals in plants due to its high solubility in water and phytotoxicity [4, 5]. Cadmium is not an essential nutrient in plants and at high concentration inhibits plant growth [6, 7]; even at relatively low concentrations it

Citation: Poornima D. Vijendra, Kavitha M. Huchappa, Roopa Lingappa, Giridhara Basappa, Sathisha G. Jayanna, and Vadlapudi Kumar, "Physiological and Biochemical Changes in Moth Bean (Vigna aconitifolia L.) under Cadmium Stress," Journal of Botany, vol. 2016, Article ID 6403938, 13 pages, 2016. doi:10.1155/2016/6403938 alters plant metabolism [8]. Detrimental effects of cadmium are manifested in inhibition of photosynthesis and in oxidative stress leading to membrane damage [9]. The intensive application of phosphate fertilizers containing traces of Cd and industrial zinc mining is expanding the soil-Cd contamination rapidly especially in the developing countries [10, 11].

Moth bean (Vigna aconitifolia L.), also called mat bean or matki bean, mout bean, or dew gram or Turkish gram, belongs to Fabaceae. Moth bean is a most popular pulse crop in India. It is cultivated for its immature pods and mature seeds and is consumed by people all around the world, especially in the developing nations [12–14] as it possesses high nutritional value and is a potential source of protein and other nutrients.

Legume crops are less tolerant to Cd toxicity than cereals and grasses and encounter strong inhibition of biomass production due to cadmium [15]. Different degrees of tolerance to cadmium stress were observed among different genotypes of the same species [16–18]. Vigna species can grow under a wide range of climates and environments [19]. Moth bean is a hot weather, drought resistant legume, particularly cultivated in hot, arid to semiarid regions. For optimum production it requires 24°C–32°C of temperature; however, the plant withstands daytime temperature of 45°C and does not tolerate water logging. Some degree of salinity and a wide pH range (3,5–10) are tolerated [15, 20]. Presence of cadmium in the soil decreases the growth of legumes like soybean [21, 22] and chickpea plants [23]. However, in the pertinent literature survey on moth bean and its heavy metal stress responses, few published reports [24, 25] are available to date revealing its cadmium stress responses.

The present study is envisaged to evaluate whether the moth bean is sensitive to cadmium like other legumes or it can withstand the cadmium stress, as this plant can withstand a wide range of environmental stresses. To test this hypothesis the present study has been taken up to evaluate the cadmium sensitivity or tolerance of moth bean plant by subjecting to different concentrations of cadmium for a short duration period of 15 days. The investigations were carried out by determining the changes in physiological growth parameters like fresh and dry mass yield, cell membrane stability, and biochemical parameters including changes in the levels of antioxidant molecules, nonprotein thiols, and antioxidant enzymes and also accumulation of cadmium in plant organs.

MATERIALS AND METHODS

Plant Material and Cadmium Treatments

Seeds of Vigna aconitifolia (moth bean) cv. PS-16 were procured from local agriculture seed stores of Davangere, Karnataka State, India. The healthy and uniform sized seeds were separated, surface-sterilized with 0.5% sodium hypochlorite (v/v in sterile distilled water) solution imbibed in sterile double distilled water for 48 hrs in the dark, and germinated on Whatman No. 1 filter paper moistened with sterile distilled water. Two-day-old sprouts were sown in germination trays filled with cleaned and dried sand for further growth, watered daily with 1x Hoagland nutrient solution [26] for 10 days. From the 11th day considered as "0" day, cadmium (Cd) in the form of cadmium chloride ($CdCl_2$) was supplemented in Hoagland nutrient solution at different concentrations, 10 µM, 50 µM, 100 µM, 200 µM, and 500 µM; simultaneously control group of seedlings were also maintained. Stress treatments were continued for 15 days, seedlings were removed from the individual treatment trays every 3-day intervals, and the experiments were conducted according to simple randomized block design. Experiments were conducted in two sets; for each set of experiments 20 healthy seedlings were taken at random, and roots and upper second fully expanded leaves were separated and subjected to physiological and biochemical analysis.

Plant Growth Measurements

Plant growth was assessed by measuring the shoot and root lengths and fresh (FW) and dry weights (DW) of the plants from individual treatment group. For the dry weight determination, samples were oven-dried at 80°C for 15 min and then vacuum-dried at 40°C to constant weight and then dry weights (DW) were recorded.

Cadmium (Cd) Uptake

Amount of cadmium (Cd) in 15-day stressed plant tissues was quantified by atomic absorption spectrometer analysis. Briefly, leaves and roots were separated from the plants, washed in deionized water for 2 min, and air-dried and further drying was carried out in a microwave station at 80°C for 2 days. Dried plant tissue was ground into fine powder; 1 g of tissue powder was digested into 3 parts of di-acid mixture containing 1 M HNO_3 and 1 M HCl (3 : 1 ratio) at 60°C. Cadmium content was determined by atomic absorption spectrometer (Chemito-AA201) analysis using cadmium standard (Sigma-Aldrich, St. Louis).

Relative Water Content (RWC)

Relative water content (RWC) was determined in fresh leaf discs that were weighed and immediately floated on double distilled water in petri dishes to saturate for the next 24 h, in the dark. The adhering water was blotted and turgor weights were recorded. Subsequently dry mass was obtained after dehydrating at 70°C for 48 h. Relative water content was calculated using the following formula:

$$\text{RWC} = \left(\frac{\text{Fresh mass} - \text{Dry mass}}{\text{Turgor mass} - \text{Dry mass}} \times 100 \right). \quad (1)$$

Cell Membrane Stability (CMS)

Cell membrane stability (% of injury) was determined by electrical conductivity measurements [27] in control and cadmium stressed leaf discs using a conductivity meter (Systronics, India). Subsequently, the leaf tissues were killed by autoclaving at 121°C and 1.06 kg cm^{-2} pressure for 15 min and electrical conductivities were measured again to determine the total electrolyte concentrations.

The CMS index (% of injury) of the leaf tissues was determined using the following equation:

Percentage of injury or CMS (%)

$$= 100 \cdot \left\{ \left\{ 1 \left[\frac{(1 \cdot T1/T2)}{(1 \cdot C1/C2)} \right] \right\} \times 100 \right\}, \qquad (2)$$

where $T1$ and $C1$ represent the electrolyte concentrations measured after incubating at 10∘ C for 24 h and $T2$ and $C2$ represent the total electrolyte concentration measured after autoclaving the leaf tissues of both treatments and controls, respectively.

Estimation of Active Oxygen Species (AOS) Levels

The amount of superoxide radicals ($O_2^{\bullet-}$) was estimated by determining the amount of nitrite formed from hydroxylamine [28] in the fresh leaves and root extracts prepared separately into 65 mM phosphate buffer (pH 7.8) and supernatants were used for the estimation of superoxide. The reaction mixture contained 0.9 mL of 65 mM phosphate buffer (pH 7.8) and 10 mM hydroxylamine hydrochloride and plant tissue extract. After incubation at room temperature 17 mM sulphanilamide and 7 mM α-naphthol were added; after incubation for 5–10 min at room temperature, diethyl ether was added and absorbance of supernatants was recorded at 530 nm. Amount of nitrite formed was determined with the help of a standard curve established with standard source of O_2^- and expressed as μM g^{-1} FW of tissue. The amount of total hydrogen peroxide (H_2O_2) was estimated in the fresh leaves and roots by ferrithiocyanate method [29]. Tissue homogenates were prepared into 5% trichloroacetic acid; supernatants were used immediately for the total peroxide estimation. The reaction mixture contained tissue extract, 50% trichloroacetic acid, and 10 mM ferrous ammonium sulphate, and colour was developed with 2.5 M potassium thiocyanate. Absorbance of reaction mixture was measured at 480 nm and amount of total H_2O_2 was calculated using the extinction coefficient 0.28 $\mu mol^{-1} cm^{-1}$.

Assay of Lipoxygenase (LOX) Activity (EC 1.13.11.12)

Lipoxygenase (EC 1.13.11.12) assay was carried out in the fresh leaf and root extracts (500 mg) as described by [30]. Tissue homogenates were prepared separately into 3 mL of 0.2 M borate buffer (pH 8.0) and supernatants used for the lipoxygenase assay. The assay mixture contained 0.2 M boric acid buffer (pH 10.0) and plant extract and linoleic acid as a substrate in a final volume of 3 mL. The reaction was carried out at 30°C for 4 min and absorption of reaction mixture was measured at 234 nm. Molar absorption coefficient of 25000 $M^{-1} cm^{-1}$ was used for determining the concentration of the hydroperoxides produced and lipoxygenase activity was expressed as absorbance increase (DA_{234}) mg^{-1} of protein min^{-1}.

Membrane Lipid Peroxidation and Protein Oxidation Products

Membrane lipid peroxidation in leaves and roots was determined as malondialdehyde (MDA) equivalent products that react with thiobarbituric acid [31]. Absorbance of reaction mixture was measured at 532 nm spectrophotometrically and concentration of MDA was calculated using extinction coefficient of 155 $mM^{-1} cm^{-1}$ and expressed as μM^{-1} g fresh weight of tissue. Protein oxidation was measured in tissue extracts as the total carbonyl group content by reaction with 2,4-dinitrophenylhydrazine (DNPH) [32]. Tissue extracts were prepared into 5 mL of 50 mM potassium phosphate buffer (pH 7.4) containing 120 mM KCl and 0.1 g PVP; aliquots of supernatants containing at least 0.5 mg protein were incubated with 0.03% Triton X-100 and 1% streptomycin sulphate for 15 min to remove the nucleic acids. After centrifugation, equal volumes of supernatant and 10 mM DNPH in 2 M HCl were mixed with each other, after incubation for 1 h, proteins were precipitated with 20% TCA (w/v), precipitate was washed three times with ethanol : ethylacetate (1 : 1) and dissolved into 6 M guanidine hydrochloride in 20 mM potassium phosphate buffer, pH was adjusted to 2.3 with trifluoroacetic acid, and absorption was measured at 380 nm. Carbonyl group content was calculated using a molar absorption coefficient of 22,000 $M^{-1} cm^{-1}$ and expressed as μM g^{-1} FW.

Estimation of Nonenzymic Antioxidants

Total ascorbate was determined according to the modified procedure of [33] in fresh leaf and root tissue homogenates prepared into 5% metaphosphoric acid and dithiothreitol (for reducing dehydroascorbate to ASC) using 5% (w/v) *N*-ethylmaleimide and 4% (w/v) 2,2′-bipyridyl. Absorbance of the reaction was measured spectrophotometrically at 525 nm and using standard ascorbate calibration, tissue total ascorbate was determined and expressed as mg^{-1} g FW. Reduced glutathione content was determined using 5,5′-dithiobis-2-nitro benzoic acid (DTNB) reagent in fresh leaf and root extracts prepared separately into 5% sulfosalicylic acid [34]. Absorbance of the reaction was measured spectrophotometrically at 412 nm and the amount of glutathione was determined using a standard calibration curve constructed with reduced glutathione (GSH). Similarly, nonprotein thiols (NPSH) were determined using DTNB reagent [35] in fresh leaf and root tissue homogenates prepared separately into 1 M HCl containing 1 mM EDTA. Absorbance of reaction mixture was measured at 412 nm and the amount of nonprotein thiols was quantified using a reduced glutathione (GSH) calibration curve. Free cysteine content was measured using acetic acid-ninhydrin reagent in the fresh leaf and root tissue extracts prepared into 5% perchloric acid [36]. Absorbance was read at 560 nm and the amount of cysteine was quantified using a calibration curve prepared with cysteine standard.

Statistical Analysis of the Data

Data from 3 trials of each experiment with at least 10 plants (tissue samples from same group) for each trial were statistically analyzed (one-way ANOVA) using GraphPad Prism software (version 3.0). Unless otherwise stated elsewhere, average values (mean) and standard errors (SE) were determined, means were compared with Dunnett's test (at 5% level of significance), and data were represented as mean ± SE.

RESULTS

Seed Germination and Seedling Growth

Seed germination was achieved within 2-3 days and the germination percentage was determined to be between 97 and 99%. Seedlings attained nearly 6.5–8.2 cms of height in 10 days on 1x Hoagland's nutrient solution.

Plant Growth Measurements

Shoot lengths and root lengths of the Cd stressed seedlings were found to be slightly affected but no significant changes were observed with the increase in stress intensity (concentration of Cd and duration of exposure) and duration compared to respective control group of plants (Figure 1). Similarly, a slight reduction in the dry weights of tissue mass in cadmium stressed plants was observed at all the concentrations of cadmium and duration (Figure 2).

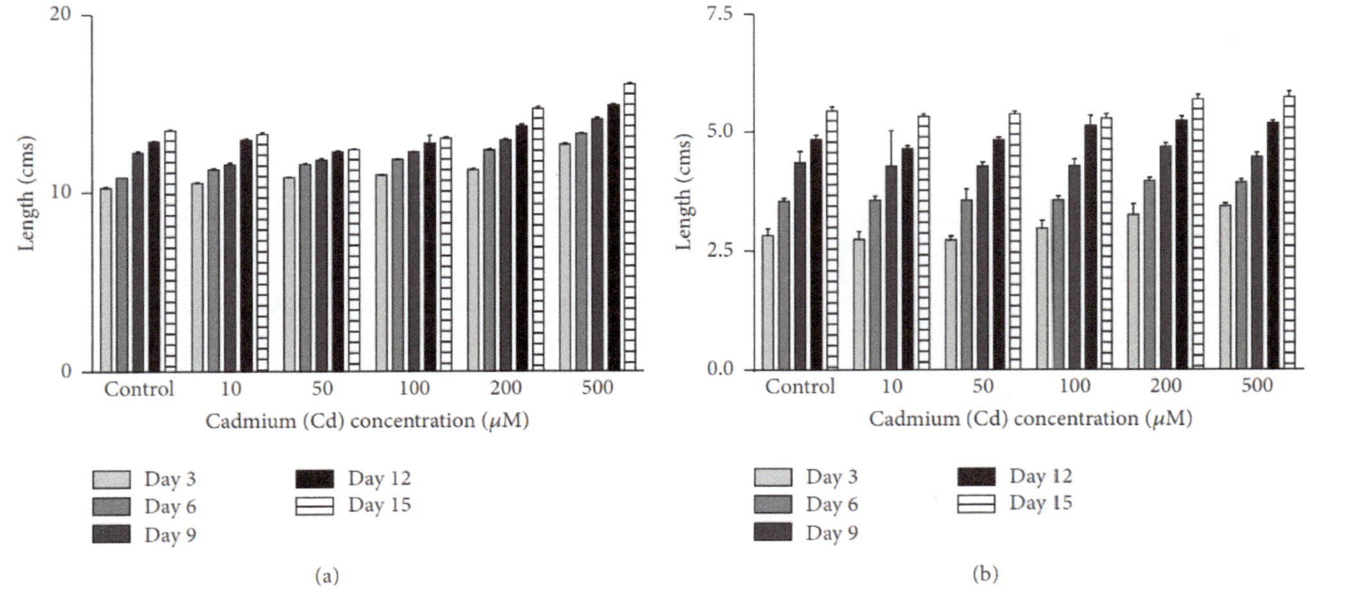

Figure 1: Changes in (a) shoot length and (b) root length. Data shown in each bar are mean ± SE of 3 trials of each experiment with at least 10 plants, analyzed by Dunnett's test (at 5% level of significance).

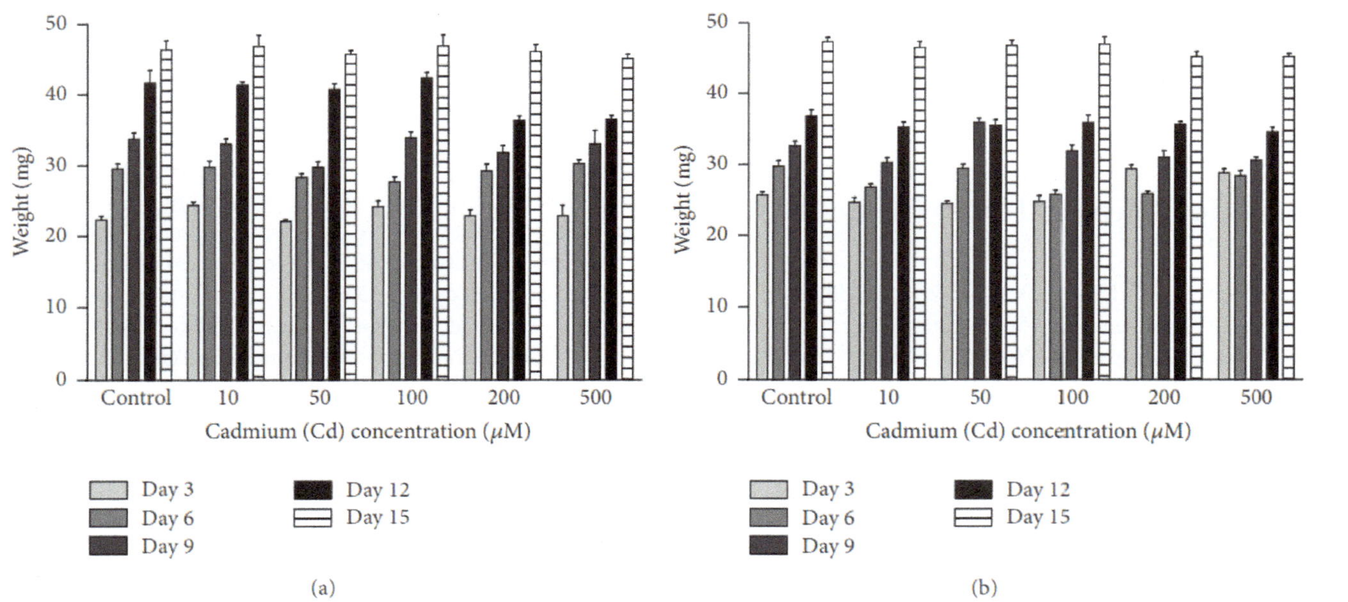

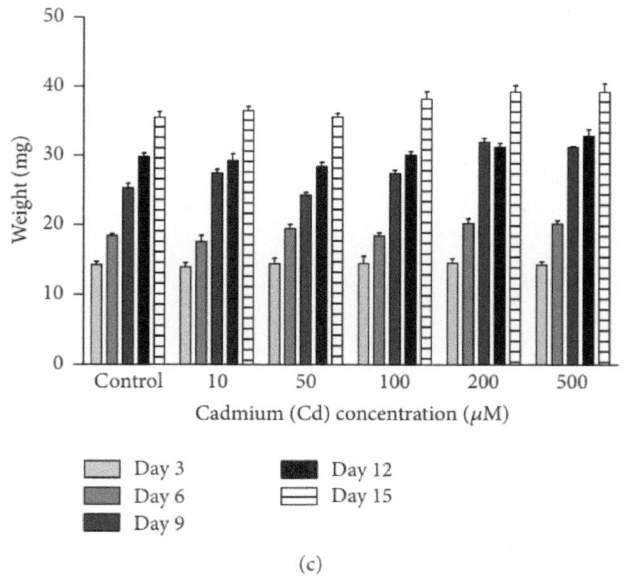

Figure 2: Changes in tissue dry weights (a) leaf, (b) shoot, and (c) root. Data shown in each bar are mean ± SE of 3 trials of each experiment with at least 10 plants, analyzed by Dunnett's test (at 5% level of significance).

Cadmium Accumulation in Tissues

Cadmium content was found to be significantly increased in both leaves and roots of stressed plants compared to respective controls. Tissue accumulation was in a concentration dependent manner supplemented in the medium. Higher amounts of cadmium were determined in leaves than in roots at 10 μM concentration supplemented in the medium, higher amounts in roots than in leaves at 200 μM concentration supplemented in the medium, and similar amounts in both leaves and roots at 500 μM concentration supplemented in the medium (Figure 3).

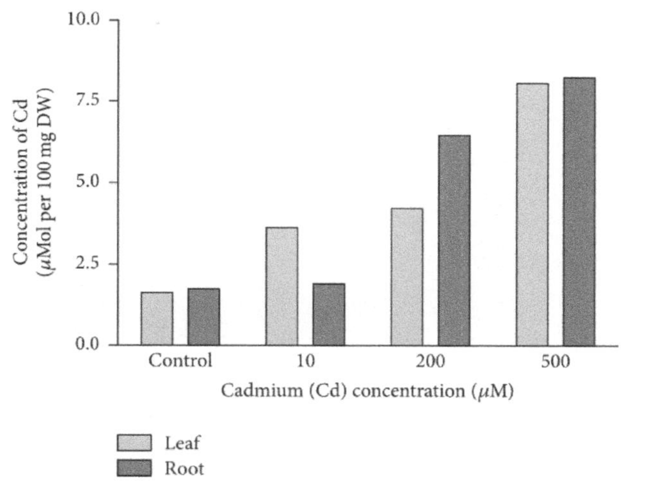

Figure 3: Cadmium content in tissues. Data shown in each bar are mean values from 3 trials of experiment with at least 10 plants/tissue from same group.

Relative Water Content (RWC)

Relative water content (RWC) was determined in fresh leaves and roots. Significant differences in relative water content were observed between the controls and cadmium stressed plants after 15 days of cadmium exposure. In the cadmium stressed plants nearly 67% decrease in RWC was observed in leaves, while in roots the decrease in RWC was nearly 56% (Figure 4) compared to control group of plants.

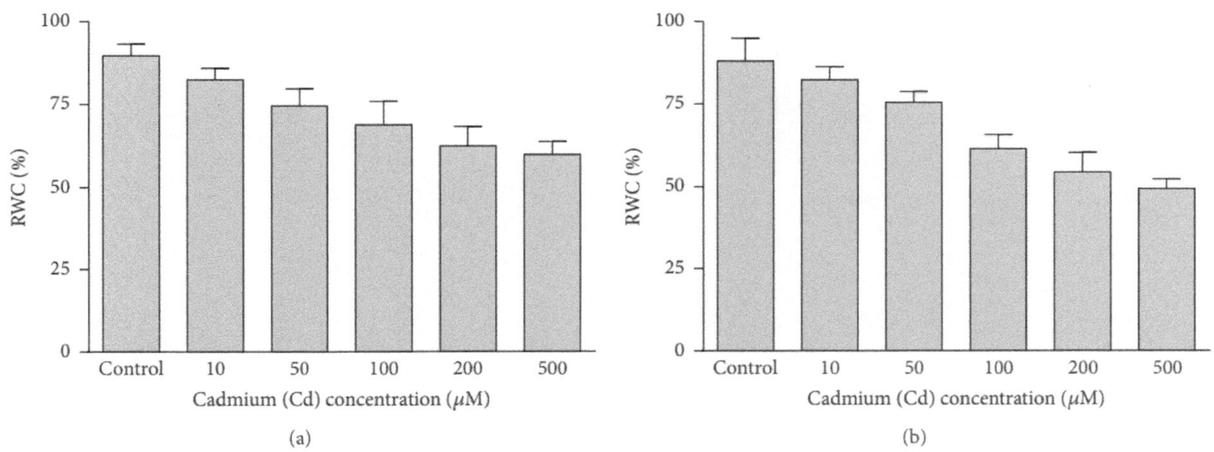

Figure 4: Changes in tissue relative water content (RWC %). (a) Leaf and (b) root. Data shown in each bar are mean ± SE of 3 trials of each experiment with at least 10 plants, analyzed by Dunnett's test (at 5% level of significance).

Cell Membrane Stability

Cell membrane stability or membrane injury (%) was assessed by an indirect measurement of electrolyte leakage of the leaf discs. Significant decrease in cell membrane stability was observed in stressed plant leaves compared to control group leaves. There was an inverse relationship observed between the intensity of cadmium stress and cell membrane stability. Greater than or equal to 100% injury was observed only at 500 μM cadmium stress, while at other concentrations of cadmium, the % of injury was observed to be less than 100% (Figure 5).

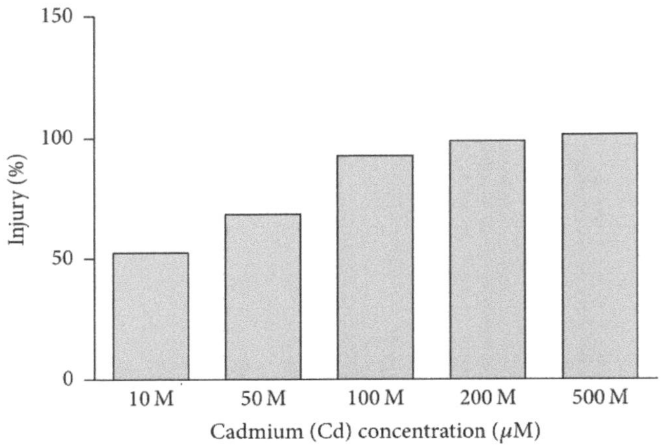

Figure 5: Changes in leaf cell membrane stability (injury %). Data shown in each bar are mean values from 3 trials of experiment with at least 10 plants/tissue from same group.

Changes in Active Oxygen Species (AOS) Levels

Cadmium stress induced an increase in superoxide radical ($O_2^{\bullet-}$) production in both leaf and root tissues, compared to controls with higher levels noticed in the stressed tissues. The levels of superoxide radical were found to be up to 4-fold higher in roots than in leaves. Cadmium at 500 μM concentration induced about 4.0-fold increase in $O_2^{\bullet-}$ production in leaves and about 6.2-fold increase in roots at the same concentration (Figures 6(a) and 6(b)). However, a consistent increase in hydrogen peroxide production was observed in both leaves and roots of stressed plants compared to controls, with higher amounts determined in leaves than in roots (Figures 6(c) and 6(d)).

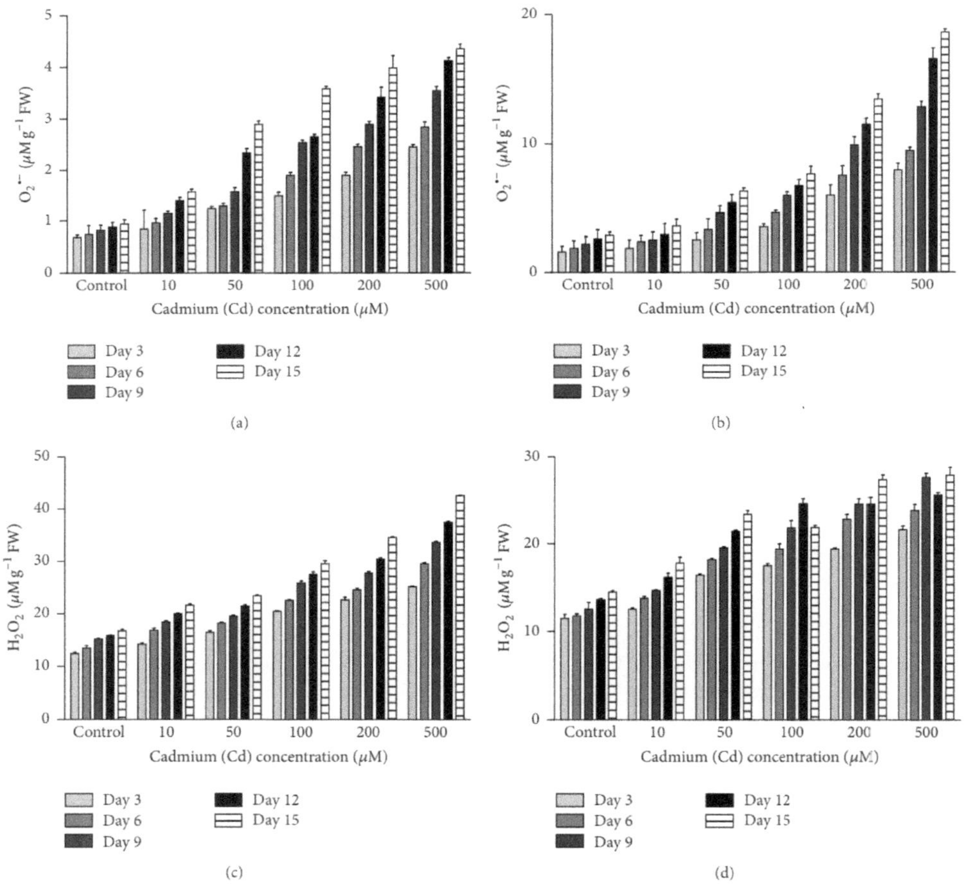

Figure 6: Changes in active oxygen species (AOS) in tissues. (a) Leaf superoxide, (b) root superoxide, (c) leaf hydrogen peroxide, and (d) root hydrogen peroxide. Data shown in each bar are mean ± SE of 3 trials of each experiment with at least 10 plants, analyzed by Dunnett's test (at 5% level of significance).

Changes in Membrane Lipid and Protein Oxidation Products

Cadmium stress caused a significant increase in the amount of lipid peroxidation in leaves and roots at all the concentrations, as measured in terms of thiobarbituric acid reactive oxygen species (TBARS) with highest amount of TBARS (MDA equivalent product); concentration was recorded in 500 µM cadmium exposed plant tissues. Compared to controls about 3-fold increase was observed in leaves and roots by day 15 (Figures 7(a)and 7(b)). Similarly, cadmium stress significantly increased carbonyl groups content in leaves and roots compared to control group of plants, with higher values recorded in leaves than in roots (Figures 7(c) and7(d)).

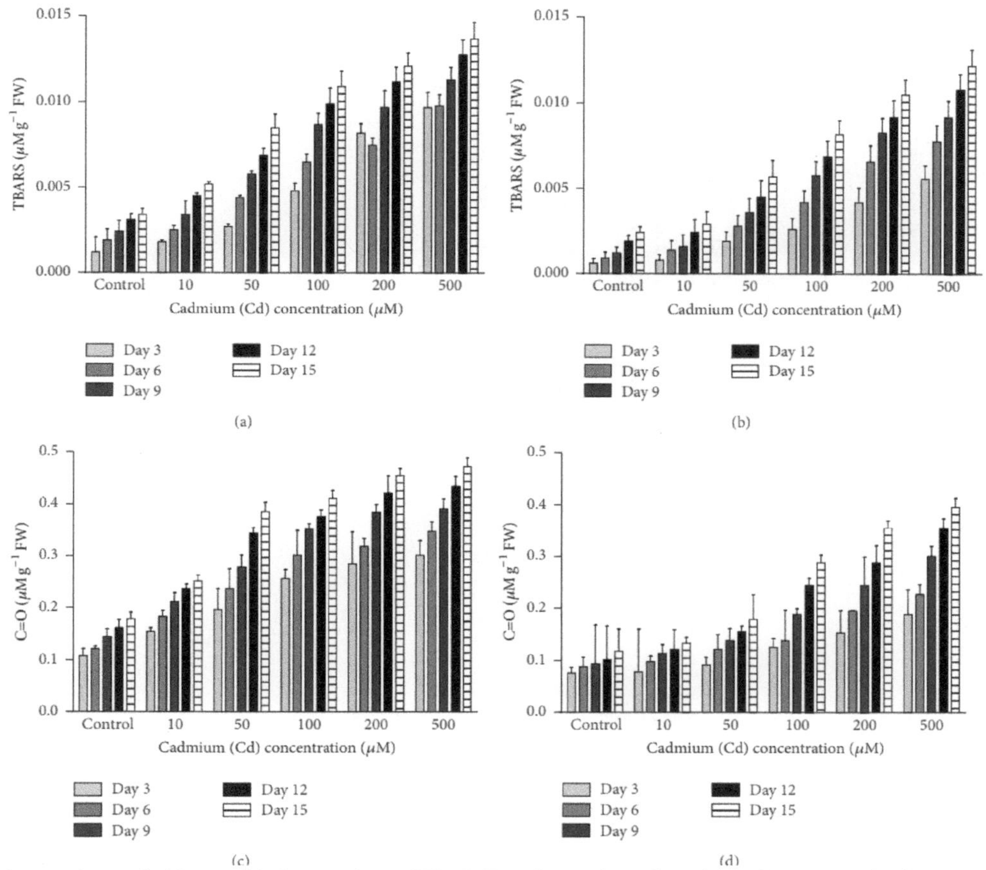

Figure 7: Changes in membrane lipid peroxidation products (TBARS) and protein carbonyl (C=O) contents in tissues. (a) Leaf TBARS, (b) root TBARS, (c) leaf carbonyls, and (d) root carbonyls. Data shown in each bar are mean ± SE of 3 trials of each experiment with at least 10 plants, analyzed by Dunnett's test (at 5% level of significance).

Changes in Lipoxygenase (LOX) Activity (EC 1.13.11.12)

Lipoxygenase activity was found to be increased in both leaves and roots of stressed plants compared to respective controls. The increase was recorded to be consistent with the intensity of cadmium stress and duration, with higher levels in leaves than in roots (Figure 8).

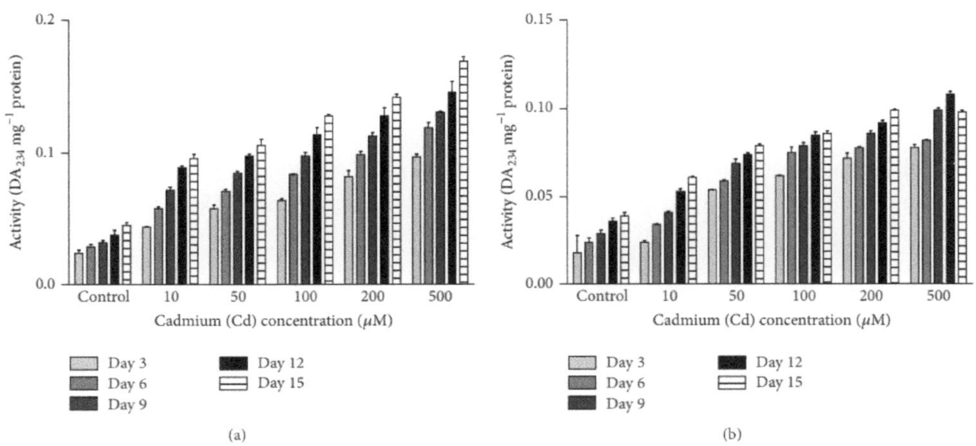

Figure 8: Changes in tissue lipoxygenase activity. (a) Leaf and (b) root. Data shown in each bar are mean ± SE of 3 trials of each experiment with at least 10 plants, analyzed by Dunnett's test (at 5% level of significance).

Changes in Nonenzymatic Antioxidants

Ascorbic acid (AsA) content was found to be slightly decreased in leaves of cadmium stressed plants compared to respective controls. However, a significant increase in ascorbate content was observed in roots, with higher levels recorded in 500 μM cadmium stressed plants on day 15 (Figures 9(a) and 9(c)). Glutathione (reduced) content was found to be significantly increased in leaves and roots of cadmium stressed plants, compared to respective controls, with higher levels recorded in leaves than in roots at all the concentrations of cadmium and duration (Figures 9(b) and 9(d)). Nonprotein thiols (NPSH) and cysteine content were found to be increased in leaves and roots of cadmium stressed plants, compared to respective controls at all the concentrations of cadmium and duration (Figures 10(a), 10(b), 10(c), and 10(d)).

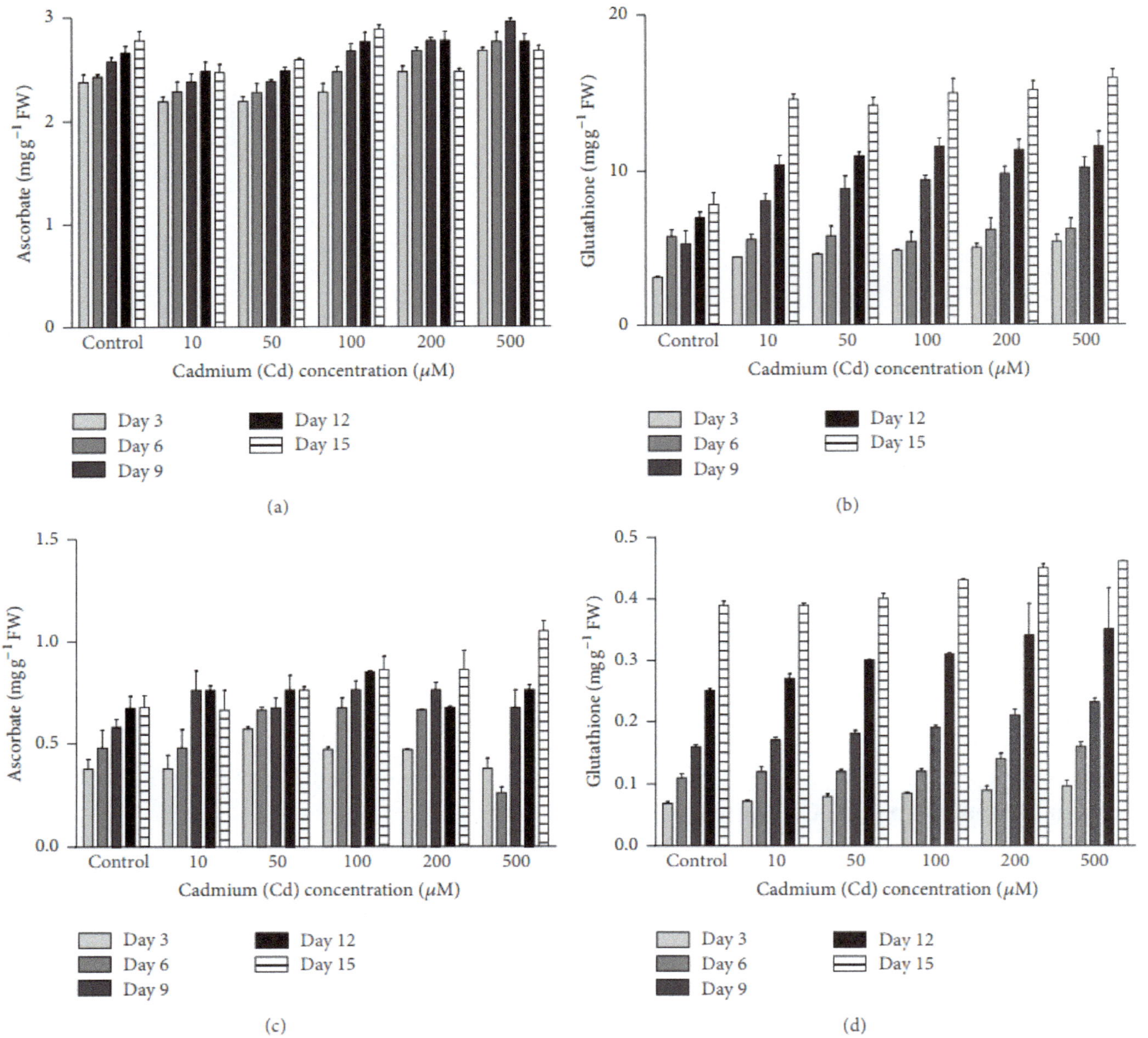

Figure 9: Changes in ascorbate (AsA) and reduced glutathione (GSH) levels. (a) Leaf AsA, (b) leaf GSH, (c) root AsA, and (d) root GSH. Data shown in each bar are mean ± SE of 3 trials of each experiment with at least 10 plants, analyzed by Dunnett's test (at 5% level of significance).

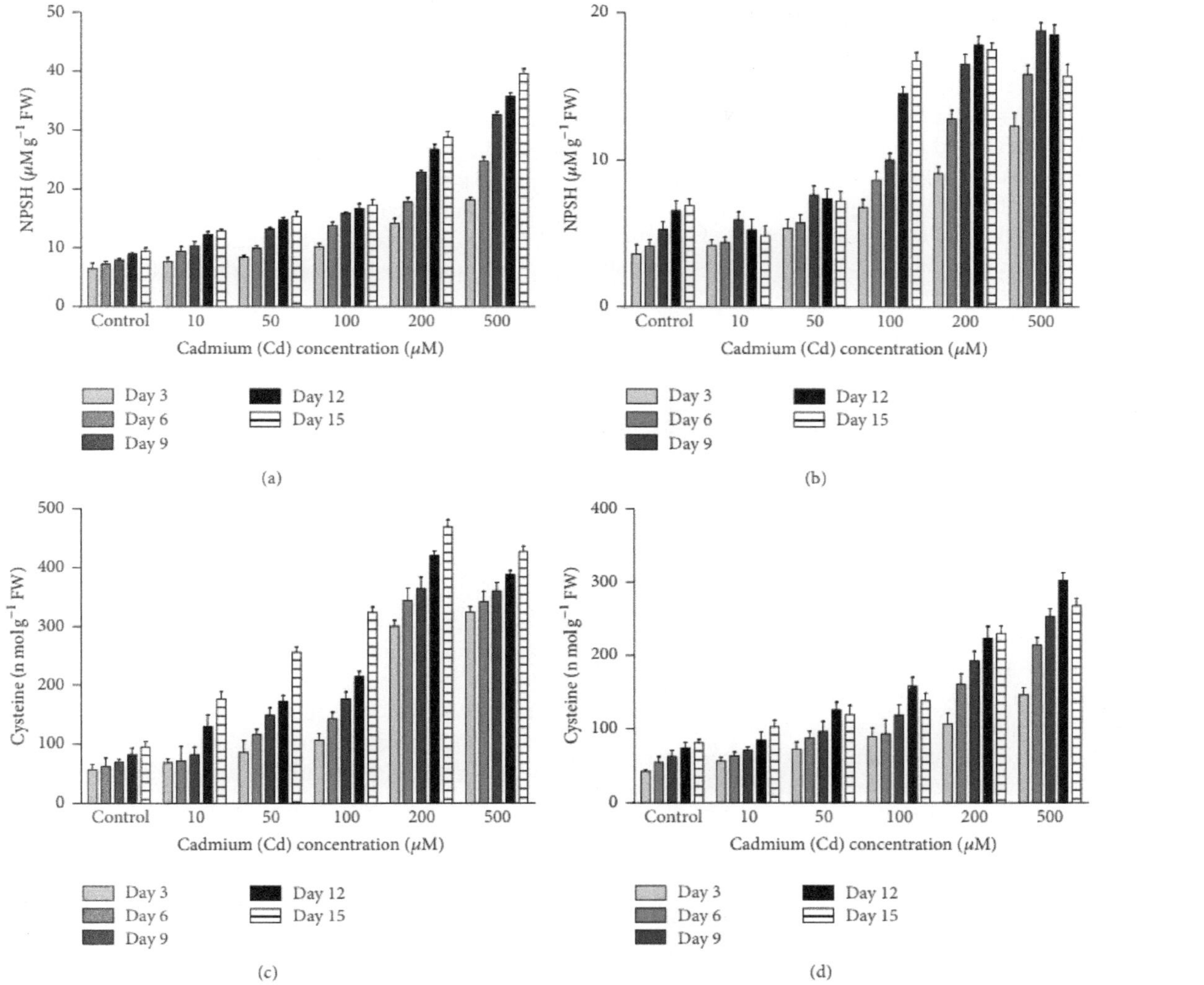

Figure 10: Changes in tissue nonprotein thiols (NPSH) and cysteine (Cys) content. (a) Leaf NPSH, (b) root NPSH, (c) leaf Cys, and (d) root Cys. Data shown in each bar are mean ± SE of 3 trials of each experiment with at least 10 plants, analyzed by Dunnett's test (at 5% level of significance).

DISCUSSION

Cadmium is one of the nonessential heavy metal pollutants, which is highly toxic to plants, because of its relatively high mobility in the soil-plant system [37, 38]. Cadmium stress impairs the physiological and biochemical processes in plants such as photosynthesis, water relations, and mineral uptake [39, 40] and also causes disruption of membrane composition and function [41, 42]. Results of the present study in cadmium stressed seedlings of V. aconitifolia revealed that cadmium stress affected shoot and root growth (Figure 1) and tissue dry matter accumulation (Figure 2). The results of cadmium quantitative analysis revealed that cadmium has accumulated in both leaves and roots on a concentration dependent manner supplemented in the medium (Figure 3), suggesting the possible translocation of cadmium from roots to aerial parts (leaves) at all the concentrations used to impose stress. Cadmium reaches the aerial parts through the xylem by easily penetrating root system of xylem [43, 44]. Cadmium quantification (Figure 3) in moth bean plants clearly suggests that leaves are also affected by cadmium stress, which might be due to the translocation of this trace metal from roots to aerial parts like shoots and leaves in moth bean plant. Cadmium affects moth bean growth and leaf pigment composition [24, 25]. The difference in cadmium uptake and tissue distribution may act as the first defense line by reducing the toxic metal level in organs or cells [45].

Leaf relative water content (RWC) and cell membrane stability were affected in the stressed seedlings compared to nonstressed control seedlings (Figures 4 and 5). Higher loss in RWC was observed in leaves (67%) compared to roots (56%); further this loss of RWC was observed to be dependent on severity of stress and duration of exposure of seedlings to the stress Figures 4(a) and 4(b). Similar findings were also reported in lettuce [46] and radish seedlings [47] that cadmium exposure reduced the relative water contents. Reference [48] demonstrated that cadmium decreases leaf conductance and interacts with stomatal regulation in a concentration dependent manner which is a manifestation of cadmium toxicity in plants that is not influenced by abscisic acid. Plant water status reflects the metabolic activity that is measured as RWC in leaves [49]. The altered leaf RWC and cell membrane damage (% injury) confirms the manifestation of cadmium toxicity in stressed V. aconitifoliaseedlings, which might be due to the elevated levels of ROS in both superoxide radicals and hydrogen peroxide (Figure 6). Published data from other plant species suggests that cadmium alters the cellular redox state, by increasing the production of reactive oxygen species (ROS) like H_2O_2 (hydrogen peroxide), $O_2^{\bullet-}$ (superoxide), and $OH^{\bullet}$ (hydroxyl radicals) [37, 50]. Accumulation of ROS such as H_2O_2 and $O_2^{\bullet-}$ in cell invariably damages the components such as DNA, protein, and lipids. Growing evidence suggests that a correlation exists between the rate of increase in H_2O_2 content and level of membrane lipid peroxidation as well as electrolyte leakage that are often used as biochemical markers to assess the extent of oxidative damage in plants under stress [51–53].

To this perspective, results of the present study strongly indicate the onset of cadmium-induced oxidative stress manifested as oxidative damage to membranes in the V. aconitifolia seedlings, as observed by the increase in amounts of TBARS (malondialdehyde), carbonyl content in leaves and roots, which are the products of membrane lipid peroxidation and protein oxidation, respectively (Figures 7(a), 7(b), 7(c), and 7(d)). Increase in TBARS content is assumed to be a common symptom of heavy metal stress [54–56]. Protein oxidation in stressed tissues is the most commonly occurring oxidative modification; accumulation of oxidized proteins reflects not only the rate of protein oxidation but also the rate of oxidized protein degradation [32, 57]. Increase in TBARS content could be correlated with the elevated levels of lipoxygenase activity (Figure 8). Lipoxygenases are a family of enzymes that catalyze oxygenation of polyunsaturated fatty acids (PUFAs) into lipid hydroperoxides (LOOHs) involved in stress responses [58]. Elevation in lipoxygenase activity during altered cellular redox state was reported in other plants also [58, 59].

Nonenzymatic mechanisms of ROS detoxification can also operate during plant stress, and the main nonenzymatic antioxidants include ascorbate and glutathione (GSH), nonprotein thiols in metal stress [60–62]. After application of heavy metals, the ascorbate-glutathione cycle seems to be a mechanism of great importance in controlling the cellular redox status [63]. It was observed during the present study that the amount of ascorbic acid (Figures 9(a) and 9(c)) remained more or less stable during the entire period of study, while glutathione (GSH) (Figures 9(b) and 9(d)), other nonprotein thiols, and cysteine concentration (Figures 10(a),10(b), 10(c), and 10(d)) were increased in cadmium stressed V. aconitifolia seedlings with increase in Cd stress intensity. Ascorbate is a primary as well as a secondary antioxidant [58, 62, 63] that may bind metals, thereby affecting their movement across biological membranes, or may act as a reducing agent, protecting the oxidation of the mercapto (-SH) groups by contributing electron or reducing power for photosystem II [64]. Similarly GSH is a well characterized antioxidant that plays a prominent role in defense system of plants [4] and plays a central role in the regeneration of ascorbate and functions as an antioxidant scavenging radical, and also GSH is the direct substrate for the synthesis of phytochelatins (PC), which chelate the metals. The nonprotein thiols (NPSH) and cysteine were measured during the present study because of their importance in polythiol synthesis (phytochelatin) and metal sequestration [65, 66].

H_2O_2 appears to be a central signal component of plant adaptation to both biotic and abiotic stresses [40]. Previous studies in other plants have given several lines of evidence indicating that cadmium exposure increased the production of H_2O_2 [57, 67, 68]. The consistent increase of H_2O_2 (Figures 6(a) and 6(b)) and ascorbate contents in leaves and roots (Figures 9(a) and 9(c)) and also elevated levels of reduced glutathione (Figures 9(b) and 9(d)) and nonprotein thiols in the leaves and roots (Figures 10(a), 10(b), 10(c), and 10(d)) ofV. aconitifolia seedlings with increasing intensity of cadmium stress in the present study suggest that H_2O_2might have induced signal transduction cascade seedlings to the imposed cadmium stress. With the increased nonenzymatic antioxidants nonprotein thiols, and translocation of cadmium from roots to leaves, these mechanisms probably confer the cadmium stress tolerance of V. aconitifolia seedlings.

CONCLUSIONS

Cadmium is one of heavy metals that accumulates in the agriculture soil at toxic levels due to anthropogenic activity. In the present study it has been observed that cadmium affected the physiological and biochemical parameters in both leaves and roots by altering redox status, increased lipoxygenase activity, and affected RWC and cell membrane stability in moth bean (V. aconitifolia) seedlings. In spite of these effects of cadmium, moth bean could maintain its growth vigour by minimizing the cadmium toxicity not only by translocation of cadmium from rhizosphere to aerial parts, but also by inducing the H_2O_2 mediated signaling, maintaining the active operation of ascorbate-glutathione cycle and increasing the concentration of nonprotein thiols and cysteine content. In the present study, it has been demonstrated that, unlike other cadmium-sensitive legumes, moth bean is tolerant to cadmium toxicity at the levels studied. Further studies are to be carried out at cellular and molecular level to confirm the possible cadmium withstanding potential of V. aconitifolia.

REFERENCES

1. Y. J. Cui, Y.-G. Zhu, R. H. Zhai, Y. Huang, Y. Qiu, and J. Z. Liang, "Exposure to metal mixtures and human health impacts in a contaminated area in Nanning, China," Environment International, vol. 31, no. 6, pp. 784–790, 2005.
2. L. Sanità Di Toppi and R. Gabbrielli, "Response to cadmium in higher plants," Environmental and Experimental Botany, vol. 41, no. 2, pp. 105–130, 1999.
3. F. Rivera-Becerril, C. Calantzis, K. Turnau et al., "Cadmium accumulation and buffering of cadmium-induced stress by arbuscular mycorrhiza in three Pisum sativum L. genotypes," Journal of Experimental Botany, vol. 53, no. 371, pp. 1177–1185, 2002.
4. S. Clemens, "Toxic metal accumulation, responses to exposure and mechanisms of tolerance in plants,"Biochimie, vol. 88, no. 11, pp. 1707–1719, 2006.
5. L. Zhao, Y.-L. Sun, S.-X. Cui et al., "Cd-induced changes in leaf proteome of the hyperaccumulator plantPhytolacca americana," Chemosphere, vol. 85, no. 1, pp. 56–66, 2011.
6. R. Anita, L. K. Chugh, V. Sawhney, and S. K. Sawhney, "Effect of cadmium and chromium on seed germination and growth of peas," in Proceedings of Environmental Pollution, S. K. Arora, M. Singh, and R. P. Aggarwal, Eds., pp. 66–77, Haryana Agricultural University Press,

Hisar, India, 1990.

7. N. C. Aery and D. K. Rana, "Growth and cadmium uptake in barley under cadmium stress," Journal of Environmental Biology, vol. 24, no. 2, pp. 117–123, 2003.

8. F. Van Assche and H. Clijsters, "Effects of metals on enzyme activity in plants," Plant, Cell & Environment, vol. 13, no. 3, pp. 195–206, 1990.

9. M. N. V. Prasad, "Cadmium toxicity and tolerance in vascular plants," Environmental and Experimental Botany, vol. 35, no. 4, pp. 525–545, 1995.

10. F. Pinot, S. E. Kreps, M. Bachelet, P. Hainaut, M. Bakonyi, and B. S. Polla, "Cadmium in the environment: sources, mechanisms of biotoxicity, and biomarkers," Reviews on Environmental Health, vol. 15, no. 3, pp. 299–323, 2000.

11. K. Peng, X. Li, C. Luo, and Z. Shen, "Vegetation composition and heavy metal uptake by wild plants at three contaminated sites in Xiangxi area, China," Journal of Environmental Science and Health-Part A Toxic/ Hazardous Substances and Environmental Engineering, vol. 41, no. 1, pp. 65–76, 2006.

12. S. S. Kadam and D. K. Salunkhe, "Nutritional composition, processing, and utilization of horse gram and moth bean," Critical Reviews in Food Science and Nutrition, vol. 22, no. 1, pp. 1–26, 1985.

13. R. Adsule, "Moth bean (Vigna aconitifolia (Jacq.) Marechal)," in Food and Feed from Legumes and Oil Seeds, E. Nwokolo and J. Smart, Eds., pp. 203–205, Chapman & Hall, London, UK, 1996.

14. L. Bravo, P. Siddhuraju, and F. Saura-Calixto, "Effect of various processing methods on the in vitro starch digestibility and resistant starch content of indian pulses," Journal of Agricultural and Food Chemistry, vol. 46, no. 11, pp. 4667–4674, 1998.

15. M. Inouhe, S. Ninomiya, H. Tohoyama, M. Joho, and T. Murayama, "Different characteristics of roots in the cadmium-tolerance and Cd-binding complex formation between mono- and dicotyledonous plants,"Journal of Plant Research, vol. 107, no. 3, pp. 201–207, 1994.

16. Y.-M. Li, R. L. Chaney, A. A. Schneiter, J. F. Miller, E. M. Elias, and J. J. Hammond, "Screening for low grain cadmium phenotypes in sunflower, durum wheat and flax," Euphytica, vol. 94, no. 1, pp. 23–30, 1997.

17. D. Ci, D. Jiang, T. Dai, Q. Jing, and W. Cao, "Variation in cadmium tolerance and accumulation and their relationship in wheat recombinant inbred lines at seedling stage," Biological Trace Element Research, vol. 142, no. 3, pp. 807–818, 2011.

18. G. Saeidi, M. Rickauer, and L. Gentzbittel, "Tolerance for cadmium pollution in a core-collection of the model legume, Medicago truncatula L. at seedling stage," Australian Journal of Crop Science, vol. 6, no. 4, pp. 641–648, 2012.

19. R. P. S. Kharb, V. P. Singh, and Y. S. Tomar, "Moth bean (Vigna aconitifolia Jacq. (Mareehal)). A review,"Forage Research Journal, vol. 13, pp. 113–132, 1987.

20. A. M. A. Mazen, "Assessment of heavy metal accumulation and performance of some physiological parameters in Zea mays L. and Vicia faba L. grown on soil amended by sewage sludge resulting from sewage water treatment in the state of Qatar," Qatar University Science Journal, vol. 15, pp. 353–359, 1995.

21. D. A. Cataldo, T. R. Garland, and R. E. Wildung, "Cadmium uptake kinetics in intact soybean plants,"Plant Physiology, vol. 73, no. 3, pp. 844–848, 1983.

22. R. H. Dowdy and G. E. Ham, "Soybean growth and elemental content as influenced by soil amendments of sewage sludge and heavy metals: seedling studies," Agronomy Journal, vol. 69, no. 2, pp. 300–303, 1977.

23. S. A. Hasan, B. Ali, S. Hayat, and A. Ahmad, "Cadmium-induced changes in the growth and carbonic anhydrase activity of chickpea," Turkish Journal of Biology, vol. 31, no. 3, pp. 137–140, 2007.

24. K. K. Bora, S. R. Mathur, and M. Lal, "Relative physiological and biochemical tolerance in moth bean cultivars to cadmium stress," Current Agriculture, vol. 27, pp. 81–84, 2003.

25. S. R. Mathur, K. B. Shukla, and M. K. Sharma, "Effect of cadmium on seedling growth, lipid peroxidation and photosynthetic pigments of mothbean cultivar," International Journal of Plant Sciences, vol. 1, no. 2, pp. 200–201, 2006.

26. D. R. Hoagland and D. I. Arnon, The Water Culture Method for Growing Plants without Soil, Circular 347, California Agricultural Experimental Station, University of California, Berkeley, Berkeley, Calif, USA, 1950.

27. Z. Wang, Y. Zhang, Z. Huang, and L. Huang, "Antioxidative response of metal-accumulator and non-accumulator plants under cadmium stress," Plant and Soil, vol. 310, no. 1-2, pp. 137–149, 2008.

28. S. Choudhury and S. K. Panda, "Toxic effects, oxidative stress and ultrastructural changes in mossTaxithelium nepalense (Schwaegr.) broth. under chromium and lead phytotoxicity," Water, Air, and Soil Pollution, vol. 167, no. 1–4, pp. 73–90, 2005.

29. S. Sugisaka, "The occurrence of peroxide in a perennial plant," Populs Gelrica, vol. 57, pp. 308–309, 1976.

30. P. E. Skorzynska, M. Drazkiewicz, and Z. Krupa, "The activity of the antioxidative system in cadmium-treated Arabidopsis thaliana," Biologia Plantarum, vol. 47, no. 1, pp. 71–78, 2004.

31. R. L. Heath and L. Packer, "Photoperoxidation in isolated chloroplasts. I. Kinetics and stoichiometry of fatty acid peroxidation," Archives of Biochemistry and Biophysics, vol. 125, no. 1, pp. 189–198, 1968.

32. L. B. Pena, L. A. Pasquini, M. L. Tomaro, and S. M. Gallego, "Proteolytic system in sunflower (Helianthus annuus L.) leaves under cadmium stress," Plant Science, vol. 171, no. 4, pp. 531–537, 2006.

33. M. Y. Law, S. A. Charles, and B. Halliwell, "Glutathione and ascorbic acid in spinach (Spinacia oleracea) chloroplasts. The effect of hydrogen peroxide and of Paraquat," Biochemical Journal, vol. 210, no. 3, pp. 899–903, 1983.

34. O. W. Griffith, "Determination of glutathione and

glutathione disulfide using glutathione reductase and 2-vinylpyridine," Analytical Biochemistry, vol. 106, no. 1, pp. 207–212, 1980.

35. S. S. Sharma, S. Kaul, A. Metwally, K. C. Goyal, I. Finkemeier, and K.-J. Dietz, "Cadmium toxicity to barley (Hordeum vulgare) as affected by varying Fe nutritional status," Plant Science, vol. 166, no. 5, pp. 1287–1295, 2004.
36. M. K. Gaitonde, "A spectrophotometric method for the direct determination of cysteine in the presence of other naturally occurring amino acids," Biochemical Journal, vol. 104, no. 2, pp. 627–633, 1967.
37. M. P. Benavides, S. M. Gallego, and M. L. Tomaro, "Cadmium toxicity in plants," Brazilian Journal of Plant Physiology, vol. 17, no. 1, pp. 21–34, 2005.
38. M. D. Groppa, M. P. Ianuzzo, E. P. Rosales, S. C. Vázquez, and M. P. Benavides, "Cadmium modulates NADPH oxidase activity and expression in sunflower leaves," Biologia Plantarum, vol. 56, no. 1, pp. 167–171, 2012.
39. U. J. López-Chuken and S. D. Young, "Modelling sulphate-enhanced cadmium uptake by Zea mays from nutrient solution under conditions of constant free Cd^{2+} ion activity," Journal of Environmental Sciences, vol. 22, no. 7, pp. 1080–1085, 2010.
40. S. S. Gill, N. A. Khan, and N. Tuteja, "Cadmium at high dose perturbs growth, photosynthesis and nitrogen metabolism while at low dose it up regulates sulfur assimilation and antioxidant machinery in garden cress (Lepidium sativum L.)," Plant Science, vol. 182, no. 1, pp. 112–120, 2012.
41. P. L. Gratão, C. C. Monteiro, M. L. Rossi et al., "Differential ultrastructural changes in tomato hormonal mutants exposed to cadmium," Environmental and Experimental Botany, vol. 67, no. 2, pp. 387–394, 2009.
42. S. M. Gallego, L. B. Pena, R. A. Barcia et al., "Unravelling cadmium toxicity and tolerance in plants: insight into regulatory mechanisms," Environmental and Experimental Botany, vol. 83, pp. 33–46, 2012.
43. D. E. Salt and W. E. Rauser, "MgATP-dependent transport of phytochelatins across the tonoplast of oat roots," Plant Physiology, vol. 107, no. 4, pp. 1293–1301, 1995.
44. M. G. Yang, X. Y. Lin, and X. E. Yang, "Impact of Cd on growth and nutrient accumulation of different plant species," Chinese Journal of Applied Ecology, vol. 9, no. 1, pp. 89–94, 1998.
45. F.-B. Wu, J. Dong, Q. Q. Qiong, and G.-P. Zhang, "Subcellular distribution and chemical form of Cd and Cd-Zn interaction in different barley genotypes," Chemosphere, vol. 60, no. 10, pp. 1437–1446, 2005.
46. G. Costa, J. Michaut, and J. Morel, "Influence of cadmium on water relations and gas exchanges, in phosphorus deficient Lupinus albus L.," Plant Physiology and Biochemistry, vol. 32, no. 1, pp. 105–114, 1994.
47. G. Costa and J. L. Morel, "Water relations, gas exchange and amino acid content in Cd-treated lettuce,"Plant Physiology and Biochemistry, vol. 32, no. 4, pp. 561–570, 1994.
48. L. Perfus-Barbeoch, N. Leonhardt, A. Vavasseur, and C. Forestier, "Heavy metal toxicity: cadmium permeates through calcium channels and disturbs the plant water status," The Plant Journal, vol. 32, no. 4, pp. 539–548, 2002.
49. D. J. Flower and M. M. Ludlow, "Contribution of osmotic adjustment to the dehydration tolerance of water-stressed pigeon pea (Cajanus cajan (L.) millsp.) leaves," Plant, Cell and Environment, vol. 9, no. 1, pp. 33–40, 1986.
50. M. C. Romero-Puertas, M. Rodríguez-serrano, F. J. Corpas, M. Gómez, L. A. del Río, and L. M. Sandalio, "Cadmium-induced subcellular accumulation of $O2^{-}$ and H_2O_2 in pea leaves," Plant, Cell & Environment, vol. 27, no. 9, pp. 1122–1134, 2004.
51. E. Bandeoğlu, F. Eyidoğan, M. Yücel, and H. A. Öktem, "Antioxidant responses of shoots and roots of lentil to NaCl-salinity stress," Plant Growth Regulation, vol. 42, no. 1, pp. 69–77, 2004.
52. E. López, C. Arce, M. J. Oset-Gasque, S. Cañadas, and M. P. González, "Cadmium induces reactive oxygen species generation and lipid peroxidation in cortical neurons in culture," Free Radical Biology and Medicine, vol. 40, no. 6, pp. 940–951, 2006.
53. S. K. Tian, L. L. Lu, X. E. Yang, H. G. Huang, K. Wang, and P. H. Brown, "Root adaptations to cadmium-induced oxidative stress contribute to Cd tolerance in the hyperaccumulator Sedum alfredii," Biologia Plantarum, vol. 56, no. 2, pp. 344–350, 2012.
54. N. Mallick, "Copper-induced oxidative stress in the chlorophycean microalga Chlorella vulgaris: response of the antioxidant system," Journal of Plant Physiology, vol. 161, no. 5, pp. 591–597, 2004.
55. A. M. Reddy, S. G. Kumar, G. Jyothsnakumari, S. Thimmanaik, and C. Sudhakar, "Lead induced changes in antioxidant metabolism of horsegram (Macrotyloma uniflorum (Lam.) Verdc.) and bengalgram (Cicer arietinum L.)," Chemosphere, vol. 60, no. 1, pp. 97–104, 2005.
56. S. Singh, S. Eapen, and S. F. D›Souza, "Cadmium accumulation and its influence on lipid peroxidation and antioxidative system in an aquatic plant, Bacopa monnieri L.," Chemosphere, vol. 62, no. 2, pp. 233–246, 2006.
57. L. Garnier, F. Simon-Plas, P. Thuleau et al., "Cadmium affects tobacco cells by a series of three waves of reactive oxygen species that contribute to cytotoxicity," Plant, Cell and Environment, vol. 29, no. 10, pp. 1956–1969, 2006.
58. G. Spiteller, "The relationship between changes in the cell wall, lipid peroxidation, proliferation, senescence and cell death," Physiologia Plantarum, vol. 119, no. 1, pp. 5–18, 2003.
59. E. Skórzyńska-Polit, B. Pawlikowska-Pawlęga, E. Szczuka, M. Drążkiewicz, and Z. Krupa, "The activity and localization of lipoxygenases in Arabidopsis thaliana under cadmium and copper stresses," Plant Growth Regulation, vol. 48, no. 1, pp. 29–39, 2006.

60. C. H. Foyer, "Ascorbic acid," in Antioxidants in Higher Plants, R. G. Alscher and J. L. Hess, Eds., pp. 31–58, CRC Press, Boca Raton, Fla, USA, 1993.

61. C. H. Foyer and G. Noctor, "Redox regulation in photosynthetic organisms: signaling, acclimation, and practical implications," Antioxidants and Redox Signaling, vol. 11, no. 4, pp. 861–905, 2009.

62. P. Ahmad, C. A. Jaleel, M. A. Salem, G. Nabi, and S. Sharma, "Roles of enzymatic and nonenzymatic antioxidants in plants during abiotic stress," Critical Reviews in Biotechnology, vol. 30, no. 3, pp. 161–175, 2010.

63. A. Cuypers, J. Vangronsveld, and H. Clijsters, "Peroxidases in roots and primary leaves of Phaseolus vulgaris Copper and Zinc Phytotoxicity: a comparison," Journal of Plant Physiology, vol. 159, no. 8, pp. 869–876, 2002.

64. K. P. Gowrinatha and V. N. R. Rao, "Reversal of heavy metal toxicity by ascorbic acid in microalgae,"Journal of Swap-Bot, vol. 9, pp. 27–29, 1992.

65. D. D. Perrin and A. E. Watt, "Complex formation of zinc and cadmium with glutathione," Biochem Biophys Acta (BBA)—General Subjects, vol. 230, no. 1, pp. 96–104, 1971.

66. M. Oven, J. E. Page, M. H. Zenk, and T. M. Kutchan, "Molecular characterization of the homo-phytochelatin synthase of soybean Glycine max: relation to phytochelatin synthase," The Journal of Biological Chemistry, vol. 277, no. 7, pp. 4747–4754, 2002.

67. Y. T. Hsu and C. H. Kao, "Toxicity in leaves of rice exposed to cadmium is due to hydrogen peroxide accumulation," Plant and Soil, vol. 298, no. 1-2, pp. 231–241, 2007.

68. K. Verma, G. S. Shekhawat, A. Sharma, S. K. Mehta, and V. Sharma, "Cadmium induced oxidative stress and changes in soluble and ionically bound cell wall peroxidase activities in roots of seedling and 3-4 leaf stage plants of Brassica juncea (L.) czern," Plant Cell Reports, vol. 27, no. 7, pp. 1261–1269, 2008.

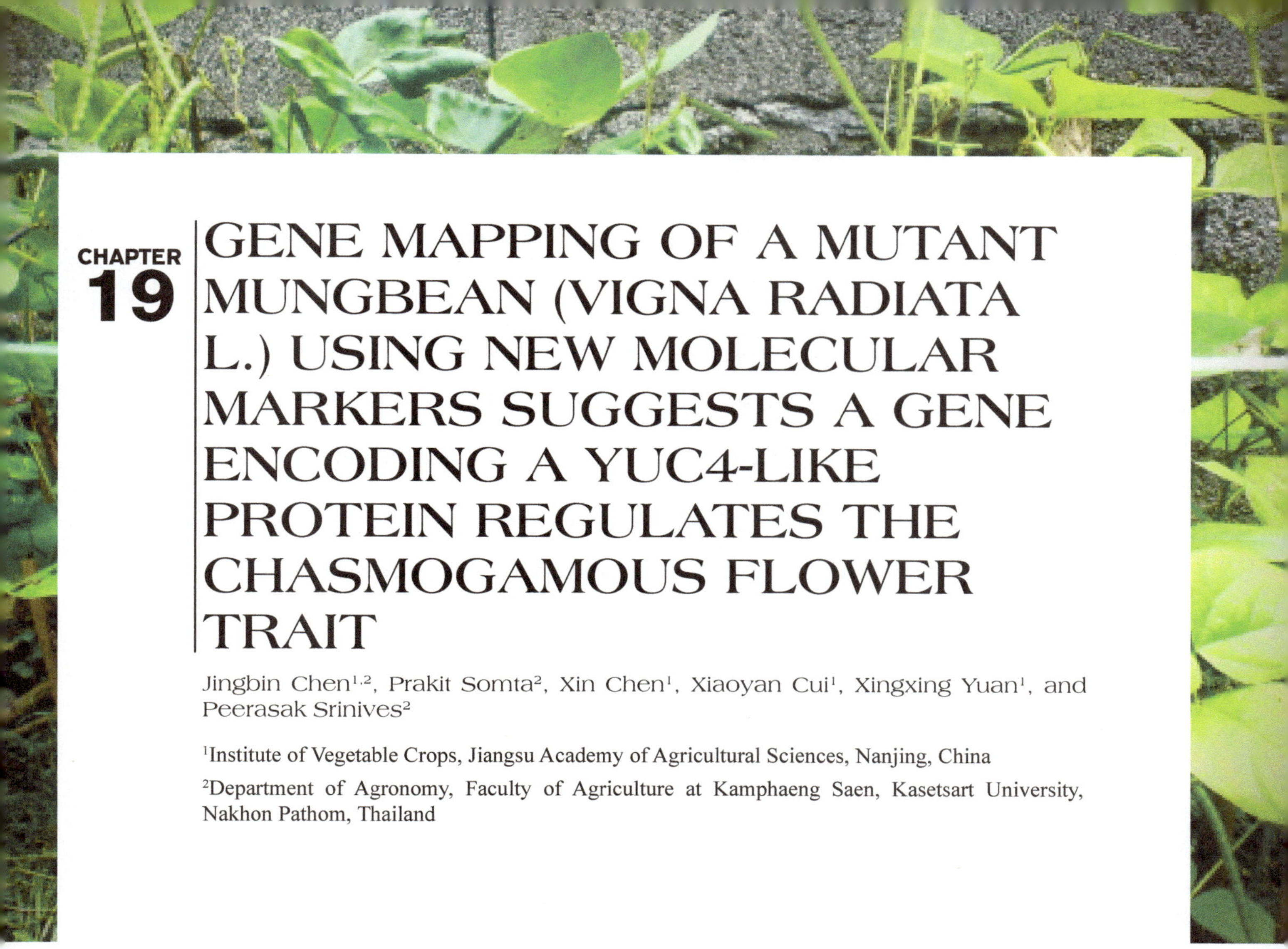

CHAPTER 19

GENE MAPPING OF A MUTANT MUNGBEAN (VIGNA RADIATA L.) USING NEW MOLECULAR MARKERS SUGGESTS A GENE ENCODING A YUC4-LIKE PROTEIN REGULATES THE CHASMOGAMOUS FLOWER TRAIT

Jingbin Chen[1,2], Prakit Somta[2], Xin Chen[1], Xiaoyan Cui[1], Xingxing Yuan[1], and Peerasak Srinives[2]

[1]Institute of Vegetable Crops, Jiangsu Academy of Agricultural Sciences, Nanjing, China

[2]Department of Agronomy, Faculty of Agriculture at Kamphaeng Saen, Kasetsart University, Nakhon Pathom, Thailand

ABSTRACT

Mungbean (*Vigna radiata* L.) is a cleistogamous plant in which flowers are pollinated before they open, which prevents yield improvements through heterosis. We previously generated a chasmogamous mutant (CM) mungbean in which open flowers are pollinated. In this study, we developed insertion/deletion (indel) markers based on the transcriptome differences between CM and Sulu-1 (i.e., normal flowering) plants. An F_2population derived from a cross between CM and Sulu-1 was used for gene mapping. Segregation analyses revealed that a single recessive gene regulates the production of chasmogamous flowers. Using newly developed indel and simple sequence repeat markers, the *cha* gene responsible for the chasmogamous flower trait was mapped to a 277.1-kb segment on chromosome 6. Twelve candidate genes were detected in this segment, including *Vradi06g12650*, which encodes a YUCCA family protein associated with floral development. A single base pair deletion producing a frame-shift mutation and a premature stop codon in *Vradi06g12650* was detected only in CM plants. This suggested that *Vradi06g12650* is a *cha* candidate gene. Our results provide important information for the molecular breeding of chasmogamous mungbean lines, which may serve as new genetic resources for hybrid cultivar development.

INTRODUCTION

Exploiting heterosis is an effective way to increase crop yield. As an important legume in Asia, several studies have been conducted on mungbean (*Vigna radiata* L. Wilczek) to investigate the importance of heterosis for seed yield and other yield-related traits (Chen et al., 2003; Soehendi and Srinives, 2005; Sorajjapinun et al., 2012; Yashpal et al., 2015). However, a major obstacle to producing mungbean hybrid seeds is floral architecture. Mungbean plants have papilionaceous flowers with five differently shaped petals, including one standard petal, two wing petals, and two keel petals (Verdcourt, 1970). The anthers and stigmas are enclosed within the pocket-like keel petals, making mungbean a cleistogamous plant whose flowers are pollinated before they open. The natural outcrossing rate of cultivated mungbean is approximately 1.68% (Sangiri et al., 2007), which is too low for the commercial production of hybrid seeds. Some accessions of cleistogamous plants exhibit a specific floral architecture that promotes natural outcrossing.

Citation: Chen J, Somta P, Chen X, Cui X, Yuan X and Srinives P (2016) Gene Mapping of a Mutant Mungbean (Vigna radiata L.) Using New Molecular Markers Suggests a Gene Encoding a YUC4-like Protein Regulates the Chasmogamous Flower Trait. Front. Plant Sci. 7:830. doi: 10.3389/fpls.2016.00830

Studies in rice revealed that exerted stigmas and anthers increase the outcrossing rate and considerably help the production of hybrid seeds (Kato and Namai, 1987). Some mutant legume crops with exerted stigmas and anthers have been described (Dahiya et al., 1984; Pundir and Reddy, 1998; Cherian et al., 2006; Srinivasan and Gaur, 2012; Dheer et al., 2014). We previously identified a chasmogamous mutant (CM), which was generated in the mungbean accession V1197 by gamma irradiation (Sorajjapinun and Srinives, 2011). The outcrossing rate of CM plants increased to 9.6%, while outcrossing was undetectable in wild-type controls. Additionally, the yield and agronomic traits of CM plants differed from those of V1197 plants, with fewer pods per plant, seeds per pod, and yield per plant. Genetic analyses using F_1, F_2, and backcross populations from the cross between CM and V1197 plants revealed that the production of chasmogamous flowers is regulated by a single gene, *cha* (Sorajjapinun and Srinives, 2011). Thus, additional research focusing on the genetic characterization of this mutant mungbean should be conducted.

In terms of molecular genetics and genomics, mungbean has not been as extensively studied as other legumes, including soybean [*Glycine max* (L.) Merr.], common bean (*Phaseolus vulgaris* L.), cowpea [*V. unguiculata* (L.) Walp.], and chickpea (*Cicer arietinum* L.). DNA markers represent important tools for genetic analyses and the mapping of genes or quantitative trait loci. Many DNA markers, especially simple sequence repeats (SSRs), have been developed for mungbean (Gwag et al., 2007; Somta et al., 2008, 2009; Tangphatsornruang et al., 2009; Gupta et al., 2014; Chen et al., 2015) or introduced from other closely related species. These markers have been used to develop linkage maps and map quantitative trait loci associated with important mungbean traits (e.g., yield and resistance to biotic and abiotic stresses; Kasettranan et al., 2010; Chankaew et al., 2011, 2013; Isemura et al., 2012; Kajonphol et al., 2012; Prathet et al., 2012; Sompong et al., 2012; Kitsanachandee et al., 2013; Alam A.K.M.M. et al., 2014; Alam A.M. et al., 2014;Chotechung et al., 2016). However, most of these markers are monomorphic or weakly polymorphic. One of the most important advances in mungbean genomics research was the release of the VC1973A whole genome sequence by Kang et al. (2014). This sequence is relevant for research related to marker development, gene mapping, and gene function analyses.

The molecular mechanisms regulating floral development can be summarized using an "ABCE" model (Krizek and Fletcher, 2005). Additionally, some hormones (e.g., auxin) are important for floral development (Okada et al., 1991;Bennett et al., 1995; Przemeck et al., 1996; Galweiler et al., 1998; Aloni et al., 2006; Cheng et al., 2006). Some*Arabidopsis thaliana* mutants have abnormal flowers, and the genes responsible for this mutation have been identified as *pin1*, *pinoid*, *mp*, and *yuc* (Cheng and Zhao, 2007). The YUCCA (YUC) family consists of flavin monooxygenases (FMOs) related to the biosynthesis of indole-3-acetic acid. The FMOs are key enzymes that convert indole-3-pyruvic acid to indole-3-acetic acid by catalyzing the hydroxylation of the amino group of tryptophan, which is the rate-limiting step in tryptophan-dependent auxin biosynthesis (Zhao et al., 2001). There are 11 *YUC* genes in the *A. thaliana* genome (Zhao et al., 2001; Cheng et al., 2006, 2007). Double, triple, and quadruple mutants of some YUC family genes exhibit severe defects in floral patterns, vascular formations, and other developmental processes (Cheng et al., 2006). Therefore, *YUC* genes play important roles during floral formation and development, with implications for flower shape.

In this study, we mapped the *cha* gene regulating the production of chasmogamous flowers in CM plants using new insertion/deletion (indel) and SSR markers. We identified a gene encoding a YUC-like protein as a likely candidate for the *cha* gene.

MATERIALS AND METHODS

Mapping Population and DNA Extraction

We previously identified a CM mungbean line in the M_2 generation of accession V1197 following gamma irradiation. The CM plants lacked wing and keel petals, which exposed the stigmas and anthers (Sorajjapinun and Srinives, 2011). A stable CM line was selected from the M_3 and M_4 generations. A CM plant was pollinated using pollen from Sulu-1, which is a mungbean line with normal flowers. Three F_1 plants were grown and the resulting flowers were morphologically analyzed. The hybrids were verified using two polymorphic indel markers. Seeds from one F_1 plant were then harvested to produce an F_2 population consisting of 127 plants. The F_2 plants and their parents were grown in a field at Kasetsart University, Kamphaeng Saen Campus, Nakhon Pathom, Thailand from May to July 2014. Flowers from each plant were individually examined to determine whether they were normal or chasmogamous.

Total genomic DNA was extracted from fresh leaf tissue of individual plants according to a slightly modified version of the method described by Lodhi et al. (1994). All DNA samples were diluted to 5 ng μl^{-1} according to lambda DNA, and analyzed by 1.0% agarose gel electrophoresis.

Transcriptome Sequencing and Development of Molecular Markers

Total RNA was extracted from young flowers of CM and Sulu-1 mungbean plants using the EasyPure Plant RNA kit (Transgene Biotech, Beijing, China). The RNA samples were used to prepare libraries for sequencing by the Illumina HiSeq 2000 sequencer at the Beijing Genomics Institute (Shenzhen, China). The resulting sequences were assembled using the Trinity program (Grabherr et al., 2011). The Sulu-1 and CM transcriptome sequences were aligned using the NCBI BLAST+ 2.2.31 program with an *E*-value cutoff of 10.0. Sequences with indels that were 5 bp or larger were randomly chosen for marker development. Primers specific for the selected indels were designed using Primer3 (Untergasser et al., 2012) with the following criteria: primer length: 18–27 nucleotides; melting temperature: 50–65°C; GC content: 40–60%; and polymerase chain reaction (PCR) product size: 100–300 bp. Transcript sequences used for indel marker development were blasted against the mungbean whole genome sequence[1] (Kang et al., 2014) to determine the physical locations of markers.

In addition to indel markers, we developed new SSR markers to fine map the *cha* locus. The mungbean chromosome 6 sequence was downloaded (Kang et al., 2014) and scanned for di-, tri-, and tetra-nucleotide repeats using SSR Hunter 1.3

(Li and Wan, 2005). Based on our initial mapping of the *cha* locus using indel markers, we focused on a 2.2-Mb genomic region of chromosome 6 (i.e., 29.9–32.1 Mb) carrying *cha*, and selected the SSRs therein. Primers for the SSRs were designed as described for the indel markers.

Molecular Marker Analysis

The newly developed indel and SSR markers were used to detect polymorphisms between CM and Sulu-1 sequences. The PCR analyses were completed using a 10-μl solution containing 2 ng genomic DNA, 1x *Taq* buffer, 2 mM $MgCl_2$, 0.2 mM dNTPs, 1 U *Taq* DNA polymerase (Thermo Scientific), and 0.5 μM forward and reverse primers. The PCR was conducted in an MJ Research PTC-200 Thermal Cycler (Bio-Rad) using the following program: 94°C for 3 min; 35 cycles of 94°C for 30 s, 55°C for 30 s, and 72°C for 30s; 72°C for 5 min. The amplicons were separated in a 6% denaturing polyacrylamide gel or a 3% agarose gel, and visualized with silver or ethidium bromide staining, respectively. The polymorphic markers were used to analyze the F_2 population.

Data Analysis and Gene Mapping

We counted the number of plants with normal and chasmogamous flowers, and analyzed the data using a χ^2 test (Mather, 1951) to confirm the monogenic inheritance in CM plants as reported earlier by Sorajjapinun and Srinives (2011). Segregation data for the floral traits and DNA markers were used to construct a linkage map with MapMaker/EXP 3.0 (Lander et al., 1987). A minimum logarithm of odds value of 3.0 and maximum recombination frequency of 4.0 were used to group the markers. The genetic map distance was calculated using the Kosambi mapping function. Linkage groups were anchored to chromosomes by the physical location of markers. The map was drawn using MapChart 2.30 (Voorrips, 2002).

Identifying and Sequencing the Candidate Gene

Based on the locations of the markers flanking *cha* on the linkage map, the predicted genes on the mungbean reference sequence (Kang et al., 2014) between the flanking markers were downloaded. Deduced protein sequences for these genes were subjected to a BLASTP search against the NCBI database to obtain information regarding their homologs and functions.

The following two pairs of primers were designed to amplify the genomic region containing a candidate gene responsible for the production of chasmogamous flowers: Seq-1-F (5′-AAGAACGAGGTTTGGCTTCA-3′)/Seq-1-R (5′-AAC TTGACCCACATCAAGGA-3′) and Seq-2-F (5′-GGAAAC ACAAATCACTATGGCA-3′)/Seq-2-R (5′-ATTGCATGTACA TGCCAGCTA-3′). The PCR was conducted using Platinum *Taq* DNA Polymerase High Fidelity (Invitrogen) following the manufacturer's instructions with some modifications (i.e., annealing temperature of 58°C and elongation time of 90 s). The PCR products were separated by electrophoresis on 1% agarose gels stained with ethidium bromide. The DNA fragments of the expected size were purified using the E.Z.N.A Gel Extraction Kit (Omega Bio-tek), and sequenced using the 3730xl DNA Analyzer (Applied Biosystems). The deduced Vr06g12650 protein and 11 *A. thaliana* YUC proteins (i.e., AT4G32540.1, AT4G13260.1, AT1G04610.1, AT5G11320.1, AT5G43890.1, AT5G25620.2, AT2G33230.1, AT4G28720.1, AT1G04180.1, and AT1G48910.1) obtained from The *Arabidopsis* Information Resource[2] database were used to construct a phylogenetic tree using the "One Click" mode and LRT statistical test of Phylogeny.fr (Dereeper et al., 2010).

RESULTS

Morphological Features of the Chasmogamous Mutant Mungbean and Inheritance of the Floral Trait

The floral architecture of CM plants differed from that of Sulu-1 plants. The CM flowers were missing the wing and keel petals (Figure 1). When the CM and Sulu-1 plants were crossed, the F_1 hybrids had normal flowers (Figure 1), which suggested a recessive gene (or genes) regulated the production of chasmogamous flowers in CM plants.

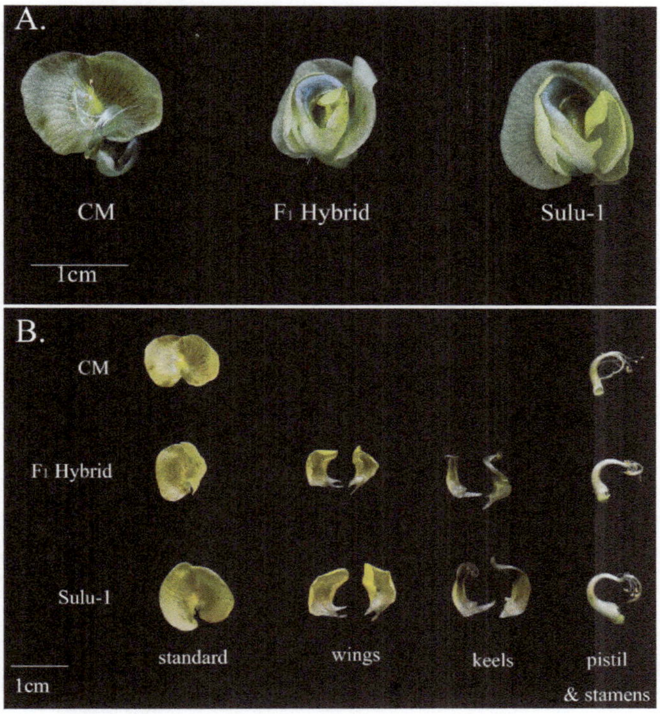

Figure 1. Floral morphology of CM and Sulu-1 plants as well as their F_1 hybrids. Whole flower (A) and individual flower parts (B). We determined the segregation ratio of the chasmogamous flower trait in F_2 plants. Out of 127 plants, 96 had normal flowers, while 31 had chasmogamous flowers. This segregation corresponded with a 3:1 ratio ($\chi^2 = 0.02$, $P = 0.88$), indicating that a single recessive gene mediated the chasmogamous flower trait in CM plants. This confirmed the results of the study by Sorajjapinun and Srinives (2011).

Development of Indel Markers

To map the gene responsible for the production of chasmogamous flowers, indel markers were developed using the transcriptomes of the Sulu-1 and CM parents. We randomly selected 140 transcript sequences carrying indels to develop markers, and determined that 84 of them (i.e., 57.1%) were polymorphic between the parents. All of the markers were co-dominant, and able to distinguish between the parents and hybrid progenies. Seventy-four markers were localized to 11 chromosomes by aligning their related transcripts with the whole mungbean genome sequence (Figure 2). Another 10 markers were located on scaffolds that could not be assembled on chromosomes.

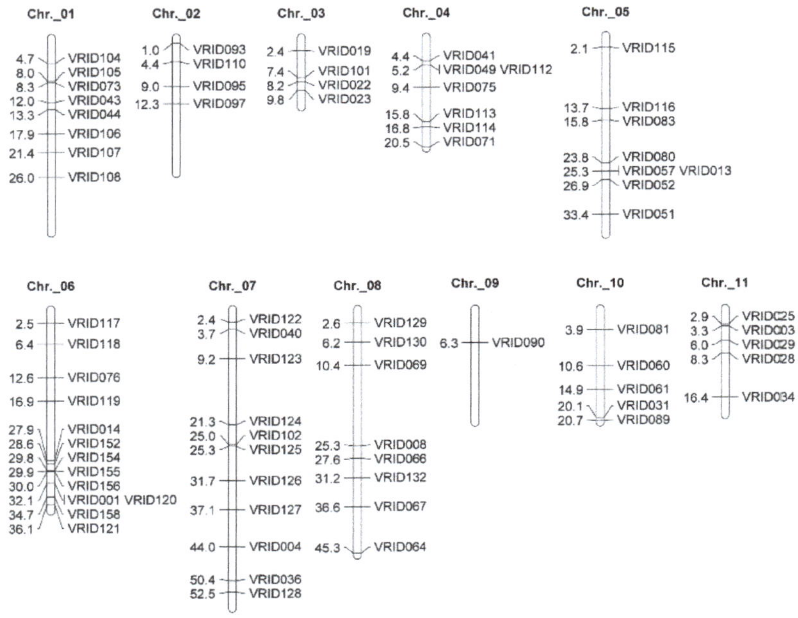

Figure 2. Physical positions on the mungbean chromosome of the indel markers polymorphic between CM and Sulu-1 plants. Unit: mega base pair (Mbp). Ten markers located on scaffolds are not shown.

Genetic Mapping of the *cha* Gene

Linkage analysis of the polymorphic indel markers and phenotyping of the flowers were completed using 127 plants from the F_2 population. The target *cha* gene was mapped to chromosome 6 between markers VRID155 and VRID120 at a distance of 1.6 and 3.4 cM, respectively (Figure 3A). To locate *cha* more precisely, seven polymorphic SSR markers were developed in the target interval by screening SSR motifs in the reference genome sequence. Eight recombinants were identified in the interval between VRID115 and VRID120. By associating the marker genotypes with floral phenotypes, the 60 and 38 recombinants restricted *cha* a segment between markers SSR09 and SSR12. These two markers were 277.1 kb apart, and were located at 30.40 and 30.68 Mb of chromosome 6, respectively (Figure 3B). Based on the mungbean whole genome sequence, 12 candidate genes were detected in this region.

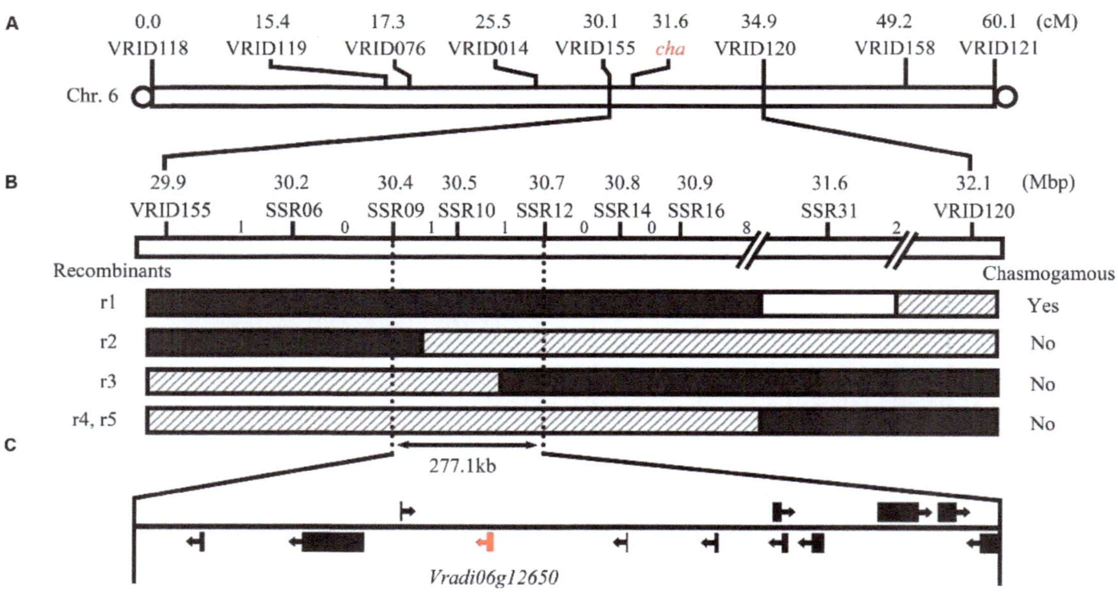

Figure 3. Mapping of the *cha* gene regulating the production of chasmogamous flowers in CM plants.Location of *cha* (A) and recombinant events on chromosome 6 (B). White, shadow, and black regions in the recombinants indicate homozygous regions for the Sulu-1 allele, and heterozygous and homozygous regions for the CM allele, respectively. The number of recombinant events between adjacent markers are indicated above the bar. Another five recombinants between the homozygous Sulu-1 genotype and heterozygous genotype are not presented because all of the plants had normal flowers. The 12 predicted genes in the candidate gene region are indicated (C).

Function Prediction and Sequencing of *cha* Candidate Genes

To predict the functions of the 12 detected *cha* candidate genes, the putative protein sequences encoded by these genes were used as queries to search the NCBI database. Their predicted functions are listed in Table S3. *Vradi06g12650*, which encoded a YUC homolog (Figure 3) related to the auxin biosynthesis pathway and floral development, was considered the most likely *cha* gene. A BLASTP search revealed that the YUC protein encoded by *Vradi06g12650* was most similar to the *A. thaliana* proteins AtYUC4 (identity: 68.13%) and AtYUC1 (identity: 62.32%). A protein sequence-based phylogenetic analysis involving Vradi06g12650 and 11 *A. thaliana* YUC proteins also revealed a close relationship between Vradi06g12650 and AtYUC1/AtYUC4 (Figure 4). We sequenced the *Vradi06g12650* coding domain in Sulu-1, CM, and wild-type V1197 mungbean plants. Comparisons among the resulting sequences and the corresponding reference sequence (i.e., from VC1973A) revealed six single nucleotide polymorphisms (SNPs) between the CM and Sulu-1 genes (Figure 5). The SNP at the 382-bp position led to an amino acid substitution from glutamine to glutamic acid, while the other SNPs resulted in non-sense mutations (Figures 5 and 6). The six SNPs were not detected between the CM and V1197 sequences. However, a 1-bp deletion at the 894-bp position was observed in the CM *Vradi06g12650*sequence (Figure 5). This deletion produced a frame-shift mutation in the 3′-terminus, resulting in a shorter CM Vradi06g12650 protein with a different C-terminus compared with the corresponding wild-type V1197, Sulu-1, and VC1973A proteins (Figure 6). Additionally, this deletion was detected in a sequence (i.e., Unigene0038420) from the CM transcriptome. These results suggested that the 1-bp deletion in *Vradi06g12650* is responsible for the production of chasmogamous flowers in CM plants.

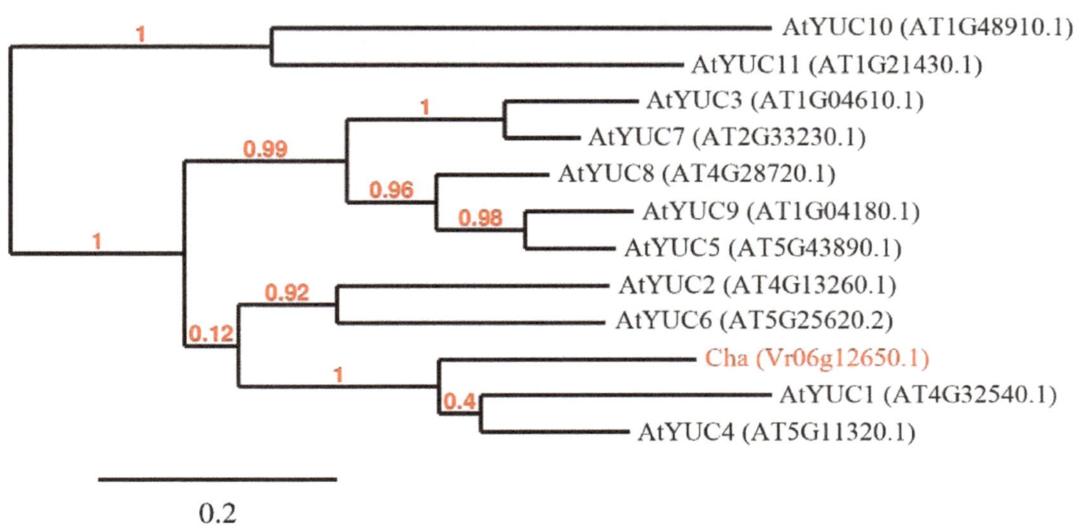

Figure 4. Phylogenetic tree of *Arabidopsis thaliana* YUCCA and Cha proteins. Sequences of 11 *A. thaliana* YUCCA proteins and the deduced mungbean Vr06g12650.1 protein were used to generate the phylogenetic tree.

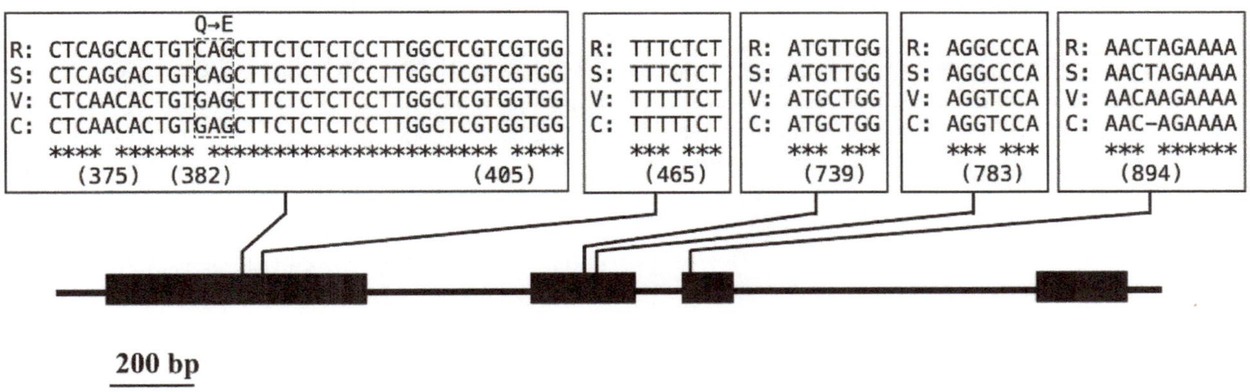

Figure 5. Variability in the *Vradi06g12650* open reading frame. Consensus sequences are marked by asterisks. Numbers under variant loci indicate the position (in bp) on the open reading frame. The R, S, V, and C correspond to V1973A (reference sequence), Sulu-1, V1197, and CM, respectively.

```
                    FAD-biding (GxGxxG)
Sulu_1    MGSCKPQQEHVHGPIIIGAGPSGLAVAACLSEHKVPFVILERSNCIASLWQHKTYDRLKL
V1197     MGSCKPQQEHVHGPIIIGAGPSGLAVAACLSEHKVPFVILERSNCIASLWQHKTYDRLKL
CM        MGSCKPQQEHVHGPIIIGAGPSGLAVAACLSEHKVPFVILERSNCIASLWQHKTYDRLKL
          ************************************************************

Sulu_1    HLPKQFCELPLKGFPHNFPKYPTKYQFISYMESYASHFNIHPRFNQTVETAHFDKASQLW
V1197     HLPKQFCELPLKGFPHNFPKYPTKYQFISYMESYASHFNIHPRFNQTVETAHFDKASQLW
CM        HLPKQFCELPLKGFPHNFPKYPTKYQFISYMESYASHFNIHPRFNQTVETAHFDKASQLW
          ************************************************************
                                     FMO-identifying motif (FxGxxxHxxxY/F)
Sulu_1    LVRTQHCQLLSPWLVVATGENAEPVLPRIHGMDHFSGSIAHTSVYKSGSEYTNQKVLVIG
V1197     LVRTQHCELLSPWLVVATGENAEPVLPRIHGMDHFSGSIAHTSVYKSGSEYTNQKVLVIG
CM        LVRTQHCELLSPWLVVATGENAEPVLPRIHGMDHFSGSIAHTSVYKSGSEYTNQKVLVIG
          *******:****************************************************
   NADP-binding motif (GxGxxG)
Sulu_1    CGNSGMEVSLDLCRHNASPYMVARNTVHVLPREMFGFSTFGIAMALYKWFPIKVVDKILL
V1197     CGNSGMEVSLDLCRHNASPYMVARNTVHVLPREMFGFSTFGIAMALYKWFPIKVVDKILL
CM        CGNSGMEVSLDLCRHNASPYMVARNTVHVLPREMFGFSTFGIAMALYKWFPIKVVDKILL
          ************************************************************

Sulu_1    LVTNFMLGNTNHYGIKRPKTGPIELKLATGKTPVLDVGQVAQIKCGNIKVMEGVKEITRK
V1197     LVTNFMLGNTNHYGIKRPKTGPIELKLATGKTPVLDVGQVAQIKCGNIKVMEGVKEITRK
CM        LVTNFMLGNTNHYGIKRPKTGPIELKLATGKTPVLDVGQVAQIKCGNIKVMEGVKEITEK
          **********************************************************.*
                      F/LATGY
Sulu_1    GAKFMDGQEKEFDAIILATGYKSNVPAWLKGCDFFTEDGMPKTPFPHGWKGEQGLYTVGF
V1197     GAKFMDGQEKEFDAIILATGYKSNVPAWLKGCDFFTEDGMPKTPFPHGWKGEQGLYTVGF
CM        VRNLWMDKKR---NLML-------------------------------------------
            :: .:::     ::*

Sulu_1    TRRGIQGTSCDAIKIAEDIASQWRTVENKNQCNSHIILLTS
V1197     TRRGIQGTSCDAIKIAEDIASQWRTVENKNQCNSHIILLTS
CM        -----------------------------------------
```

Figure 6. Alignment of the deduced Cha protein sequences from Sulu-1, V1197, and CM plants by ClustalX. Conserved flavin monooxygenase motifs are indicated (on top). Asterisk indicates conserved amino acid.

DISCUSSION

Advantages of Developing Markers from Transcriptomes and the Utility of *cha* for Mungbean Improvement

Mungbean is not a model plant. Therefore, its molecular genetics and genome have not been as comprehensively studied as in many other crops. The available molecular markers are insufficient for genetics research and mungbean breeding programs. Additionally, some reports have indicated that the polymorphic information content of mungbean SSR markers is low (Tangphatsornruang et al., 2009; Chen et al., 2015). Thus, it is important to develop more mungbean genomic resources. In this study, we developed indel markers by comparing the transcriptomes of two mungbean accessions, which resulted in the detection of several polymorphic markers between the parents of the mapping population. The differences in indel sizes between Sulu-1 and CM plants enabled the indels to be distinguished by PCR followed by 3% agarose gel electrophoresis analysis, which is easier to complete than polyacrylamide gel electrophoresis. This resulted in a more efficient genetic mapping procedure.

With the newly developed markers, the *cha* gene responsible for the production of chasmogamous flowers in mungbean was mapped to chromosome 6. Several newly developed markers were identified closely linked to the target gene. Our data revealed that the marker SSR10 was completely linked to *cha*. However, this finding is based on a relatively small population of 127 F_2 plants. Because the CM floral phenotype was observed to be regulated by a single recessive gene, these linked markers may be useful for the marker-assisted selection of mungbean plants producing chasmogamous flowers.

Functions of the *cha* Candidate Gene

Vradi06g12650 is the most likely *cha* gene. This gene putatively encodes a YUC homolog involved in auxin biosynthesis and floral development (Zhao et al., 2001; Cheng et al., 2006). The YUC proteins constitute a family of FMOs containing several conserved sequence motifs, including the FAD-binding motif, FMO-identifying motif, NADPH-binding motif, and F/LATGY motif (Schlaich, 2007). The protein encoded by *Vradi06g12650* is highly similar to *A. thaliana* YUC4. The CM *Vradi06g12650* coding sequence differed from that of the wild-type V1197 by a 1-bp deletion at the 894-bp position. This deletion was only detected in CM plants, and results in a frame-shift of the coding sequence leading to the absence of the LATGY motif in the C-terminus of the predicted YUC protein (Figure 6). There are 11 *A. thaliana* genes encoding YUC proteins, suggesting there may be some functional redundancy among these proteins. Mutational inactivation of a single *YUC* family gene in *A. thaliana* caused no obvious developmental defects, while a double *yuc1 yuc4* mutant and a quadruple *yuc1 yuc4 yuc10 yuc11* mutant exhibited severe defects in the formation of floral organs (Zhao et al., 2001; Cheng et al., 2006). These findings differed from our observation that a defect in a single *YUC4*-like gene causes a dramatic morphological abnormality in mungbean floral organs. However, it is worth noting that the mutation detected in our study produced an immature YUC4-like protein lacking the F/LATGY motif, which is highly conserved among YUC proteins. Therefore, this defective YUC4-like protein may be responsible for the abnormal floral development in CM plants. Additional studies are required to characterize how *cha* affects the production of chasmogamous flowers in mungbean.

AUTHOR CONTRIBUTIONS

JC designed the InDel and SSR markers and prepared the manuscript. PS performed gene mapping and reviewed the manuscript. XChen involved in bioinformatics analysis and reviewed the manuscript. XCui sequenced the candidate gene. XY conducted the hybrids and developed the populations. PS designed the study and refined the manuscript.

ACKNOWLEDGMENTS

We are thankful to the Center for Agricultural Biotechnology, Kasetsart University, Joint Legume Research Center Kasetsart University and Jiangsu Academy of Agricultural Sciences, and Institute of Vegetable Crops, Jiangsu Academy of Agricultural Sciences/Jiangsu Key Laboratory for Horticultural Crop Genetic Improvement, for molecular lab facilities.

REFERENCES

1. Alam, A. K. M. M., Somta, P., and Srinives, P. (2014). Identification and confirmation of quantitative trait loci controlling resistance to mungbean yellow mosaic disease in mungbean [*Vigna radiata* (L.) Wilczek]. *Mol. Breed.* 34, 1497–1506. doi: 10.1007/s11032-014-0133-0
2. Alam, A. M., Somta, P., Muktadir, M., and Srinives, P. (2014). Quantitative trait loci associated with seed weight in mungbean (*Vigna radiata* (L.) Wilczek). *Kasetsart J. (Nat. Sci.)* 48, 197–204.
3. Aloni, R., Aloni, E., Langhans, M., and Ullrich, C. I. (2006). Role of auxin in regulating *Arabidopsis* flower development. *Planta* 223, 315–328. doi: 10.1007/s00425-005-0088-9
4. Bennett, S. R., Alvarez, J., Bossinger, G., and Smyth, D. R. (1995). Morphogenesis in pinoid mutants of *Arabidopsis thaliana*. *Plant J.* 8, 505–520. doi: 10.1046/j.1365-313X.1995.8040505.x
5. Chankaew, A., Somta, P., Isemura, T., Tomooka, N., Kaga, A., Vaughan, D. A., et al. (2013). QTL mapping reveals conservation of major and minor loci for powdery mildew resistance in four sources of resistance in mungbean [*Vigna radiata* (L.) Wilczek]. *Mol. Breed.* 32, 121–130. doi: 10.1007/s11032-013-9856-6
6. Chankaew, S., Somta, P., Sorajjapinun, W., and Srinives, P. (2011). Quantitative trait loci mapping of Cercospora leaf spot resistance in mungbean, *Vigna radiata* (L.) Wilczek. *Mol. Breed.* 28, 255–264.
7. Chen, H., Qiao, L., Wang, L., Wang, S., Blair, M. W., and Cheng, X. (2015). Assessment of genetic diversity and population structure of mung bean (*Vigna radiata*) germplasm using EST-based and genomic SSR markers. *Gene* 566, 175–183. doi: 10.1016/j.gene.2015.04.043
8. Chen, X., Sorajjapinun, W., Reiwthongchum, S., and Srinives, P. (2003). Identification of parental mungbean lines for production of hybrid varieties. *CMU J.* 2, 97–106.
9. Cheng, Y., Dai, X., and Zhao, Y. (2006). Auxin biosynthesis by the YUCCA flavin monooxygenases controls the formation of floral organs and vascular tissues in *Arabidopsis*. *Genes Dev.* 20, 1790–1799. doi: 10.1101/gad.1415106
10. Cheng, Y., Dai, X., and Zhao, Y. (2007). Auxin synthesized by the YUCCA flavin monooxygenases is essential for embryogenesis and leaf formation in*Arabidopsis*. *Plant Cell* 19, 2430–2439. doi: 10.1105/tpc.107.053009
11. Cheng, Y., and Zhao, Y. (2007). A role for auxin in flower development. *J. Integr. Plant Biol.* 49, 99–104. doi: 10.1111/j.1744-7909.2006.00412.x
12. Cherian, C. A., Mallikarjuna, N., Jadhav, D., and Saxena, K. (2006). Open flower segregants selected from *Cajanus platycarpus* crosses. *J. SAT Agric. Res.*2, 1–2.
13. Chotechung, S., Somta, P., Chen, J., Yimram, T., Chen, X., and Srinives, P. (2016). A gene encoding a polygalacturonase-inhibiting protein (PGIP) is a candidate gene for bruchid (Coleoptera: Bruchidae) resistance in mungbean (*Vigna radiata*). *Theor. Appl. Genet.* doi: 10.1007/s00122-016-2731-1 [Epub ahead of print].
14. Dahiya, B., Lather, V., Solanki, I., and Kumar, R. (1984). Genetic resources: useful spontaneous mutants in chickpea. *Int. Chick. News.* 11, 5–7.
15. Dereeper, A., Audic, S., Claverie, J. M., and Blanc, G. (2010). BLAST-EXPLORER helps you building datasets for phylogenetic analysis. *BMC Evol. Biol.*10:8. doi: 10.1186/1471-2148-10-8
16. Dheer, M., Punia, S. S., Ram, B., Jain, N. K., Koli, N. R., Bhatnagar, V., et al. (2014). Morphological features of an open flower mutant plant and characterization of their progenies in lentil (*Lens culinaris* Medik.). *Genet. Resour. Crop Evol.* 61, 879–886. doi: 10.1007/s10722-014-0100-y
17. Galweiler, L., Guan, C., Muller, A., Wisman, E., Mendgen, K., Yephremov, A., et al. (1998). Regulation of polar auxin transport by AtPIN1 in *Arabidopsis*vascular tissue. *Science* 282, 2226–2230. doi: 10.1126/science.282.5397.2226
18. Grabherr, M. G., Haas, B. J., Yassour, M., Levin, J. Z., Thompson, D. A., Amit, I., et al. (2011). Full-length transcriptome assembly from RNA-Seq data without a reference genome. *Nat. Biotechnol.* 29, 644–652. doi: 10.1038/nbt.1883
19. Gupta, S. K., Bansal, R., and Gopalakrishna, T. (2014). Development and characterizatiom of genic SSR markers for mungbean (*Vigna radiata* (L.) Wilczek). *Euphytica* 195, 245–258.
20. Gwag, J. G., Chung, J. W., Chung, H. K., Lee, J. H., Ma, K. H., Dixit, A., et al. (2007). Characterization of new

microsatellite markers in mung bean, *Vigna radiata* (L.). *Mol. Ecol. Resour.* 6, 1132–1134. doi: 10.1007/s11033-014-3436-7

21. Isemura, I., Kaga, A., Tabata, S., Somta, P., Srinives, P., Shimizu, T., et al. (2012). Construction of a genetic linkage map and genetic analysis of domestication related traits in mungbean (*Vigna radiata*). *PLoS ONE* 7:e41304. doi: 10.1371/journal.pone.0041304

22. Kajonphol, T., Sangsiri, C., Somta, P., Toojinda, T., and Srinives, P. (2012). SSR map construction and quantitative trait loci (QTL) identification of major agronomic traits in mungbean (*Vigna radiata* (L.) Wilczek). *SABRAO J. Breed. Genet.* 44, 71–86.

23. Kang, Y. J., Kim, S. K., Kim, M. Y., Lestari, P., Kim, K. H., Ha, B. K., et al. (2014). Genome sequence of mungbean and insights into evolution within*Vigna* species. *Nat. Commun.* 5, 5543. doi: 10.1038/ncomms6443

24. Kasettranan, W., Somta, P., and Srinives, P. (2010). Mapping of quantitative trait loci controlling powdery mildew resistance in mungbean (*Vigna radiata*(L.) Wilczek). *J. Crop Sci. Biotechnol.* 13, 155–161. doi: 10.1007/s12892-010-0052-z

25. Kato, H., and Namai, H. (1987). Floral characteristics and environmental factors for increasing natural outcrossing rate for F1 hybrid seed production of rice *Oryza sativa* L. *Jpn. J. Breed.* 37, 318–330. doi: 10.1270/jsbbs1951.37.318

26. Kitsanachandee, R., Somta, P., Chatchawankanphanich, O., Akhtar, K. P., Shah, T. M., Nair, R. M., et al. (2013). Detection of quantitative trait loci for mungbean yellow mosaic India virus (MYMIV) resistance in mungbean (*Vigna radiata* (L.) Wilczek) in India and Pakistan. *Breed. Sci.* 63, 367–373. doi: 10.1270/jsbbs.63.367

27. Krizek, B. A., and Fletcher, J. C. (2005). Molecular mechanisms of flower development: an armchair guide. *Nat. Rev. Genet.* 6, 688–698. doi: 10.1038/nrg1675

28. Lander, E. S., Green, P., Abrahamson, J., Barlow, A., Daly, M. J., Lincoln, S. E., et al. (1987). MAPMAKER: an interactive computer package for constructing primary genetic linkage maps of experimental and natural populations. *Genomics* 1, 174–181. doi: 10.1016/0888-7543(87)90010-3

29. Li, Q., and Wan, J.-M. (2005). SSRHunter: development of a local searching software for SSR sites. *Hereditas* 27, 808–810.

30. Lodhi, M. A., Ye, G.-N., Weeden, N. F., and Reisch, B. I. (1994). A simple and efficient method for DNA extraction from grapevine cultivars and *Vitis*species. *Plant Mol. Biol. Rep.* 12, 6–13. doi: 10.1007/BF02668658

31. Mather, K. (1951). *The Measurement of Linkage in Heredity.* London: Methuen & Co. LTD.

32. Okada, K., Ueda, J., Komaki, M. K., Bell, C. J., and Shimura, Y. (1991). Requirement of the auxin polar transport system in early stages of *Arabidopsis*floral bud formation. *Plant Cell* 3, 677–684. doi: 10.1105/tpc.3.7.677

33. Prathet, P., Somta, P., and Srinives, P. (2012). Mapping QTL conferring resistance to iron deficiency chlorosis in mungbean [*Vigna radiata* (L.) Wilczek].*Field Crops Res.* 137, 230–236. doi: 10.1016/j.fcr.2012.08.002

34. Przemeck, G. K., Mattsson, J., Hardtke, C. S., Sung, Z. R., and Berleth, T. (1996). Studies on the role of the *Arabidopsis* gene MONOPTEROS in vascular development and plant cell axialization. *Planta* 200, 229–237. doi: 10.1007/BF00208313

35. Pundir, R., and Reddy, G. (1998). Two new traits-open flower and small leaf in chickpea (*Cicer arietinum* L.). *Euphytica* 102,357–361. doi: 10.1023/A:1018307709788

36. Sangiri, C., Kaga, A., Tomooka, N., Vaughan, D., and Srinives, P. (2007). Genetic diversity of the mungbean (*Vigna radiata*, Leguminosae) genepool on the basis of microsatellite analysis. *Aust. J. Bot.* 55, 837–847. doi: 10.1071/Bt07105

37. Schlaich, N. L. (2007). Flavin-containing monooxygenases in plants: looking beyond detox. *Trends Plant Sci.* 12, 412–418. doi: 10.1016/j.tplants.2007.08.009

38. Soehendi, R., and Srinives, P. (2005). Significance of heterosis and heterobeltiosis in F1 hybrid of mungbean (*Vigna radiata* (L.) Wilczek) for hybrid seed production. *SABRAO J. Breed. Genet.* 37, 97–105.

39. Sompong, U., Somta, P., Raboy, V., and Srinives, P. (2012). Mapping of quantitative trait loci for phytic acid and phosphorus contents in seed and seedling of mungbean (*Vigna radiata* (L.) Wilczek). *Breed. Sci.* 62, 87–92. doi: 10.1270/jsbbs.62.87

40. Somta, P., Musch, W., Kongsamai, B., Chanprame, S., Nakasathien, S., Toojinda, T., et al. (2008). New microsatellite markers isolated from mungbean (*Vigna radiata* (L.) Wilczek). *Mol. Ecol. Resour.* 8, 1155–1157. doi: 10.1111/j.1755-0998.2008.02219.x

41. Somta, P., Seehalak, W., and Srinives, P. (2009). Development, characterization and cross-species amplification of mungbean (*Vigna radiata*) genic microsatellite markers. *Conserv. Genet.* 10, 1939–1943. doi: 10.1007/s10592-009-9860-x

42. Sorajjapinun, W., Chankaew, S., and Srinives, P. (2012). Comparison of hybrid vigor based on parental distance in SSR markers and agronomic traits in mungbean (*Vigna radiata* (L.) Wilczek). *Sonklanakarin J. Sci. Technol.* 34, 143–151.

43. Sorajjapinun, W., and Srinives, P. (2011). Chasmogamous mutant, a novel character enabling commercial hybrid seed production in mungbean. *Euphytica*181, 217–222. doi: 10.1007/s10681-011-0408-z

44. Srinivasan, S., and Gaur, P. M. (2012). Genetics and characterization of an open flower mutant in chickpea. *J. Hered.* 103, 297–302. doi: 10.1093/jhered/esr125

45. Tangphatsornruang, S., Somta, P., Uthaipaisanwong, P., Chanprasert, J., Sangsrakru, D., Seehalak, W., et al. (2009). Characterization of microsatellites and gene contents from genome shotgun sequences of mungbean (*Vigna radiata* (L.) Wilczek). *BMC Plant Biol.* 9:137. doi: 10.1186/1471-2229-9-137

46. Untergasser, A., Cutcutache, I., Koressaar, T., Ye, J., Faircloth, B. C., Remm, M., et al. (2012). Primer3-new

capabilities and interfaces. *Nucleic Acids Res.*40:e115. doi: 10.1093/nar/gks596

47. Verdcourt, B. (1970). Studies in the Leguminosae-Papilionoïdeae for the 'Flora of Tropical East Africa': IV. *Kew Bull.* 24, 507–569. doi: 10.2307/4103051
48. Voorrips, R. E. (2002). MapChart: software for the graphical presentation of linkage maps and QTLs. *J. Hered.* 93, 77–78. doi: 10.1093/jhered/93.1.77
49. Yashpal, Singh, M. N., Pathak, N., and Saroj, S. K. (2015). Combining ability, heterosis and inbreeding depression in inter specific hybrids involving greengram [*Vigna radiata* (L.) Wilczek] and blackgram [*Vigna* mungo (L.) Hepper]. *Electron. J. Plant Breed.* 6, 87–92.
50. Zhao, Y., Christensen, S. K., Fankhauser, C., Cashman, J. R., Cohen, J. D., Weigel, D., et al. (2001). A role for flavin monooxygenase-like enzymes in auxin biosynthesis. *Science* 291, 306–309. doi: 10.1126/science.291.5502.306

CHAPTER 20

A NATURAL FRAMESHIFT MUTATION IN CAMPANULA EIL2 CORRELATES WITH ETHYLENE INSENSITIVITY IN FLOWERS

Line Jensen, Josefine Nymark Hegelund, Andreas Olsen, Henrik Lütken, and Renate Müller

Department of Plant and Environmental Sciences, Faculty of Science, University of Copenhagen, Højbakkegård Allé 9-13, 2630 Taastrup, Denmark

ABSTRACT

Background

The phytohormone ethylene plays a role in development and senescence of climacteric flowers. In ornamental plant production, ethylene sensitive plants are usually protected against negative effects of ethylene by application of chemical inhibitors. In *Campanula,* flowers are sensitive to even minute concentrations of ethylene.

Results

Monitoring flower longevity in three *Campanula* species revealed *C. portenschlagiana* (*Cp*) as ethylene sensitive, *C. formanekiana* (*Cf*) with intermediate sensitivity and *C. medium* (*Cm*) as ethylene insensitive. We identified key elements in ethylene signal transduction, specifically in *Ethylene Response Sensor 2* (*ERS2*),*Constitutive Triple Response 1* (*CTR1*) and *Ethylene Insensitive 3- Like 1 and 2* (*EIL1* and *EIL2*) homologous. Transcripts of *ERS2, CTR1* and *EIL1* were constitutively expressed in all species both throughout flower development and in response to ethylene. In contrast, *EIL2* was found only in *Cf* and *Cm*. We identified a natural mutation in *Cmeil2* causing a frameshift which resulted in difference in expression levels of *EIL2*, with more than 100-fold change between *Cf* and *Cm* in young flowers.

Conclusions

This study shows that the naturally occurring 7 bp frameshift discovered in *Cmeil2*, a key gene in the ethylene signaling pathway, correlates with ethylene insensitivity in flowers. We suggest that transfer of the *eil2*mutation to other plant species will provide a novel tool to engineer ethylene insensitive flowers.

BACKGROUND

Ethylene is a gaseous phytohormone involved in regulating processes of horticultural importance encompassing flower development, fruit ripening, abscission and leaf and flower senescence [1]. In the ethylene signal transduction pathway,

Citation: Line Jensen, Josefine Nymark Hegelund, Andreas Olsen, Henrik Lütken, and Renate Müller. A natural frameshift mutation in Campanula EIL2 correlates with ethylene insensitivity in flowers, BMC Plant Biology BMC series – open, inclusive and trusted (2016) 16:117, DOI: 10.1186/s12870-016-0786-4 ethylene perception is facilitated via a copper co-factor present in receptor proteins integrated in the endoplasmic reticulum (ER) [2, 3]. In *Arabidopsis*, receptor proteins comprising EThylene Response 1 (ETR1) and Ethylene Response Sensor 1 (ERS1) or ETR2, ERS2 and Ethylene INsensitive 4 (EIN4) have been characterised. They differ by the functionality of their kinase domains [4, 5, 6]. Ethylene receptors exist as dimers and physically interact with the negative regulator Constitutive Triple Response 1 (CTR1) [7]. The kinase activity of CTR1 is directed towards the C-terminal of Ethylene INsensitive 2 (EIN2), a positive regulator of the ethylene response [8, 9, 10]. Ethylene binding to receptors deactivates CTR1 and results in a dephosphorylation of EIN2 [10]. Subsequently, the C-terminal of EIN2 is cleaved and translocated from ER to the nucleus [11, 12] where Ethylene INsensitive 3/Ethylene Insensitive 3-Like (EIN3/EIL)-dependent transcription and activation of the ethylene response occur [13, 14, 15].

Postharvest quality of many ornamental plants is sensitive to ethylene during production and distribution [16,17]. In climacteric plants, flower development is controlled by intrinsic rise in ethylene production and respiration which promotes flower development and senescence. Plant species with climacteric flower senescence are sensitive to exogenous ethylene and may exhibit accelerated petal or flower wilting upon exposure. Commercially important climacteric ornamental plants include carnations, orchids, *Kalanchöe, Campanula* and roses [18, 19, 20, 21]. Endogenous ethylene production may arise due to natural floral development but also in response to stress, elevated CO_2 production [22] or increased auxin production [23]. Hence, ornamental plants are often treated with chemical inhibitors blocking ethylene signaling to improve postharvest quality and prolong flower longevity [24].

Genetic approaches designed to reduce ethylene sensitivity in flowers have modified signaling via the ethylene signal transduction pathway. The *etr1-1* ethylene receptor mutant from *Arabidopsis* fails to bind ethylene [25]. Expression of *etr1-1* in *Petunia* and *Campanula carpatica* flowers [26, 27, 28] results in ethylene insensitivity, delayed senescence and postponed flower abscission. Also, transgenic *Petunia* expressing reduced levels of*PhEIN2* displayed delayed flower senescence [29]. However, to date genetic approaches successfully prolonging flower longevity have resulted in transgenic plants [30].

Campanula is an economically important ornamental plant, used as indoor potted plant, garden plant, as well as cut flower. The *Campanula* genus consists of approximately 415 species [31]. Here, we characterise expression patterns of flower expressed *ERS, CTR* and *EIL* genes in response to floral development and exogenous ethylene in three ornamental species of *Campanula; C. portenschlagiana (Cp), C. formanekiana (Cf)* and *C. medium (Cm)*. The ethylene insensitivity identified in flowers of *C. medium* correlates with the occurrence of a natural mutation in the open reading frame of *EIL2*. This finding holds promise for new breeding strategies towards ethylene insensitive ornamental plants.

RESULTS

Campanula Sensitivity to Ethylene

To understand the physiological variation in ethylene sensitivity among *Campanula* species, we used ethylene exposure tests in a postharvest environment. *C. portenschlagiana, C. formanekiana* and *C. medium* were selected due to their relevance as ornamental plants.

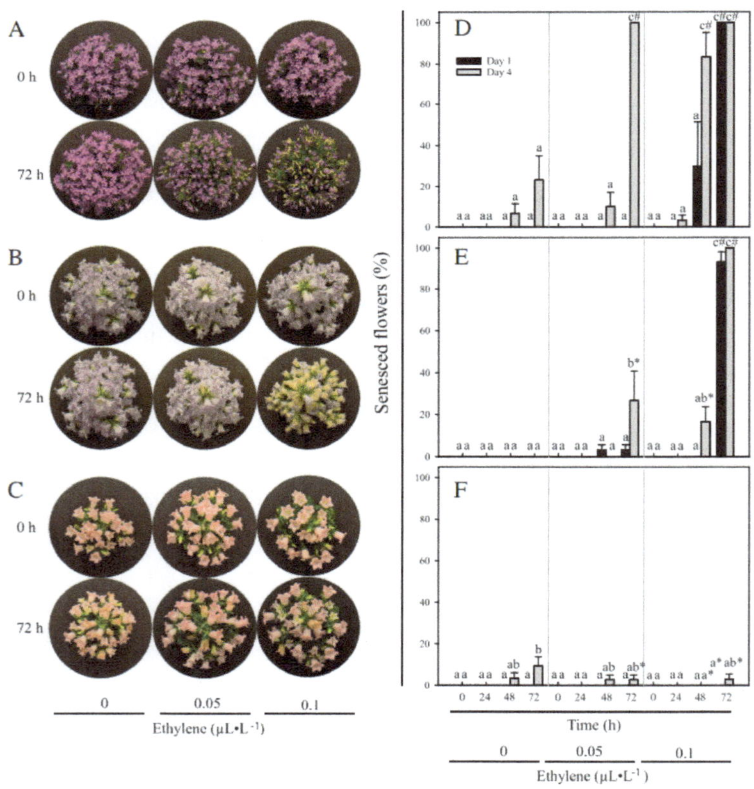

Figure 1. Floral development in response to ethylene exposure in *Campanula*. Visual presentation of flower responses to ethylene concentrations of 0, 0.05 or 0.1 $\mu L \cdot L^{-1}$ ethylene for 72 h **a** *C. portenschlagiana (Cp)*, **b** *C. formanekiana (Cf)*,**c** *C. medium (Cm)*. Graphical presentation of flower senescence (%) in responses to ethylene concentrations of 0, 0.05 or 0.1 $\mu L \cdot L^{-1}$ ethylene for 0–72 h **d** *Cp*, **e** *Cf*, **f** *Cm*. Flower senescence were monitored every 24 h for 1-day (black) and 4-day (grey) old flowers. Data are means ± SE. Values with same letters within species and are not significantly different ($P > 0.05$). Values with same symbols between species are not significantly different ($P > 0.05$).

Cp was found to be sensitive to ethylene from concentrations of 0.05 $\mu L \cdot L^{-1}$. Individual flower sensitivity increased with flower age. In *Cp*, old flowers did not survive 0.05 $\mu L \cdot L^{-1}$ ethylene treatment for 48 h whereas young flowers maintained longevity in a 0.1 $\mu L \cdot L^{-1}$ethylene environment for more than 48 h. *Cp* flowers, regardless of age, did not survive after 72 h of the high ethylene treatment (Fig. 1a, d). Less pronounced ethylene sensitivity was found in *Cf* where 26 % of old flowers wilted in response to 72 h of 0.05 $\mu L \cdot L^{-1}$ ethylene as opposed to 100 % of old *Cp* flowers. Increased ethylene concentration for the same period resulted in complete senescence of 4-day old *Cf* flowers (Fig. 1b, e). As in*Cp*, young *Cf* flowers were less ethylene sensitive than old flowers, however, 93 % of young flowers wilted in response to 0.1 $\mu L \cdot L^{-1}$ ethylene (Fig. 1e). Thus, flowers of neither *Cp* nor *Cf* could tolerate 72 h of 0.5 $\mu L \cdot L^{-1}$ethylene, regardless of flower age. In contrast, *Cm* flowers were non-responsive to ethylene, they maintained both colour and turgor for 72 h in the 0.1 $\mu L \cdot L^{-1}$ ethylene environment (Fig. 1c, f).

Identification of Key Genes in Ethylene Signal Transduction

Degenerate primers were used to identify expressed components in the ethylene signal transduction pathway in *Campanula* flowers. This allowed identification of partial homologs for *ERS2*, *CTR1* and *EIL*. *Campanula*transcript fragments of *ERS2* and *CTR1* were translated to protein and named according to their closest relative in *Arabidopsis* whereas *Campanula EIN3/EIL* homologs were named *EIL1* and*EIL2*. Protein alignment of *Campanula* ERS2 showed high similarity within species in the identified region whereas *Campanula* CTR1 proteins differed. In *Cf* and *Cm*, *ERS2* was encoded by a single gene whereas *Cp ERS2* was represented by two loci containing different introns but resulting in identical partial transcripts. Sequencing showed some polymorphisms among the partial *CTR1*transcripts. These could not be separated in RT-PCR reactions and may represent different alleles in the same locus.

Also *EIL* transcripts were identified by degenerate primers using flower cDNA as template. Sequencing identified two partial *EIL1* homologs *EIL1a* and *EIL1b* in *Cp* and only one partial *EIL1* homolog in *Cf* and *Cm*. In*Cf* the cDNA pool contained an additional *EIL* homolog, *EIL2*, however this transcript was not readily detectable in *Cm* cDNA using the degenerate primers. As *Cf* and *Cm* have different sensitivities towards ethylene (Fig. 1e, f), primers were designed to separate and amplify both *EIL1* and *EIL2* fragments from *Cf* and *Cm* genomic DNA. Interestingly, *Cmeil2* was found to contain a deletion of 7 bp in the *EIL2* ORF resulting in a frame shift in the corresponding protein (Fig. 2). The 7 bp deletion in *Cmeil2* was verified from independent gDNA extractions (data not shown). At the nucleotide level *CfEIL2* and *Cmeil2* shared 96 % identity to each other and 76 % identity to *CfEIL1* and *CmEIL1* respectively (Table 1). PCR reactions specific for *EIL2* using *Cp* gDNA or cDNA did not amplify a product.

```
AtEIN3    TAATAACCCGATTGGACCGACTCCTCATACCTTGCAAGAGCTTCAAGACACGACTCTTGG   560
CfEIL2    GAGTAATTCGATTGGCCCGACCCCACATACTCTCCAAGAGCTTCAAGATACGACCCTCGG   420
Cmeil2    GAGTAATTCGATTGGCCCTACCCCACATACTCTCCAAGAGCTTCAAGATACGACCCTCGA   420
           * ***  ******* ** ** ** *****  * ************** ***** ** *

AtEIN3    ATCGCTTTTGTCTGCGTTGATGCAACACTGTGATCCTCCTCAGAGACGTTTTCCTTTGGA   620
CfEIL2    TTCGTTATTATCTGCTCTCATGCAGCATTGTGATCCTCCCCAGAGGCGTTTCCCTCTGGA   480
Cmeil2    TTTGTTATTATCTGCTCTC-------ATTGTGATCCTCTACAGAGGCGTTTCCCTCTCGA   473
           * * * ** *****  *        * **********  ***** ***** *** * **

AtEIN3    GAAAGGAGTTCCTCCTCCGTGGTGGCCTAATGGGAAAGAGGATTGGTGGCCTCAACTTGG   680
CfEIL2    GAAAGGTGTTCCACCCCCGTGGTGGCCCATAGGGACGGAGGAATGGTGGCCTCAATTGGG   540
Cmeil2    GAAAGGTGTTCCGCCCCCGTGGTGGCCCACGAGGACGGAGGAATGGTGGCCTCAATTGGG   533
          ****** ***** ** *********** *   ***  ***** ************ * **
```

Figure 2. Alignment of *EIL2* gDNA from *Campanula*. The 65 bp fragments of *EIL2* gDNA span the region where *C. medium*(*Cm*) contains a 7 bp deletion and *C. formanekiana* (*Cf*) does not. As a reference the closest ortholog *AtEIN3* from*Arabidopsis thaliana* is included [Genbank: O24606.1]. The position of the deletion in *Cmeil2* is underlined. The alignment was produced in Clustal Ω [62].

Table 1. Percent identity matrix of partial *EIL* nucleotide sequences from *Campanula*

	CpEIL1a	*CpEIL1b*	*CfEIL1*	*CmEIL1*	*CfEIL2*	*Cmeil2*
CpEIL1a	100					
CpEIL1b	93.06	100				
CfEIL1	88.49	86.95	100			
CmEIL1	88.83	87.29	96.28	100		
CfEIL2	74.24	75.13	75.80	76.48	100	
Cmeil2	73.08	73.80	74.14	75.51	96.31	100

The identity matrix was produced by Clustal Ω [62]

Abbreviations: *Cp C. portenschlagiana*, *Cf C. formanekiana*, *Cm C. medium*

Expression Analysis of Putative *ERS2, CTR1, EIL1* and *EIL2* Homologues

To address the transcriptional regulation of the ethylene signal transduction pathway in *Campanula*, flower tissues were harvested at five developmental stages (from bud to fully expanded flower on day 4, Fig. 3). Transcripts of putative *ERS2, CTR1* and *EIL1* were expressed constitutively during flower development (Fig. 4). To determine whether *Campanula ERS2, CTR1* or *EIL1* were responsive to ethylene, transcriptional analysis were performed in young (1-day old) flowers exposed to 0.025 $\mu L \cdot L^{-1}$ or 0.050 $\mu L \cdot L^{-1}$ ethylene for 24 h. However, neither *ERS2*, *CTR1* nor *EIL1* transcripts were responsive to the applied ethylene treatments (Fig. 5).

Figure 3. Developmental stages of *Campanula* flowers. Developmental stages are unripe bud (B), day 0 (one day before flowering) and 1, 2 and 4-day old flowers of **a** *Campanula portenschlagiana* (*Cp*), **b** *C. formanekiana* and **c** *C. medium* (*Cm*).

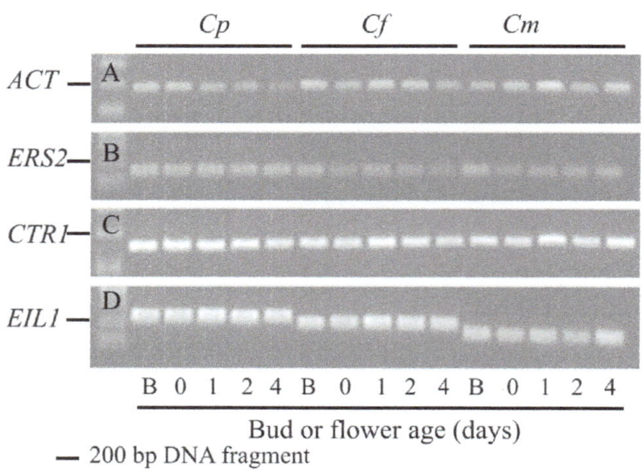

Figure 4. Expression of *ERS2* (**b**), *CTR1* (**c**) and *EIL1* (**d**) in *Campanula* during flower development. As reference gene *Actin* (*ACT*(**a**)) from *Campanula* was used. The developmental stages from bud to flower were; bud (B), flower the day before opening (0), 1-day old flower (1), 2-days old flower (2) and 4-days old flower (4). The presented *Campanula* species are *C. portenschlagiana* (*Cp*), *C. formanekiana* (*Cf*) and *C. medium* (*Cm*).

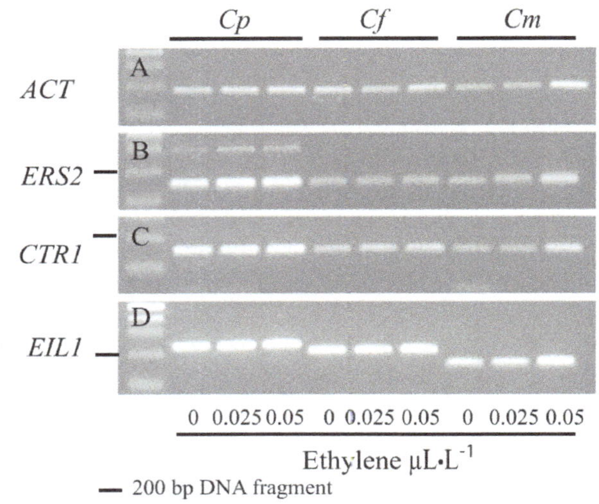

Figure 5. Expression of *ERS2* (**b**), *CTR1* (**c**) and *EIL1* (**d**) in *Campanula* in response to transient ethylene exposure. As reference gene *Actin* (*ACT* (**a**)) from *Campanula* was used. Ethylene was supplied to sealed glass tanks in the following concentrations; 0 $\mu L \cdot L^{-1}$ (0), 0.025 $\mu L \cdot L^{-1}$ (0.025) and 0.050 $\mu L \cdot L^{-1}$ (0.050) ethylene. The presented *Campanula*species are *C. portenschlagiana* (*Cp*), *C. formanekiana* (*Cf*) and *C. medium* (*Cm*). Flowers were marked the day before flower opening; the experiments started the following day (with 1-day old flowers) and were allowed to proceed for 24 h.

Expression patterns of *CfEIL2* and *Cmeil2* transcripts were analyzed by classic RT-PCR and RT-qPCR. *CfEIL2*transcripts showed expression pattern and levels similar to those of *CfEIL1* throughout flower development and in response to 0.025 $\mu L \cdot L^{-1}$ and 0.050 $\mu L \cdot L^{-1}$ ethylene for 24 h (Fig. 6). In contrast, *Cmeil2* was detectable in trace amounts when analysed by RT-PCR. Quantitative analysis by RT-qPCR showed consistently very low levels of *Cmeil2* through flower development and no transcriptional response to ethylene. Expression levels of*CfEIL2* and *Cmeil2* differed by more than 100-fold in young flowers (day 0, day 1) whereas the same comparison in old flowers (day 4) yielded only 40-fold changes. The variation in fold change was primarily due to non-significant increases in *Cmeil2* transcript levels. Expression levels of *CfEIL2* and *Cmeil2* in response to ethylene were not found to be significantly different due to the large variation in *CfEIL2* expression levels (Fig. 6b).

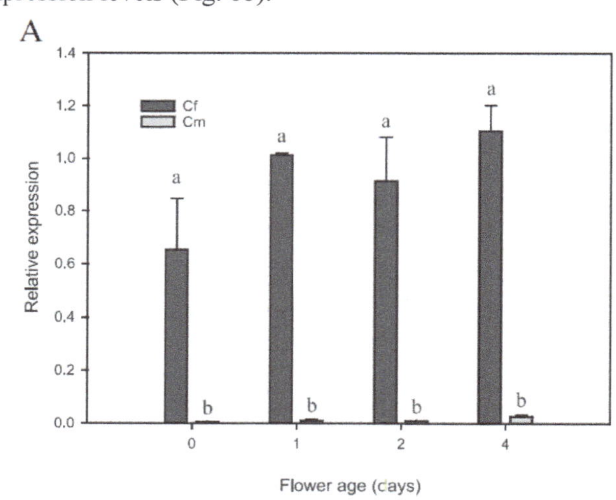

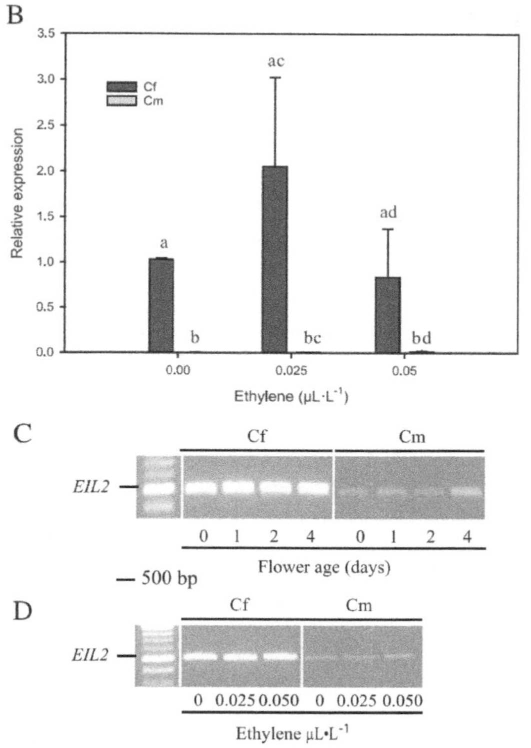

Figure 6. Expression of *EIL2* in large flowered *Campanula* during flower development and following ethylene exposure. *Campanula* species are *C. formanekiana* (*Cf*) and *C. medium* (*Cm*). In **a**, **c** the floral developmental stages were; day before opening (0), 1-day old flower (1), 2-days old flower (2) and 4-days old flower (4). In **b**, **d** ethylene was supplied to sealed glass tanks at the following concentrations; 0.00 $\mu L \cdot L^{-1}$ (0), 0.025 $\mu L \cdot L^{-1}$ (0.025) and 0.050 $\mu L \cdot L^{-1}$ (0.050) ethylene. Results from RT-qPCR are presented as relative expressions in (**a**) and (**b**). The corresponding results of*EIL2* RT-PCR loaded on agarose gels are presented in (**c**) and (**d**). The values of 1-day flowers and 0 $\mu L \cdot L^{-1}$ ethylene were set to the value 1 in (**a**) and (**b**), respectively. As reference gene *Actin* (*ACT*) from *Campanula* was used, these RT-PCR results are presented in Figs. 4a and 5a. RT-qPCR data were normalized to transcripts of the same *ACT*. Data are means ± SE. Values with same letters are not significantly different ($P > 0.05$).

The *eil2* Frameshift Mutation is Unique for *C. Medium*

Alignment of the putative EIL2 protein fragment from *Campanula* with EIL protein sequences from other plants confirmed the presence of three conserved domains found in other EILs. These domains comprise the two basic amino acids binding domains (BD I and BD II) and the proline-rich region (PR) (Fig. 7). At the protein level the putative EILs from *Campanula* were closely related to each other when compared to other EILs except for Cmeil2. Cmeil2 showed high homology to other EILs until the position of the frameshift. The sequence following downstream of the frameshift was only observed in *Cm* and did not show any homology to previously reported EIL proteins. Omission of the deletion from the *Cmeil2* reading frame resulted in a protein that perfectly aligned with other EIL2 proteins (data not shown). Phylogenetic analysis using other plant EILs indicated a close relation among *Campanula* EILs and a clear phylogenetic separation of *Campanula*EIL1 and EIL2 proteins (Fig. 8). Some branch points in the phylogenetic analysis yielded low bootstrap values due to the size of the aligned fragment (200 amino acids) and the high identity among all the EILs (Fig. 8).

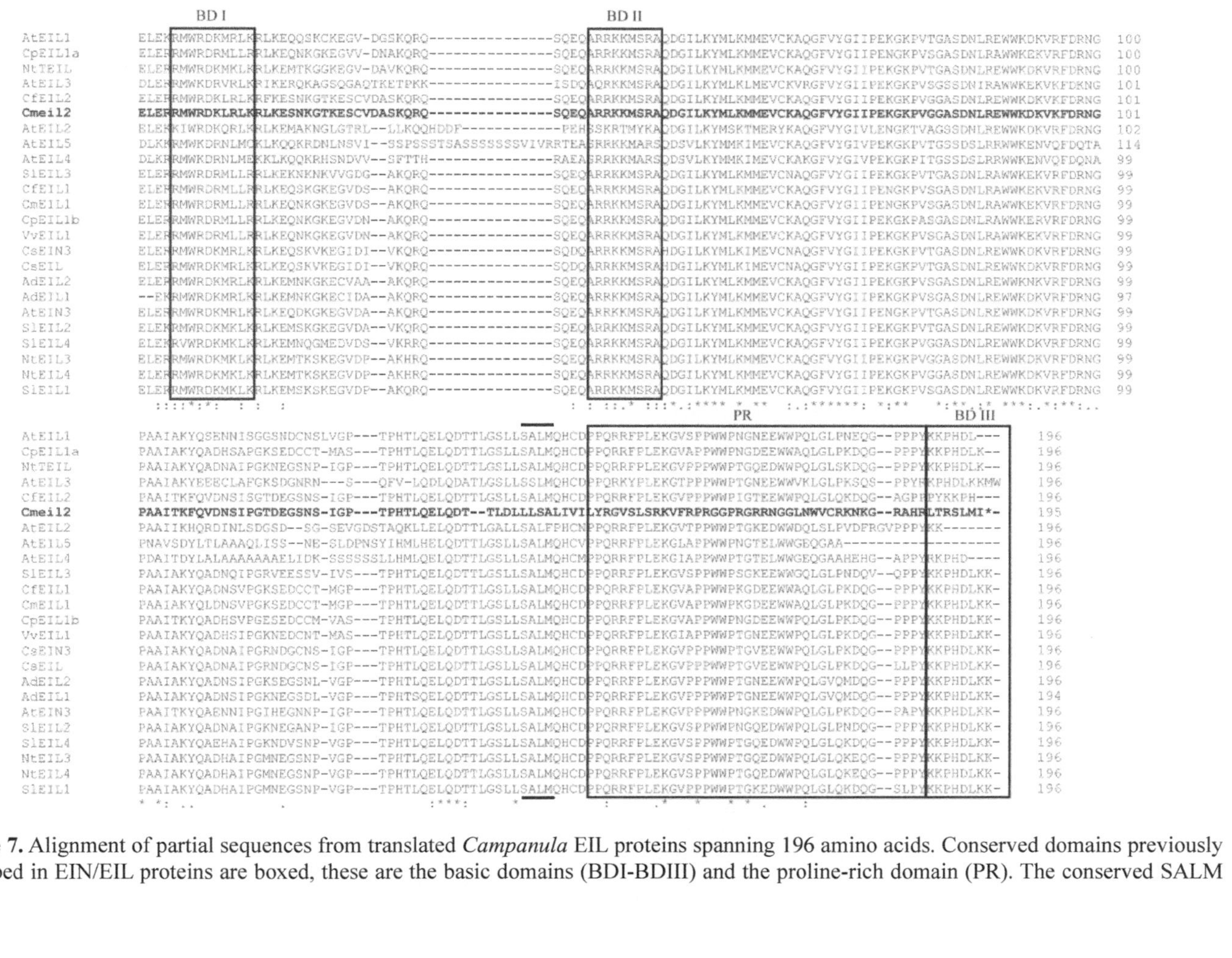

Figure 7. Alignment of partial sequences from translated *Campanula* EIL proteins spanning 196 amino acids. Conserved domains previously described in EIN/EIL proteins are boxed, these are the basic domains (BDI-BDIII) and the proline-rich domain (PR). The conserved SALM

motif in which Cmeil2 is mutated is marked with (····). Conserved aa among all EIN/EILs are marked below with an asterisk. Cmeil2 is presented in bold. The alignment were produced from partial EIN/EIL protein sequences of *C. portenschlagiana* (*Cp*), *C. formanekiana* (*Cf*), *C. medium* (*Cm*), *Actinidia deliciosa* (*Ad*), *Arabidopsis thaliana* (*At*), *Cucumis sativus* (*Cs*), *Nicotiana tabacum* (*Nt*), *Solanum lycopersicum* (*Sl*) and*Vitis vinifera* (*Vv*). Previously named proteins are presented by their species abbreviation followed by their name. The alignment was produced in Clustal Ω [62].

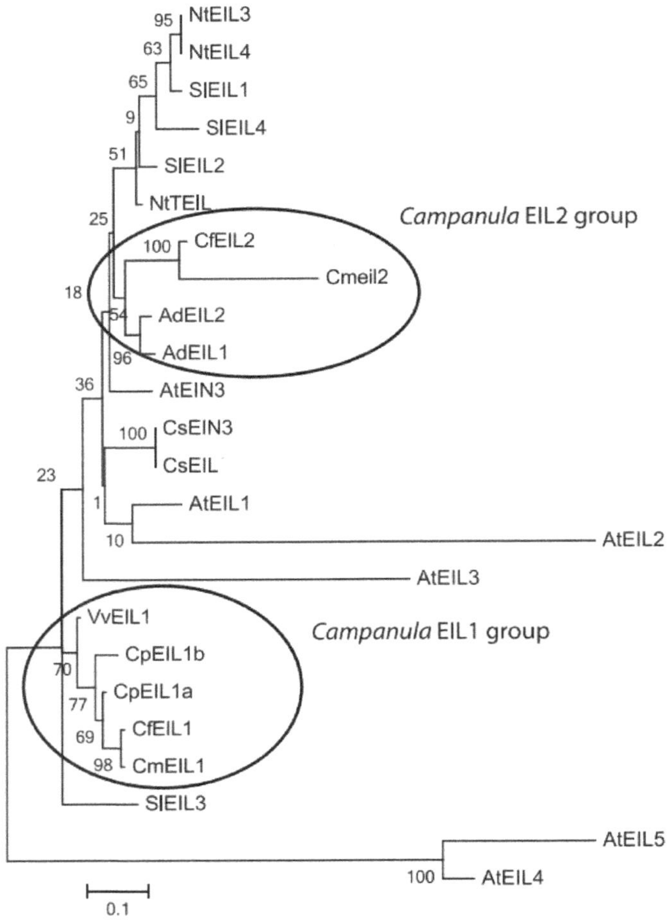

Figure 8. Phylogenetic tree of EILs in plants. The phylogenetic tree was produced from partial EIN/EIL protein sequences spanning 196 amino acids. The clustering in two groups of EIL1 and EIL2 proteins from *Campanula* are highlighted in ellipses. Low bootstrap values in some parts of the phylogenetic analysis are due to the size of the partial EIN/EIL proteins (196 aa) and the high level of aa identity among them. The phylogenetic tree were produced from partial EIN/EIL protein sequences of *C. portenschlagiana* (*Cp*), *C. formanekiana* (*Cf*), *C. medium* (*Cm*), *Actinidia deliciosa*(*Ad*), *Arabidopsis thaliana* (*At*), *Cucumis sativus* (*Cs*), *Nicotiana tabacum* (*Nt*), *Solanum lycopersicum* (*Sl*) and *Vitis vinifera* (*Vv*). Previously named proteins are presented by their species abbreviation followed by their name. The phylogenetic analysis were produced from MEGA version 6 [63].

Characterization of *eil2* in *Campanula* Species and Cultivars

To elucidate the natural occurrence of *eil2* in *Campanula*, close relatives to *C. medium* were identified as *C. hofmannii*, *C. alpina*, *C. alpestris* and *Edraianthus graminifolius* and *C. incurva* as close relative to *C. formanekiana* [32]. As the *Cmeil2* mutation disrupts an *Nla*III restriction site in *EIL2* a simple screen for the presence of the mutation was developed (Fig. 9). *EIL2* PCR products digested with *Nla*III resulted in either two or three DNA fragments depending on the presence or lack of the *eil2* mutation, respectively. Results obtained via *Nla*III digests were verified by sequencing. Interestingly, *eil2* was found to be specific for *Cm* and did not occur in related *Campanula* species (Fig. 9a). Intraspecific *Nla*III restriction analysis among *Cm* cultivars confirmed the occurrence of *eil2* regardless of cultivar origin (Fig. 9b). Thus the reported frameshift mutation in *Cmeil2* is specific for *Cm* and occurs in all tested *Cm* both among non domesticated and domesticated cultivars.

```
A                                                      NlaIII
A.thaliana                 ACTCTTGGATCGCTTTTGTCTGCGTTGATGCAACACTGTGATCCTCCTCAGAGACGTTTT
C.alpina                   ACGCTTGGTTCGTTATTATCTGCTCTCATGCAACATTGTGATCCTCCCCAGAGGCGTTTC
C.medium non domesticated  ACCCTCGGTTTGTTATTATCTGCTCTC-------ATTGTGATCCTCCACAGAGGCGTTTC
C.medium domesticated      ACCCTCGGTTTGTTATTATCTGCTCTC-------ATTGTGATCCTCCACAGAGGCGTTTC
C.medium calycanthema      ACCCTCGGTTTGTTATTATCTGCTCTC-------ATTGTGATCCTCCACAGAGGCGTTTC
C.alpestris                ACCCTCGGTTCGTTATTATCTGCTCTCATGCAGCATTGTGATCCTCCCCAGAGGCGTTTC
C.incurva                  ACCCTCGGTTCGTTATTATCTGCTCTCATGCAGCATTGTGATCCCCCCCAGAGGCGTTTC
C.formanekiana             ACCCTCGGTTCGTTATTATCTGCTCTCATGCAGCATTGTGATCCTCCCCAGAGGCGTTTC
C.hofmannii                ACCCTCGGTTCGTTATTATCTGCTCTCATGCAGCATTGTGATCCTCCCCAAAGGCGTTTC
E.graminifolius            ACCCTCGGTTCGTTATTATCTGCTCTCATGCAGCATTGTGATCCTCCGCAAAGGCGTTTC
                           ** ** ** * * * ** *****  *        * ******** ** ** ** *****
B
C.medium line 2            ACCCTCGGTTCGTTATTATCTTCTCTCATTGTGATCCTCCCCAGAGGCGTTTCCCTCTGG
C.medium line 1            ACCCTCGATTTGTTATTATCTGCTCTCATTGTGATCCTCTACAGAGGCGTTTCCCTCTCG
C.medium non domesticated  ACCCTCGGTTTGTTATTATCTGCTCTCATTGTGATCCTCCACAGAGGCGTTTCCCTCTCG
C.medium domesticated      ACCCTCGGTTTGTTATTATCTGCTCTCATTGTGATCCTCCACAGAGGCGTTTCCCTCTCG
C.medium calycanthema      ACCCTCGGTTTGTTATTATCTGCTCTCATTGTGATCCTCCACAGAGGCGTTTCCCTCTCG
                           ******* ** ********** *****************  ***************** *
```

Figure 9. Alignment of genomic *EIL2* DNA sequences. The aligned area presented is centered on the seven bp deletion in *C. medium* (represented with -). Nucleotides bordering the mutation are boxed. Conserved nucleotides are marked with asterisk. **a** *EIL2* from six *Campanula* species, three *C. medium* cultivars, *Edraianthus graminifolius*, and*Arabidopsis EIN3*. The restriction site of *Nla*III found in *EIL2 Campanula* sequences not having the mutation is marked with bold line **b** *eil2* from three *C. medium* cultivars and two PKM breeding lines (line 1 (Sweet Mee®) and line 2).

Characterization of *eil2* in Hybrids of *Cf* × *Cm*

As *Cm* is homozygote for *eil2* whereas *Cf* is homozygote for *EIL2* the performance of *eil2* in heterozygote plants were evaluated by ethylene exposure tests in *C. formanekiana* × *C. medium* hybrids. The presence of*eil2* in *Cf* × *Cm* hybrids (A-E) was verified using the *Nla*III screening system (Fig. 10). Young flowers were exposed to high ethylene concentrations of 5.0 $\mu l \cdot L^{-1}$ for 72 h and flower responses were scored in categories of no response, signs of senescence and complete senescence (Table 2). Interestingly, four heterozygote *Cf* × *Cm* hybrids showed phenotypes indistinguishable from that of *Cf* with 87–100 % of flowers showing complete senescence in response to ethylene. A single hybrid (E) exhibited an intermediate phenotype with 57 % senesced flowers. In contrast, two *Cm* breeding lines maintained flower longevity longer and only 0–5 % of flowers senesced as a result of 72 h of 5.0 $\mu l \cdot L^{-1}$ ethylene exposure.

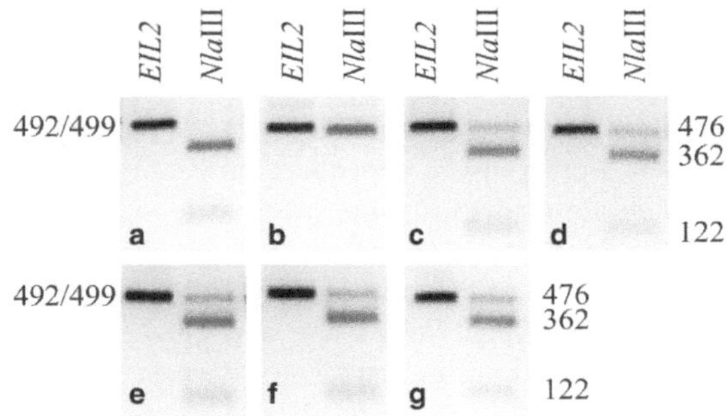

Figure 10. Detection of *EIL2* in hybrids of *C. formanekiana* (*Cf*) and *C. medium* (*Cm*). The original PCR products (*EIL2*/*eil2*) are 499 bp or 492 bp in *Cf* and *Cm*, respectively. Upon digestion with the restriction enzyme *Nla*III *CfEIL2* produces DNA fragments of 362 bp, 122 bp and 16 bp (**a**). In contrast *Cm* containing the 7 bp deletion in *eil2* produces DNA fragments of 476 bp and 16 bp (**b**). The 16 bp DNA fragments are not detected. *Nla*III restriction analysis of 5 heterozygote *Cf* × *Cm* hybrids shows that all hybrids contain both *Cf* and *Cm* specific DNA fragments (**c–g**, hybrids A–E). Hybrids were produced according to [57].

Table 2. Flower responses to ethylene exposure in *C. medium*, *C. formanekiana* and *C. formanekiana* × *C. medium* hybrids. Plants were exposed to concentrations of 0 or 5 $\mu l \cdot L^{-1}$ ethylene for 72 h. Flower responses were grouped in three categories (no senescence, signs of senescence and complete senescence). Data are in % and presented as means ± SE (*n* = 2, except *Cm*line 1; *n* = 4). Values with same letters within treatments in same response are not significantly different (*P* > 0.05)

Species/hybrid	No senescence	Signs of senescence	Complete senescence
	0 $\mu l \cdot L^{-1}$ ethylene, 72 h		
C. formanekiana	87 ± 4.7 [a]	13 ± 4.7[d]	0.0 ± 0.0[a]
C. medium line 1[1]	3.3 ± 2.9[d]	97 ± 2.9 [a]	0.0 ± 0.0[a]
C. medium line 2[1]	53 ± 19 [ab]	47 ± 19[cd]	0.0 ± 0.0 [a]
Hybrid A	53 ± 0.0 [ab]	47 ± 0.0[cd]	0.0 ± 0.0 [a]
Hybrid B	17 ± 2.4[bcd]	83 ± 2.4 [ab]	0.0 ± 0.0 [a]
Hybrid C	40 ± 4.7[bc]	60 ± 4.7[bc]	0.0 ± 0.0 [a]
Hybrid D	10 ± 7.1[cd]	90 ± 7.1 [ab]	0.0 ± 0.0 [a]

Hybrid E	3.3±2.4[cd]	97±2.4 [ab]	0.0±0.0 [a]
	5 µl·L⁻¹ ethylene, 72 h		
C. formanekiana	0.0±0.0 [a]	0.0±0.0[c]	100±0.0 [a]
C. medium line 1[1]	0.0±0.0 [a]	95±4.3 [a]	5.0±4.3[c]
C. medium line 2[1]	0.0±0.0 [a]	100±0.0 [a]	0.0±0.0[c]
Hybrid A	0.0±0.0 [a]	13±9.4[c]	87±9.4 [a]
Hybrid B	0.0±0.0 [a]	3.3±2.4[c]	97±2.4 [a]
Hybrid C	0.0±0.0 [a]	0.0±0.0[c]	100±0.0 [a]
Hybrid D	0.0±0.0 [a]	10±7.1[c]	90±7.1 [a]
Hybrid E	0.0±0.0 [a]	43±7.1[b]	57±7.1[b]

[1] *C. medium* breeding lines 1 (Sweet Mee®) and 2

DISCUSSION

Ethylene Sensitivity in *Campanula*

Ethylene sensitivity of flowers is a recurring problem affecting breeders, producers and costumers of ornamental plants [17]. Thus characterisation of physiological and molecular variations in economically important *Campanula* species are much needed. In the present study, ethylene sensitivity among *C. portenschlagiana*, *C. formanekiana* and *C. medium* were found to depend on physiological and genetic factors. Ethylene sensitivity in flowers were dependent on genotype as *Cp* was highly sensitive, *Cf* had an intermediate level of sensitivity and *Cm* was found to be insensitive to high concentrations of ethylene (Fig. 1, 10). Labelling of flower developmental stage showed a link between flower age and ethylene sensitivity, where a higher proportion of flowers senesced among the older flowers. Only old flowers of *Cp* were sensitive to low amounts of 0.05 µL·L⁻¹ ethylene, whereas all flowers of *Cp* were sensitive to 0.1 µL·L⁻¹ ethylene. Doubling the ethylene concentration resulted insignificant increase in senesced *Cf* flowers from 26 to 100 % and also in complete loss of flower longevity in young *Cf* flowers (Fig. 1). In the same experimental settings all flowers of *Cm* were insensitive to ethylene. A similar correlation between flower age and ethylene sensitivity has been observed in *Pelargonium peltatum* [33]. This age dependent increase in sensitivity may also be connected to pollination as some plant species induce flower senescence upon pollination [34, 35, 36].

Whereas *Cf* flowers are ethylene sensitive in a concentration dependent manner (Fig. 1e, Table 2) not even a 50-fold increase in ethylene concentration reduced flower longevity in *Cm*. This indicates that ethylene insensitivity of *Cm* is independent of ethylene concentrations (Fig. 1f, Table 2).

Constitutive Expression of *ERS2* and *CTR1* in Flowers

Expression of *ERS2* was found to be constitutive in *Campanula* during floral development and transcripts in young flowers were also unresponsive to low concentrations of ethylene (Figs. 4b and 5b). In both *Arabidopsis* and roses ethylene receptors are encoded by five genes [6, 37, 38, 39], some of which exhibit differential expression in response to exogenous ethylene and are regulated during flower development tissues [40, 41]. Also, exogenously applied ethylene does not affect levels of *Dianthus caryophyllus ERS2* in petals but this gene is regulated by flower development [42]. Collectively, results obtained in other plants indicate that ethylene receptor families comprise multiple members. The genome of *Cp* was found to contain two homologs of *ERS* receptors whereas only one gene/transcript was identified in *Cf* and *Cm*. As gDNA was also used as template in the cloning reactions mRNA levels in flower tissues should not be the determining factor. Thus all three *Campanula* species likely encode additional ethylene receptors which were not identified here due to primer specificities.

Similarly as for *ERS2*, all *Campanula CTR1* transcripts were constitutively expressed during flower development and did not respond to application of exogenous ethylene (Figs. 4c and 5c). The same pattern was observed in roses where *RhCTR1* and *RhCTR2* were constitutively expressed throughout flower development; however both *RhCTR* transcript levels increased in response to ethylene [43]. In some plant species the genomic structure of *CTR1* is highly complex. Banana and tomato *CTR1* exist in a 15 exon 14 intron structure yielding complete ORFs of 11.5 and 12 kb, respectively [44, 45]. Even with a long elongation time in PCR reactions we were not able to identify the full-length sequence of *Campanula CTR1*. However small polymorphisms detected in the partial *Cp*, *Cf* and *Cm CTR1* transcripts indicate that these plants may be heterozygote in the *CTR1* locus or that an additional copy of *CTR1* exists. Two and four *CTR* homologs have been identified in roses and tomato, respectively [43, 46]. Hence an additional *CTR1* in *Campanula* is not unlikely.

Occurrence and Expression of EILs in *Campanula*

The *EIN3/EIL* family encodes transcription factors mediating the initialization of the physiological ethylene response [13, 47]. In the three *Campanula* species investigated here approximately 600 bp of a flower expressed *EIL* homologue were identified, *EIL1*. All *EIL1* transcripts were consistently expressed through flower development and did not respond to applied ethylene (Figs. 4d and 5d). In the small flowered *Cp* two close homologs of *EIL1* were identified by sequencing as *EIL1a* and *EIL1b*, however the two could not be separated during expression analysis. In contrast, the two large flowered

Campanula (*Cf* and *Cm*) both contained the ORF of *EIL2*, a gene not detected in *Cp*. *CfEIL2* gene expression was like *CfEIL1* constitutive throughout flower development and nonresponsive to exogenous ethylene (Fig. 6c, d). In *Cm*, *eil2* was not expressed due to the deletion of 7 bp in the ORF (Figs. 2 and 6). Translational analysis of the partial Cmeil2 protein indicated that the 7 bp deletion would introduce a frameshift prior to what should have been the proline rich domain in *Cmeil2*(Fig. 7). On nucleotide level *CfEIL2* and *Cmeil2* share 96 % identity (Table 1), showing that they are close homologs. The close homology among *CfEIL2* and *Cmeil2* was supported by phylogenetic analysis (Fig. 8). Translation of the putative *Cmeil2* indicate that the frameshift mutation in *eil2* disrupts the putative major DNA binding domain in EIL2 in front of the proline rich region and simultaneously introduces a stop codon 50 amino acids further downstream in the eil2 protein sequence. Hence, if translated the protein would be truncated to approximately 40 % of the expected size when compared to AtEIN3 (Fig. 7). The position and functionality of EIL DNA binding domains have been characterised in *Arabidopsis* EIL3 and in cucumber EIN3 [48, 49].

Whether the promoters of *CfEIL2* and *Cmeil2* share the same specificity remains to be shown, however *Cmeil2*may have been expressed in flowers at one point as traces of the transcript was observed in RT-PCR and in RT-qPCR (Fig. 6). Finally, additional *EIL* homologs may be present in *Campanula* as 4–6 homologs have been identified in *Arabidopsis*, tomato and tobacco [13, 50, 51, 52].

Previous studies of *EIN3/EIL* homologs have shown constitutive expression in *Paeonia*, however in *Dianthus caryopyllus DcEIL1/2* and *DcEIL3* transcripts in petals and styles increased rapidly after ethylene treatment of flowers and then gradually declined [53,54]. The constitutive expressions of *ERS2*, *CTR1* and *EILs* in the ethylene signal transduction pathway in *Campanula* indicate that flowers are capable of a fast physiological response in the presence of ethylene. This correlate well with earlier results where *Campanula* has been described as an ethylene sensitive species [20]. Furthermore, the lack of transcriptional response to ethylene exposure could be a combination of regulatory steps on the protein level [55, 56].

The *Cmeil2* Phenotype was Inherited as a Recessive Trait

In the present study, the *eil2* frameshift mutation was identified not only in Danish domesticated *Cm* but also in non domesticated specimens of *Cm* and in the closely related *C. medium* var. *calycanthema* (Fig. 9). This suggests that the *eil2* frameshift must have occurred in an earlier ancestor of *Cm*. None of the closely related*Campanula* species *C. alpina*, *C. alpestris*, *C. hofmannii* or *E. graminifolius* contained *eil2* (Fig. 9). In hybrids of*Cf*× *Cm*, heterozygote *eil2* resulted in very similar ethylene responses as observed in *Cf* as only one of five hybrids showed an intermediate phenotype shifted towards increased ethylene insensitivity when compared to *Cf* (Table 2). This indicated that the *eil2* phenotype was restored by the presence of a wild type *EIL2* and was inherited as a recessive trait. Homozygote *eil2* in *Cf* could not be obtained via crossings as both male and female parts of *Cf*× *Cm* hybrids were sterile. Results obtained here are the first to describe the effects of an*ein3*/*eil* mutation in flowers. Also, no reports exist of approaches where *EIN3/EIL* genes have been knocked out or silenced via gene modification. Therefore the *eil2* phenotype described in *Cm* cannot be directly compared to related phenotypes in flowers of other plant species. In *Arabidopsis*, the closest homolog to*Campanula EIL2* is *EIN3* (*AtEIN3*). In *Arabidopsis*, *ein3* mutants are well characterised and they too are inherited in a recessive manner [13]. Also phenotypes in *ein3* or *eil* mutants in *Arabidopsis* are only described in seedlings or mature rosettes. To our knowledge ethylene sensitivity and floral development has not been described in *Arabidopsis ein3* or *eil* mutants. However, in support of our results are data from tomato where expression of *LeEIL1*, *LeEIL2* or *LeEIL3* antisense transcripts results in ethylene insensitive buds [51]. Collectively, this study is the first to correlate ethylene insensitivity in flowers to an *ein*/*eil* mutant phenotype.

Future Approaches to Achieve Ethylene Insensitive Plants

Previous studies have indicated that there may be specific functions for the individual *EILs*. Thus to alter the physiological response of flowers to ethylene, the right ortholog in each plant species has to be identified. In the framework of this study, the *Cmeil2* frameshift mutation could potentially be transferred to *EIL2* homologs of related species by conventional crossing or the deletion could be introduced by wide hybridisation among related species, assisted by embryo rescue techniques when necessary [57].

The frameshift in *Cmeil2* is positioned in a highly conserved region, and it therefore holds potential for molecular breeding towards ethylene insensitive plants. However, the identification of the full genomic sequence in *Cm* and *Cf* and the full sequence of the translated gene product in *Cf* are essential steps in this process. Ultimately, we propose that the identified 7 bp deletion in *Cmeil2* may be used to confer ethylene insensitivity to other plant species. This may be feasible via targeted mutagenesis techniques utilizing ZNF, TALENs [58] or CRISPR/Cas9 [59]. As a result, ethylene insensitivity may be transferred from *Cm* to other important climacteric ornamentals e.g., roses, *Petunia*, carnations, or even to edible climacteric crops such as broccoli and tomato.

CONCLUSIONS

We characterised the physiological and molecular responses among three *Campanula* species to exogenous ethylene. Key genes in the ethylene signal transduction pathway *ERS*, *CTR* and *EIL1* were found to be constitutively expressed in *Campanula* and unresponsive to exogenous ethylene. However, *EIL2* was found to be specific for the large flowered species *C. formanekiana* and *C. medium*, but was not expressed in *Cm* due to a 7 nucleotide frameshift in the coding region of *Cmeil2*. The natural mutation identified here in *Cmeil2*correlates with the observed ethylene insensitivity in this species. This finding holds great potential for future breeding strategies towards ethylene insensitive plants.

METHODS

Plant Materials and Growth Conditions

Campanula portenschlagiana Schultes 'Blue GET MEE®' (*Cp*), *Campanula formanekiana* Degen & Doefler 'Blue MARY MEE®' (*Cf*), *Campanula medium* L. 'Sweet MEE®' (*Cm*), *C. medium* breeding line 2 and *C. medium* var.*calycanthema* were received from the nursery Gartneriet PKM A/S (Odense, Denmark) in a developmental stage with young flower buds. Hybrids of *C. formanekiana* × *C. medium* (A-E) were produced by ovule culture [57] or at Gartneriet PKM A/S. Upon arrival, plants were transferred to a greenhouse with 18 °C day/15 °C night and a 16-h photoperiod of natural light. Seeds of the non domesticated *C. medium* were provided by the Alpine Staudengärtnerei (Leisning, Germany). Seeds of *C. alpina*, *C. alpestris*, *C. hofmannii*, *C. incurva* and*Edraianthus graminifolius* were obtained from B & T World Seeds (Aigues-Vives, France).

Identification of Putative *ERS2*, *CTR1* and *EIL* Genes

Based on the NCBI GenBank [60] sequence from *Campanula carpatica* (GenBank: AF413669) intron spanning primers were designed to amplify partial fragments of *ERS2*. *ERS2* PCR on gDNA produced two *ERS2* products in *Cp* and one *ERS2* genomic fragment in *Cf* and *Cm*. In *Cp*, The two *CpERS2* products were derived from two genes containing different intron sizes but coding for very similar transcripts. A partial *CTR1* was produced using degenerate primers aligning to conserved areas among*Musa acuminata, Solanum lycopersicon, Arabidopsis thaliana* and *Rosa hybrida CTR1* sequences (GenBank:JF430422, GenBank:AF096250, GenBank:NM_180429, and GenBank:AY032953). Sequencing showed some polymorphisms in *CTR1* of *Cp*, *Cf* and *Cm*. This could indicate that more than one copy of *CTR1* exist in*Campanula*. Partial *EIL* genes cloned via degenerate primers produced from *Malus* x *domestica* and *Solanum lycopersicon EIL* sequences (GenBank:GU732486 and GenBank:NM_001247617). *Campanula ACT* was amplified using degenerate primers, sequenced and from this sequence specific primers were designed for expression analyses.

Genomic DNA from *Cp*, *Cf*, and *Cm* was isolated from flowers with DNeasy Plant Mini Kit (Qiagen) using 300 mg of plant material following manufacturer's recommendations. PCR reactions used 100–250 ng gDNA, 2 % (v/v) DMSO and polymerase LaTaq (Takara Bio Inc.) as manufacturer recommends. Reactions followed the program; 4 min 94 °C, 33–35 cycles of [30 s 94 °C, 1 min 60 °C, 1 min 72 °C] and a final 7 min elongation step at 72 °C in a MyCycler (Biorad). Cloning of PCR-products was via TOPO TA Cloning® kit (Life Technologies Corp, Invitrogen) as recommended by manufacturer. Plasmids were purified by QIAprep Spin Miniprep kit (Qiagen) and sequenced by Eurofins MWG Operon. *EIL2* PCR products were purified with QIAquick PCR purification Kit (Qiagen) and restriction analyses using *Nla*III were done as supplier recommends (New England Biolabs).

Ethylene Exposure Experiments

To monitor flower development, individual buds were labelled one day before flower opening. This stage was termed day 0. In the following days newly opened flowers (day 1) and 4 days old flowers (day 4) were identified, tagged and used in subsequent experiments. This allowed two morphologically different stages to be monitored simultaneously throughout the ethylene exposure experiments. Ethylene exposure were conducted in a climate chamber in glass tanks with postharvest growth conditions; 20 °C day/18 °C night, 16-h photoperiod at 10–12 $\mu mol\ m^{-2} \cdot s^{-1}$ provided by cool-white fluorescent tubes (Philips Master TL-D-36 W/830). Each glass tank had a volume of 128 L and contained three plants. Flower labeling resulted in each glass tank containing three plants with a total of 15 labelled flowers for each developmental stage (day 1 and day 4). Except for *Cm* where 8–13 labeled flowers pr. growth stage were used. Ethylene concentrations of 0 $\mu L \cdot L^{-1}$, 0.05 $\mu L \cdot L^{-1}$ or 0.1 $\mu L \cdot L^{-1}$ were obtained by injection of gaseous ethylene (Mikrolab Aarhus A/S) into sealed glass tanks. Flowers were monitored, tanks ventilated and ethylene reinjected every 24 h. For Fig. 1 a senescent flower was defined as a flower showing twisted or closed corolla or wilted. For Table 2, ethylene sensitivity of *Cf* × *Cm* hybrid flowers were classified in three categories; no symptoms, signs of senescence (partial wilting and discoloration of corolla) and complete senescence (complete wilting and full discoloration of corolla). The latter experiments were done in glass tanks with 5 $\mu l \cdot L^{-1}$ ethylene for 72 h. *Cp* and *Cf*experiments were repeated twice whereas *Cm* experiments were in three replicates.

Gene expression analyses were done at developmental stages: bud, day 0 (one day before flowering), day 1, day 2 and day 4 (Fig. 3). Plants were subjected to low ethylene concentrations of 0.025 $\mu L \cdot L^{-1}$ and 0.050 $\mu L \cdot L^{-1}$ ethylene for 24 h. Each tank contained three plants with labeled 1-day flowers. Flowers from each tank were pooled, harvested in liquid nitrogen, grinded and used for RNA extraction. Each experiment was performed in three replicates.

RNA Extraction, cDNA Synthesis and Expression Analysis

RNA was extracted using RNeasy Plant Mini kit (Qiagen) with the following change to manufacturer's protocol: Cell lysis were done using RLT buffer with 0.01 % β-mercaptoethanol (v/v) for 1 min at 56 °C. RNA yield and purity (A260/A280 ratio > 2.0) was estimated by a Nanodrop™ 1000 Spectrophotometer (Thermo Fisher Scientific Inc.). RNA integrity was evaluated on 1.2 % agarose gels. Purified RNA was stored at -80 °C. RNA was DNase treated with Amplification Grade DNase I (Invitrogen) and cDNA synthesis was done using iScript cDNA Synthesis kit as recommended (Biorad). In 20 µl reactions 0.8 µg RNA was used. No contamination of DNA in cDNA was verified in non RT samples. Expression analyses were done using 5-fold diluted cDNA and ExTaq as polymerase as recommended (Takara Bio Inc.) in MyCycler (Biorad). Primers and gene specific reaction settings used in expression analysis. RT-PCR program; 4 min 94 °C, 25–32 cycles of [30 s 94 °C, 1 min 55 °C, 1 min 72 °C] and 7 min 72 °C. Reactions for RT-qPCR were performed on an ICycler instrument (Bio-Rad) by using the iQ SYBR Green Supermix (Bio-Rad) according to supplier's instructions [61] using the program; 95 °C for 10 min, 50 cycles of [30 s at 95 °C, 1 min at 57.5 °C and 1 min at 72 °C]. To evaluate the efficiency of qPCR, serial dilutions of cDNA were used to generate a standard curve. This resulted in R^2 values of 0.998 and 0.994 for *ACT* and *EIL2* primer sets, respectively. Threshold cycles (Ct), (defined as cycle were the signal exceeds ten times the standard deviation of the baseline), for *CfEIL2* and*Cmeil2* were standardized to the corresponding *Actin* Ct (ΔCt). The relative quantification of target gene *CfEIL2*and *Cmeil2* between the different treatments was determined as $2^\wedge(-\Delta\Delta Ct)$. Values are based on three replicates.

Bioinformatics and Statistics

Sequence identification and analysis were done using CLC sequence viewer (CLC bio), BLAST and Clustal Ω [62]. Phylogenetic analysis were conducted using MEGA version 6.06 [63]. Statistical analyses were done in SigmaPlot v. 13 by one way analysis of variance using the Holm-Sidak method.

ACKNOWLEDGEMENTS

Gartneriet PKM A/S is greatly acknowledged for providing plant material and Christian Hald Madsen is thanked for fruitful discussions.

REFERENCES

1. Abeles FB, Morgan PW, Saltveit ME. Ethylene in plant biology. 2nd ed. San Diego: Academic; 1992.
2. Bleecker AB, Estelle MA, Somerville C, Kende H. Insensitivity to ethylene conferred by dominant mutation in *Arabidopsis thaliana*. Science. 1988;241:1–25.
3. Rodriguez FI, Esch JJ, Hall AE, Binder BM, Schaller GE, Bleecker AB. A copper cofactor for the ethylene receptor ETR1 from*Arabidopsis*. Science. 1999;283(5404):996–8.
4. Gamble RL, Coonfield ML, Schaller GE. Histidine kinase activity of the ETR1 ethylene receptor from *Arabidopsis*. Proc Natl Acad Sci U S A. 1998;95(13):7825–9.
5. Moussatche P, Klee HJ. Autophosphorylation activity of the *Arabidopsis* ethylene receptor multigene family. J Biol Chem. 2004;279(47):48734–41.
6. Hua J, Sakai H, Nourizadeh S, Chen QG, Bleecker AB, Ecker JR, et al. EIN4 and ERS2 are members of the putative ethylene receptor gene family in Arabidopsis. Plant Cell. 1998;10(8):1321–32.
7. Clark KL, Larsen PB, Wang X, Chang C. Association of the *Arabidopsis* CTR1 Raf-like kinase with the ETR1 and ERS ethylene receptors. Proc Natl Acad Sci U S A. 1998;95(9):5401–6.
8. Kieber JJ, Rothenberg M, Roman G, Feldmann KA, Ecker JR. CTR1, a negative regulator of the ethylene response pathway in arabidopsis, encodes a member of the Raf family of protein kinases. Cell. 1993;72(3):427–41.
9. Alonso JM, Hirayama T, Roman G, Nourizadeh S, Ecker JR. EIN2, a bifunctional transducer of ethylene and stress responses in *Arabidopsis*. Science. 1999;284(5423):2148–52.
10. Ju C, Yoon GM, Shemansky JM, Lin DY, Ying ZI, Chang J, et al. CTR1 phosphorylates the regulator EIN2 to control ethylene hormone signaling from the ER membrane to the nucleus in *Arabidopsis*. Proc Natl Acad Sci. 2012;109(47):19486–91.
11. Wen X, Zhang C, Ji Y, Zhao Q, He W, An F, et al. Activation of ethylene signaling is mediated by nuclear translocation of the cleaved EIN2 carboxyl terminus. Cell Res. 2012;22(11):1613–6.
12. Qiao H, Shen Z, Huang SC, Schmitz RJ, Urich MA, Briggs SP, et al. Processing and subcellular trafficking of ER-tethered EIN2 control response to ethylene gas. Science. 2012;338(6105):390–3.
13. Chao Q, Rothenberg M, Solano R, Roman G, Terzaghi W, Ecker JR. Activation of the ethylene gas response pathway in arabidopsis by the nuclear protein ETHYLENE-INSENSITIVE3 and related proteins. Cell. 1997;89(7):1133–44.
14. Chang KN, Zhong S, Weirauch MT, Hon G, Pelizzola M, Li H, et al. Temporal transcriptional response to ethylene gas drives growth hormone cross-regulation in *Arabidopsis*. eLife. 2013;2, e00675.
15. Li Z, Peng J, Wen X, Guo H. *Ethylene-insensitive3* is a senescence-associated gene that accelerates Age-dependent leaf senescence by directly repressing *miR164* transcription in *Arabidopsis*. Plant Cell. 2013;25(9):3311–28.
16. Ferrante A, Trivellini A, Scuderi D, Romano D, Vernieri P. Post-production physiology and handling of ornamental potted plants. Postharvest Biol Tec. 2015;100:99–108.
17. Rogers HJ. From models to ornamentals: how is flower senescence regulated? Plant Mol Biol. 2013;82(6):563–574.
18. Serek M, Woltering EJ, Sisler EC, Frello S, Sriskandarajah S. Controlling ethylene responses in flowers at the receptor level. Biotechnol Adv. 2006;24(4):368–81.
19. Sun Y, Christensen B, Liu F, Wang H, Müller R. Effects of ethylene and 1-MCP (1-methylcyclopropene) on bud and flower drop in mini *Phalaenopsis* cultivars. Plant Growth Regul. 2009;59(1):83–91.
20. Woltering EJ. Effects of ethylene on ornamental pot plants: a classification. Sci Hort. 1987;31(3–4):283–94.
21. Müller R, Andersen AS, Serek M. Differences in display life of miniature potted roses (*Rosa hybrida* L.). Sci Hort. 1998;76(1–2):59–71.
22. Finlayson SA, Reid DM. Influence of CO2 on ACC oxidase activity from roots of sunflower (*Helianthus annuus*) seedlings. Phytochemistry. 1994;35(4):847–51.
23. Hansen H, Grossmann K. Auxin-induced ethylene triggers abscisic acid biosynthesis and growth inhibition. Plant Physiol. 2000;124(3):1437–48.
24. Lütken H, Clarke J, Müller R. Genetic engineering and sustainable production of ornamentals: current status and future directions. Plant Cell Rep. 2012;31(7):1141–57.
25. Chang C, Kwok S, Bleecker A, Meyerowitz E. Arabidopsis ethylene-response gene ETR1: similarity of product to two-component regulators. Science. 1993;262(5133):539–44.
26. Wilkinson JQ, Lanahan MB, Clark DG, Bleecker AB, Chang C, Meyerowitz EM, et al. A dominant mutant receptor from*Arabidopsis* confers ethylene insensitivity in heterologous plants. Nat Biotech. 1997;15(5):444–7.

27. Clevenger DJ, Barrett JE, Klee HJ, Clark DG. Factors affecting seed production in transgenic ethylene-insensitive petunias. J Amer Soc Hort Sci. 2004;129(3):401–6.

28. Sriskandarajah S, Mibus H, Serek M. Transgenic *Campanula carpatica* plants with reduced ethylene sensitivity. Plant Cell Rep. 2007;26(6):805–13.

29. Shibuya K, Barry KG, Ciardi JA, Loucas HM, Underwood BA, Nourizadeh S, et al. The role of PhEIN2 in ethylene responses throughout plant development in Petunia. Plant Physiol. 2004;136(2):2900–12.

30. Olsen A, Lütken H, Hegelund JN, Müller R. Ethylene resistance in flowering ornamental plants – improvements and future perspectives. Horticulture Res. 2015;2:15038.

31. Heywood VH, Brummitt RK, Culham A, Seberg O. Flowering plant families of the world. Richmond: Kew Publishing; 2007.

32. Mansion G, Parolly G, Crowl AA, Mavrodiev E, Cellinese N, Oganesian M, et al. How to handle speciose clades? mass taxon-sampling as a strategy towards illuminating the natural history of *Campanula* (Campanuloideae). PLoS ONE. 2012;7(11), e50076.

33. Cameron AC, Reid MS. 1-MCP blocks ethylene-induced petal abscission of *Pelargonium peltatum* but the effect is transient. Postharvest Biol Tec. 2001;22:169–77.

34. Porat R, Halevy AH, Serek M, Borochov A. An increase in ethylene sensitivity following pollination is the initial event triggering an increase in ethylene production and enhanced senescence of *Phalaenopsis* orchid flowers. Physiol Plant. 1995;93(4):778–84.

35. Giblin DE. Variation in floral longevity between populations of *Campanula rotundifolia* (*Campanulaceae*) in response to fitness accrual rate manipulation. Am J Bot. 2005;92(10):1714–22.

36. Tanase K, Onozaki T, Satoh S, Shibata M, Ichimura K. Effect of Age on the auto-catalytic ethylene production and the expression of ethylene biosynthetic gene *DcACS1* in petals of long-life carnations. Jpn Agric Res Q. 2011;45(1):107–16.

37. Hua J, Chang C, Sun Q, Meyerowitz E. Ethylene insensitivity conferred by Arabidopsis ERS gene. Science. 1995;269(5231):1712–4.

38. Ma N, Tan H, Liu X, Xue J, Li Y, Gao J. Transcriptional regulation of ethylene receptor and CTR genes involved in ethylene-induced flower opening in cut rose (*Rosa hybrida*) cv. Samantha. J Exp Bot. 2006;57(11):2763–73.

39. MöLler R, Lind-Iversen S, Stummann BM, Serek M. Expression of genes for ethylene biosynthetic enzymes and anethylene receptor in senescing flowers of miniature potted roses. J Hortic Sci Biotechnol. 2000;75(1):12–8.

40. Hua J, Meyerowitz EM. Ethylene responses Are negatively regulated by a receptor gene family in *Arabidopsis thaliana*. Cell. 1998;94(2):261–71.

41. Müller R, Stummann BM, Serek M. Characterization of an ethylene receptor family with differential expression in rose (*Rosa hybrida* L.) flowers. Plant Cell Rep. 2000;19(12):1232–9.

42. Shibuya K, Nagata M, Tanikawa N, Yoshioka T, Hashiba T, Satoh S. Comparison of mRNA levels of three ethylene receptors in senescing flowers of carnation (Dianthus caryophyllus L.). J Exp Bot. 2002;53(368):399–406.

43. Müller R, Owen CA, Xue Z-T, Welander M, Stummann BM. Characterization of two CTR-like protein kinases in *Rosa hybrida* and their expression during flower senescence and in response to ethylene. J Exp Bot. 2002;53(371):1223–5.

44. Hu H-L, Do Y-Y, Huang P-L. Expression profiles of a MhCTR1 gene in relation to banana fruit ripening. Plant Physiol Biochem. 2012;56:47–55.

45. Leclercq J, Adams-Phillips LC, Zegzouti H, Jones B, Latché A, Giovannoni JJ, et al. *LeCTR1*, a tomato *CTR1*-like gene, demonstrates ethylene signaling ability in Arabidopsis and novel expression patterns in tomato. Plant Physiol. 2002;130(3):1132–42.

46. Adams-Phillips L, Barry C, Kannan P, Leclercq J, Bouzayen M, Giovannoni J. Evidence that CTR1-mediated ethylene signal transduction in tomato is encoded by a multigene family whose members display distinct regulatory features. Plant Mol Biol. 2004;54(3):387–404.

47. Cho Y-H, Yoo S-D. Novel connections and gaps in ethylene signaling from the ER membrane to the nucleus. Front Plant Sci. 2015;5(733):1–7.

48. Bie BB, Pan JS, He HL, Yang XQ, Zhao JL, Cai R. Molecular cloning and expression analysis of the ethylene insensitive3 (EIN3) gene in cucumber (*Cucumis sativus*). Genet Mol Res. 2013;12(4):4179–91.

49. Yamasaki K, Kigawa T, Inoue M, Yamasaki T, Yabuki T, Aoki M, et al. Solution structure of the major DNA-binding domain of *Arabidopsis thaliana* ethylene-insensitive3-like3. J Mol Biol. 2005;348(2):253–64.

50. Rieu I, Mariani C, Weterings K. Expression analysis of five tobacco EIN3 family members in relation to tissue-specific ethylene responses. J Exp Bot. 2003;54(391):2239–44.

51. Tieman DM, Ciardi JA, Taylor MG, Klee HJ. Members of the tomato LeEIL (EIN3-like) gene family are functionally redundant and regulate ethylene responses throughout plant development. Plant J. 2001;26(1):47–58.

52. Yokotani N, Tamura S, Nakano R, Inaba A, Kubo Y. Characterization of a novel tomato EIN3-like gene (LeEIL4). J Exp Bot. 2003;54(393):2775–6.

53. Iordachescu M, Verlinden S. Transcriptional regulation of three EIN3-like genes of carnation (Dianthus caryophyllus L. cv. Improved White Sim) during flower development and upon wounding, pollination, and ethylene exposure. J Exp Bot. 2005;56(418):2011–8.

54. Wang Y, Zhang C, Jia P, Wang X, Wang W, Dong L. Isolation and expression analysis of three EIN3-like genes in tree peony (*Paeonia suffruticosa*). Plant Cell Tiss Org. 2013;112(2):181–90.

55. Xu J, Zhang S. Regulation of ethylene biosynthesis and signaling by protein kinases and phosphatases. Mol Plant. 2014;7(6):939–42.

56. Kendrick MD, Chang C. Ethylene signaling: new levels of complexity and regulation. Curr Opin Plant Biol. 2008;11(5):479–85.

57. Röper AC, Lütken H, Christensen B, Boutilier K, Petersen KK, Müller R. Production of interspecific *Campanula* hybrids by ovule culture: exploring the effect of ovule isolation time. Euphytica. 2015;203(3):643–57.

58. Lusser M, Parisi C, Plan D, Rodríguez-Cerezo E. Deployment of new biotechnologies in plant breeding. Nat Biotechnol. 2012;30:231–9.

59. Belhaj K, Chaparro-Garcia A, Kamoun S, Nekrasov V. Plant genome editing made easy: targeted mutagenesis in model and crop plants using the CRISPR/Cas system. Plant Methods. 2013;9(1):39.

60. Benson DA, Cavanaugh M, Clark K, Karsch-Mizrachi I, Lipman DJ, Ostell J, et al. GenBank. Nucleic Acids Res. 2013;41(D1):D36–42.

61. Lütken H, Jensen LS, Topp SH, Mibus H, Müller R, Rasmussen SK. Production of compact plants by overexpression of AtSHI in the ornamental *Kalanchoë*. Plant Biotech J. 2010;8(2):211–22.

62. McWilliam H, Li W, Uludag M, Squizzato S, Park YM, Buso N, et al. Analysis Tool Web Services from the EMBL-EBI. Nucleic Acids Res. 2013;41:W597–600.

63. Tamura K, Stecher G, Peterson D, Filipski A, Kumar S. MEGA6: Molecular Evolutionary Genetics Analysis version 6.0. Mol Biol Evol. 2013;30:2725–9.

CHAPTER 21

PLANT-BASED VACCINES: PRODUCTION AND CHALLENGES

Erna Laere[1], Anna Pick Kiong Ling[1], Ying Pei Wong[1], Rhun Yian Koh[1], Mohd Azmi Mohd Lila[2], and Sobri Hussein[3]

[1]Department of Applied Biomedical Sciences and Biotechnology, School of Health Sciences, International Medical University, 57000 Bukit Jalil, Kuala Lumpur, Malaysia

[2]Department of Veterinary Pathology and Microbiology, Faculty of Veterinary Medicine, Universiti Putra Malaysia (UPM), 43400 Serdang, Selangor, Malaysia

[3]Agrotechnology and Bioscience Division, Malaysian Nuclear Agency, Bangi, 43000 Kajang, Selangor, Malaysia

ABSTRACT

Plant-based vaccine technologies involve the integration of the desired genes encoding the antigen protein for specific disease into the genome of plant tissues by various methods. Agrobacterium-mediated gene transfer and transformation via genetically modified plant virus are the common methods that have been used to produce effective vaccines. Nevertheless, with the advancement of science and technology, new approaches have been developed to increase the efficiency of former methods such as biolistic, electroporation, agroinfiltration, sonication, and polyethylene glycol treatment. Even though plant-based vaccines provide many benefits to the vaccine industry, there are still challenges that limit the rate of successful production of these third-generation vaccines. Even with all the limitations, continuous efforts are still ongoing in order to produce efficient vaccine for many human and animals related diseases owing to its great potentials. This paper reviews the existing conventional methods as well as the development efforts by researchers in order to improve the production of plant-based vaccines. Several challenges encountered during and after the production process were also discussed.

INTRODUCTION

Vaccines help in stimulating the antibodies production in human and animals and provide immune protection against several diseases [1]. However, the unavailability of vaccines for the treatment of fatal diseases has caused problems and driven global attention towards production of safer, easier, and more effective vaccines. Generally, there are three types of vaccine production methods, namely, the egg-based vaccines, cell-based vaccines, and vaccines produced using investigational-manufacturing systems. The most common example of egg-based vaccine is the influenza vaccine produced in 9-to-12-day-old embryonated eggs [2, 3]. This conventional method has been applied for over 60 years and it involves the injection of virus particles into the eggs and further incubation for several days to allow the replication of virus particles. In order to produce a vaccine, the antigen isolated from the purification process of the eggs containing vaccine virus particle would undergo additional procedures. However, the selection of most appropriate influenza virus strains to be replicated for vaccine production remains to be the main limitation in this method as not all strains of influenza virus are able to replicate in embryonated eggs, hence affecting the amount of vaccine produced in the eggs [2]. Apart from that, a large number of eggs are required to produce a vaccine while the regulatory

Citation: Erna Laere, Anna Pick Kiong Ling, Ying Pei Wong, Rhun Yian Koh, Mohd Azmi Mohd Lila, and Sobri Hussein, "Plant-Based Vaccines: Production and Challenges," Journal of Botany, vol. 2016, Article ID 4928637, 11 pages, 2016. doi:10.1155/2016/4928637

approval for vaccines produced from egg-based method is rather time-consuming [2, 4]. Meanwhile, the main limitation in producing the cell-based vaccines would be the requirement of high-priced fermentation facilities. With various limitations in the former two conventional methods, particularly in relation to time, the expensive manufacturing process due to the need of cold storage for the temperature-sensitive vaccines [5], and the risk of unwanted immune response and developing the disease, investigational-manufacturing systems, which utilize the biological systems such as plant, insect cells, or bacteria culture to manufacture vaccines have recently gained the attention of researchers. Among these, plant-based vaccine production has received particular attention due to the numerous advantages it may offer.

The attempt to produce vaccines in plants was made by Hiatt and coworkers in 1989 [6]. The concept of utilizing transgenic plants to produce and deliver subunit vaccines was introduced by Dr. Arntzen and his colleagues and proved that this concept can overwhelm the limitations in traditional vaccine production [6]. The first subunit vaccine was produced by them in tobacco plants by expressing surface protein antigen of Streptococcus mutants. They also initiated the production of hepatitis B and heat-labile toxin B subunit in potato tubers as well as potato plants. In 1998, it was proven, for the first time, by National Institute of Allergy and Infectious Diseases (NIAID) that significant immunogenicity can be induced safely by an edible vaccine [6], utilizing the concept of plants as bioreactor. Due to the fact that the plant-based vaccine is easy to handle as it does not require complicated storage and its production is cost-effective and easy to scale up for large production, this method may provide a cheaper alternative for vaccine production [5, 7–10]. Moreover, plant-based edible vaccines produced through this method are able to provide a needleless, convenient, and easy route of administration [9–11]. Among the plants that have been commonly used as bioreactor are tobacco, potato, tomato, corn, and rice. To date, there are many transgenic plants that have been used to produce four different types of vaccines: bacterial vaccines, viral vaccines, parasite vaccines, and immunocontraceptive vaccines [9].

There are several plant-based vaccines that have been produced, with some of them being currently at the clinical trial phase. Among them, the most common types of vaccines are against virus and bacteria that cause fatal illness in human and animals and usually Nicotiana plants are utilized as the bioreactor. However, to date, only two products have been licenced: (a) plant made scFV mAB used in the production of a recombinant HBV vaccine in Cuba and (b) Newcastle disease virus (NDV) vaccine for poultry approved by the US Department of Agriculture (USDA) [12]. There is no plant-based vaccine that has received the license from US Food and Drug Administration (FDA). This is due to the fact that plant-based vaccines are classified under the genetically modified crop category [13]. In view of this exciting yet challenging research, the first part of this paper focuses on the conventional and refined expression technologies for improved plant-based vaccines production, while the latter part discusses challenges encountered during and after the production process.

PRODUCTION OF PLANT-BASED VACCINES

Plant-based vaccine production mainly involves the integration of transgene into the plant cells. The target sequence of the selected antigen is integrated with the vector before being transferred into the expression system. The transgene can then be expressed in the plants either through a stable transformation system or through transient transformation system, depending on the location where the transgene has been inserted in the cells. Stable transformation system can be achieved through nuclear or plastid integration [14]. It is called stable or permanent due to the permanent changes occurring in recipient cells' genetics as the target transgene is integrated into the genome of host plant cells [15]. Biolistic and genetically modified Agrobacterium strain can lead to the formation of stable transfection. However, as Agrobacterium tumefaciens is not infecting many plant species naturally, it limits the application of Agrobacterium strain for stable transformation of the desired gene. Generally, stably transgenic plant cells produce a lower amount of subunit antigen, in the range of 0.01 to 0.30% of total soluble plant protein.

On the other hand, transient transformation system involves the production of desired protein or antigen soon after the heterologous gene resides transiently in the host cells [14, 16]. The transgene is not incorporated into the genome of the plant cells. In this plant expression system, the regeneration of whole plant is not required and the frequency of its occurrence is higher. These characteristics overcome the pitfalls related to the stable integration [14]. Two most commonly used methods that would achieve transient expression of a desired protein in plants are the Agrobacterium-mediated transformation of genetically modified plant virus and particle bombardment [14].

Direct Gene Delivery Method

As mentioned earlier, there are several methods that can be used to produce plant-based vaccines. Basically, these methods are divided into two categories, which are direct and indirect gene delivery [16, 17]. The direct gene delivery method simply means the direct introduction of DNA or RNA into the plant cells [12]. In this section, the most common direct gene delivery approach, biolistic method, will be further discussed.

Biolistic method is a vector-independent method and it is also known as gene gun or microprojectile bombardment method [12]. This is an alternative method of gene transfer for nuclear transformation ifAgrobacterium-mediated transformation is not feasible [6, 18, 19]. It involves the use of gold or tungsten as microcarrier to coat the DNA [6, 20]. The coated DNA will then be placed on top of macrocarrier, inserted into gene gun, and subjected to high pressure of helium gas [6, 20]. Due to the high pressure, the coated DNA will travel at a high speed within a vacuum and penetrate into the cells of targeted plant [21]. The advantages of this method are that it forms a stable integration of the transgene into the plant genome and it can be applied to transfer foreign DNA into a variety of types of plant host species as well as various cell types [14]. There is no vector requirement for this method and it will aid in cotransformation [14]. However, it requires a costly particle gun device, it is labour intensive, and it can cause severe damage to the plant tissues [16, 20].

Biolistic method can be used to achieve two types of antigen expression in the transgenic plants: nuclear and chloroplast transformation. Nuclear transformation is done by integrating the desired gene into the nucleus of the plant cells via nonhomologous recombination [22, 23]. The transgene might be inserted at the same locus or different loci to create the stable transgenic plants [22]. Even though the plants can inherit the transgene to the offspring, nuclear transgenic plants show a low expression level of antigens which lead to the requirement of a huge quantity of plant material to produce the right dose of administration and it might cause pleiotropic and position effects due to the random integration of the transgene [23–25].

A fascinating alternative to nuclear transformation with the aim to increase the yield of recombinant protein production from a single transformation step is known as chloroplast transformation [6, 17, 26]. The formation of chloroplast transgenic plants involves the use of biolistic process, which will deliver the desired DNA into chloroplasts, followed by the integration of gene of interest into the chloroplast genome from plastid transformation vector at flanking sequence via homologous recombination [18, 26]. The advantages of chloroplast transformation compared to the nuclear transformation are the ability to eliminate gene silencing effect, a rapid and low cost production due to its high copy number in a plant cell, its potential to express multiple genes in plastids and less technical work, natural transgene containment, and ensuring a site-specific insertion of the transgene in the chloroplast genome [6, 17, 25, 27, 28]. As plastid will be generally maternally inherited to the next offspring, gene flow through pollination to the transgenic plant's weedy or wild relatives in the same environmental location can be prevented as the transgene is inserted and expressed in the plastid genome [29]. However, as no glycosylation process happens in plastids, thus this method is not an option for production of functional heterologous proteins that require a complex posttranslational modification.

Among nuclear and chloroplast transformation, most of the recently reported plant-based vaccines are produced through chloroplast transformation. Some examples of vaccines that were derived from chloroplast to fight against bacterial diseases are cholera, Lyme disease, anthrax, tetanus, and plague, while vaccines to fight against viral diseases are rotavirus and canine parvovirus (CPV) [26]. Production of most of these vaccines utilized tobacco as the model plant. The cholera toxin B subunit, Bacillus anthracis protective antigen, and tetanus toxin Fragment C genes were all expressed utilizing the transgenic tobacco model [30–32]. Similarly, a protective peptide, namely, 2L21, which prevents dogs from CPV infection, was also successfully expressed in the tobacco model [33]. Some plant derived vaccines were also further tested in animal models to validate their efficacy. For instance, results showed that mice immunized with the chloroplast-derived anthrax vaccine survived through the anthrax toxin challenge [34] while mucosal immunisation of mice with the chloroplast-derived tetanus vaccine increased the antibodies against tetanus in the body, indicating the safe use of the vaccine applied through nasal or oral route [32].

Apart from tobacco model, two plant-based vaccines against dengue and rabies viruses have been reported to be produced in Lactuca sativa (lettuce) and Zea mays, respectively, through biolistic method. The dengue virus vaccine targeting dengue-3 serotype polyprotein (DENV3prM/E) was produced in Lactuca sativa via chloroplast transformation [35]. Results showed that L. sativa expressed polyprotein of the antigen in different forms as monomers, heterodimers, or multimers in Western blot. The plant grew normally and the transgenes were inherited by the next progeny without any segregation. In another example, the rabies vaccine was produced in Zea mays whereby the embryogenic callus was first transformed with the pregenerated construct with a constitutive promoter from cauliflower mosaic virus (CaMV) by biolistics, and the regenerated plants were grown in a greenhouse [36]. Then, the transgenic Zea mays expressing the rabies virus glycoprotein was fed to mice through oral route. Results showed that the treatment protected the animals from the rabies virus challenge, conferring the beneficial effect of the plant-based vaccine as a potent oral immunogen.

Based on the current evidences, antigen expression through chloroplast transformation confers several benefits. First of all, the introduced genes were inherited stably in subsequent generations ensuring the continuous supply of the source [30]. Furthermore, high yield of antigens in the chloroplasts would lessen the amount of plant material required for vaccination, and this would make the encapsulation of freeze-dried material or pill formation easier [33]. Even though production of plant-based vaccine from chloroplast transformation had been identified as an alternative method to overcome the weakness in nuclear transformation, more studies need to be carried out. This is due to the fact that chloroplast transformation method is yet to be applied on many plant species apart from tobacco plant. In addition, it is difficult to create plants that will have uniformly transformed plastid (homoplasmic). Due to the later limitation, generation of transplastomic plants that are genetically stable is hindered.

Indirect Gene Delivery Methods

Despite using direct gene delivery method, indirect gene delivery methods show more significant efficacy in vaccines production as indirect gene delivery involves the utilization of plant bacteria, particularly theAgrobacterium species and plant viruses, which naturally infect the plant cells and are able to integrate the gene of interest into plant genome [16].

Agrobacterium-Mediated Gene Transfer

Agrobacterium is a Gram-negative soil pathogenic bacterium that naturally will infect the plants and transfer their genes (T-DNA) to the nucleus of the plant cells [17, 19]. Two strains of Agrobacterium species that have been commonly used as a biological vector are Agrobacterium tumefaciens (A. tumefaciens) and Agrobacterium rhizogenes (A. rhizogenes). The main difference between these two species is the plasmid that they carry. A. tumefaciens carries tumour-inducing plasmid (Ti-plasmid), while A. rhizogenes carries root-inducing plasmid (Ri-plasmid) [18, 37]. However, A. tumefaciens is the most preferred strain by researchers for stable expression of the desired protein. In the Ti-plasmid, there are genes encoding for plant hormones such as auxin and cytokinin synthesis, which will induce tumour tissue in plants. However, for vaccine production, these genes will be deleted to form disarmed Ti-plasmid and heterologous gene is inserted forming a recombinant plasmid vector [37]. The recombinant plasmid vector is transformed into A. tumefaciens and with the help from vir gene of the bacterium, the introduced heterologous gene is transferred by the transformed bacterium and integrated into the host plant nuclear genomic DNA by nonhomologous recombination at random sites [17, 19, 37]. The transformed bacteria are transferred into the plant leaves by soaking the leaves in the A. tumefaciens culture. This method is able to yield a stable integration of the transgene into the genome of the plant [33, 38].

Much research conducted prior to 2000 and early 2000 produced various vaccines through Agrobacterium-mediated transformation system in dicotyledonous plant models. For instance, Arakawa et al. successfully expressed gene encoding for cholera toxin B subunit protein in potato leaf explants using A. tumefaciens [39]. Similarly, potato has been transformed to produce VP60 protein against rabbit hemorrhagic disease virus, in which rabbits immunized with the potato's leaf extract showed increased anti-VP60 antibody titers and were protected against the hemorrhagic disease [40]. These discoveries have led to the production of more diverse antigens in various crop species. Helicobacter pylori TonB protein was expressed in transgenic Arabidopsis thaliana through this method [41]. The antigen produced was recognizable by rabbit anti-TonB antiserum and suitable to be used as vaccine against Helicobacter infections by oral administration. In addition, Li et al. showed that hepatitis B surface antigen gene was able to be introduced into tomato plants mediated by A. tumefaciens [42]. Agrobacterium-mediated transformation has also been used for the production of antigens such as heat-labile enterotoxin from Escherichia coli, Norwalk virus capsid protein, hepatitis B surface antigen, and transgenic alfalfa expressing proteins from the foot and mouth disease virus by using potato model [43].

Most of the preliminary Agrobacterium-mediated transformation studies have been focussing on dicotyledon plants. However, based on recent reports, this method is not limited to dicotyledon plants alone. Yoshida et al. have successfully introduced the Aβ42 gene into rice using the Agrobacterium method [44]. When this transgenic brown rice expressing Aβ was orally administered to mice, their serum anti-Aβ antibody titers was elevated [44]. In addition, similar levels of CTB-specific systemic IgG and mucosal IgA antibodies with toxin-neutralizing activity were induced in mice and macaques orally immunized with rice-based oral cholera toxin B subunit vaccine, MucoRice-CTB/Q, or MucoRice-CTB/N produced through Agrobacterium-mediated method [45].

Generally, this method allows a large fragment of foreign DNA to be inserted into the Ti-plasmid of A. tumefaciens. It is also a simple and cost-effective gene transfer method with higher efficiencies [20, 22]. On the contrary, this method is considered as a rather slow process and produced a low yield [20]. The yield of desired protein is low due to the limitation during the transferring of transformed bacterium into the plant tissues. The soaking method will create the "position effect," in which only the cell layers located at the edge of the explants will receive the transgene and not the whole cells of the explants [16]. Hence, researchers have been refining this expression system by incorporating various possible approaches to improve the production of plant-based vaccines. Among them is incorporation of two agroinfiltration approaches: syringe agroinfiltration as well as vacuum agroinfiltration in the Agrobacterium-mediated transformation system.

In an effort to overcome the problems related to the stable expression of the transgene, researchers also use agroinfiltration method to insert the desired gene into plants cells. Agroinfiltration is a method that involves the infiltration of A. tumefaciens suspension into the intracellular spaces of desired parts of the plants by using a syringe and results in transient expression of desired protein or transgene [46, 47]. It can improve the expression level of antigen protein in the plant cells [48]. There are two methods of agroinfiltration, which are syringe infiltration and vacuum infiltration. Syringe infiltration is the simplest method where, by using needleless syringe, the transformed A. tumefaciens is injected into the leaf. However, creation of a small nick by using a needle has to be done before that and researcher has to ensure that it does not pierce through both sides of the leaf. The advantages offered by this method are multiple transgene constructs which might be introduced into different parts of a single leaf and it can be utilized for a variety of recombinant protein production applications [16]. Despite that, this method allows a rapid and cost-effective analysis on transgene expression level as the procedures involved are relatively simple and do not require sophisticated instruments [16, 49, 50]. On top of that, agroinfiltration can be applied to many plant species after the protocols have been optimised [50].

Another method of agroinfiltration is vacuum infiltration which involves the submerging of the leaves in the infiltration buffer containing transgene-carrying A. tumefaciens. A negative atmospheric pressure is subjected to the submerged leaves in the vacuum chamber with the aim of withdrawing the air present in the interstitial spaces of the leaves and occupying the space with the transformed A. tumefaciens [51]. Compared to syringe infiltration, this method is more complicated as there is a need for vacuum equipment and reducing the flexibility in the introduction of multiple transgenes on a single leaf. However, it benefits the recombinant protein production through the scalable production process as well as the robustness and rapidness of this method to infiltrate a large number of plants with Agrobacterium [51]. Agrobacterium tumefaciens has also been utilized to deliver the "deconstructed" viral vectors to lettuce cells and agroinfiltration with geminiviral replicon vectors was found to be able to produce high level of virus-like-particles (VLPs) derived from the Norwalk virus capsid protein (NVCP) and therapeutic mAbs against Ebola (EBV) or West Nile (WNV) viruses in lettuce [51]. Chen et al. further concluded that lettuce is an excellent host for agroinfiltration with deconstructed viral vectors [51].

Table 1 summarises some of the host plants as well as transformation methods used to produce the plant-based vaccines for human and animal diseases.

Table 1: The plant-based vaccines production for human and animal diseases.

Diseases	Pathogens	Plants	Transformation method	References
Diarrheal	Norwalk virus	Nicotiana benthamiana	Agrobacterium tumefaciens	[8]
Diarrheal	ETEC	Corn		[61]
Tuberculosis	Mycobacterium tuberculosis	Arabidopsis thaliana	Agrobacterium	[62]
Avian H5N1 influenza	HA protein of H5N1	Nicotiana benthamiana	Agrobacterium	[2]
Dengue	Dengue virus type 2 E glycoprotein (EIII)	Nicotiana tabacum cv. MD609	Agrobacterium tumefaciens	[10]
Rabies	Rabies virus	Nicotiana benthamiana, tomato	Agroinfiltration	[63]
Hepatitis B	HBsAg	Tomato	Agrobacterium tumefaciens	[42]
Foot and mouth disease	FMDV	Stylosanthes guianensis cv. Reyan II		[64]
Gaucher disease	Taliglucerase alfa	Carrot	Stable transformation	[65]
Nerve agents attack	Acetylcholinesterase	Tobacco	PEGylated	[66]
Diabetics	Insulin	Safflower	Agrobacterium tumefaciens	[5]
Human immunodeficiency	HIV	Tobacco	Agroinfiltration	[67]
Bluetongue	Bluetongue virus	Nicotiana benthamiana	Agroinfiltration	[68]
Ebola	ebola virus	Nicotiana benthamiana	Agroinfiltration	[69]

ETEC: enterotoxigenic Escherichia coli; HA: hemagglutinin; HBsAg: hepatitis B surface antigen; FMDV: foot and mouth disease virus.

Genetically Engineered Plant Virus

In this method, a suitable plant virus is modified in order to create chimeric gene for viral coat protein. Thus, it acts as a vector to deliver genetic materials into the plant cells [6, 38]. This method results in transient expression of antigen in plants [6, 38]. The recombinant virus will express the desired protein or peptide as a by-product of viral replication activity during viral infection in the plants [28, 38]. In addition, the synthesis and accumulation of vaccine epitopes can be achieved by modifying the viral capsid proteins [38]. Several advantages of plant virus mediated infection include a high level of recombinant protein expression within a short period of time after infection, easiness to generate multiple antigen copies on the viral particle's surface, and allowing large-scale viral infections in plants [6, 9, 28, 38]. However, products from the viral replication have to be purified first from the infected plants before being used for vaccination [38]. This production method will also cause the death of the plants after infection. Thus, once the vaccine has been harvested, another plant needs to be infected with the recombinant virus and this reinfection procedure has to be done repeatedly for continuous vaccine production [38].

In the earlier development, plant virus expression system involves mostly the engineered RNA viruses such as tobacco mosaic virus (TMV), potato virus X (PVX), alfalfa mosaic virus (AlMV), cucumber mosaic virus (CMV), and cowpea mosaic virus (CPMV) as expression vector [52]. These viruses are not known to replicate in mammalian cells; hence they act as an excellent alternative replicating vaccine vectors for development of both human and veterinary vaccines. Moreover, most of the expression systems have demonstrated that the vaccine antigens produced are protective against challenge infection. For example, the modified CPMV was used to infect Vigna unguiculata in the production of a few antigens including VP2 capsid protein of mink enteritis virus (MEV), VP2 capsid protein of canine parvovirus, Pseudomonas aeruginosa outer membrane protein F, and Staphylococcus aureus D2 domain of fibronectin-binding protein. Subcutaneous injection of the chimeric CPMV expressing the enteritis virus antigen has been shown to protect minks against the challenge with virulent MEV [53].

TMV has also been used to produce various antigens in plant model. Hybrids of TMV containing epitope from murine hepatitis virus (MHV) were propagated in tobacco plants, followed by purification of the virus particles. Mice immunized with the purified hybrid viruses developed serum IgG and IgA specific for the epitope and TMV coat protein. In the study, immunogen administered and protection against MHV infection were found to have a positive correlation [54]. Foot and mouth disease virus (FMDV) epitopes were also expressed in tobacco by the TMV-based vector and guinea pigs, mice, and swine were used to evaluate the protective effects of the recombinant virus. Most of the animals were protected against the FMDV challenge [55]. Pseudomonas aeruginosa outer membrane protein F was expressed in tobacco using TMV as well. The chimeric vaccine produced offered immunoprotection against chronic pulmonary infection by Pseudomonas aeruginosa in a mouse model [56]. To date, more and more RNA viruses, which also include papaya mosaic virus (PapMV), bamboo mosaic virus (BaMV), tomato bushy stunt virus (TBSV), plum pox, and Potyvirushave been added into the list of expression vectors used in plant-based vaccine production [57].

With the advances in plant virus molecular biology, DNA viruses such as geminiviruses have been further developed as one of the state-of-the-art plant expression systems [58]. geminiviruses have a small, single-stranded DNA genome that replicates in the nucleus of host cells by a rolling circle replication mechanism using a double-stranded DNA intermediate [59]. Geminiviral vectors based on bean yellow dwarf virus have been constructed to express its replication initiator protein (Rep) to produce a vaccine against Staphylococcus enterotoxin B (SEB) [60]. In another example, geminiviral replicon system was used to produce an Ebola immune complex (EIC) in leaves of Nicotiana benthamiana. Subcutaneous immunization of BALB/C mice with purified EIC resulted in anti-Ebola virus antibody production at levels comparable to those obtained with Ebola glycoprotein (GP1) virus-like particle [59]. Apart from the two examples above, beet curly top virus (BCTV) and tobacco yellow dwarf virus (TYDV) are the other geminiviruses vectors that have been used in the production of various vaccines and therapeutic proteins such as SEB, Norwalk virus VLP, HBVcAg, WNVE protein Mab, HPV-1L1 protein, HIV-1 type C p24, HAV VP1, and vitronectin [58].

METHODS TO INCREASE THE EFFICIENCY OF GENE DELIVERY

Additional methods can be applied in the production of some vaccines with the aim of improving the efficiency of transgene delivery into the host plant tissues. These are by using chemical stimulant and sonication.

DNA Uptake by Chemical Stimulation

A well-known chemical stimulant which has been used by researchers to accelerate DNA uptake by plant protoplasts is polyethylene glycol (PEG). PEG works by precipitating ionic macromolecules (in this case the DNA), promoting the uptake of the DNA to the protoplasts by endocytosis as well as allowing the transient expression of desired genes [70]. The efficiency of PEG-mediated gene transformation to enhance DNA uptake by protoplasts is depending on several factors such as the concentration of PEG, the inoculation period, and the amount of plasmids [70–72]. The advantages of this method are as follows: (a) it can be incorporated with other gene delivery methods to enhance their efficiency, (b) it will not cause damage to protoplast, and (c) this method is cost-effective as there is no requirement for expensive technical equipment and not much adaptation is required for different protoplasts [70]. However, it can be very difficult to conduct as it requires expertise to carry out the procedures and the successful transformation is highly depending on the genotype of the protoplasts [73]. Furthermore, PEG can also be toxic to the protoplasts and result in a low survival rate and ceased cell division [71]. Due to the complexity of this method, PEG transformation has not been widely used in plant-based vaccine production. Nevertheless, in generating transplastomic tobacco expressing a 25 kda protein antigen against Mycobacterium leprae and Mycobacterium avium, Hassan et al. successfully transformed the mmpI gene along with an adjuvant lymphotoxin-beta (LTB) into tobacco chloroplast by the Polyethylene glycol (PEG) mediated transformation method [74].

Sonication

Sonication is a technique that utilizes sound waves to agitate particles in solution and, aiming to mix solution, increase the rate of dissolution and remove dissolved gases from liquid. In plant-transformation, sonication will cause the formation of microwounds on plant tissue and enhance the delivery of naked DNA into the plant protoplast [75]. However, it has been reported that sonication alone is not a preferable method for gene delivery in soybean and kidney beans as it caused more negative effects on the plant cells and resulted in low transformation efficiency [75, 76]. In many studies, the sonication assisted Agrobacterium-mediated transformation (SAAT) would be used to induce mechanical disruption and formation of wounds on plant cells by ultrasonic waves [77]. The wounded cells will then allow the penetration of Agrobacterium into the deeper part of plant tissues, thus increasing the likelihood of plant cells being infected [76]. Being a very easy method, of low cost, and significant to enhance Agrobacterium-mediated gene transfer are the benefits of SAAT method. Apart from successful application of SAAT in the transformation of Chenopodium rubrum L. [78] and Beta vulgaris L. [79], this approach has also been applied in the production of recombinant Escherichia coli wild-type heat-labile holotoxin and Escherichia coli mutant LT vaccine adjuvants in Nicotiana tabacum, in which the highest systemic LT-B-specific IgG titres were detected in birds [80].

Recently, another approach called the combination of sonication plus vacuum infiltration assistedAgrobacterium-mediated transformation had evolved to optimise the efficiency of Agrobacterium-mediated gene transfer [76]. This approach has successfully transformed the Fraxinus pennsylvanica hypocotyls to harbour binary vector pq35GR containing the neomycin phosphate transferase (nptII) and β-glucuronidase (GUS) fusion gene and an enhanced green fluorescent protein gene [81]. Nevertheless, the application of this combined approach is still scarce and remains at the developmental stage in plant-based vaccines research.

CHALLENGES OF PLANT-BASED VACCINES

Although many plant-based vaccines that have been produced are still in phase 1 clinical trials, some vaccines have proceeded or completed phases II and III trials [82]. These therapeutics were produced in various transgenic plants such as insulin in transgenic safflower (SemBioSys), growth factor in transgenic barley (ORF Genetics), taliglucerase alfa in transgenic carrot (Protalix BioTherapeutics), avian influenza vaccine in transgenic tobacco (Medicago), and Ebola Vaccine in transgenic tobacco (Mapp Biopharmaceutical) [82, 83]. Nevertheless, up till today, there is no plant made vaccine that has been approved to be marketed for human consumption. Thus, it is worthwhile to note that even though the production of plant-based vaccines had been initiated almost two decades since 1989 [77], a few challenges still have to be overcome in order to develop them into highly efficacy vaccines. The issues that need to be addressed could start from the upstream processes to the implementation of the vaccines. Generally, three main challenges are the selection of antigen and plant expression host, consistency of dosage, and manufacturing of vaccines according to Good Manufacturing Practice (GMP) procedures.

Selection of Antigen and Plant Expression Host

The first issue is the selection of an antigen and the right plant expression host [25, 37]. This stage is very important in developing a vaccine that is able to fulfil all the requirements needed because not all antigens are compatible with the selected host plants [37]. The proper and careful selection will not only help to determine the safeness of the vaccine produced, it can also be used to produce thermal-stable vaccine [37]. Meanwhile, identification of antigen candidate of poorly characterised pathogen with promising characteristics can be done by applying genomics or proteomics approaches [23].

Consistency of Dosage

The consistency of dosage is another challenge that the researchers have to face as dosage produced may vary within the plants of the same species, from fruit to fruit and from generation to generation due to the size and ripeness of the fruits or plants [18, 37]. The transgenic plants show intrinsic variability in the antigen expression due to the position and pleiotropic effects caused by nonspecific integration of the transgene into the host plant genome [25]. On top of that, it is also quite difficult to evaluate the required dosage for every patient. Levels of innate and adaptive immune responses generated in different individuals may vary based on the types of antigens being exposed in the body. Between two patients with different body weight as well as their age, the dosage of plant-based vaccine required will be different. If this issue is not monitored carefully, an immunological tolerance will be induced when the patient is overdosed while reduction in antibody production will occur when the patient is underdosed [37]. Besides that, gene silencing might be induced due to the accumulation of mRNA in the transgenic plant cells as the growth of the plants is stopped and the fruit formation is reduced while the antigen content is increased [84]. In such case, consumption of plant-based vaccines may induce allergic reaction and few side effects such as toxicity on central nervous system, cytokine-induced sickness, and autoimmune diseases [37].

Manufacturing of Vaccines according to GMP Procedures

The ultimate goal of plant-based vaccines is to produce stable transgenics vaccines which are safe for consumption while reducing the production cost. Besides all the underlying issues that may affect the efficacy of plant-based vaccines, the regulatory guideline regulates by U.S. Department of Agriculture (USDA) and FDA especially the growth of transgenic plants, production and purification of plant-based vaccines, and all phases of clinical trial until marketable stage shall be strictly implemented [63]. Therefore, the manufacturers shall ensure their responsibility to follow the Good Agricultural Practices (GAP) and Good Manufacturing Practice (GMP) so that the upstream to downstream production of plant-based vaccines is strictly controlled for quality management.

Generally, to produce a plant product that could meet the quality standard, the biomanufacturing facilities must be well equipped so that complete processing cycles of the plant vaccines could be accomplished. The facilities include equipment for plant and bacterium cultivation, infiltration, plant harvest, and protein purification [85]. Takeyama et al. also summarised a few GMP plants that produce various vaccines such as influenza HA antigen, Norovirus capsid protein subunit vaccine, and rice-based cholera vaccine [85]. Concurrently, Kashima et al. reported that in order to produce a plant vaccine that meets the governmental regulatory requirements, a lot of steps and precautions need to be taken into consideration. During the production of a rice-based oral cholera vaccine, MucoRice-CTB, the biomanufacturing agency successfully established specific techniques to maintain the seed of MucoRice-CTB. The agency further evaluated the seed's propagation and stored seeds were renewed periodically to maintain the good quality. Furthermore, cultivation of the plant using a closed hydroponic system helps to minimise the variations in vaccine production. The rice produced was polished, powdered, and packaged to make the MucoRice-CTB drug substance. Final check on the identity, potency, and safety of MucoRice-CTB product must be conducted and only the products that met the quality requirements will be released [86].

It remains a great challenge to maintain the GMP standard for the product in plant-based vaccine industry. Besides the equipment, facilities, and method used to produce the vaccine, other considerations that have to be taken into account are those stated in the GMP guideline published by WHO [87, 88]. GMP for biological products guideline stated that some particular precautions are necessary for the manufacture, control, and administration of biological products as procedures and processes used in the production usually lead to high variation in the quality of products. Thus, the precautious steps should start from the very beginning of the production processes. However, in-process control is also important during the manufacturing of the biological products. Skillful staff are required to run the production processes and thus the biomanufacturing agency should provide necessary training to the staff. Buildings for the vaccine production must be designed in a way that operations can be carried out smoothly. A special design is required for plant vaccine production, in which the seedlots should be stored separately from other materials. Some other general rules of GMP shall be followed to maintain the quality standard of the vaccine products. These include the facts that standard operating procedures shall be implemented for all manufacturing operations, all products shall be clearly labelled, lot processing and distribution records shall be properly kept, and quality assurance and control shall be in place in monitoring the product quality.

FUTURE PROSPECTS

Although there are several challenges in the production and application of plant-based vaccines, the development of a better and widely acceptable plant-based vaccine among researchers still remains intact. Research in plant-based vaccine production currently has focused on the development of methods that can increase the amount of antigen produced in the transgenic plants, thus enhancing a significant immune response. The first strategy that can be applied to increase the amount of antigen in the transgenic plant tissues is by optimizing the bacterial or viral genes coding sequence so that the expression is similar to plant nuclear genes [1]. It is also important to determine the suitable subcellular compartment in the plant cells that can yield optimal quantity and quality of antigen.

Besides that, fusion of genes encoding the antigenic protein to immunomodulatory (mucosal adjuvants) proteins can be carried out more extensively as this approach has been shown to have the potential to increase the immunogenicity on the desired antigen. The examples of mucosal adjuvants are bacterial enterotoxins (B subunit of cholera toxin, CTB), secondary metabolites derived from plant, and mammalian and bacterial immunomodulators [89]. This technique is proven to prevent the diarrhoea disease caused by cholera toxin. Through this method, development of multicomponent vaccines is initiated. It has been demonstrated that the fusion of CTB to an enterotoxin protein of rotavirus and E. coli adhesion protein has provided defence towards cholera, rotavirus, and enterotoxigenic E. coli through the production of tricomponent subunit vaccine in the transgenic potato [89].

Other than development of methods to increase the immunogenicity of plant-based vaccines, more studies are anticipated in overcoming the problem related to dose variability in the transgenic plants. This is following the discoveries by Rigano and colleagues, who showed that some food-processing techniques (batch-processing and freeze drying) could maintain the normal conformation and native antigenicity of material in transgenic plants, such as tomato, potato, and Arabidopsis [25], thus standardizing the concentration of antigens in the plants.

CONCLUDING REMARKS

Plant-based vaccines are the emerging type of vaccines that have a higher therapeutic value to treat many human and animal diseases. A stable and transient gene expression can be obtained based on the gene delivery methods used. By far, chloroplast transformation via biolistic or particle bombardment gene delivery method has been considered as a very promising alternative for better production of plant-based vaccines. However, the development and improvement of suitable gene delivery methods for efficient and optimum vaccine production shall be continued. There are also some bioethical issues arising from the production of plant-based vaccines such as the risk of transferring allergens from transgenic plants to human and animals. As some of the plant-based vaccines use bacteria and virus as the vectors, the pathogens might be reactivated and infect other organisms that consume them. The benefits and advantages of plant-based vaccines shall be able to overwhelm the challenges faced by this interesting biological product. Thus, it is anticipated that regulatory approval will be granted ultimately to help in the global disease control.

ACKNOWLEDGMENTS

The study is funded by the Ministry of Higher Education, Malaysia, under the Exploratory Research Grant Scheme (ERGS) [ERGS/1/2012/SKK10/IMU/03/4].

REFERENCES

1. V. Doshi, H. Rawal, and S. Mukherjee, "Edible vaccines from GM crops: current status and future scope,"Journal of Pharmaceutical and Scientific Innovation, vol. 2, no. 3, pp. 1–6, 2013.
2. A. L. Greer, "Early vaccine availability represents an important public health advance for the control of pandemic influenza," BMC Research Notes, vol. 8, no. 1, article 191, 2015.
3. T. Huda, H. Nair, E. Theodoratou et al., "An evaluation of the emerging vaccines and immunotherapy against staphylococcal pneumonia in children," BMC Public Health, vol. 11, supplement 3, article S27, 2011.
4. N. Landry, B. J. Ward, S. Trépanier et al., "Preclinical and clinical development of plant-made virus-like particle vaccine against avian H5N1 influenza," PLoS ONE, vol. 5, no. 12, Article ID e15559, 2010.
5. C. A. Penney, D. R. Thomas, S. S. Deen, and A. M. Walmsley, "Plant-made vaccines in support of the Millennium Development Goals," Plant Cell Reports, vol. 30, no. 5, pp. 789–798, 2011.
6. J. Saxena and S. Rawat, "Edible vaccines," in Advances in Biotechnology, pp. 207–226, 2014.
7. A.-A. E. Aboul-Ata, A. Vitti, M. Nuzzaci et al., "Plant-based vaccines: novel and low-cost possible route for mediterranean innovative vaccination strategies," Advances in Virus Research, vol. 89, pp. 1–37, 2014.
8. H. Lai and Q. Chen, "Bioprocessing of plant-derived virus-like particles of Norwalk virus capsid protein under current Good Manufacture Practice regulations," Plant Cell Reports, vol. 31, no. 3, pp. 573–584, 2012.
9. Z.-J. Guan, B. Guo, Y.-L. Huo, Z.-P. Guan, J.-K. Dai, and Y.-H. Wei, "Recent advances and safety issues of transgenic plant-derived vaccines," Applied Microbiology and Biotechnology, vol. 97, no. 7, pp. 2817–2840, 2013.
10. M.-Y. Kim, M.-S. Yang, and T.-G. Kim, "Expression of dengue virus e glycoprotein domain III in non-nicotine transgenic tobacco plants," Biotechnology and Bioprocess Engineering, vol. 14, no. 6, pp. 725–730, 2009.
11. C. O. Tacket, "Plant-based oral vaccines: results of human trials," in Plant-produced Microbial Vaccines, A. V. Karasev, Ed., vol. 332 of Current Topics in Microbiology and Immunology, pp. 103–117, 2009.
12. S. Naderi and B. Fakheri, "Overview of plant-based vaccines," Research Journal of Fisheries and Hydrobiology, vol. 10, no. 10, pp. 275–289, 2015.
13. S. S. Korban, "Opportunities and challenges for plant-based vaccines," in Agricultural Biotechnology: Beyond Food and Energy to Health and the Environment, pp. 71–77, National Agricultural Biotechnology Council, 2005.

14. F. Altpeter, N. Baisakh, R. Beachy et al., "Particle bombardment and the genetic enhancement of crops: myths and realities," Molecular Breeding, vol. 15, no. 3, pp. 305–327, 2005.

15. H. Ma and G. Chen, "Gene transfer technique," Nature and Science, vol. 3, no. 1, pp. 25–31, 2005.

16. Q. Chen and H. Lai, "Gene delivery into plant cells for recombinant protein production," BioMed Research International, vol. 2015, Article ID 932161, 10 pages, 2015.

17. E. Gómez, S. C. Zoth, E. Carrillo, and A. Berinstein, "Developments in plant-based vaccines against diseases of concern in developing countries," Open Infectious Diseases Journal, vol. 4, no. 1, pp. 55–62, 2010.

18. N. Mishra, P. N. Gupta, K. Khatri, A. K. Goyal, and S. P. Vyas, "Edible vaccines: a new approach to oral immunization," Indian Journal of Biotechnology, vol. 7, no. 3, pp. 283–294, 2008.

19. T.-G. Kim and M.-S. Yang, "Current trends in edible vaccine development using transgenic plants,"Biotechnology and Bioprocess Engineering, vol. 15, no. 1, pp. 61–65, 2010.

20. C. P. Shah, M. N. Trivedi, U. D. Vachhani, and V. J. Joshi, "Edible vaccine: a better way for immunization,"Clinical Trial, vol. 3, no. 1, pp. 1–4, 2011.

21. I. K. Vasil and V. Vasil, "Transformation of wheat via particle bombardment," Methods in Molecular Biology, vol. 318, pp. 273–283, 2006.

22. E. Gómez, S. C. Zoth, and A. Berinstein, "Plant-based vaccines for potential human application," Human Vaccines, vol. 5, no. 11, pp. 738–744, 2009.

23. S. J. Streatfield, "Plant-based vaccines for animal health," Revue Scientifique et Technique de l Office International des Epizooties, vol. 24, no. 1, pp. 189–199, 2005.

24. A. Molina, J. Veramendi, and S. Hervás-Stubbs, "Induction of neutralizing antibodies by a tobacco chloroplast-derived vaccine based on a B cell epitope from canine parvovirus," Virology, vol. 342, no. 2, pp. 266–275, 2005.

25. M. M. Rigano and A. M. Walmsley, "Expression systems and developments in plant-made vaccines,"Immunology and Cell Biology, vol. 83, no. 3, pp. 271–277, 2005.

26. D. Verma and H. Daniell, "Chloroplast vector systems for biotechnology applications," Plant Physiology, vol. 145, no. 4, pp. 1129–1143, 2007.

27. S. Rosales-Mendoza, D. O. Govea-Alonso, E. Monreal-Escalante, G. Fragoso, and E. Sciutto, "Developing plant-based vaccines against neglected tropical diseases: where are we?" Vaccine, vol. 31, no. 1, pp. 40–48, 2012.

28. L. Santi, "Plant derived veterinary vaccines," Veterinary Research Communications, vol. 33, supplement 1, pp. 61–66, 2009.

29. S. H. Wani, N. Haider, H. Kumar, and N. B. Singh, "Plant plastid engineering," Current Genomics, vol. 11, no. 7, pp. 500–512, 2010.

30. H. Daniell, S.-B. Lee, T. Panchal, and P. O. Wiebe, "Expression of the native cholera toxin B subunit gene and assembly as functional oligomers in transgenic tobacco chloroplasts," Journal of Molecular Biology, vol. 311, no. 5, pp. 1001–1009, 2001.

31. J. Watson, V. Koya, S. H. Leppla, and H. Daniell, "Expression of Bacillus anthracis protective antigen in transgenic chloroplasts of tobacco, a non-food/feed crop," Vaccine, vol. 22, no. 31-32, pp. 4374–4384, 2004.

32. J. S. Tregoning, P. Nixon, H. Kuroda et al., "Expression of tetanus toxin Fragment C in tobacco chloroplasts," Nucleic Acids Research, vol. 31, no. 4, pp. 1174–1179, 2003.

33. A. Molina, S. Hervás-Stubbs, H. Daniell, A. M. Mingo-Castel, and J. Veramendi, "High-yield expression of a viral peptide animal vaccine in transgenic tobacco chloroplasts," Plant Biotechnology Journal, vol. 2, no. 2, pp. 141–153, 2004.

34. V. Koya, M. Moayeri, S. H. Leppla, and H. Daniell, "Plant-based vaccine: mice immunized with chloroplast-derived anthrax protective antigen survive anthrax lethal toxin challenge," Infection and Immunity, vol. 73, no. 12, pp. 8266–8274, 2005.

35. A. P. Kanagaraj, D. Verma, and H. Daniell, "Expression of dengue-3 premembrane and envelope polyprotein in lettuce chloroplasts," Plant Molecular Biology, vol. 76, no. 3–5, pp. 323–333, 2011.

36. E. Loza-Rubio, E. Rojas, L. Gómez, M. T. J. Olivera, and M. A. Gómez-Lim, "Development of an edible rabies vaccine in maize using the Vnukovo strain," Developments in Biologicals, vol. 131, pp. 477–482, 2008.

37. M. Sharma and B. Sood, "A banana or a syringe: journey to edible vaccines," World Journal of Microbiology and Biotechnology, vol. 27, no. 3, pp. 471–477, 2011.

38. J. Yu and W. H. Langridge, "Novel approaches to oral vaccines: delivery of antigens by edible plants,"Current Infectious Disease Reports, vol. 2, no. 1, pp. 73–77, 2000.

39. T. Arakawa, D. K. X. Chong, J. L. Merritt, and W. H. R. Langridge, "Expression of cholera toxin B subunit oligomers in transgenic potato plants," Transgenic Research, vol. 6, no. 6, pp. 403–413, 1997.

40. S. Castañón, M. S. Marín, J. M. Martín-Alonso et al., "Immunization with potato plants expressing VP60 protein protects against rabbit hemorrhagic disease virus," Journal of Virology, vol. 73, no. 5, pp. 4452–4455, 1999.

41. I. Kalbina, L. Engstrand, S. Andersson, and Å. Strid, "Expression of Helicobacter pylori TonB protein in transgenic Arabidopsis thaliana: toward production of vaccine antigens in plants," Helicobacter, vol. 15, no. 5, pp. 430–437, 2010.

42. T. Li, J. K. Sun, Z. H. Lu, and Q. Liu, "Transformation of HBsAg (hepatitis B surface antigen) gene into tomato mediated by Agrobacterium tumefaciens," Czech Journal of Genetics and Plant Breeding, vol. 47, no. 2, pp. 69–77, 2011.

43. A. Ziemienowicz, “Agrobacterium-mediated plant transformation: factors, applications and recent advances,” Biocatalysis and Agricultural Biotechnology, vol. 3, no. 4, pp. 95–102, 2014.

44. T. Yoshida, E. Kimura, S. Koike et al., “Transgenic rice expressing amyloid β-peptide for oral immunization,” International Journal of Biological Sciences, vol. 7, no. 3, pp. 301–307, 2011.

45. Y. Yuki, M. Mejima, S. Kurokawa et al., “Induction of toxin-specific neutralizing immunity by molecularly uniform rice-based oral cholera toxin B subunit vaccine without plant-associated sugar modification,” Plant Biotechnology Journal, vol. 11, no. 7, pp. 799–808, 2013.

46. L. Srinivas, G. B. Sunil Kumar, T. R. Ganapathi, C. J. Revathi, and V. A. Bapat, “Transient and stable expression of hepatitis B surface antigen in tomato (Lycopersicon esculentum L.),” Plant Biotechnology Reports, vol. 2, no. 1, pp. 1–6, 2008.

47. Y. Yang, R. Li, and M. Qi, “In vivo analysis of plant promoters and transcription factors by agroinfiltration of tobacco leaves,” Plant Journal, vol. 22, no. 6, pp. 543–551, 2000.

48. M.-Y. Kim, Y.-S. Jang, M.-S. Yang, and T.-G. Kim, “High expression of consensus dengue virus envelope glycoprotein domain III using a viral expression system in tobacco,” Plant Cell, Tissue and Organ Culture, vol. 122, no. 2, pp. 445–451, 2015.

49. M. W. Lee and Y. Yang, “Transient expression assay by agroinfiltration of leaves,” Methods in Molecular Biology, vol. 323, pp. 225–229, 2006.

50. B. M. Leckie and C. N. Stewart Jr., “Agroinfiltration as a technique for rapid assays for evaluating candidate insect resistance transgenes in plants,” Plant Cell Reports, vol. 30, no. 3, pp. 325–334, 2011.

51. Q. Chen, H. Lai, J. Hurtado, J. Stahnke, K. Leuzinger, and M. Dent, “Agroinfiltration as an effective and scalable strategy of gene delivery for production of pharmaceutical proteins,” Advanced Techniques in Biology & Medicine, vol. 1, no. 1, article 103, 2013.

52. M. Fujiki, J. F. Kaczmarczyk, V. Yusibov, and S. Rabindran, “Development of a new cucumber mosaic virus-based plant expression vector with truncated 3a movement protein,” Virology, vol. 381, no. 1, pp. 136–142, 2008.

53. K. Dalsgaard, Å. Uttentha, T. D. Jones et al., “Plant-derived vaccine protects target animals against a viral disease,” Nature Biotechnology, vol. 15, no. 3, pp. 248–252, 1997.

54. M. Koo, M. Bendahmane, G. A. Lettieri et al., “Protective immunity against murine hepatitis virus (MHV) induced by intranasal or subcutaneous administration of hybrids of tobacco mosaic virus that carries an MHV epitope,” Proceedings of the National Academy of Sciences of the United States of America, vol. 96, no. 14, pp. 7774–7779, 1999.

55. L. Wu, L. Jiang, Z. Zhou et al., “Expression of foot-and-mouth disease virus epitopes in tobacco by a tobacco mosaic virus-based vector,” Vaccine, vol. 21, no. 27–30, pp. 4390–4398, 2003.

56. J. Staczek, M. BEndahmane, L. B. Gilleland, R. N. Beachy, and H. E. Gilleland, “Immunization with a chimeric tobacco mosaic virus containing an epitope of outer membrane protein F of Pseudomonas aeruginosa provides protection against challenge with P. aeruginosa,” Vaccine, vol. 18, no. 21, pp. 2266–2274, 2000.

57. M. Lebel, K. Chartrand, D. Leclerc, and A. Lamarre, “Plant viruses as nanoparticle-based vaccines and adjuvants,” Vaccines, vol. 3, no. 3, pp. 620–637, 2015.

58. K. L. Hefferon, “DNA virus vectors for vaccine production in plants: spotlight on geminiviruses,”Vaccines, vol. 2, no. 3, pp. 642–653, 2014.

59. S. H. Bhoo, H. Lai, J. Ma et al., “Expression of an immunogenic Ebola immune complex in Nicotiana benthamiana,” Plant Biotechnology Journal, vol. 9, no. 7, pp. 807–816, 2011.

60. K. L. Hefferon, “Plant virus expression vectors set the stage as production platforms for biopharmaceutical proteins,” Virology, vol. 433, no. 1, pp. 1–6, 2012.

61. C. O. Tacket, M. F. Pasetti, R. Edelman, J. A. Howard, and S. Streatfield, “Immunogenicity of recombinant LT-B delivered orally to humans in transgenic corn,” Vaccine, vol. 22, no. 31-32, pp. 4385–4389, 2004.

62. M. M. Rigano, M. L. Alvarez, J. Pinkhasov et al., “Production of a fusion protein consisting of the enterotoxigenic Escherichia coli heat-labile toxin B subunit and a tuberculosis antigen in Arabidopsis thaliana,” Plant Cell Reports, vol. 22, no. 7, pp. 502–508, 2004.

63. I. Perea Arango, E. Loza Rubio, E. Rojas Anaya, T. Olivera Flores, L. Gonzalez de la Vara, and M. A. Gómez Lim, “Expression of the rabies virus nucleoprotein in plants at high-levels and evaluation of immune responses in mice,” Plant Cell Reports, vol. 27, no. 4, pp. 677–685, 2008.

64. D. M. Wang, J. B. Zhu, M. Peng, and P. Zhou, “Induction of a protective antibody response to FMDV in mice following oral immunization with transgenic Stylosanthes spp. as a feedstuff additive,” Transgenic Research, vol. 17, no. 6, pp. 1163–1170, 2008.

65. R. B. Malabadi, N. T. Meti, G. S. Mulgund, K. Nataraja, and S. Vijaya Kumar, “Recent advances in plant derived vaccine antigens against human infectious diseases,” Research in Pharmacy, vol. 2, no. 2, pp. 8–19, 2012.

66. J. Atsmon, E. Brill-Almon, C. Nadri-Shay et al., “Preclinical and first-in-human evaluation of PRX-105, a PEGylated, plant-derived, recombinant human acetylcholinesterase-R,” Toxicology and Applied Pharmacology, vol. 287, no. 3, pp. 202–209, 2015.

67. R. Strasser, A. Castilho, J. Stadlmann et al., “Improved virus neutralization by plant-produced anti-HIV antibodies with a homogeneous β1,4-galactosylated N-glycan profile,” The Journal of Biological Chemistry, vol. 284, no. 31, pp. 20479–20485, 2009.

68. E. C. Thuenemann, A. E. Meyers, J. Verwey, E. P. Rybicki, and G. P. Lomonossoff, “A method for rapid production of heteromultimeric protein complexes in

plants: assembly of protective bluetongue virus-like particles," Plant Biotechnology Journal, vol. 11, no. 7, pp. 839–846, 2013.

69. W. Phoolcharoen, S. H. Bhoo, H. Lai et al., "Expression of an immunogenic Ebola immune complex inNicotiana benthamiana," Plant Biotechnology Journal, vol. 9, no. 7, pp. 807–816, 2011.

70. J. M. Jeon, N. Y. Ahn, B. H. Son et al., "Efficient transient expression and transformation of PEG-mediated gene uptake into mesophyll protoplasts of pepper (Capsicum annuum L.)," Plant Cell, Tissue and Organ Culture, vol. 88, no. 2, pp. 225–232, 2007.

71. A. Hassanein, L. Hamama, K. Loridon, and N. Dorion, "Direct gene transfer study and transgenic plant regeneration after electroporation into mesophyll protoplasts of Pelargonium × hortorum, 'Panaché Sud.'," Plant Cell Reports, vol. 28, no. 10, pp. 1521–1530, 2009.

72. F. Locatelli, C. Vannini, E. Magnani, I. Coraggio, and M. Bracale, "Efficiency of transient transformation in tobacco protoplasts is independent of plasmid amount," Plant Cell Reports, vol. 21, no. 9, pp. 865–871, 2003.

73. W. Craig, D. Gargano, N. Scotti et al., "Direct gene transfer in potato: a comparison of particle bombardment of leaf explants and peg-mediated transformation of protoplasts," Plant Cell Reports, vol. 24, no. 10, pp. 603–611, 2005.

74. S. W. Hassan, Z. Mehmood, M. T. Waheed, and A. G. Lossl, "Towards generation of transplastomic tobacco, expressing a 35 kda protein as an antigen: a step towards affordable plant made vaccine againstMycobacterium," Cloning & Transgenesis, vol. 3, no. 1, article 117, 2013.

75. E. R. Santarém, H. N. Trick, J. S. Essig, and J. J. Finer, "Sonication-assisted Agrobacterium-mediated transformation of soybean immature cotyledons: Optimization of transient expression," Plant Cell Reports, vol. 17, no. 10, pp. 752–759, 1998.

76. Z. Liu, B.-J. Park, A. Kanno, and T. Kameya, "The novel use of a combination of sonication and vacuum infiltration in Agrobacterium-mediated transformation of kidney bean (Phaseolus vulgaris L.) with leagene," Molecular Breeding, vol. 16, no. 3, pp. 189–197, 2005.

77. E. P. Rybicki, "Plant-based vaccines against viruses," Virology Journal, vol. 11, article 205, 2014.

78. J. I. F. Solís, P. Mlejnek, K. Studená, and S. Procházka, "Application of sonication-assisted Agrobacterium-mediated transformation in Chenopodium rubrum L.," Plant, Soil and Environment, vol. 49, no. 6, pp. 255–260, 2003.

79. M. Klimek-Chodacka and R. Baranski, "A protocol for sonication-assisted Agrobacterium rhizogenes-mediated transformation of haploid and diploid sugar beet (Beta vulgaris L.) explants," Acta Biochimica Polonica, vol. 61, no. 1, pp. 13–17, 2014.

80. T. Miller, M. Fanton, S. Nickelson, H. Mason, and S. Webb, "Safety and immunogenicity of bacterial and tobacco plant cell line derived recombinant native and mutant Escherichia coli heat-labile toxin in chickens," Avian Pathology, vol. 41, no. 5, pp. 441–449, 2012.

81. N. Du and P. M. Pijut, "Agrobacterium-mediated transformation of Fraxinus pennsylvanica hypocotyls and plant regeneration," Plant Cell Reports, vol. 28, no. 6, pp. 915–923, 2009.

82. L. Faye and V. Gomord, "Success stories in molecular farming—a brief overview," Plant Biotechnology Journal, vol. 8, no. 5, pp. 525–528, 2010.

83. M. McCarthy, "US signs contract with ZMapp maker to accelerate development of the Ebola drug,"British Medical Journal, vol. 349, Article ID g5488, 2014.

84. P. Lal, V. Ramachandran, R. Goyal, and R. Sharma, "Edible vaccines: current status and future," Indian Journal of Medical Microbiology, vol. 25, no. 2, pp. 93–102, 2007.

85. N. Takeyama, H. Kiyono, and Y. Yuki, "Plant-based vaccines for animals and humans: recent advances in technology and clinical trials," Therapeutic Advances in Vaccines, vol. 3, no. 5-6, pp. 139–154, 2015.

86. K. Kashima, Y. Yuki, M. Mejima, et al., "Good manufacturing practices production of a purification-free oral cholera vaccine expressed in transgenic rice plants," Plant Cell Reports, vol. 35, no. 3, pp. 667–679, 2016.

87. World Health Organisation, "Good manufacturing praticies for biological products," WHO Technical Report Series 822, World Health Organisation, Geneva, Switzerland, 1992.

88. World Health Organization, Report WHO Informal Consultation on Scientific Basis for Regulatory Evaluation of Candidate Human Vaccines from Plants, WHO Quality Assurance and Safety of Biologicals, 2005.

89. J. Yu and W. H. R. Langridge, "A plant-based multicomponent vaccine protects mice from enteric diseases," Nature Biotechnology, vol. 19, no. 6, pp. 548–552, 2001.

INDEX